ESSENTIALS OF PHYSICAL GEOGRAPHY

THIRD EDITION

ESSENTIALS OF PHYSICAL GEOGRAPHY

THIRD EDITION

ROBERT E. GABLER
Western Illinois University

ROBERT J. SAGER
Los Angeles Harbor College

SHEILA M. BRAZIER
Golden West College

DANIEL L. WISE
Western Illinois University

SAUNDERS COLLEGE PUBLISHING
Philadelphia New York Chicago
San Franciso Montreal Toronto
London Sydney Tokyo

Text Typeface: Baskerville
Compositor: The Clarinda Company
Acquisitions Editor: John J. Vondeling
Project Editor: Sally Kusch
Copyeditor: Becca Gruliow
Art Director: Carol C. Bleistine
Text Designer: Edward A. Butler
Cover Designer: Lawrence R. Didona
Layout Artist: James Gross
New Text Artwork: J & R Technical Services
Production Manager: Tim Frelick
Assistant Production Manager: JoAnn Melody

Cover Credit: Worthington Glacier. Tina Waisman.
Frontispiece: Water erosion in arid lands, Bryce Canyon
National Park. (R. Gabler)

Library of Congress Cataloging-in-Publication Data

Essentials of physical geography

   Includes index.
   1. Physical geography—Text–books—1945–
I. Gabler, Robert E.
GB54.5.E84 1986   551   86-17658
ISBN 0-03-008958-1

ESSENTIALS OF PHYSICAL GEOGRAPHY 3/e   0-03-008958-1

Requests for permission to make copies of any part of the work should be mailed to: Permissions, Holt, Rinehart and Winston, Inc., Orlando, Florida 32887.

  89 071 9876543

Printed in the United States of America

PREFACE

Physical geographers are often asked a somewhat provocative question by their students, colleagues and friends. Are natural disasters presently occurring in greater numbers and causing problems of more significant dimensions than during previous periods of earth history—and are humans, in some unknown way, partly responsible? It does seem that with each passing week the newspaper headlines or the television newscasts report some new tragedy based on cataclysmic events in the natural environment. Hurricanes in Florida are followed by volcanic eruptions and mudslides in Colombia, earthquakes in Mexico, forest fires in California, typhoons and flooding in Bangladesh, drought and starvation in Africa, the dying of forests in Germany, and tornadoes in Texas. But there is no simple answer to the query concerning the role of humans in natural disasters. In some they may contribute to the cause, and, along with other living things, in all cases they are the victims.

It is also unlikely that the frequency of unusual events within the natural environment is on the increase. But the amount of reporting and the thoroughness of coverage are certainly greater now than they were a generation ago. In addition, the rapid expansion of human populations has forced more and more people to live in the "hazard" zones where natural disasters are likely to occur. Hence, although humans may play no role in the environmental processes leading up to the event, they are often unwittingly or involuntarily responsible for the dimensions of any disaster stemming from such processes.

Most important for those who are concerned about the loss of life and human suffering associated with disastrous natural events, is the understanding that the processes involved are *not* unusual and similar events have been associated with these processes throughout earth history. Only a knowledge of their origins and a thorough understanding of the events themselves can lessen their impact on individuals and society. The authors of *Essentials of Physical Geography* believe that the study of physical geography can provide the information and explanations necessary to comprehend the changes, great and small, unusual and routine, that take place within the physical environment. It is only through a knowledge and understanding of physical geography that humans can learn to establish a compatible relationship with the earth upon which they live.

COVERAGE

Physical geography provides a broad view of the earth and its component systems; it identifies physical phenomena and stresses their distribution and relationships. This text provides an introduction to all major aspects of the earth system. It covers a wide range of topics, from atmospheric elements to the earth's interior. It helps to explain the origins, development, significance, and distribution of processes and events that occur within, on, or above the surface of the earth. *Essentials of Physical Geography* contains ideas and information about the earth as a planet and as a human environment. It is a companion piece to any course for which physical geography is a major component. It is designed to provide the broadest possible coverage of the earth's physical patterns and processes and it is written primarily for the individual who is unlikely to study these topics again in depth. Although phenomena as diverse as glaciers, soils, midlatitude cyclones, and tides are considered, there has been a conscious effort throughout to emphasize interrelationships—to focus on all phenomena, including those associated with humans, as interdependent parts of one integrated earth system.

NEW FEATURES

In keeping with the increasing trend to use models as a tool in geographic education, the theme of earth systems has been expanded and emphasized in this third edition of *Essentials of*

Physical Geography. It is introduced in Chapter 1, reoccurs throughout the book, and is the organizational theme for the discussion of terrestrial and marine ecosystems in Chapters 9 and 17. Although every effort has been made to inform students concerning the application and value of system models in understanding the physical environment, the theoretical approach has been deliberately integrated with real world examples. The authors have also reinforced their point of view that any introductory physical geography text, designed for a general rather than a specialized education, should include the human element, especially by citing the ways in which physical elements and earth systems interact with humans and their activities. Both students and faculty alike are urged to utilize the Viewpoint essays as points of departure for meaningful classroom discussion concerning the human element in physical geography. The essays from previous editions of the text have been revised and retained, and new Viewpoints have been added to Chapter 1, entitled "The Geographer" and to Chapter 15, entitled "Yellow Earth and Yellow River."

Numerous other additions and improvements will be evident to previous users of this text. The illustration program has been thoroughly revised and expanded. Full color photographs and line drawings are found throughout the book wherever the use of color enhances student appreciation or understanding of the illustration. This is especially the case when the subjects are earth-sun relationships (Chapter 1), remote sensing imagery (Chapter 2), cloud types (Chapter 5), climatic regions (Chapters 7 and 8), ecosystems and major biomes (Chapter 9), soil profiles (Chapter 10), landforms and rock types (Chapter 11), tectonic plate theory (Chapter 12), stream-modified features (Chapter 14), and coastal zones (Chapter 17). The text utilizes the metric system, the one which is most widely accepted throughout the world. However, English system equivalents for all important measurements are included in parentheses because it is recognized that conversion to metric is not complete in the United States, and both stu-

dents and teachers may benefit from the mental exercise of comparing the numbers as they appear side by side.

INITIAL CHAPTERS

The most significant changes included in the current revision of *Essentials of Physical Geography* will be noted in the fundamental reorganization of topics and chapter outlines and the addition of new, expanded, or updated text. As previously mentioned, Chapter 1 has been lengthened in order to introduce students to the theory of models and systems and to demonstrate the practical application of models to an understanding of the real world. The new discussion assists in an examination of the earth as an integrated system, and it precedes an overview of the important role that physical geography plays in helping humans to understand and appreciate their environment. Chapter 1 fulfills the objective of introducing the earth as a planet by concluding with a review of earth motions, earth-sun relationships, and the seasonal fluctuation that occurs on earth as a result.

Chapter 2 provides the student with a thorough discussion of earth location, globes, maps, and the products of remote sensing. In order to include the explanations essential to an understanding of the laboratory exercises which usually accompany physical geography courses, the chapter provides detailed information on the relationships of longitude to clock time, the Public Lands Survey System, map essentials and map projections, aerial photography and satellite imagery. The chapter is greatly enhanced by eight pages of full color plates that include digital terrain models, topographic map and aerial photograph comparison, and LANDSAT imagery.

ATMOSPHERIC ELEMENTS

Because the distributional patterns associated with the atmospheric elements and controls owe so much to the earth-sun relationships introduced in Chapter 1, these elements are examined sequentially in Chapters 3, 4, and 5. The discussion of solar energy and temperature in Chapter 3 leads logically to a discussion of atmospheric pressure and winds in Chapter 4 and atmospheric moisture and precipitation in Chapter 5. The interrelationships of the various atmospheric elements and controls are emphasized, and the importance of energy and moisture budgets is stressed. In this edition, the treatment of several topics has been expanded and more clearly illustrated. These topics include the Coriolis effect and frictional drag, upper air winds, and the discussion of stability and instability as conditions of air.

WEATHER

Chapter 6 examines weather and the events associated with changes in the atmospheric elements. Significant revision occurs in the sections on North American air masses, the movement of cyclones, and the relationship of upper air flow to surface weather. The discussion of midlatitude cyclones and local weather has been written to provide realistic examples of weather change, and the text is accompanied by diagrams that illustrate a model situation in the central United States. The major purpose of the chapter is to provide students with sufficient background to fully comprehend the daily weather map and better understand the frequently unreliable science of weather prediction.

TOPIC REORGANIZATION

Chapters 7 through 10 are the result of the major reorganization associated with the third edition. The topics normally associated with climate, vegetation, and soils have been reordered, and extensive new coverage of ecosystems has been included. Climate is introduced in Chapter 7 and is immediately followed by a review of the distribution of climatic types and climatic regions of the humid tropics and arid lands. Climatic regionalization is concluded in Chapter 8 prior to the detailed examination of ecosystems and the world's major biomes in Chapter 9 and soils in Chapter 10. Because physical geography texts should be comprehensive but must still match the time contraints of courses as limited as a semester or a quarter in length, the authors believe that the new organization should provide instructors with an interesting option. Those who have severe restrictions on class time could assign either the material on climatic regions or the detailed discussions of ecosystems and soils as supplementary rather than required reading. However, we urge our colleagues to emphasize the importance of climatic regions as they exist in the real world, and we urge students to read all of Chapters 7 and 8 even if the material is not assigned by their instructor. Physical geography is, after all, *geography* and it includes an examination of both the earth and the humans that occupy its major regions. Geography, as a discipline, has been accused of abandoning the regional study which brings the current world into focus, and the authors of this text are reluctant to add to the problem. It is in physical geography courses where students must learn about the Amazon Basin, the Mediterranean lands, the Sahel, the midlatitude prairies, the tundra and the Gobi—there is no other subject in which you can expect such information to be included.

CLASSIFICATION

As in past editions the authors regard methods of classification as an essential element in any science course. A discussion of classification and classification of climate appear in Chapter 7, classification of major biomes is included in Chapter 9, classification of soils is located in Chapter 10 and classification of landforms is introduced in Chapter 11. Climatic classification is a modified version of the Köppen System; vegetation classification is patterned primarily after Kuchler; soils are classified utilizing both the USDA Great Soil Groups and the Comprehensive Soil Classification System (the 7th Approximation); landform classification is based on the work of Edwin Hammond. The world distribution of the various classes in each system is stressed and full color maps have been included in the appropriate chapters to reinforce the presentation.

LANDFORMS AND THE OCEAN

Chapters 11 through 16 provide a detailed study of the development, appearance and distribution of the earth's landforms. A second major instance of chapter reorganization occurs in this portion of the third edition with the introduction of separate chapters dealing with landforms and earth structure (Chapter 11) and tectonic processes (Chapter 12). Chapter 12 provides a thorough discussion of plate tectonics and permits the special emphasis which this significant theory deserves. Not only has the material on plate tectonics been rewritten and reinforced with new diagrams but there has been significant revision as well throughout the six chapters—in the discussion of rocks and minerals, diastrophism, volcanism, the availability of ground water, the erosion cycle in deserts, and glacial systems. There are also numerous opportunities for students to recognize the close relationship between the earth's landform features and phenomena from the other earth spheres: atmosphere, biosphere, and hydrosphere. Separate chapters on gradation (Chapter 13), water in humid lands (Chapter 14), water and wind in arid lands (Chapter 15), and glaciers (Chapter 16) focus on the processes associated with the development of various physical landscapes. Landforms are considered as transitory features that are constantly in the process of change and are the products of earth systems.

Chapter 17 provides an examination of the global ocean and its increasing significance, as humankind faces an uncertain future with an ever-changing resource base. This seems an appropriate conclusion to a physical geography text. An understanding and appreciation of coastal regions and the global ocean are essential to an educated citizenry, as land-based societies turn increasingly toward the sea.

ANCILLARIES

The ancillaries available with this textbook include an instructor's manual, a computerized test bank, and a set of overhead transparencies. The instructor's manual offers an unusual innovation because it contains within its pages sufficient material for a complete course syllabus keyed to the textbook. The student syllabus contains lists of major ideas, detailed outlines, key terminology, and study questions for each chapter in the textbook. In addition, the manual contains a series of essays addressed to the student which should assist the student in developing better learning skills. Elements provided specifically for the instructor include suggested instructional activities and both recommended references and teaching aids also organized by chapter.

The test bank is available in both Apple and IBM versions and contains hundreds of multiple choice, short answer, and true/false test items which may be utilized in both chapter or unit evaluation. The transparency set comprises over 100 2- and 4-color transparencies selected from those line drawings that illustrate the major topics in the text.

ACKNOWLEDGEMENTS

Once again the remarkable staff of Saunders College Publishing has successfully shepherded *Essentials of Physical Geography* through the rigors of preparing another edition. We wish to express our sincere appreciation to Don Jackson, Publisher; John Vondeling, Associate Publisher; and both Margaret Mary Kerrigan and Kate Pachuta, Assistant Editors, for their efforts, counsel, assistance, and support. Special commendation is also due Sally Kusch, Senior Project Editor; Art Director Carol Bleistine; and Manager of Editing, Design and Production, Tim Frelick. Without their encouragement, unlimited patience, and competent guidance, this edition would not be blessed with a wealth of new art work, a splendid design, and a totally professional appearance.

The analysis and review preceding the third edition of *Essentials of Physical Geography* have been more painstaking and extensive than for any previous edition of this text. In the early stages of revision a number of individuals provided helpful criticism of the second edition and offered a number of excellent suggestions which have been incorporated in the present book. Among the early reviewers were Professors Kang-tsung Chang, University of North Dakota; M. Stanley Dart, Kearney State College; Michael DeMers, Mankato State University; Thomas P. Grimes, Ball State University; David W. Icenogle, Auburn University; Richard S. Jarvis, State University of New York at Buffalo; Carol W. McCarty, Aquinas Junior College; Harold McConnell, Florida State University; Gerald E. Nelson, Casper College; Earl J. Senninger, Charles Stewart Mott Community College; Rodman E. Snead, University of New Mexico; Sten A. Taube, Northern Michigan University; Harold L. Throckmorton, San Diego Mesa College; and Vernon O. Walton, Essex Community College.

During the preparation of final manuscript for the third edition three of the original reviewers and several additional physical geographers accomplished the monumental task of reading through each new or revised chapter in its initial form. It was their helpful advice, constructive suggestions for chapter reorganization, and extensive marginal notes which have most influenced the third edition, and for which the authors wish to publicly express their heartfelt thanks. Included among those who devoted so many long and tedious hours are the previously mentioned M. Stanley Dart, David W. Icenogle, and Vernon O. Walton and Professors Richard A. Boerckel, United States Military Academy; Ralph Hannon, Santa Ana College; William J. Reynolds, United States Military Academy; and Stephen Stadler, Oklahoma State University.

At each campus represented by one of the authors there were untold numbers of friends and colleagues who contributed generously of their time and counsel and in many instances provided contributions in the form of manuscript suggestions and specific illustrations. To each one, both known and unknown, we owe a debt of gratitude. And a special word of thanks is due Mrs. Pamela Hines without whose assistance the final manuscript would never have appeared at the offices of the publisher in an organized, coherent, and thoroughly professional format.

Water erosion, Badlands, Bryce Canyon. (R. Gabler)

CONTENTS OVERVIEW

U-shaped glacial troughs, Mt. Rainier. (R. Gabler)

Wilbur Creek, Glacier National Park. (R. Gabler)

CONTENTS

Yellowstone National Park. (R. Gabler)

Wizard Island, Crater Lake, Oregon. (R. Gabler)

Sandstone weathering, Zion National Park. (R. Gabler)

Travertine deposit, Yellowstone National Park. (R. Gabler)

Chapman's Bay, South Africa. (R. Sager)

1

EARTH SYSTEMS AND MOVEMENTS

An astronaut orbiting in a spacecraft, occupying a space laboratory, or flying the Space Shuttle has an unparalleled view of planet earth. It is easy for the astronaut to see the earth's component parts and the various objects which appear on the planet's surface. Continents, oceans, forests, deserts, cloud patterns, cities and other evidence of civilization can be observed independently or as they relate to one another. Utilizing special cameras and other scientific equipment, many attributes or characteristics of these objects, such as size, shape, and temperature can be mapped and studied (Fig. 1.1). Thus, the astronaut is in an ideal position to understand and appreciate the **earth system.**

An acceptable definition of a **system** is a set of objects and/or characteristics of objects that are linked in some way to one another. The objects, which are often described as **variables,** are interrelated in such a way that they function together as an organized whole. Parts of the earth system affect other parts, and they appear in worldwide patterns that, when examined, demonstrate clear interconnections. The location and position of mountains help to determine the distribution and amounts of rainfall, which, in turn, affect the amount and types of vegetation. Vegetation, moisture, and underlying rock affect the kind of soil. Vegetation and soil type influence the runoff of water, and the circle is complete when the amount of runoff is a major factor in the stream erosion, which reduces mountains.

Certainly the most important attribute of the earth system is that it is a **life support system.** Like the space vehicle that supports the astronaut orbiting the earth, the earth system provides the necessary combination of elements and characteristics to permit life, as we know it, to exist. If any critical parts of a life support system are changed, living things may no longer be able to survive. For instance, if all the oxygen is used up in a spacecraft, the crew in it will die. Or if there is no way to keep the craft at the right temperature, its occupants will burn or freeze. If food supplies run out, the astronauts will starve. On the earth, natural processes provide a constant supply of oxygen, the sun maintains temperatures at a tolerable level, and there is a continuous cycle of creation of new food supplies for living things.

LANDSAT image of Tokyo and Tokyo Bay, Japan (NASA)

1

FIGURE 1.1 Thermal infrared image of Typhoon Vera, 1979. The highest clouds, representing coolest temperatures, appear in shades of blue and white, while the lower clouds, with warmer temperatures, appear in shades of red. (NOAA)

The earth, then, is a "set" of interrelated components that are vital and necessary for the existence of all living creatures. As we move toward the last decade of the twentieth century, we have come to realize that important parts of our life support system, which may be called **natural resources,** can be abused and over- used, thereby threatening the functioning of the whole system.

We are aware that some of the earth's resources, such as air and water, can be polluted to the point where they are unusable or even lethal to some life forms (Fig. 1.2). By polluting the oceans, we may be killing off some impor-

FIGURE 1.2 Toxic chemicals, such as the ones discovered in this solid waste dump in southeast Chicago, pose a life-endangering threat to local water supplies. (Illinois EPA)

tant fish species, while less desirable species might increase in number. Acid rain, caused by industries, power plants, and automobiles releasing pollutants into the atmosphere, is damaging forests and killing fish in lakes hundreds of miles away from the source of the pollution. We may be using up some other resources, especially those we need for fuel, too rapidly. While we still have enough coal to last several hundred years, we have frequently been warned about future shortages in our petroleum supplies. When nonrenewable resources such as mineral fuels are gone, the alternative resource is invariably less desirable or more expensive.

FIGURE 1.3 This crowded city street in Delhi, India, is typical of the major metropolitan areas of densely populated south and southeast Asia. (S. Brazier)

We are learning that there are limits to the amount of space on the earth and that we must use it wisely. In the search for living space, we occasionally construct buildings in places that are not safe, and many places where we live are overcrowded (Fig. 1.3). Also, we sometimes plant crops in areas that are ill suited to agriculture because there is not always enough good farmland to fill our needs.

As we continue to explore space, we are learning more and more about the world in which we live. With the use of cameras and other scientific instruments mounted on manned and unmanned spacecraft, we can see larger parts of the earth than we could before, and we have come to a fuller realization that there are limits to the support that can be given to humankind by the earth. All citizens of the earth must understand the effects of their actions on the complex earth system. It is to physical geography and other earth sciences that we look in order to learn the consequences of our activities upon the world in which we live.

THE EARTH SYSTEM

There are four major divisions of our earth system, and the interactions between the various subsystems within these divisions create our physical landscape (Fig. 1.4). The **atmosphere** is the blanket of air that envelops and insulates the earth. Composed of many gases, its movements and processes create the changing conditions we know as weather and climate. Within the atmosphere is the earth. The earth's landforms—together with its crust, rocks, minerals, and soils—make up the **lithosphere.** And the processes operating in and on the lithosphere, the changes in landforms, create a part of the physical landscape that in turn affects the development of the **biosphere.** All living things—people, animals, and plants—make up the biosphere. The fourth major subsystem is the **hydrosphere,** which is composed of the waters of the earth and the atmosphere.

We can see many interactions between these major divisions of the earth. For exam-

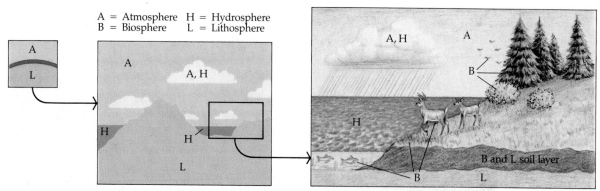

A = Atmosphere H = Hydrosphere
B = Biosphere   L = Lithosphere

FIGURE 1.4   The four interdependent subsystems of the earth's physical environment.

ple, the hydrosphere acts as a supply of water for humans and as a home for many types of animal life and vegetation. It affects the lithosphere as countless streams wear down land formations, and it influences the atmosphere through evaporation, condensation, and the effects of ocean temperatures on climate.

Further, there are many instances of overlap between the four divisions. Soil can be examined as part of the biosphere or the hydrosphere as well as part of the lithosphere. The water stored in plants and animals is part of both the biosphere and the hydrosphere, and water in clouds belongs to the atmosphere as well as the hydrosphere. That we cannot draw definite dividing lines between the divisions underscores the interrelatedness of the various parts of the earth system. Like an engine or like the human body, the earth is a system that functions only when all its parts work together harmoniously.

MODELS AND SYSTEMS

Since the earth is so large and complex, physical geographers must find some way to simplify it for purposes of description, study, and understanding. To do this they develop representations of the real world called **models.** When we describe the earth as a huge system or a complex group of interrelated systems, we are really using models, which help us to organize what we are observing. Models also assist us in

explaining the processes that are involved. Throughout the chapters that follow, we will use the concept of the earth system as a model as well as many other models to help us simplify a complex earth and focus on the most important elements in the physical environment.

It has already been pointed out that the earth system operates as a unit and that each part has connections with all the other parts. But the earth is simply too complex to allow construction of a single satisfactory model of the entire system or even a substantial part of it. Hence, it is necessary to identify and study within the physical environment smaller subsystems that demonstrate strong internal connections. Examples of earth subsystems usually examined by physical geographers include the atmospheric water cycle, cyclonic storm systems, stream systems, the systematic heating of the atmosphere, and ecosystems.

Systems can be described in various ways, and among the most popular with physical geographers are those systems involving the movement of energy. In **energy systems** the geographer can trace the movement of energy into the system (input), the effects of the energy on the matter within the system, and the movement of energy out of the system (output). The subsystem involved in the heating of the atmosphere is an example of an energy subsystem. Because energy cannot be contained within the boundaries of a system, all energy systems are also described as **open systems.**

A **closed system** is one in which no substantial amount of either energy or matter crosses its boundaries. Only the solar system is a truly closed system. Although energy moves in and out of the open earth system, some earth *subsystems* can be considered closed in respect to the movement of materials alone. It can even be said that, except for meteoric remains reaching the earth's surface or the escape of occasional gas molecules and manmade satellites leaving the outer atmosphere, the entire earth system is closed to the movement of material. The atmospheric subsystem that involves the movement of water from atmosphere to earth to ocean and back to the atmosphere again is another good example of a **materials system** that is closed. Water may appear in the system in all three of its major states, as liquid, gas, or solid ice, and may be transformed from one to the other many times, but there is no gain or loss of water by the system.

Most earth subsystems are open systems and both energy and material move freely across system boundaries. A stream system is an excellent illustration of an open earth subsystem in which both material and energy in the form of soil, weathered rock, heat, and precipitation enter the stream while water and sediments leave the stream where it empties into the ocean or some other standing body of water.

The various characteristics, or *variables*, of an open system have a tendency to reach a balance with one another and with the factors that influence the system from outside its boundaries. If the amount of material entering the system balances that which leaves, the system is said to have reached a state of **dynamic equilibrium.** In systems involving plant and animal life we often hear this called the "balance of nature." Each animal population will in the long run adjust naturally to the food supply of its habitat whether the food is in the form of vegetation or other animal populations. Sometimes we humans intrude on habitats, and our activities can disturb the natural balance, destroy dynamic equilibrium, and force the life forms in the habitats to seek a new and different balance.

There is a simple but important mechanism that helps a system regulate itself and maintain equilibrium. It is labeled **feedback.** This means that when one of the elements in a system changes, often because of outside influence, there is a sequence of changes in the other elements of the system. Feedback occurs when this sequence of changes ultimately affects the first element once again. More often than not the result is **negative feedback,** whereby the sequence of changes serves to counteract the direction of change in the initial element. Let us look once more at our example of a stream and examine more closely its relationship to its channel. If the amount of water in the stream *increases* suddenly, as during a heavy rainstorm, the velocity of the stream will *increase,* which may in turn cause greater erosion of the stream channel. Greater erosion often *increases* the width of the stream channel, which would then *increase* external friction by enlarging the amount of surface over which the water flows. External friction wastes energy and *decreases* stream erosion, and negative feedback has taken place (Fig. 1.5).

Some earth subsystems exhibit **positive feedback;** that is, the sequence of changes serves to reinforce and not counteract the direction of initial change. For example, during several periods in the planet's history, the earth has experienced significant cooling on a global scale. This cooling of the atmospheric system led to the growth of great ice sheets, which covered large portions of the earth's surface. The massive ice sheets increased greatly the amount of solar energy that was reflected back to space from the earth's surface, thus increasing the cooling trend in the atmosphere and the further growth of the ice sheets. The result over considerable periods of time was positive feedback.

There are both advantages and disadvantages in viewing the earth as a vast system, which in the great scheme of things is only one small subsystem in the solar system and the universe, but which, itself, is made up of count-

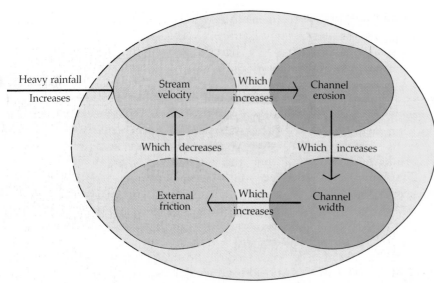

FIGURE 1.5 Heavy rainfall *increases* stream velocity, which can place into motion a sequence of changes, which may ultimately result in *decreased* stream velocity; an example of negative feedback.

STREAM SYSTEM

less subsytems. As we have mentioned, systems are models. And like all models they are not the same as reality. They are products of the human mind and are only a way of looking at the real world. The examination of various earth subsystems will help us to understand the natural processes involved in the development of the atmosphere, lithosphere, hydrosphere, and biosphere as we know them today. They may even help to reconstruct past events and predict future change. But we must be careful not to substitute the model for the real world.

ENVIRONMENT—ECOLOGY—ECOSYSTEM

One of the great advantages of considering subsystems when studying physical geography is that systems serve to illustrate the relationships among the various elements within the systems. As scientists, geographers are keenly interested in relationships, and they pay special attention to the relationships between humans and the physical environment. In no way can the human-environment relationship be better illustrated than through the examination of the human impact on ecosystems.

In the last half of the twentieth century people have become more conscious of their environment than ever before. We talk about the environment and worry about ecology. Popular news magazines often devote whole sections to discussions of environmental issues. But what are we really talking about when we use words like *environment? ecology? ecosystem?*

In the broadest sense, our environment can be defined as our surroundings; it is made up of all the physical, social, and cultural aspects of our world that affect our growth, our being, and our way of living (Fig. 1.6). Humans also share environments with plants and animals. We can speak of a plant's environment and include in our discussion the soil in which the plant grows, the amount of sunlight and rainfall it receives, the gases that surround it, the range of temperature in the air, and the various plants that grow nearby and serve to block winds, sunlight, and rain.

Just as humans interact with their environment, so do the other animals and plants. The study of these relationships between organisms, whether animal or plant, and their environments is a science known as **ecology.** Ecological relationships are complex but naturally balanced "webs of life." Disrupting the natural

FIGURE 1.6 The physical and cultural attributes of a site combine to form a unique environment. The humans who occupy the site, such as these boat dwellers of Aberdeen, Hong Kong, are both influenced by the environment and are an integral part of it. (Donald Marshall collection, WIU International Programs)

ecology of a community of organisms may have negative results (although this is not always so). For example, the filling in or pollution of coastal marshlands may disrupt the natural ecology of such areas. As a result, fish spawning grounds may be destroyed and the food supply of some marine animals and migratory birds could be greatly depleted. The end product is the destruction of valuable plant and animal life.

Ecosystem is a contraction of **ecological system.** That is, an ecosystem refers to a community of organisms and the relationships of those organisms to their environment. An ecosystem is dynamic in that its various parts are always in flux. For instance, plants grow, rain falls, animals eat, and soil matures, all changing the environment of a particular ecosystem. Since each member of the ecosystem belongs to the environment of every other part of that system, any change in one alters the environment for all the others. And as those components react to the alteration, they in turn continue to transform the environment for the others. A change in the atmosphere from conditions of sunshine to those of rain affects plants, soils, and animals. Heavy rain may carry away soils and plant nutrients, so the plants may not be

able to grow as well and the animals then may not be able to eat as much. On the other hand, the addition of moisture to the soil may help some of the plants grow, increasing the amount of shade beneath them and thus keeping other plants from growing.

The concept of an ecosystem can be applied at almost any scale, in a wide variety of geographic locations, and under all environmental conditions where life is possible. Hence, a farm pond, a grass-covered field, a marsh, a forest, or a portion of a desert can be viewed as an ecosystem. Even the earth itself may be considered one large ecosystem. Ecosystems are found wherever there is an exchange of materials among living organisms and functional relationships between the organisms and their natural environment. Ecosystems are open systems with movement of energy and material across their boundaries. Although some ecosystems, such as a small lake or a desert oasis, have clear-cut boundaries, the limits of many others are not as precisely defined. Often the variation from one ecosystem to another is obscure and the transition occurs slowly over distance.

Since human beings first walked the earth, they have been part of each ecosystem they have inhabited. However, with the advanced technology of the twentieth century people today have a greater ability than ever before to alter the world's natural ecosystems. The Aswan High Dam provides a classic example of the effects of human technology on an ecosystem. Built on the Nile River in Egypt to provide a reservoir of water for the irrigation of arid lands, the Aswan High Dam had unexpected results because it disturbed a natural ecosystem (Fig. 1.7). Because of the dam, silt was no longer carried downstream. The lands below Aswan in the Nile Delta had always been enriched by the deposit of these silts when the Nile overflowed its banks each spring, but now more and more fertilizers have to be applied because of a lack of soil nutrients contained in the silt. In fact, Egypt, once a land with rich soils along the Nile, has become a large importer of fertilizers. Nutrients for fish are also being blocked by the dam, and this has resulted in a decrease in the fish catch of the eastern Mediterranean Sea. Furthermore, the water below the dam moves at a faster rate than before because it does not carry as much silt and other materials, and it now threatens buildings and piers constructed along the

banks of the river and may even undermine the banks themselves. The changes and problems that the Aswan High Dam has created in the ecosystem of the Nile Delta should serve as a useful lesson. It is during the planning stages of activities that alter the environment, such as building a dam or a freeway or changing the course of a river, that the geographer's view of the whole system and his or her sensitivity to its components can be of service.

POLLUTION

When examining our effect upon our environment, we cannot ignore the problem of pollution. But what exactly is pollution? First, there are many varieties, including air pollution, water pollution, noise pollution, visual pollution, and solid waste pollution. Is amount involved? Can we say that one car in the middle of the desert or alone on a New Hampshire mountain road pollutes the atmosphere? Or would emptying the dregs of our soft drink into the Mississippi alone constitute polluting those waters? Technically yes, but not to any significant degree, since pollution does not occur simply be-

FIGURE 1.7 The Nile River at Aswan, Egypt. Although this quiet scene is reminiscent of Egypt in past centuries, the completion of the high dam on the Nile at Aswan brought profound change to the environment and ecosystems of the river from the dam to the river's mouth near Alexandria. (Courtesy of Samuel E. Turner)

FIGURE 1.8 **(a) On a clear day it is not difficult to understand why Los Angeles, California was named "The City of Angels." (R. Gabler) (b) But far too often the great metropolis justifies its designation as "Smog Capital" of the United States. (R. Sager)**

cause of the addition of foreign material to a system like the atmosphere or the hydrosphere.

Pollution does occur, however, when more foreign material is put into a system than the system can tolerate. In a large city more pollutants may get blown, pushed, and exhausted into the air than the atmosphere can handle, thus creating a "smoggy" day (Fig. 1.8). Pollution is the accumulation, to a level intolerable to humans, of undesirable elements in any one of the diverse aspects of the physical environment. In the strictest sense, there is natural pollution (lime, iron, or sulphur in water supplies, smoke from forest fires, or dust from the eruption of volcanoes). But in our current usage, *pollution* goes beyond the natural elements to include those wastes in the water, air, or other aspects of the environment for which humans are responsible.

Humans always leave evidence of their presence, but this evidence becomes a problem and is considered pollution when it becomes too great—that is, when it significantly alters the natural environment or when it threatens normal growth and reproduction or the normal functioning of all lifeforms, including human beings.

THE STUDY OF PHYSICAL GEOGRAPHY

The word *geography* comes from two Greek roots. *Geo-* refers to the earth, and *-graphy* means picture or writing. Geography examines, describes, and explains the earth—its variation from time to time and from place to place. Geography is often called the "spatial science" because it includes the recognition, analysis, and explanation of likenesses and differences or of variations in phenomena as they are distributed on the earth's surface (through *earth* space).

Geography is both a physical and a social science. In its concern with the natural environment, geography is very much a physical science. Yet geography also examines humanity's two-way relationship with the earth and is thus a social science as well.

Cultural or **human geography** is the study of human activity and of the results of that activity. The human geographer studies such subjects as population distribution, cities and urbanization, natural resource utilization, industrial location, and transportation net-

FIGURE 1.9 Chicago, midwestern U.S. focal point for industry, commerce, financial institutions, interstate highways, air traffic, and millions of housing units, represents fertile ground for study by cultural geographers. (Photo by Kee Chang, Chicago Association of Commerce and Industry)

works (Fig. 1.9). When a geographical study concentrates primarily on the physical and human features of a specific region, such as Canada or the Middle East, we call this **regional geography.**

Physical geography encompasses the study of the natural aspects of the human environment. That is, physical geographers look at the atmospheric elements that affect the surface of the earth and that together make up weather and climate. They examine the variations in soil and in natural vegetation. The varieties of water bodies on earth, their movements, effects, and other characteristics are subjects of physical geography, as are the landforms of the earth, their formation, and their modification (Fig. 1.10). Yet although physical geographers emphasize the elements that make

FIGURE 1.10 Colorado National Monument near Grand Junction, Colorado. Natural environments, with their component rock structures, landforms, soils, vegetation, moisture conditions, and atmospheric elements, are the subjects of study in physical geography. (R. Gabler)

up our physical environment, they do not ignore the effect of people on those elements.

Learning about our environment and the processes that govern it is the major function of physical geography. The knowledge learned can be of help to us in analyzing problems such as whether we should continue to build nuclear power plants, allow offshore oil development, or drain coastal marshlands. Army intelligence must predict the effects that weather and terrain may have on military operations. Industries must evaluate how the development of a new plant site may alter the surrounding environment. Closer to home, there are smaller problems that the principles and processes of physical geography can help to analyze. For example, should you plant your new lawn before or after the spring rains? What sort of effects can be expected from a proposed shopping center? Will it make your home more or less valuable and livable? What are the advantages and disadvantages of a particular home site? What hazards—flooding, landslide, earthquake—might your house be subject to? What can you do to minimize the potential damage to your house from a natural disaster?

It is apparent then that the study of physical geography and an understanding of the natural environment are of major value to all of us. But perhaps you have wondered, what do those scientists who call themselves physical geographers do? What kinds of jobs do they hold? Physical geography sounds interesting and exciting, but can I make a living at it? By applying their knowledge, skills, and research techniques to problems of the real world, physical geographers make major contributions to human well-being and the economic development of society. Such *applied* physical geography takes many forms and the *Viewpoint* on pages 12–13 serves to introduce the reader to only a few.

Finally, not only is physical geography a source of useful knowledge and an opportunity for employment, it is also first and foremost a visual science. Even if you forget every fact discussed in the following pages, you will have been shown new ways to look at, to see, even to evaluate the world around you. Just as you see a painting differently after an art course, so, too, will you look at sunsets, waves, storms, prairies, and mountains with a different, more informed eye. You will see greater variety in the landscape, not because there is any more there, but because your eye will have been trained to see it differently.

THE EARTH AS A PLANET

The color television shots of the earth from the Apollo spacecrafts on their lunar missions gave us some unforgettable views of our planet. The cameras showed the earth as a sphere of blue oceans, green and brown landmasses, and swirls of white clouds. One astronaut has described the earth as it appears to one who has traveled close to the moon:

> The earth looked so tiny in the heavens that there were times during the *Apollo 8* mission when I had trouble finding it. If you can imagine yourself in a darkened room with only one clearly visible object, a small blue-green sphere about the size of a Christmas tree ornament, then you can begin to grasp what the earth looks like from space. I think that all of us subconsciously think that the earth is flat or at least almost infinite. Let me assure you that, rather than a massive giant, it should be thought of as the fragile Christmas tree ball which we should handle with considerable care.

This Island Earth,
edited by Oran W. Nicks,
NASA, SP-250, 1970

SIZE AND SHAPE OF THE EARTH

For most of his history man has pondered the size and shape of the world in which he lives. Though it was not until the 1960s that man was able to travel deep enough into space to see the shape of the earth, ancient Greeks as early as Pythagoras in 540 B.C. theorized that the earth was a sphere. However, it was not until about 200 B.C. that a philosopher named Eratosthenes made a fairly accurate estimate of the

(Text continues on page 14)

VIEWPOINT: THE GEOGRAPHER

It was 7:30 P.M. and Sue, a *meteorologist* for the National Weather Service, was observing the weather radar in anticipation of heavy rainfall associated with an approaching cold front. As she observed the radar screen, a slight chill ran down her spine as she saw a "hook echo," the characteristic image of a tornado. Within minutes, her fear was confirmed by a state patrolman who reported a large funnel cloud aloft. Sue quickly put into motion the procedures to issue a tornado warning for those communities in the path of this approaching storm. Although property damage was extensive and some injuries were reported, miraculously no one was killed by this tornado. Sue's warning was credited with saving many lives.

Rob is an *intelligence officer* for an infantry battalion of the U.S. Army. In this position he must post current weather reports and forecasts and keep an up-to-the-minute situation map of his battalion's area of military operations using state-of-the-art graphics and map overlays. On a typical day he coordinates air-photo reconnaissance, receives results of SLAR (Side-Looking Airborne Radar) and satellite imagery, and directs the gathering of information and the writing of intelligence reports. He and his staff must be prepared at all times to provide an up-to-date assessment of the military situation in their area of operation.

Sue and Rob both work for the government, but, the common thread between their occupations is more than that. Before talking about that common denominator, let us look briefly at a half dozen or so other occupations.

Daily weather map preparation by meteorologists provides an essential public service. (WIU Visual Production Center)

Traveling by boat across coastal marsh, a scientist conducts an inventory of wildlife resources and then helps to design and promote legislation to protect tideland areas. She is *project manager* for a state Department of Natural Resources. Another scientist analyzes high-altitude photographs and other remotely sensed data produced from orbiting satellites as an *intelligence operation specialist* for a private corporation. A third individual examines passenger data, explores industrial sites, analyzes roadbed problems, and recommends routes to be upgraded or discontinued as a *railroad planner* for a state Department of Transportation. A fourth identifies and plans the development of new resort areas, leads holiday tours, and organizes vacation cruises as *contract administrator* for a large U.S. recreation firm. Another individual supervises field studies of selected ecosystems occupied by declining and threatened animal species as *regional manager* for a major conservation and environmental protection society. A sixth individual produces detailed maps and elaborate reports on strategic areas of the world as a *cartographer* for the U.S. Department of Defense. Yet another studies the patterns of movement and frequency of destructive tornadoes, recommends sites for public building construction, and suggests programs to lessen the impact of flood and drought as a *national disaster planner* for a state Division of Emergency Preparedness.

What do Sue, Rob, and these other individuals have in common? Of course the title of this Viewpoint gives the answer away. Each of the above-mentioned men and women are geographers. Their positions represent only a few from the long list of jobs filled each year by college and university graduates who hold a Bachelor's or Master's Degree in the discipline of geography. Each one has successfully completed the academic requirements in geographic tools and methodology, physical geography, human geography, regional studies, and related fields leading to the degree. Each has gained the knowledge, field experience, and laboratory skills essential to earning a living in the real world.

The variety of tasks performed by geographers may come as a surprise. We have been led to believe that such topics as countries and capitals, rivers and mountains, the products of a nation, or the major regions where cotton and steel are produced occupy the majority of the geographer's attention. Place names, locations, and distributions are as inseparable from geography as dates from history or elements and compounds from chemistry. But knowledge of location, dates, or chemical elements comprises only the framework of information around which the disciplines of geography, history, and chemistry, respectively, are organized. Much more important, in the *applied* or job-related aspects of each discipline, are the research tools, the techniques of analysis, and the special view of a problem that the scientist in each discipline represents.

The traditional tools of geography are the map and the photograph. To these have been added, in the last half of the twentieth century, the products of remote sensing from satellites and the computer as an essential tool for storing, organizing, and analyzing the vast amount of new information about the earth that has become available. Geographers use these tools to describe the location of things on the earth's surface and to determine why these things are located where they are. Although geographers may specialize by focusing on certain types of problems, they remain generalists with a broad but integrated point of view. They have a working knowledge of associated fields of study and are often able to utilize and interrelate the discoveries of other scientists. They are concerned with patterns of distribution of, and interrelationships between, both physical and cultural phenomena. They are unlikely to take a biased or prejudiced view of a problem because they have been educated to see the various aspects of the earth in their true relationships and to identify those elements in the physical and cultural environment that are the key to a problem's solution.

Military intelligence relies heavily upon topographic analysis. (WIU Visual Production Center)

Physical geographers working in the field gather research data, which lead to a more complete assessment of environmental change. (R. Gabler)

Now that we have discussed what geographers do, let us look at who employs geographers. As our occupational descriptions imply—and correctly so—a large number of applied geographers are employed in local and state government offices, in urban and regional planning agencies, in agencies of the Federal Government, in the Armed Forces, in the Foreign Service, and in international organizations. In other words, geography has a major role in government, and geographers are widely employed by various government agencies.

There are also increasing employment opportunities in the private sector, in companies performing tasks similar to those undertaken by government agencies. Real estate development corporations, insurance companies, weather consultant firms, communications industries, environmental consultant agencies, and a host of other organizations employ geographers.

In its applied aspects geography is an exciting, challenging, and highly contributive field of study. Jobs in geography are filled by individuals who enjoy solving problems, who are deeply concerned about the future of the earth on which we live, and who have a knack for recognizing the whole as greater than the sum of its parts. Keep in mind that we have discussed only a few of the career opportunities that are available in the physical aspects of the discipline. Many more opportunities exist in other facets of physical geography as well as in the area of human geography. As one geographer stated when defining the field, "Geography *is* what geographers *do!*"

circumference of the earth. He accomplished this by measuring the shadows cast by vertical poles at noon in two different cities in Egypt; his estimate was within a few hundred miles of the earth's actual cirumference. The accuracy of Eratosthenes' calculation is all the more amazing if we imagine ourselves trying to estimate the shape and size of our planet without the benefit of today's maps, globes, or navigational devices. What kind of description could we provide of the earth, its landforms and oceans, its shape and dimensions, if we knew nothing but what we can see around us? What things can you think of, for example, that might prove the earth is a sphere?

The apparent boundary line between the sky and the earth is called the **horizon.** If you hold a basketball to represent the earth in front of you with the sky as a background, you can see a similar boundary line, which is curved no matter which way you turn the ball. If you were far enough from the earth's surface, the horizon would similarly appear to be curved as indicated in Figure 1.11.

There are many aspects of our world that are related to the curvature of the earth. For example, the curvature affects the intensity and duration of solar radiation received at different locations on the earth. Differences in temperature from place to place and currents in the oceans and atmosphere are also related to the earth's near sphericity. Further, humans determine time by a system based on the earth as a sphere. They have devised a full system of direction and location by means of a grid based on the shape of the globe, and part of their navigation system is based on the earth's spherical shape.

For most purposes the earth can be considered a perfect sphere with an equatorial circumference of 39,840 kilometers (24,900 mi). However, due to forces associated with earth rotation, the area near the earth's equator actually bulges out somewhat, and the two poles are accordingly flattened slightly. So instead of a perfect shape, the earth is more properly an **oblate spheroid** or **ellipsoid of rotation.**

The earth's deviation from a perfect sphere is exceedingly minor. Nevertheless, these irregularities do affect navigation, mapping, and distance accuracies. People working in the areas of navigation, surveying, aeronau-

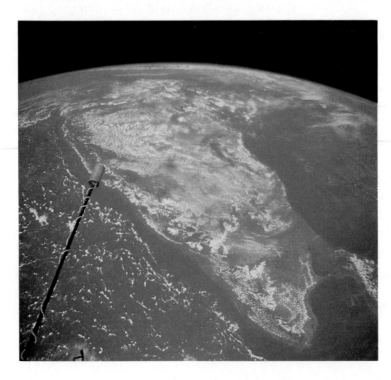

FIGURE 1.11 A satellite view of earth showing the subcontinent of India. The spherical shape of the planet can be clearly seen. (NASA)

tics, and cartography must include in their calculations the deviations of the earth's shape from true sphericity. Scientists have been able to identify the extent of such deviations from measurements of variations in gravitational pull acting on satellites. They have found that the diameter of the earth at the equator is 12,758 kilometers (7927 mi), while from pole to pole it is 12,714 kilometers (7900 mi). On a globe with a diameter of 12 inches this difference of 44 kilometers would be 4/100 of an inch, a deviation of about 1/3 of 1 percent and not noticeable to the naked eye.

Other deviations from the earth's true sphericity are caused by its landforms. Mount Everest in the Himalayas is the highest point of land on the earth at 8847 meters (29,028 ft) above sea level. The lowest known point on the earth's surface is in the Challenger Deep, a part of the Mariana Trench in the Pacific Ocean southwest of Guam. This spot is 11,033 meters (36,198 ft) below sea level. The difference between these two points, 19,880 meters or just over 12 miles, is insignificant when reduced in scale to a globe with a 12-inch diameter.

MOVEMENTS OF THE EARTH

The earth has three basic movements: **galactic movement, rotation,** and **revolution.** The first of these is the movement of the earth with the sun and the rest of the solar system in an orbit around the center of the Milky Way Galaxy. This movement has limited effect upon the changing environment of the earth and is generally the concern of astronomers rather than of geographers. The other two movements of the earth, rotation on an axis and revolution around the sun, are of vital interest to the physical geographer. The consequences of these movements are the phenomena of day and night, variations in the length of day, and to a major extent, the changing seasons.

ROTATION Rotation refers to the turning of the earth on its own axis, an imaginary line extending from the North Pole to the South Pole. The earth rotates on its axis at a uniform rate, making one complete turn with respect to the sun in 24 hours.

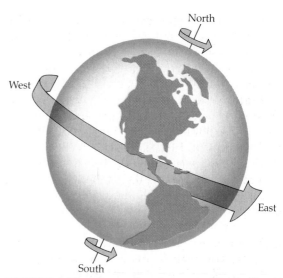

FIGURE 1.12   The earth spins around a tilted axis as it follows its orbit around the sun. The earth's rotation is from west to east, making the stationary sun appear to rise in the east and set in the west.

The earth turns in an eastward direction (Fig. 1.12). The sun "rises" in the east and apparently moves westward across the sky, but it is actually the earth, not the sun, that is moving, and it rotates toward the morning sun (that is, toward the east).

The earth, then, rotates in a direction opposite to the apparent movement of the sun, moon, and stars across the sky. Looking down on a globe from above the North Pole, the direction of rotation is counterclockwise. This eastward direction of rotation not only defines the movement of the zone of daylight on the earth's surface but also helps define the circulatory movements of the atmosphere and oceans.

The velocity of rotation on the earth varies with the distance of a given place from the **equator** (the imaginary circle around the earth halfway between the two poles). All points on the globe take 24 hours to make one complete rotation. However, the rotational velocity depends on how great a distance must be covered in that 24 hours. The rotational velocity at the poles is nearly zero. You can see this by spinning a globe with a finger touching the North Pole. Your finger does not move while the rest of the globe rotates. The greatest velocity of rotation is found at the equator, where the distance traveled by a point in 24 hours is largest. For points along the equator, such as Kampala, Uganda, the velocity is about 460 meters (1500 ft) per second or approximately 1660 kilometers (1038 mi) per hour (Fig. 1.13). In comparison, at Leningrad, USSR, where the distance traveled during one complete rotation of the earth is about half that at the equator, the earth rotates about 830 kilometers per hour.

We are unaware of the speed of rotation, however, because (1) the rate is constant for each place on the earth's surface; (2) the atmosphere rotates with the earth; and (3) there are no nearby objects, either stationary or moving at a different rate with respect to the earth, to which we can relate the earth's movement. Thus, without references we are unable to perceive the speed of rotation.

Rotation accounts for our alternating days and nights. This can be demonstrated by shining a light at a globe while rotating the globe slowly toward the east. You can see that half the sphere is illuminated while the other half is not and that new points are continually moving into the illuminated section of the globe while others are moving into the darkened sector. This corresponds to the earth's rotation and the sun's energy striking the earth. While one half of the earth receives the light and energy of solar radiation, the other half is in darkness. The line around the earth separating the light and dark halves is known as the **circle of illumination.**

REVOLUTION   While the earth rotates on its axis, it also revolves around the sun in an elliptical, almost circular, orbit at an *average* distance from the sun of about 150,000,000 ki-

FIGURE 1.13 The speed of rotation of the earth varies with the distance from the equator.

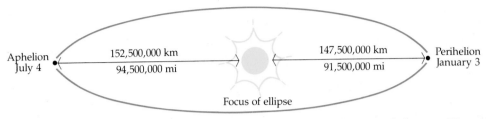

FIGURE 1.14 Oblique view of the elliptical orbit of the earth around the sun. The earth is closest to the sun at perihelion and farthest away at aphelion. Note that in the Northern Hemisphere summer (July), the earth is farther from the sun than at any other time of year.

lometers (93,000,000 mi) (Fig. 1.14). About January 3 the earth is closest to the sun and is said to be at **perihelion** (from Greek: *peri*, close to; *helios*, sun); its distance then from the sun is approximately 147,500,000 kilometers.

Around July 4 the earth is about 152,500,000 kilometers from the sun. It is then that the earth has reached its furthest point from the sun and is said to be at **aphelion** (Greek: *ap*, away; *helios*, sun). Five million kilometers is insignificant in space, and these varying distances from the earth to the sun do not materially affect the receipt of energy on earth. Hence they have no relationship to the seasons.

The period of time the earth takes to make one revolution around the sun determines the length of one year. Because the earth makes $365\frac{1}{4}$ rotations on its axis during the time it takes to complete one revolution of the sun, a year is said to have $365\frac{1}{4}$ days. Because of the difficulty of dealing with a fraction of a day, it has been decided that a year would have 365 days and that in every fourth year, called *Leap Year*, an extra day would be added in February.

PLANE OF THE ECLIPTIC, INCLINATION, AND PARALLELISM The earth in its orbit around the sun moves in a constant plane. This plane is called the **plane of the ecliptic**. The earth's axis is tilted at an angle of $23\frac{1}{2}°$ from the vertical to the plane of the ecliptic and thus has a constant **angle of inclination**, as it is called, of $66\frac{1}{2}°$ with the plane (Fig. 1.15).

In addition to a constant angle of inclination, the earth's axis maintains another characteristic called **parallelism**. As the earth revolves around the sun, the earth's axis remains parallel to its former positions. That is, at every position in the earth's orbit the axis remains pointed toward the same spot in the sky. For the North Pole that spot is close to the star we call the North Star, or Polaris. Thus, the earth's axis is fixed with respect to the stars outside our solar system but not with respect to the sun.

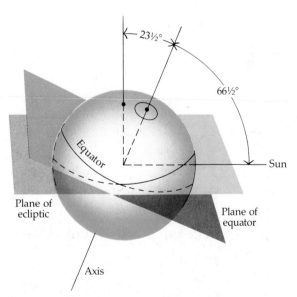

FIGURE 1.15 The plane of the ecliptic is defined by the orbit of the earth around the sun. The $23\frac{1}{2}°$ inclination of the earth's rotational axis causes the plane of the equator to cut across the plane of the ecliptic.

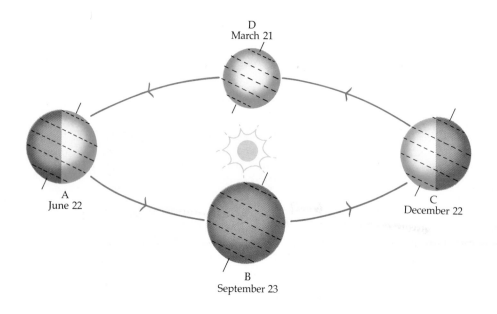

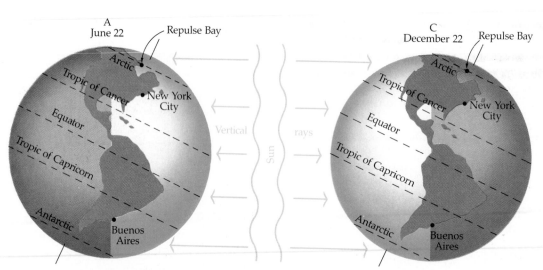

FIGURE 1.16 Geometric relationships betweeen the earth and the sun at the solstices. Note the differing day lengths at the summer and winter solstices in the Northern and Southern Hemispheres.

To get a better picture of what happens to the earth in its movement around the sun, pick up a globe and carry it around an imaginary or substitute sun. Keep certain facts in mind: (1) the earth in its orbit around the sun and the sun itself lie in the plane of the ecliptic, (2) the earth's axis is inclined with respect to that plane so as to make a constant angle of inclination with it of $66\frac{1}{2}°$, and (3) the earth's axis as the earth moves around the sun remains parallel to itself in all former positions. As you will see as you walk the globe around the sun under these conditions, the earth's axis changes position with respect to the sun as the earth re-

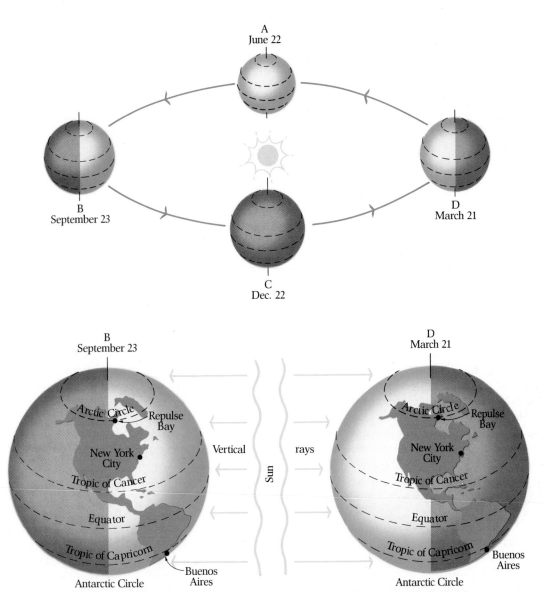

FIGURE 1.17 Geometric relationships between the earth and the sun at the equinoxes. Day length is 12 hours everywhere, since the circle of illumination crosses the equator at right angles and cuts through both poles.

volves around the sun. Sometimes one pole of the axis is tilted toward the sun, sometimes away, and sometimes neither. These changing positions have a distinct bearing on seasonal changes on the earth's surface and on variations in intensity of the sun's rays from place to place and from time to time.

THE SEASONS

About June 22 the earth is in a position in its orbit so that the northern tip of its axis is inclined toward the sun at an angle of 23½° from a line perpendicular to the plane of the ecliptic. This is called the summer

solstice (from Latin: *sol*, sun; *sistere*, to stand) in the Northern Hemisphere. We can best see what is happening if we refer to Figure 1.16, position A. On that diagram we can see that unequal parts of the Northern and Southern Hemispheres receive light from the sun. That is, as we imagine rotating the earth under these conditions, a larger portion of the Northern Hemisphere remains in daylight than of the Southern Hemisphere. Conversely, a larger portion of the Southern Hemisphere remains in darkness than of the Northern Hemisphere.

Thus, referring again to Figure 1.16, position A, a person living at Repulse Bay, Canada, north of the Arctic Circle, has a full 24 hours of daylight at the June solstice and can go hunting at 1:00 A.M. We can also see that someone living in New York City will experience a longer period of daylight than of darkness. And someone living in Buenos Aires, Argentina, will have a longer period of darkness on that day. This day is called the winter solstice in the Southern Hemisphere.

Now let us imagine the movement of the earth from its position at the June solstice toward a position a quarter of a year away in September. As the earth moves toward that new position, we can imagine the changes that will be taking place in our three cities. First, in Repulse Bay, from no darkness at all on the June solstice there will be an increasing amount of darkness through July, August, and September. In New York sunset will be getting earlier, although it will still be light enough to play softball after dinner. And in Buenos Aires the situation will be reversed. As the earth moves toward its position in September, we can see that the periods of daylight in the Southern Hemisphere will begin to get longer, the nights shorter.

Finally, on or about September 23 the earth will reach a position known as an **equinox** (Latin: *equus*, equal; *nox*, night). On this date all over the earth day and night will be of equal length. Thus, on the equinox conditions are identical for both hemispheres. As you can see on page 19, Figure 1.17, position B, the earth's axis points neither toward nor away from the sun; the circle of illumination passes through both poles, and it cuts the earth in half along its axis.

Imagine again the revolution and rotation of the earth while moving from September 23 toward a new position another quarter of a year later in December. We can see that in Repulse Bay the nights will be getting longer and longer until on the winter solstice, which occurs on or about December 22, this northern town will experience 24 hours of darkness (Figure 1.16, position C). The only natural light at all in Repulse Bay will be a faint glow at noon refracted from the sun below the horizon. And in New York, too, the days will get shorter, the sun will set earlier, until by the time you do your Christmas shopping, it's dark before the stores close at 5:30. Again, we can see that in Buenos Aires the situation is reversed, and on December 22 that city will experience its summer solstice and conditions will be much as they were in New York City in June. It may be a sweltering day in Buenos Aires and everyone will go to the beach for the Christmas holidays.

Moving from late December through another quarter of a year to late March, Repulse Bay will have longer and longer periods of daylight, as will New York, while in Buenos Aires the nights will be getting longer, though they still will not be as long as the days. Then on or about March 21, the earth will again be in an equinox position similar to the one in September (Figure 1.16, position D). Again, days and nights will be equal all over the earth. In other words, it will be 12 hours from sunrise to sunset and from sunset to sunrise. Finally, moving through another quarter of the year toward the June solstice where we began, Repulse Bay and New York City are both experiencing longer periods of daylight than of darkness, and the sun is setting earlier and earlier in Buenos Aires until on or about June 22 Repulse Bay and New York City will have their longest day of the year, Buenos Aires its shortest. Further, we can see that on June 22 a point on the Antarctic Circle in the Southern Hemisphere will experience a winter solstice similar to that which Repulse Bay had on December 22, with no daylight in 24 hours except what will appear at noon as a glow of twilight in the sky.

LINES RELATED TO EARTH REVOLUTION

Looking at the diagrams of the earth in its various positions as it revolves around the sun, we can see that the angle of inclination is important. For, on June 22, because the earth's axis is tilted 23½° toward the sun with respect to a line drawn perpendicular to the plane of the ecliptic, the sun's rays can reach that far (23½°) beyond the North Pole. The **Arctic Circle,** an imaginary line drawn around the earth 23½° from the North Pole (or 66½° north of the equator) marks this limit. We can see from the diagram that all points on or north of the Arctic Circle will experience no darkness on the June solstice, and further that all points south of the Arctic Circle will have some darkness on that day. The **Antarctic Circle** in the Southern Hemisphere (23½° north of the South Pole, or 66½° south of the equator) marks a similar limit.

Furthermore, it can be seen from the diagrams that the sun's **vertical,** or **direct, rays** (rays that strike the earth's surface at right angles) also shift position in relation to the poles and the equator as the earth revolves around the sun. At the time of the June solstice, the sun's rays are vertical or directly overhead at noon at all points located 23½° *north* of the equator. This imaginary line around the earth marks the northernmost position that the solar rays will ever be directly overhead during a full revolution of our planet around the sun. The imaginary line marking this limit is called the **Tropic of Cancer.** Six months later at the time of the December solstice, the solar rays are vertical and the noon sun is directly overhead at all points 23½° *south* of the equator. The imaginary line marking this limit is known as the **Tropic of Capricorn.** At the times of the March and September equinoxes, the vertical solar rays will strike directly only at the equator, and the noon sun is directly overhead at all points on that line.

Note also that on any day of the year the sun's rays can strike the earth directly at only one position on or between the two tropics. All other positions that day will receive the sun's rays at an angle of less than 90° (or will receive no sunlight).

INSOLATION AND THE SEASONS

Solar radiation received at the earth's surface is known as **insolation** (for *in*coming *sol*ar radia*tion*), and it is the main source of energy on our planet. The seasonal variations in temperature that we experience are due primarily to fluctuations in insolation.

What causes these variations in insolation and thus in the seasons? One logical answer might be that the radiation given off by the sun fluctuates greatly and on a regular basis throughout the year. But it would be improbable for such cycles to correspond exactly with the time that it takes for the earth to revolve around the sun. Also, any fluctuations in solar radiation are not very large. We must thus look elsewhere for the causes of changes in amounts of insolation.

It is true that the earth's atmosphere affects the amount of insolation received. Heavy cloud cover, for instance, will keep more solar radiation from reaching the earth's surface than will a clear blue sky. However, cloud cover is irregular and unpredictable, and it affects total insolation to only a minor degree over long periods of time.

The real answer to the question of what causes variations in insolation can be found by once again studying Figures 1.16 and 1.17. Two major phenomena vary regularly for a given position on the earth as our planet rotates on its axis and revolves around the sun: the duration of daylight and the angle of solar rays. The amount of daylight controls the duration of solar radiation, and the angle of the sun's rays directly affects the intensity of solar radiation received. Together, the intensity and the duration of radiation are the major factors that affect the amount of insolation.

This situation is like an oven in which a roast is being cooked. The roast will cook faster and get browner if (1) the temperature is turned up and/or (2) someone leaves the oven on longer than usual. Likewise, a spot on the earth will receive more insolation if (1) the sun shines more directly, or (2) the sun shines longer, or (3) both. One reason that places along the Tropics of Cancer and Capricorn are so hot during their summer solstice is that the sun's rays are intense and the day is long (there are many hours of daylight).

The intensity of solar radiation received at any one time varies from place to place because the earth presents a spherical surface to insolation. Therefore, only a portion of the earth's surface can receive radiation at right angles, while the rest is struck at varying oblique angles. As we can see from Figure 1.18, a bundle of rays of solar energy that strikes the earth at a vertical angle covers less area than an equal amount striking the earth at an oblique angle. And we can see that the closer to a right angle that such a group of rays strikes the earth, the smaller will be the area covered. Since the amount of energy is the same no matter what kind of angle the rays make with the surface, it follows that the smaller the area that is struck, the greater will be the intensity per unit area. Conversely, the more oblique the angle at which the sun's rays strike the earth, the greater the area over which those rays will be spread, and so less energy will strike the earth per unit area. In addition, the atmosphere limits to some extent the amount of insolation that reaches the earth's surface, and oblique rays must pass through a greater thickness of atmosphere than vertical rays.

The duration of solar energy is related to the amount of daylight received at a particular point on the earth, since no insolation is received at night. Obviously, the longer the period of daylight, the greater the amount of solar radiation that will be received at that location. And, as we have seen, periods of daylight vary in length through the seasons of the year as well as from place to place on the earth's surface.

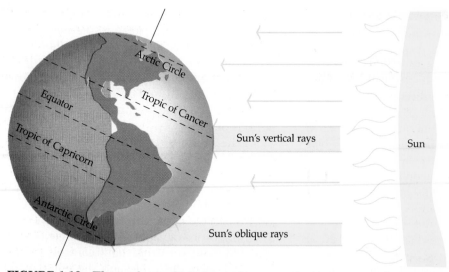

FIGURE 1.18 **The angle at which the sun's rays strike the earth determines the amount of solar energy received per unit of surface area. This amount in turn controls the seasons. The diagram represents the June condition, in which solar radiation strikes the surface perpendicularly only in the Northern Hemisphere, creating summer conditions there. In the Southern Hemisphere, oblique rays are spread over large areas, producing less receipt of energy per unit of area and making this the winter hemisphere.**

VARIATIONS OF INSOLATION WITH LATITUDE

Neglecting for the moment the influence of the atmosphere on variations in insolation during a 24-hour period, a place will receive its greatest insolation at solar noon when the sun has reached its zenith or highest point in the sky for that day. At any location, no insolation will be received during the hours of darkness. The amount of energy received after daybreak increases as the earth rotates until the time of solar noon. The amount of insolation then decreases until the next period of darkness begins.

We also know that the amount of daily insolation received at any one location on the earth varies with the seasons. There are three distinct patterns in the distribution of the seasonal receipt of solar energy in each hemisphere. These patterns serve as the basis for recognizing six latitudinal zones, or bands, of insolation and temperature that circle the earth (Fig. 1.19).

If we look first at the Northern Hemisphere, we may take the Tropic of Cancer and the Arctic Circle as the dividing lines for three of these distinctive zones. The area between the equator and the Tropic of Cancer can be called the north **tropical zone.** Here, insolation is always high but is greatest at the two times during the year that the sun is directly overhead at noon. These dates vary according to latitude. The north **midlatitude zone** is the wide band between the Tropic of Cancer and the Arctic Circle. In this belt, insolation is greatest on the June solstice, when the sun reaches its highest noon altitude and the period of daylight is long, and it is least when the sun is low in the sky and the period of daylight is short at the December solstice. The north **polar zone,** or **arctic zone,** extends from the Arctic Circle to the pole. In this region, as in the midlatitude zone, insolation is greatest at the time of the June solstice, but it ceases during the period that the sun's rays are blocked entirely by the tilt of the earth's axis. This period lasts for 6

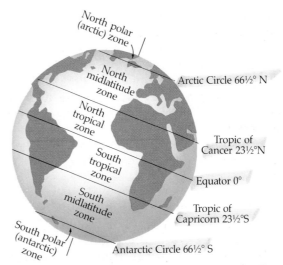

FIGURE 1.19   The line of the equator, the Tropics of Cancer and Capricorn, and the Arctic and Antarctic Circles define six latitudinal zones that have distinctive insolation characteristics.

months at the North Pole, but it is as short as 1 day directly on the Arctic Circle.

Similarly, there are the south tropical zone, the south midlatitude zone, and the south polar or **Antarctic zone** separated by the Tropic of Capricorn and the Antarctic Circle in the Southern Hemisphere. These areas get their greatest amounts of insolation at opposite times of the year from the northern zones.

Despite various patterns in the amount of insolation received in these zones, there are generalizations that we can make. For example, total annual insolation at the top of the atmosphere at a particular latitude remains constant from year to year. Furthermore, annual insolation tends to decrease from lower latitudes to higher latitudes. And the closer to the poles a place is located, the greater will be its seasonal variations caused by fluctuations in insolation.

The amount of insolation received by the earth is an important concept in the understanding of atmospheric dynamics and the distribution of climate and vegetation. Such climatic elements as temperature, precipitation, and winds are controlled in part by the amount of insolation received by the earth. People are dependent upon certain levels of insolation for physical comfort, and plant life is especially sensitive to the amount of available insolation. You may have noticed plants that have wilted in too much sunlight or that have grown brown in a dark corner away from a window. Over a longer period of time, deciduous plants have an annual cycle of budding, flowering, leafing, and losing their leaves. This cycle is apparently determined by the fluctuations of increasing and decreasing solar radiation that mark the changing seasons. Even animals respond to seasonal changes; some animals hibernate, while many North American birds fly south toward warmer weather as winter approaches, and many animals breed at such a time that their offspring will be born in the spring, when warm weather is approaching.

QUESTIONS FOR DISCUSSION AND REVIEW

1. How do geographers define a system? What is meant by the earth system?

2. What is a life support system?

3. List the four major divisions of the earth system. Give examples of how the divisions interact with one another.

4. What are open and closed systems; energy and materials systems? How does feedback affect the dynamic equilibrium of a system?

5. Define *ecosystem*. Give reasons why the study of ecosystems often serves to illustrate the close relationships between humans and the environment.

6. Give an example of an ecosystem in your local area that has been affected by man. In your opinion, was the change good or bad? What values are you using in making such a judgement?

7. Define *pollution*. How have various kinds of pollution affected your own life? List some sources of pollution in your city or town.

8. Define *geography*, and discuss how it fits into the realm of the physical as well as the social sciences.

9. In the twentieth century we tend to accept without proof that the world is round. What are some of the ways to prove that it really is round?

10. Is the earth an absolutely perfect sphere? Why not? How are the imperfections important in the study of geography?

11. Describe briefly how the earth's rotation and revolution affect life on earth.

12. If the sun is closest to the earth on January 3, why isn't winter in the Northern Hemisphere warmer than winter in the Southern Hemisphere?

13. Define *plane of the ecliptic, angle of inclination, parallelism.*

14. Identify the two major factors that cause regular variation in insolation throughout the year. How do they combine to cause the seasons?

15. Given what you know of the sun's relation to life on earth, explain why the solstices and equinoxes have been so important to cultures all over the world. What are some of the major festivals associated with these times of the year?

16. Describe in your own words the relationship between insolation and latitude.

2

REPRESENTATIONS OF THE EARTH

LOCATION ON THE EARTH

The determination of the location of particular features or of distributions of phenomena on the earth's surface is a necessary part of physical geography, and methods developed to find and record locations have become an important part of the collection of tools used by geographers.

The principle of location is essential to geographers in their attempts to describe and analyze different aspects of what we call the *earth system*. Locational relationships between two phenomena on the earth's surface, whether they are close together or thousands of miles apart, may be relevant to an understanding of how they fit within the entire system. The particular location of a mountain, for instance, can affect the amount of rainfall in a neighboring area, and rainfall in turn can affect nearby rivers and lakes, soil, vegetation, animals, and the human land use.

It is probable that almost as soon as people began to communicate, they began to look for and develop a common language of location. They probably used features of the landscape and would tell someone to "Go along the river until it forks, then follow the left branch until you come to the campsite." Just as we use streets, traffic lights, and city blocks, primitive humans used a river, or trees, hills, and other obvious landmarks.

It was when ancient people began to sail across the open seas that they began to look for more permanent direction signals and ways to describe location. They found that if they used the stars and the sun and their patterns of rising, setting, and apparent circling in the heavens, they could give more accurate directions, long before the development of the first crude compass. The stars and the sun and their relationship to the earth still form a basic part of navigation, the science of location.

No one knows when or where the first maps were made as their origin is lost in antiquity. Probably the first crude maps were developed by different cultures in several different parts of the world, such as China, Egypt, the Pacific Islands, Greece, Mexico, and Peru. Such early maps were made of sticks, or drawn on clay tablets, stone slabs, papyrus, linen, and silk. These were the early ancestors of our present maps and were fundamental to the beginnings of geography.

Map furnished by Dr. Donald E. Luman, Department of Geography, Laboratory for Cartography and Spatial Analysis, Northern Illinois University.

Maps and globes are visual representations of all or part of the earth. These representations carry a great deal of information, much of it in symbolic form. So, not only has a standard language of location been developed, but there are also symbols and signs that need to be understood before we can learn all that these visual representations tell us.

Globes, maps, and atlases (collections of maps) are more than a fundamental part of geography. In fact, an entire science of mapmaking, known as **cartography,** has developed. Today, sophisticated computer-assisted data-collection methods have been developed using remote sensing from aircraft and space satellites to help us map and monitor the earth (and other planets). In addition, we can all think of numerous fields, such as navigation, political science, planning, surveying, history, meteorology, and geology, in which representations of the earth, either globes or maps, are vital. In our everyday lives—in school, newspapers, television, hiking, and all kinds of traveling—we all have had experience with representations of the earth.

GLOBES AND GREAT CIRCLES

The globe is a nearly perfect representation of the earth. It shows the shape of the earth, the spatial relationships between the different land and water forms, comparative distances between different features, comparative sizes, and accurate direction. It is an extremely useful geographic tool, for it enables us to see, instead of trying to imagine, many things about the earth. For example, learning about the seasons or understanding how the duration of daylight in one place changes as the earth revolves around the sun are both made a great deal easier. Because the globe is spherical like the earth, it can represent almost without distortion various features and relationships on the earth's surface (Fig. 2.1).

Globes have their limitations, however. They are clumsy to carry and expensive to produce. And because they show the entire earth, they cannot show the detail of a particular area in which we might be interested. For instance, it is impractical to take a globe with us to go hiking. We would rather use a map that

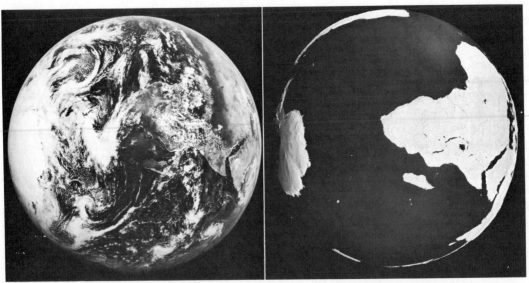

FIGURE 2.1   The earth as seen from space, showing a portion of Africa and surrounding oceans (left), and a manufactured globe representing the same general viewpoint (right). (NASA)

showed elevations and the pattern of trails and rivers and that could be folded up and put in a pocket. Despite these limitations, globes, as the most perfect representations of the earth, serve many vital functions. Furthermore, since maps are actually attempts to depict the spherical earth on a flat surface, the understanding of the globe helps us understand maps.

An imaginary circle on the surface of the earth (like the equator or any other you want to imagine) whose plane passes through the center of the earth is called a great circle (Fig. 2.2a,b). It is "great" because it is the largest possible circle that can be drawn around the earth. It has several useful characteristics: (1) Every great circle cuts the earth in half and each half is known as a hemisphere; (2) every great circle is a circumference of the earth; and (3) great circles provide the shortest routes of travel on the earth's surface. Circles whose planes do not pass through the center of the earth are called small circles (Fig. 2.2c). A small circle does not cut the earth in half, and it is not a circumference of the earth.

Find the great circle that connects two places, and you will have found the shortest route between them. Put a rubber band around a globe to help visualize this. Connect any two cities such as Moscow and New York, San Francisco and Tokyo, New Orleans and Paris, Kansas City and Singapore, by stretching the rubber band around the globe so that it touches both cities and divides the surface of the globe in half. The rubber band then forms a great circle, and the route marked by it between the two cities is the great circle route. This is often the route that navigators chart for planes and ships to follow because it saves time and fuel. The greater the distance between two points, the larger the savings will be in distance by following the great circle route.

It may seem strange that the old saying about a straight line being the shortest distance between two points is not always true. This *is* true on a flat surface. But the earth is a spherical body, and the arc of a great circle is the shortest distance between two points *on that surface*.

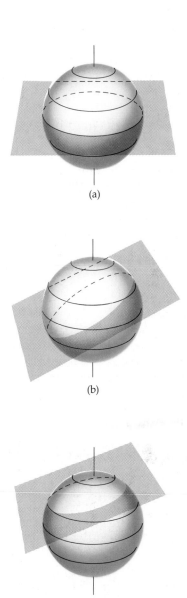

(a)

(b)

(c)

FIGURE 2.2 Planes slicing through a globe's center (a) at the equator and (b) obliquely. In either case the globe is cut into equal halves, and the line of intersection of the plane with the globe is a great circle having the same circumference as the globe itself. In (c) the plane slices the globe into unequal portions. The line of intersection of such a plane with the globe is a small circle.

One other circle that we have already encountered is an example of a great circle. The circle of illumination always divides the earth into a light half and a dark half, a day hemisphere and a night hemisphere.

LATITUDE AND LONGITUDE

Imagine that as you travel across the United States during summer vacation, you pass through the state of Colorado. A friend suggests you stay at a spectacular campground near the town of Cimarron. You will pull the Colorado road map out of the glove compartment and look up Cimarron in the index. It says "D2." Turning the map over, you run your left index finger down the side of the map until you reach D. Holding this finger on the D, you run the index finger of your other hand along the bottom of the map until you reach 2. Then you bring your left index finger across from D while you bring your right one up from 2. They meet in a box marked on the map. Scanning the area within the box, you locate Cimarron (Fig. 2.3).

What you have used is a coordinate system of intersecting lines, which on the map make a grid system of boxes. It is more difficult to describe location on a sphere like the earth than on a flat surface such as the roadmap. Imagine, for example, a perfectly smooth spherical object like a playground ball. Let us say that there is a leak in the ball. You know where it is, and you want to describe its location in a note to the person who's going to fix it. However, you soon reach a state of frustration as it becomes apparent that there are no fixed reference points. You cannot say that the leak is so many inches from something if the sphere is the same everywhere with no corners, points, lines, or trademark.

MEASUREMENT OF LATITUDE The earth (globe), however, unlike the playground ball, does have two fixed reference points. As the earth rotates uniformly on an imaginary axis, the ends of that axis are fixed points called the **North Pole** and **South Pole.** These two points can be used to locate a great circle that lies equidistant between them. This is the equator. Now we can measure the angular distance north of the equator or south of it; this distance north or south is called **latitude** and is measured in degrees.

Let us try to locate the latitude of a place on the spherical surface of the earth. For example, we can attempt to describe the latitude of Los Angeles without mentioning any physical or political features of the earth.

Imagine a line that goes from the center of the earth to Los Angeles and another line that goes from the center of the earth to a point on the equator directly south of Los Angeles. These two lines make an angle and, as indicated in Figure 2.4a, the arc of this angle is the distance on the spherical surface that Los Angeles is north of the equator. That is, the arc of that angle is the latitude of Los Angeles, the distance of the city north of the equator in degrees. The angle made by the two imaginary lines is just over 34°; hence the latitude of Los Angeles is about 34°N (that is, north of the equator).

According to this system the equator is designated as 0° latitude. As we go farther and farther north or south of the equator, the angles and their arcs will increase until we reach

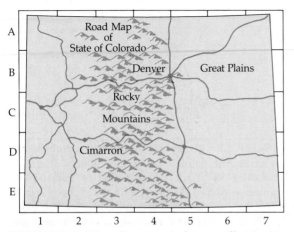

FIGURE 2.3 Use of a rectangular coordinate system to locate a position accurately.

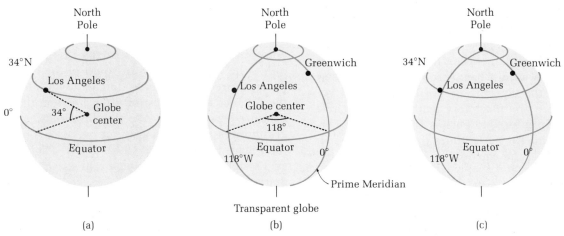

FIGURE 2.4 (a) The geometric basis for the latitude of Los Angeles. Latitude is the angular distance north or south of the equator. (b) The geometric basis for the longitude of Los Angeles. Longitude is the angular distance east or west of the prime meridian, which by international agreement passes through Greenwich, England. (c) The location of Los Angeles is 34°N, 118°W.

the North or South Pole. At the poles then we reach our maximum latitude of 90° north or south.

Since the circumference of the earth is approximately 40,000 kilometers (25,000 mi) and there are 360° in that circumference, we can determine by simple division that a degree of latitude is equal to about 112 kilometers (70 mi). Since a degree of latitude covers a relatively large area, it is conventional to break degrees into smaller parts representing shorter distances. Thus, Los Angeles is actually located at 34°03′N (34 degrees, 3 minutes north). There are 60 minutes in every degree. We can get even more accurate: 1 minute is made up of 60 seconds, which are indicated as ″. So we could have a place at latitude 23°34′12″S, which we would read as 23 degrees, 34 minutes, 12 seconds south. A minute of latitude is 1.9 kilometers (1.2 mi) long, and a second is about 30 meters (102 ft).

When visually determining latitude for navigation purposes, a **sextant** is used (Fig. 2.5). This instrument measures the angle between the horizon and a celestial body, such as the noonday sun or the North Star (Polaris). However, no matter how accurately we pin-

point the latitude of a place like Los Angeles, we have given only half the story. For we have only described the location of Los Angeles as being approximately 34° north of the equator. There are, however, an infinite number of points that are 34° north of the equator.

MEASUREMENT OF LONGITUDE In order to describe the location of Los Angeles we must show where on the latitude of 34°N we can find Los Angeles. Remembering the grid system we used with the road map, we can see

FIGURE 2.5 **For the purpose of determining the latitude of the observer, a sextant is used to measure celestial angles.**

that if we could also measure the east-west position of Los Angeles, we could pinpoint its location.

However, in order to describe east-west position, we must have a starting point, just as we had the equator for north and south. We could use any imaginary half of a great circle drawn from pole to pole as an east-west reference point for all latitudes (or distances north or south of the equator), as it would cut all latitudes.

The starting point of east-west measurement on the globe is arbitrary and, in the past, was a matter of national pride. So many countries had their own beginning of a half great circle for east-west measurement that confusion was rampant and locations were often dependent upon the nationality of the mapmaker. This confusion was finally resolved by international agreement in 1884, when the half of a great circle that passed through Greenwich, England (near London), was generally accepted as the starting point. We call this half-circle the **prime meridian** and distance east or west of the prime meridian **longitude.**

Like latitude, longitude is measured in degrees, minutes, and seconds. Returning to Los Angeles, let us imagine that same line drawn from the center of the earth to the point where the half-circle passing through Los Angeles crosses the equator. This time, a second imaginary line will go from the center of the earth to a point where the prime meridian crosses the equator. Figure 2.4b shows that these two lines drawn from the center of the earth make an angle, the arc of which is the distance that Los Angeles is east or west of the prime meridian, in degrees. Los Angeles is 118° west of Greenwich.

The prime meridian is at 0° longitude, and as we go farther and farther east or west of the prime meridian, our longitude will increase. As we travel toward the east, for example, we will finally reach the middle of the Pacific Ocean, which is halfway around the world from Greenwich, or at 180°E. This place is also 180°W. Thus, longitude is measured in degrees east or west from 0° at Greenwich to a maximum of 180° in the Pacific Ocean.

THE GLOBAL GRID

We have seen, then, that we can locate any point on the globe (earth's surface) by its latitude—the distance north or south of the equator measured in degrees—and its longitude—the distance east or west of the prime meridian measured in degrees. We do this by using a grid system similar to the one we used on a road map. However, the global grid is a special grid system that is applicable to the earth itself, that can be used on all globes, that can be transferred to maps of the earth, and that can be understood by all people all the time (i.e., it is not dependent upon special times of the year, special features of the land, or on use of a particular globe or map). The global grid will serve us long into the future too. It is the basis for determining location on, and the mapping of, the moon, planets, and other celestial bodies.

PARALLELS AND MERIDIANS The global grid is made up of lines that run east and west around the globe and lines that run north and south from pole to pole (Fig. 2.6). The east-west lines mark the distance north or south of the equator. These lines of latitude all circle the

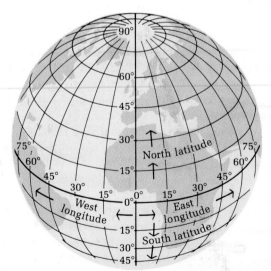

FIGURE 2.6 Globelike representation of the earth, showing the global grid with parallels and meridians at 15° intervals.

Plate 1   Standard USGS Topographic Map Symbols

Primary highway, hard surface . . . .

Secondary highway, hard surface . . . .

Light-duty road, hard or improved surface . . . .

Unimproved road . . . .

Road under construction, alinement known . . . .

Proposed road . . . .

Dual highway, dividing strip 25 feet or less . . . .

Dual highway, dividing strip exceeding 25 feet . . . .

Trail . . . .

Railroad: single track and multiple track . . . .

Railroads in juxtaposition . . . .

Narrow gage: single track and multiple track . . . . . .

Railroad in street and carline . . . .

Bridge: road and railroad . . . .

Drawbridge: road and railroad . . . .

Footbridge . . . .

Tunnel: road and railroad . . . .

Overpass and underpass . . . .

Small masonry or concrete dam . . . .

Dam with lock . . . .

Dam with road . . . .

Canal with lock . . . .

Buildings (dwelling, place of employment, etc.) . . . .

School, church, and cemetery . . . .   Cem

Buildings (barn, warehouse, etc.) . . . .

Power transmission line with located metal tower . . . .

Telephone line, pipeline, etc. (labeled as to type) . . . .

Wells other than water (labeled as to type) . . . .   oOil . . . . . oGas

Tanks: oil, water, etc. (labeled only if water) . . . .   ● ● ● ⊘Water

Located or landmark object; windmill . . . .   o . . . . . . !

Open pit, mine, or quarry; prospect . . . .   ⋈ x

Shaft and tunnel entrance . . . .   ■ . . . . . Y

Horizontal and vertical control station:

    Tablet, spirit level elevation . . . .   BM △ 5653

    Other recoverable mark, spirit level elevation . . . .   △ 5455

Horizontal control station: tablet, vertical angle elevation   VABM △ 95I9

    Any recoverable mark, vertical angle or checked elevation   △3775

Vertical control station: tablet, spirit level elevation . . . . .   BM × 957

    Other recoverable mark, spirit level elevation . . . .   × 954

Spot elevation . . . .   × 7369   × 7369

Water elevation   670   670

Boundaries: National . . . .

    State . . . .

    County, parish, municipio . . . .

    Civil township, precinct, town, barrio . . . .

    Incorporated city, village, town, hamlet . . . .

    Reservation, National or State . . . .

    Small park, cemetery, airport, etc. . . . .

    Land grant . . . .

Township or range line, United States land survey . . . . .

Township or range line, approximate location . . . .

Section line, United States land survey . . . .

Section line, approximate location . . . .

Township line, not United States land survey . . . .

Section line, not United States land survey . . . .

Found corner: section and closing . . . .

Boundary monument: land grant and other . . . .   □ . . . . □

Fence or field line . . . .

Index contour . . . .   Intermediate contour . . . .

Supplementary contour . . . .   Depression contours . . . .

Fill . . . .   Cut . . . .

Levee . . . .   Levee with road . . . .

Mine dump . . . .   Wash . . . .

Tailings . . . .   Tailings pond . . . .

Shifting sand or dunes . . . .   Intricate surface . . . .

Sand area . . . .   Gravel beach . . . .

Perennial streams . . . . .   Intermittent streams . . .

Elevated aqueduct . . . . .   Aqueduct tunnel . . . . .

Water well and spring . o . . . . . . ⌒   Glacier . . . .

Small rapids . . . .   Small falls . . . .

Large rapids . . . .   Large falls . . . .

Intermittent lake . . . . . .   Dry lake bed . . . .

Foreshore flat . . . .   Rock or coral reef . . . .

Sounding, depth curve . .   Piling or dolphin . . . . .

Exposed wreck . . . .   Sunken wreck . . . .

Rock, bare or awash; dangerous to navigation . . . .

Marsh (swamp) . . . .   Submerged marsh . . . .

Wooded marsh . . . .   Mangrove . . . .

Woods or brushwood . . . .   Orchard . . . .

Vineyard . . . .   Scrub . . . .

Land subject to
controlled inundation . . . .   Urban area . . . .

Plate 2   World Map of Population Density

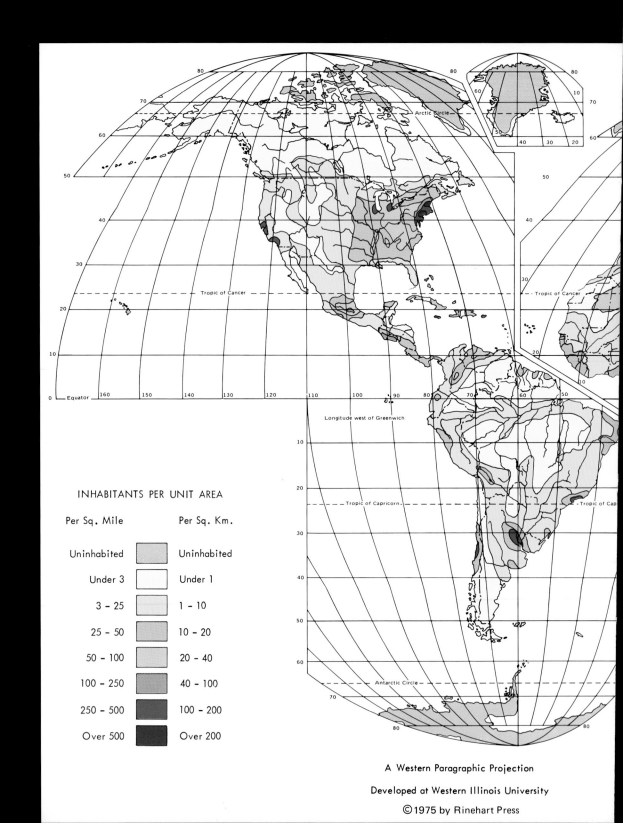

INHABITANTS PER UNIT AREA

Per Sq. Mile		Per Sq. Km.
Uninhabited		Uninhabited
Under 3		Under 1
3 – 25		1 – 10
25 – 50		10 – 20
50 – 100		20 – 40
100 – 250		40 – 100
250 – 500		100 – 200
Over 500		Over 200

A Western Paragraphic Projection

Developed at Western Illinois University

©1975 by Rinehart Press

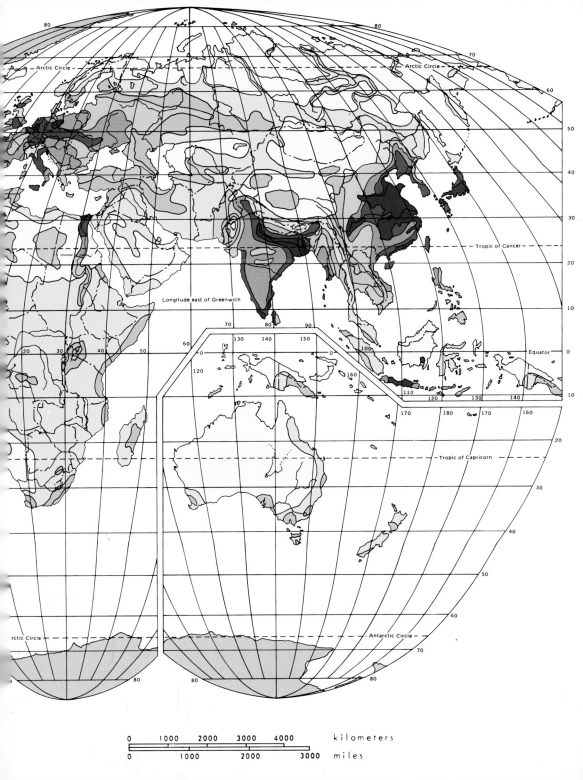

Arctic Circle   Arctic Circle

80   80
                                              70
                                                   60
                                                        50
                                                        40
Tropic of Cancer

Longitude east of Greenwich

70   80   90
60   130   140   150
120   0   0
                                100
                           160
                      110
                 120   130   140

170   180   170   160
Equator

20   30   40   50

20   30   40

10

Tropic of Capricorn

30

40

50

60

rctic Circle   Antarctic Circle
                                              70

80   80   80

kilometers
0   1000   2000   3000   4000

0   1000   2000   3000   miles

Top: 1 : 250,000 scale,
1 inch = nearly 4 miles.
Area shown,
107 square miles.

Center: 1 : 62,500 scale,
1 inch = nearly 1 mile.
Area shown,
6¾ square miles.

Bottom: 1 : 24,000 scale,
1 inch = 2000 feet.
Area shown,
1 square mile.

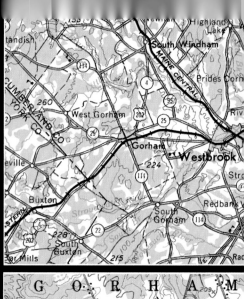

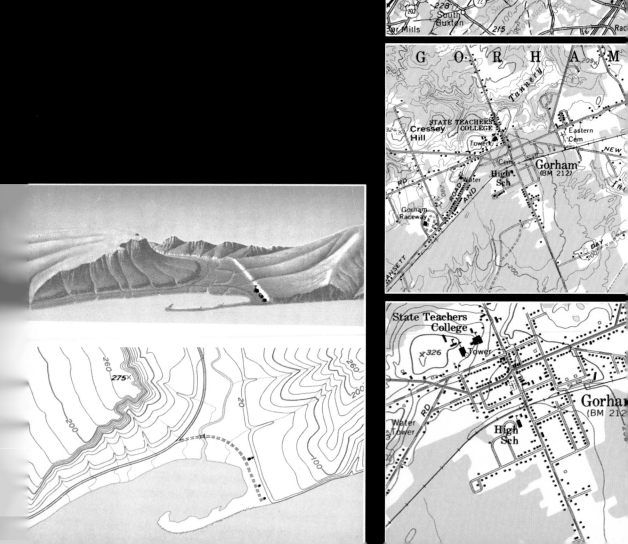

Plate 4 Digital Terrain Modeling

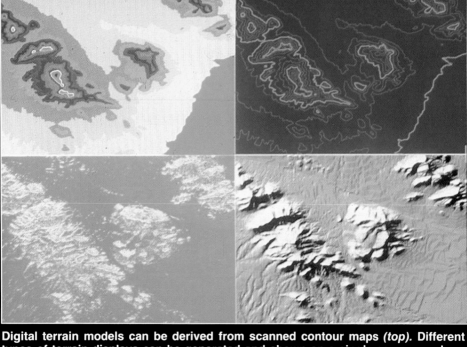

Digital terrain models can be derived from scanned contour maps *(top).* **Different types of terrain displays can be generated and shown on a single screen, such as color-scaled contour maps, conventional contour maps, color-coded shaded relief maps, and 3-D stereographic shaded relief maps** *(bottom). (Intergraph Corporation, Huntsville, Alabama)*

Plate 6   Color Aerial Photograph of Laguna Beach, California

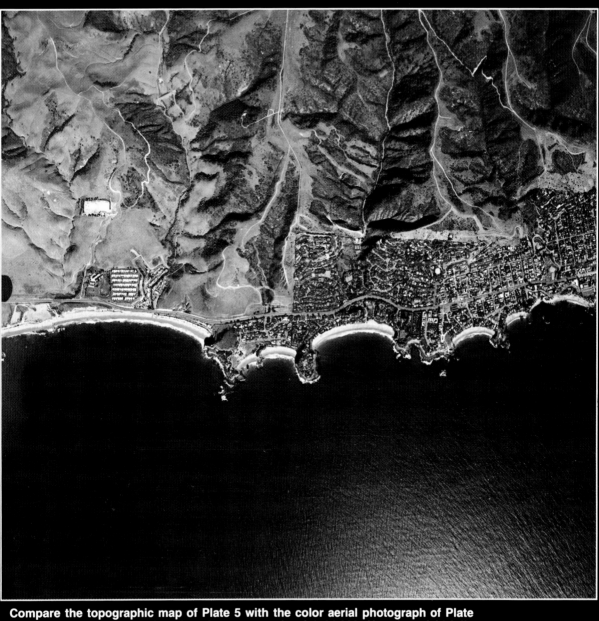

Compare the topographic map of Plate 5 with the color aerial photograph of Plate 6, which has been reproduced at the same scale of 1:24,000. Note how faithfully the topographic map represents both the physical and cultural features shown on the aerial photograph. Can you recognize any differences resulting from the different dates associated with the map (1972) and the aerial photograph (1979)? *(Map: USGS; Photo: Aerial Eye, Inc. Irvine, CA)*

Note the extreme detail of this false-color imagery of Glacier Bay National Park, Alaska. The Brady Icefield covers much of the Fairweather Range and is the source of many glaciers. Large amounts of sediment are deposited in the sea by glacial meltwaters. The red areas are forested. *(NASA, LANDSAT 1057-19542)*

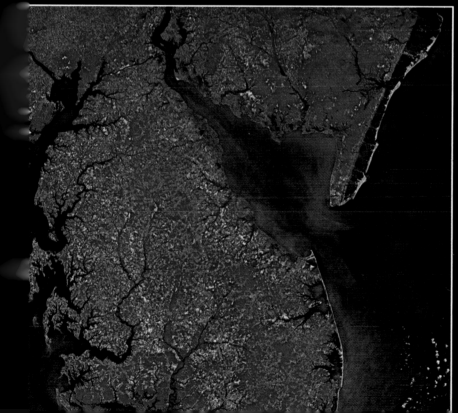

A false-color image of sections of New Jersey, Delaware, and Maryland. Delaware Bay is in the upper center and Chesapeake Bay on the far left. The light-colored patches are cleared and urbanized areas. Note the clarity of the barrier islands, wetlands, and estuaries along the Atlantic coast. Coastal states use LANDSAT data for mapping and monitoring their coastal environments. *(NASA, Landsat 1079-15133)*

globe, are evenly spaced, and are parallel to the original line of latitude, the equator. Hence, they are known as **parallels.** Parallels separated by 1° of latitude are everywhere about 112 kilometers (70 mi) apart. The only parallel that is a great circle is the equator; all the others are small circles.

The lines running north and south are half great circles that converge at the poles. These half great circles mark longitude, or the distance east or west of the prime meridian. Known also as **meridians,** each of these lines of longitude, when joined with its mate halfway around the globe, forms a great circle. The meridians, too, at any particular latitude are evenly spaced around the globe, although meridians do get closer together as we move poleward from the equator. At the equator, meridians separated by 1° of longitude are about 112 kilometers (70 mi) apart, and at 60°N or S they are about 56 kilometers (35 mi) apart.

LONGITUDE AND TIME

TIME ZONES

The relationship of longitude to time was used to set up the time zones that we know today. Until about a hundred years ago each town or area used what was known as "local time." That is, **solar noon** was determined for a location by finding the precise moment at which a vertical stake cast its shortest shadow. This meant that the sun had reached its highest altitude at that location; thus it was noon there, and all clocks were set to that time. Because of the earth's rotation, noon in a nearby town to the east took place a little earlier. A town to the west experienced noon a little later.

This system of using local time worked satisfactorily until the development during the nineteenth century of improved (i.e., faster) transportation, such as the railroad, and communication, such as the telegraph.

With these technological changes, the use of local time by each community became impractical. Hence, in 1884, the International Meridian Conference was held in Washington, D.C., to set up a system of time zones. The earth was divided into 24 zones, one for each hour of the day; thus, each time zone spans 15° of longitude. The prime meridian was made the central meridian of its time zone, and the precise noon at the prime meridian established the time for all places between $7^1/_2$°E and $7^1/_2$°W of that meridian. This pattern followed around the earth. Every fifteenth degree of longitude is the central meridian for a time zone of $7^1/_2$° of longitude to either side. However, as shown in Figure 2.7, the lines separating time zones do not follow meridians exactly. This discrepancy exists because jogs in the lines have been established for political or geographical reasons. Thus, in the United States, time zones often follow state boundaries. Imagine, for instance, the confusion that would result if Chicago were in one time zone and its suburbs in another.

The time of the Greenwich meridian time zone (sometimes called *Universal Time, U.T.,* or *Zulu Time*) is used for reference, and times to the east or west can be distinguished by comparison with it. Thus, time zones to the west of the prime meridian are said to have slow time and places to the east are on fast time. So a place 90° to the east of the prime meridian would be 6 hours fast, while time in the Pacific Time Zone in the United States, whose central meridian is 120°W, is said to be 8 hours slow.

For navigation, longitude is determined by the use of the *chronometer,* an extremely accurate clock. Two chronometers are used, one set on Greenwich, the other on local time. The number of hours between them, fast or slow, determines longitude (1 hour equals 15° of longitude).

Until the advent of electronic navigation by ground and satellite-based radio beams, the sextant and chronometer were the navigator's basic tools for determining location.

INTERNATIONAL DATE LINE

Exactly halfway around the globe from the Greenwich Meridian is the **International Date Line.** It is an imaginary line that follows the 180th meridian, except for jogs to separate

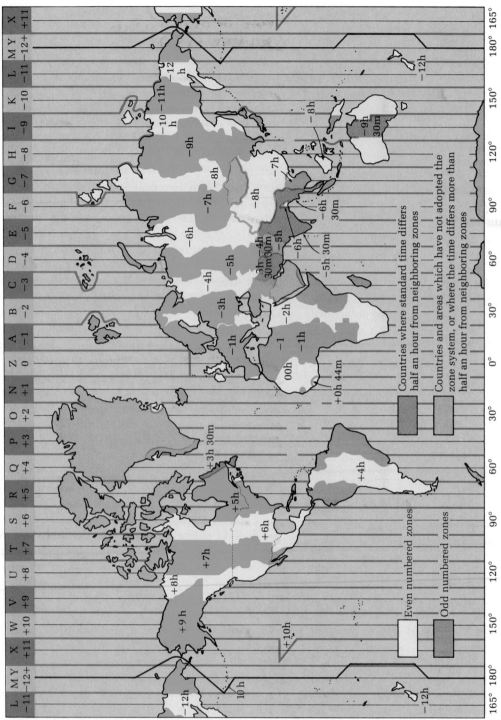

FIGURE 2.7   World time zones reflect the fact that the earth turns through 15° of longitude each hour. Thus time zones are approximately 15° wide. Political boundaries usually prevent the time zones from being perfectly longitudinal. (After U.S. Navy Oceanographic Office, No. 5192)

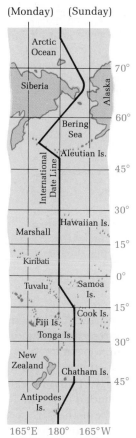

(Monday)   (Sunday)

Arctic Ocean
70°
Siberia
Alaska
60°
Bering Sea
International Date Line
Aleutian Is.
45°
30°
Hawaiian Is.
Marshall
15°
Kiribati
0°
Tuvalu
Samoa Is.
15°
Cook Is.
Fiji Is.
Tonga Is.
30°
New Zealand
Chatham Is.
45°
Antipodes Is.

165°E   180°   165°W

FIGURE 2.8 The new day officially issues from the International Date Line and then sweeps westward around the earth to disappear when it again reaches the line. Thus west of the line is always a day later than east of the line.

Alaska and Siberia and to keep together some Pacific island groups (Fig. 2.8).

At the International Date Line we turn our calendar a full day back if we are traveling east and a full day forward if we are traveling west. Thus, if we are traveling east from Tokyo to San Francisco and it is 4:30 P.M. Tuesday when we cross the International Date Line, it will be 4:30 P.M. Monday on the other side. Or, if we are traveling west from Alaska to Siberia, and it is 10:00 A.M. Wednesday when we reach the International Date Line, it will be 10:00 A.M. Thursday on the other side.

THE PUBLIC LANDS SURVEY SYSTEM

We have already seen that the global grid, made up of parallels and meridians, is universal and can be applied to *any* representation of the earth. There is another system of lines intersecting at right angles that is of interest because of its effect on the landscape of the United States. This system is referred to as the **Public Lands Survey System,** or the **Township and Range System.** Thought to have been suggested by Thomas Jefferson, it was proposed as an aid to parceling out the public land for sale west of Pennsylvania. By the Township and Range System, land was divided by means of occasional selected north-south lines called **principal meridians** and east-west lines called **base lines** (Fig. 2.9). The base lines were surveyed and laid out according to astronomical calculations. The north-south meridians, though perpendicular to the base lines, had to be adjusted periodically to counteract the effects of the curvature of the earth. If adjustments were not made, the north-south lines, like all meridians, would tend to converge, and the block laid out by the grid system would be smaller to the north.

The result of this system is a pattern of nearly square blocks, called *townships,* laid out in horizontal tiers north and south of the base lines, and vertical columns ranging east and west of the principal meridians. A **township** is an area of 6 miles on a side (36 square miles or 93 square kilometers). As illustrated in Figure 2.10, townships are first labeled by their position north or south of the surveyed base line. So a township in the third row south of a baseline will be called *Township No. 3 South,* which is abbreviated T3S.

However, just as providing only the latitude of Los Angeles was insufficient description of its location, we must also name a township according to which column or **range** it is located in to the east or west of the principal meridian of the survey area. Thus, if Township No. 3 South is in the second range to the east

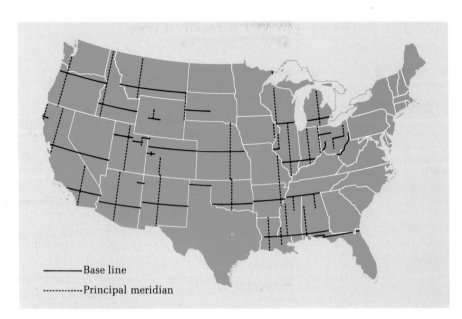

FIGURE 2.9 Principal baselines and meridians of the United States Public Lands Survey System (Township and Range System).

of the principal meridian, its full location can be given as T3S, R2E (Range No. 2 East).

The Public Lands Survey System further divided townships in 36 **sections** of 1 square mile or 640 acres (2.6 square kilometers). These sections are numbered from 1 to 36, beginning in the northeast corner and going back and forth across the section, ending in the southeast corner with 36. Each section then was divided into four quarters or corners, the northeast, northwest, southeast, and southwest quarters, each with 160 acres. Each corner could then be further subdivided into four quarters of 40 acres each, sometimes known as *forties*. These, too, were named by their position in the quarter: northeast, northwest, southeast, and southwest forty. Thus, we can describe the location of the shaded forty in Figure 2.10 as being in the SW ¼ of the NE ¼, Sec. 14, T3S, R2E, which we can find if we know the identity of both the principal meridian and the base line for the survey.

The Public Lands Survey System has had an enormous impact upon the landscape of the

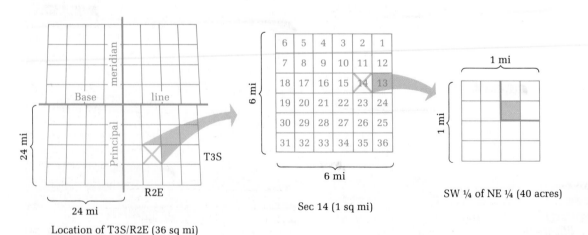

FIGURE 2.10   Method of location according to the Public Lands Survey System.

United States; it is what gives the great portion of our Midwest and West its patchwork-quilt appearance from the air (Fig. 2.11). The land was staked out, fenced, and farmed in blocks formed by the divisions of the Public Lands Survey System. Even road maps in many states reflect the use of this grid, as many roads follow boundaries laid down by the government survey system.

MAPS AND MAP PROJECTIONS

Although globes are the most accurate representation of the earth, they do have their limitations, as mentioned earlier in this chapter. Maps, on the other hand, as representations of the earth, or parts of the earth, on flat paper can make up for many of these limitations. They can be reproduced easily and inexpensively, they can depict the entire earth or show a small area in great detail, and they are far easier to handle, transport, and use than a globe.

ADVANTAGES OF MAPS

As graphic representations of the earth, maps supply an enormous amount of information that would take pages and pages to describe—probably less successfully—in words. Just imagine trying to describe in words all that a map of a small area like your city or campus can tell you—size, area, distances, street patterns, railroads, bus routes, locations of special buildings like hospitals, libraries, museums, routes to freeways and major highways, locations of business districts and residential areas, population centers, and so forth. Maps can show true courses for navigation and true shapes of earth features; they can be used to measure areas or distances, and they can show optimal routes.

In addition, because they are graphic representations and use symbolic language, maps can show many relationships. For example, they can show the distribution of landform varieties on the earth's surface, or they can reveal the relationship between varieties of climate and vegetation, or they can identify possible

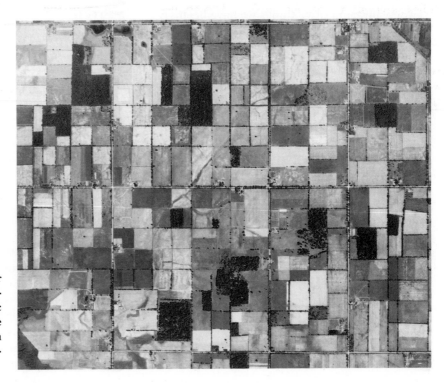

FIGURE 2.11 Rectangular road and field patterns resulting from the Public Lands Survey System in the agricultural midwestern United States. (U.S. Department of Agriculture)

routes and their comparative distances. The possibilities of maps are practically endless.

For many reasons, then, the map is the geographer's most important tool. The geographer can draft a map to show almost any relationship in the environment with which he or she is concerned. As the geographer's interest is essentially spatial—geographers are interested in the relationships among the subsystems of the environment and their changes through time and space—he or she can use the map to give a pictorial description of these relationships.

Besides serving as a vital tool to the geographer, maps are useful in every walk of life. They are valuable to tourists, political scientists, historians, geologists, pilots, soldiers, sailors, hikers, and even burglars. Maps are important because they can be drawn to fit any of an endless variety of special needs.

LIMITATIONS OF MAPS

Despite all the positive things we have been saying about maps, they can never depict the earth as accurately as can globes. Every map ever drawn distorts the earth in some way. This distortion is a basic problem of transforming a representation of the spherical earth onto a flat surface. Mathematically, it is impossible to create a completely perfect representation of the earth, or globe, on a flat surface. You just cannot make a sphere flat or two-dimensional and still maintain all its properties accurately.

On a single globe we can compare size, shape, and area of all the features of the earth, and we can measure distance, direction, shortest routes, and true courses, all with great accuracy. But because of the inherent distortion of maps, we can never compare and measure all these properties in one map with complete accuracy. There is, therefore, no such thing as a "perfect map."

We must accept this distortion of the earth as represented in maps. We should realize, though, that when a map is of a small area,

the distortion will be so slight as to be negligible. Thus, if we use a map of a state park for a day's hike, the distortion will be too small to affect us, and we can have full confidence in the map's information. It is when we attempt to show large portions of the earth's surface, or the whole earth, that its curvature has to be considered, and the distortion in maps becomes most apparent.

Maps can obviously be useful to us despite their limitations and distortions. We must, however, be aware of each map's particular characteristics. We should know what properties it depicts accurately, what features it distorts, and for what purpose the map was designed. Given this information, we can make accurate comparisons and measurements on maps.

The grid system of latitude and longitude that we have examined was designed for a spherical surface. On the globe this grid system has certain properties. Of these there are four primary ones: (1) parallels of latitude are always parallel, (2) parallels are evenly spaced, (3) meridians of longitude converge (meet) at the poles, and (4) meridians and parallels meet everywhere at right angles.

There are thousands of different ways to project the grid of the globe onto a flat surface in order to make a map. However, in none of these projections will all four of the properties of the global grid system listed above be maintained. Thus, checking the grid system of a map projection for these four properties can help us to discover the area or areas of distortion for that particular projection.

PROPERTIES OF MAP PROJECTIONS

SHAPE No map can depict the true shape of large areas of the earth, such as continents. What we are really saying when we state that it is impossible to make the spherical globe flat is that we cannot preserve the shape of the globe; we have to make compromises. So a flat map cannot depict large areas of the globe without distortion of shape.

Small areas, regions, lakes, islands, bays, can be depicted in their true shape, with little apparent distortion. And, in fact, a map can have the property of showing true shape for small areas. Large areas, like countries and continents, will still appear in distorted shapes. Maps that *do* maintain true shape for small areas are said to be **conformal.**

In order to preserve the shape of small areas, we would expect that the parallels and meridians, when transferred from a globe to a flat surface conformal map, would maintain their global relationship. This they do: On all conformal maps, maps that preserve true shape for small areas, the meridians and parallels always cross at right angles, just as they do on the globe.

AREA   It is possible for a mapmaker to create a map that has true area for all parts of the earth depicted. That is, areas shown on a flat map can have the same proportions throughout as they have in reality. Thus, if we cover any two parts of the map with, say, a penny, we will be covering up equivalent areas on the earth's surface. When a map is drawn with this property, it is said to be an **equal-area** map. This ability to show equal area is an especially useful property for showing things like population distribution. As long as the map has equal area and a symbol represents the same quantity throughout the map, we can get a good idea of the distribution of anything from churches to cornfields to volcanic craters.

The only way to show equal area on a flat map, however, is to pull areas out of shape. It is, in fact, impossible on a flat map to show both equal area and true shape. Most of us are familiar with the Mercator map, which is commonly used in schools and textbooks and is often found on office walls (Fig. 2.12). In this particular map projection, the shape of areas is true, so it is conformal; it therefore cannot have the property of equal area, and areas do appear greatly out of proportion. Greenland, for instance, is shown to be about the size of South America, whereas in reality South America is about nine times the size of Greenland

(see Fig. 2.7)! A coin placed on Greenland in this conformal map would cover an area of land far smaller than that covered by the same coin placed on South America. So the shape of areas on this map has been preserved at the expense of showing equal area.

This, then, is a major problem of making maps. It is impossible to show both true shape and equal area on the same map. The cartographer must decide if he wants to show true shape at the expense of equal area or if he wants to maintain equal area while sacrificing true shape. Or he can compromise and create a map that approaches both properties but has neither.

DISTANCE   Just as no flat map can show true shape all over the earth or for large areas, neither can it maintain constant scale of distance all over the earth. The scale on a map cannot be applied correctly everywhere on a map depicting a large area. On a map of a small area, the distortion in distance will be so small as to be insignificant, and the accuracy will usually be sufficient for our purposes.

It is possible for a map to have the property of **equidistance** in specific instances. That is, on a map of the earth the equator may have equidistance or the same constant scale all along its length. Or all the meridians, but not the parallels, may have equidistance. On another map all straight lines drawn from the center may have equidistance, but the scale will not be constant for lines not drawn from the center.

DIRECTION   Because the compass directions on the earth are actually curved around the sphere, not all flat maps can show true compass directions. Thus, a given map may be able to show true north, south, east, and west, but the directions between those points may not appear accurately. So, if we are sailing toward an island, that island's location may be shown correctly according to its longitude and latitude, but the direction in which we must sail to get there may not be accurate, and we may pass right by it.

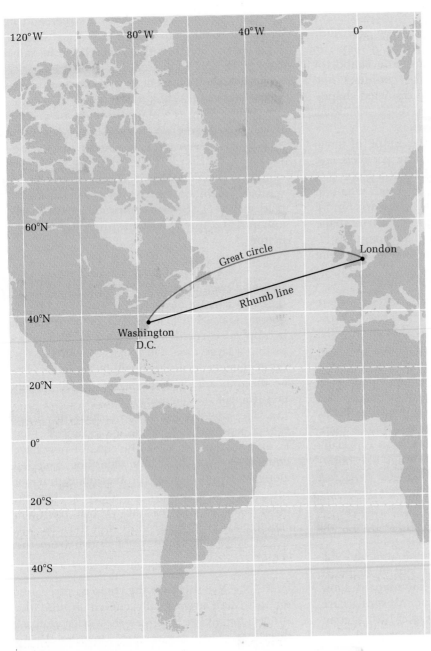

FIGURE 2.12 The Merca-tor projection, often mis-used for general-purpose maps, was designed for the purpose of navigation. Its one useful property is that the lines of constant com-pass heading, or rhumb lines, are straight lines. The Mercator is developed from a cylindrical projection. Compare the sizes of Green-land and South America and check their proportional sizes on a globe to observe the great amount of polar distortion on this projection.

Maps that are able to show true direction are called **azimuthal** map projections. These are drawn with a center or a focus, and all lines drawn from that center maintain true compass direction (Fig. 2.13).

MAP PROJECTIONS

We have been using the term *map projection* without ever stopping to explain what is really meant by it. Stated simply, a map projection is

FIGURE 2.13 Azimuthal map centered on the North Pole. Although this is the conventional orientation of such a map, it could be centered anywhere on the earth. It shows true direction to all other points. Azimuthal maps can show only one hemisphere of the earth at one time.

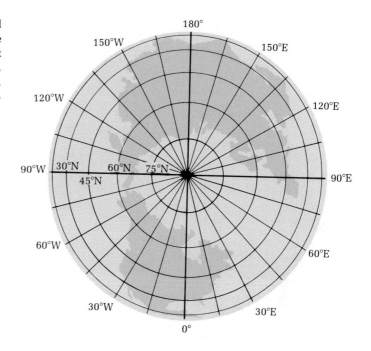

a representation of the three-dimensional globe onto a two-dimensional flat surface. And we have seen that we cannot retain all the properties of the globe when we flatten it out into a map.

However, there is one aspect of the globe that all map projections must maintain. That is the property of location. All places shown on a map must be in the same location with respect to latitude and longitude as they are on the globe. No matter how much we change the appearance of the grid of parallels and meridians when we project it onto a flat surface, we must be sure that all places still have their same location, that all bays, cities, lakes, and mountains are still at their accurate latitude and longitude.

Although maps are not actually made by light projections, some projections can be demonstrated by putting a light inside a transparent globe so that the grid lines are projected onto a plane (**planar projection),** a cylinder (**cylindrical projection),** or a cone (**conic projection),** which can be rolled out flat (Fig. 2.14).

Many map projections today are actually produced by complex mathematical computations and are plotted by a computer on printout or CRT (cathode ray tube) screen.

THE MERCATOR PROJECTION One of the most well-known of all projections is the Mercator, named for its inventor Gerardus Mercator, who devised this projection by mathematical calculation in 1569. The mathematical computations are complex, but the theory behind the Mercator projection is easy to understand.

The **Mercator projection,** shown in Figure 2.12, is actually a cylindrical projection that has been mathematically adjusted. Imagine a transparent globe with a light inside moving along its equator that casts the shadows of the grid onto a sheet of paper wrapped around the globe so as to form a cylinder. If the paper is touching the globe everywhere at the equator, then the equator on the flat map will be exactly as it is on the globe; it will be in scale. The rest of the grid, however, will not be in scale (Fig.

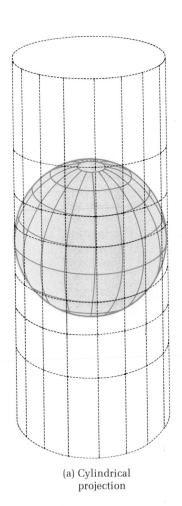

(a) Cylindrical
projection

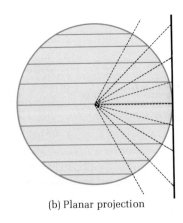

(b) Planar projection

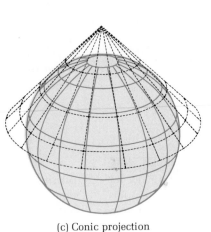

(c) Conic projection

FIGURE 2.14 The theory behind the development of the (a) cylindrical, (b) planar, and (c) conic projections. These projections are not actually made in this manner, but they can be demonstrated by light projection from a transparent globe.

2.14a). The meridians will appear as parallel lines instead of converging at the poles. Obviously, then, there will be an enormous amount of distortion east and west for the small land areas of the polar regions as they are stretched to the same width as the equator. This is where Mercator's mathematical adjustments come in. The spacing of the parallels on the Mercator projection is not even as on the globe but is mathematically calculated to be proportional to the increased spacing of the meridians as they approach the poles. The result is a grid that is made up of rectangles that increase in size toward the poles.

The mathematical adjustments of the Mercator projection provide for conformality.

Small areas on this map keep their true shape, but this means that this projection cannot also have equal area. Since the Mercator projection is used so widely in schools, it has led generations of students into wrongly believing that Greenland is as large as South America.

Besides conformality, the Mercator does have one important property that no other projection has. A straight line drawn anywhere on a Mercator projection is a true compass heading. Such a line is called a **rhumb line,** and it is of great value to navigators (Fig. 2.12). Although the distance of a rhumb line drawn from place to place may not be accurate or to scale, the compass heading will always be accurate no matter where on the map it is drawn.

THE GNOMONIC PROJECTION This is one of the oldest map projections. It is a planar projection, for it can be made by projecting the grid lines onto a flat or plane surface (Fig. 2.14b). If we put a flat sheet of paper, which is a plane, tangent to (touching) the globe at the equator, the grid will be projected with great distortion. The parallels and meridians will be unevenly spaced, and what is plotted—the land and water areas—will also be badly distorted.

Despite its distortion, the gnomonic projection does have one valuable characteristic. Looking at it in Figure 2.15 we can see that all the parallels appear curved except the equator, which appears as a straight line. In addition, *all* the meridians appear as straight lines. The one thing we have learned that the equator and meridians have in common is that they are all great circles (or halves of great circles). Thus, if the equator and meridians appear as straight lines on the gnomonic projection, then all great circles must appear as straight lines on a map using this projection. In fact, the gnomonic is the only map projection on which all arcs of great circles emerge as straight lines. This one characteristic makes the gnomonic projection vitally important to navigators who are interested in knowing the great circle route between any two places. All a navigator needs to do is draw a straight line between where he is and where he wants to go and he has drawn the great circle route, the shortest route, between the two places. People who work with long-distance communications and radar also find this projection useful, since radio waves follow arcs of great circles.

There is an interesting relationship between the gnomonic and Mercator projections. Great circles on the Mercator projection appear as curved lines, while rhumb lines appear

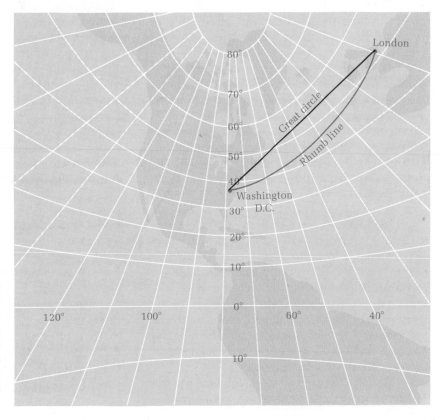

FIGURE 2.15 The gnomonic projection produces extreme distortion of distances, shapes, and areas. Yet, it is extremely valuable for navigation, for it is the only projection that shows all great circles as straight lines. It is developed from a planar projection.

straight (Fig. 2.12). On the gnomonic projection the situation is reversed. There great circles appear as straight lines, while rhumb lines are curves (Fig. 2.15).

THE LAMBERT CONFORMAL CONIC PROJECTION

In a simple conic projection a cone is fitted over the globe with its pointed top centered over a pole. The sides of the cone are tangent to the earth at one prechosen parallel, known as the *standard parallel* (Fig. 2.14c). The standard parallel on the resulting projection will be exactly like its corresponding parallel on the globe. The other parallels of a simple conic projection, like the lines of latitude they represent, are concentric circles that become smaller the closer they are to the pole. The meridians on a simple conic projection appear as straight lines radiating from the pole. This projection is a good example of a compromise projection. It is neither conformal nor does it have equal area. However, the distortions are slight, especially near the standard parallel, although they increase with distance from the standard parallel.

The simple conic projection can be improved by having the cone intersect the globe at *two* parallels, thus giving the projection two standard parallels whose scales are the same as their corresponding parallels on the globe and are therefore in scale with each other. This increases the area of the projection in which there is only slight distortion.

The Lambert conformal conic projection is a mathematically adjusted conic projection based on two standard parallels (Fig. 2.16). By some adjusting of the spacing of the parallels, Lambert was able to make his conic projection conformal.

Two standard parallels are used to make a Lambert conformal conic projection of the United States, and the result is a highly accurate and useful map. The distortion on this map is only 1 percent from the East Coast to the West Coast and from the Canadian to the Mexican borders. The distortion is so slight that intercontinental aircraft navigating across the United States can use a Lambert conformal conic map. On this projection a straight line will be an almost perfect great circle route and will have almost no distortion in scale.

THE WESTERN PARAGRAPHIC EQUAL-AREA PROJECTION

This projection was developed in the cartographic laboratory at Western Illinois University in conjunction with the preparation of the colored map plates found in this book (Plate 1). It is a combination of two equal-area projections. The objective of a **combination projection** is the retention of the desirable properties of both separate projections. In this case one projection is used for the lower latitudes and another for the higher latitudes, to minimize shape distortion in all areas of the world (Fig. 2.17a).

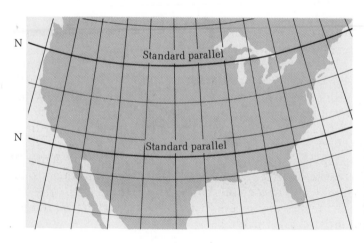

FIGURE 2.16 The Lambert conformal conic projection is a mathematical projection having two standard parallels. This projection is used when angles and shapes of midlatitude areas are to be kept as accurate as possible.

FIGURE 2.17 The paragraphic projection has been developed by joining two different projections, one for the lower latitudes and another for the higher latitudes (a), to minimize shape distortion in all areas of the map. Distortion by projections can also be reduced by interruption (b), that is, by having a central meridian for each segment of the map. (Western Illinois University Cartography Laboratory)

(a)
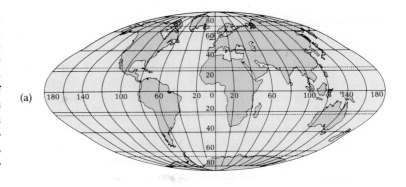

(b)
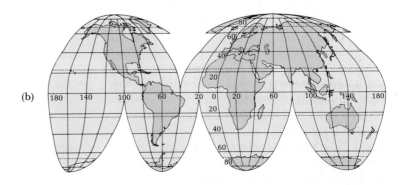

Distortion can also be reduced by using an **interrupted projection** to produce several smaller segments. Each segment has its own central meridian, and therefore no part of the projection suffers from extreme distortion of shape (Fig. 2.17b). For the interrupted paragraphic projection each major continental area is based on its own central meridian, and the oceans are interrupted. The only common element to all segments of the projection is the line representing the equator. If our interest were centered on the ocean areas of the world, we could interrupt the projection in the continental areas instead.

The final step in the construction of the Western paragraphic projection was condensation, to conserve space while preserving detail. Since the projection is composed of individual segments having only the equator as a common element, the distance between the land areas is meaningless, and, in fact, large gaps occur in the ocean areas. To increase the scale of the representation while retaining important information in a limited space, the map is condensed along the equator in the ocean areas (see the map plates throughout this book). The Western paragraphic can be considered a *combination, interrupted, condensed, equal-area* projection.

MAP ESSENTIALS

There are, as we have mentioned, a great number of ways to project the global grid onto a flat surface, though we have examined only a small sample of these projections. In addition, the map grid is only one element in the development of a useful map. There is a great deal more the map should include if it is to be an accurate, informative, and indispensable tool of the geographer. Among the essential items are title, date, legend, scale, and direction. In addition, the cartographer must select the best way to depict the phenomena the map is designed to show.

TITLE, DATE, AND LEGEND   Every map should have a title. The title should tell what area is depicted, what the map is about, what relationship it shows. For example, a map that we can use for hiking in Yellowstone National Park should obviously have a title that tells us that it is a map of the park, and if it shows trails, it might say something like "Yellowstone National Park: Trails and Camping Areas."

Each map should indicate the date of its construction or the date to which its information applies. That is, if the map shows the distribution of population in the United States, it should tell when these census data were gathered in order to let us know whether the information is current or outdated.

A map should also have a **legend** or a key to any symbols used on it. For example, if one dot on a map represents 1000 people or a symbol of a pine tree represents a roadside park, the key should tell us so. Likewise, if color is used on the map to represent, say, different climatic regions, then the color coding used should be given in the legend. Map symbols can show a wide variety of features on the earth's surface (Plate 1).

SCALE   A map is a representation of all or part of the earth and depicts its features as smaller than reality. In order to determine size or distance on a map we need to know what the **scale** is (Plate 3). That is, we need to know the relationship or ratio between distance on the actual earth and distance on the map. Some indication must be provided on every map to show to what scale the map is drawn and to what parts of the map that scale is applicable. Such a representation of scale enables us to measure distance, determine area, and make comparisons of size.

Because making a flat map of the spherical earth involves stretching some places more than others, the scale given can never be totally accurate all over the map. However, if the map is of a small area of the earth's surface, say of a city or a state, then the distortion of the map will be so slight that the scale can be accepted as applicable everywhere. Because globes differ from the earth in size only, that is, because they are not distorted in any way like a map, the scale of a globe is applicable everywhere.

One way to show scale on a map is with a **verbal scale.** This is a statement on the map that says something like "1 inch to 1 mile" (1 inch on the map represents 1 mile on the ground), or "1 inch to 1000 miles" (1 inch represents 1000 miles). A scale like this, however, is not applicable to a map after it has been reduced or enlarged. Nor can it be used by people who do not understand what a mile is, or an inch. That is, in order to use a verbal scale we must understand the units of measurement in which it is stated.

The **representative fraction (R.F.)** is presented as a ratio that is free of units of measurement. Thus, it can be used with any unit of measurement—feet, inches, meters, centimeters—as long as the same unit is used on both sides of the ratio. As an example, a map may have a scale of 1:63,360, which can also be expressed $\frac{1}{63,360}$. This means that 1 inch on the map represents 63,360 inches on the ground. It can also mean that 1 cm on the map represents 63,360 cm on the ground. In the case of our example above where we used inches, knowing that 1 inch on the map represents 63,360 inches on the ground may be difficult to conceptualize, unless we realize that 63,360 inches is the same as 1 mile. Thus, the representative fraction 1:63,360 means the map has the same scale as a map with a verbal scale of 1 inch to 1 mile.

A third kind of scale, a graduated line or **linear scale,** is sometimes provided on a map and is the most useful scale for measuring distances. The graduated line is marked off with specific map distances proportionate to distances on the earth. To use such a scale you take a straight edge such as the end of a piece of paper, measure and mark on it the distance between the two points on the map you are interested in, and then compare that distance to the graduated scale on which you can read off the equivalent distance on the earth's surface.

The major advantage of such a scale is that it is still applicable even if the map is reduced or enlarged when it is reproduced.

Maps may be said to be of small, medium, or large scale. **Small-scale** maps show large areas, include little detail, and have large denominators in their representative fractions. **Large-scale** maps, on the other hand, show small areas of the earth's surface in greater detail and have small denominators in their representative fractions. (To help avoid confusion, remember that ½ is a larger fraction than ¹⁄₁₀₀.) Maps with representative fractions between 1:100,000 and 1:1,000,000 can be said to be medium-scale maps. Maps with larger representative fractions than this (remember that 1:24,000 is larger than 1:100,000) can be said to be large-scale maps. Small-scale maps would have representative fractions less than 1:1,000,000.

DIRECTION Direction can be shown on a map by the lines of latitude and longitude, since lines of latitude are east-west lines and lines of longitude run north-south. In addition, a map may have an arrow pointing north from which we can orient ourselves with respect to any direction. Such a north arrow may be marked true north or magnetic north, or two different north arrows may be given, one for true north and one for magnetic north.

The earth has magnetic forces that radiate from its interior and make the earth into a giant bar magnet. Like a bar magnet, the earth has a magnetic north pole and south pole and fields of force that radiate in a pattern into space. Although they shift position from time to time, the magnetic poles of the earth are close to the geographic poles. A compass needle (a magnet) aligns itself with the magnetic force fields of the earth, and the north-seeking end of the needle aligns itself with the magnetic north pole. If we know the **magnetic declination,** or the angle that represents the difference in direction between magnetic north and true geographic north, we can then adjust our direction accordingly (Fig. 2.18). Thus, if

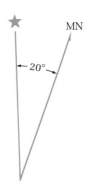

FIGURE 2.18 Map symbol showing true north, symbolized with a star representing Polaris or the North Star, and magnetic north, symbolized by an arrow. The example indicates 20° east magnetic declination.

our compass is pointing toward north and we know that where we are the magnetic declination is 20°E, we can adjust our course knowing that our compass is pointing 20° east of true north. So we should turn 20° west from the direction pointed out by our compass in order to face true north.

Magnetic declination varies from place to place and through time over the earth's surface. For this reason, maps that show magnetic declination are drawn and revised periodically. This is a case in which the date on a map is very important.

A map of magnetic declination is known as an **isogonic map** (Fig. 2.19). On such a map, points that have the same magnetic declination are connected by lines called **isogonic lines**. It is possible to be in a position where magnetic north and true north are in alignment. The line on an isogonic map that connects these points is called the **agonic line** (line of no magnetic declination).

LOCATION Having examined globes and maps and learned some of their special language and symbols, we now have the ability to communicate our location or understand the language of location. We have seen that we can give location on a map or globe by means of

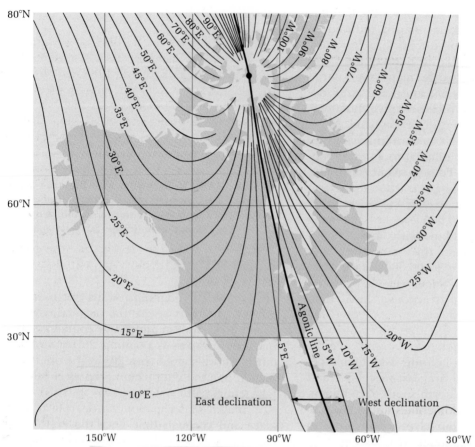

FIGURE 2.19 Isogonic map of North America, showing magnetic declination that must be added (west declination) or subtracted (east declination) from the compass reading to determine true directions. (After Hydrographic Chart 1706)

the coordinate system of latitude and longitude. We can also show location by providing our bearing from a known position and by providing the distance between the known and unknown positions. The bearing may be given by using **azimuth.** In the azimuth system, direction is given in degrees of a circle with respect to north. That is, if we imagine a circle of 360° with north at 0° (and at 360°) and read the degrees clockwise, we can give the direction of any location with respect to north in terms of the degrees of a circle. For instance, straight east would have an azimuth of 90°, while south would be 180°. Azimuths are used when giving precise directional instructions for military and navigational purposes.

DIFFERENTIATION AND DISTRIBUTION ON MAPS

There is an enormous variety of specialized information that may be shown on maps, but for each map there is a specific purpose. To fulfill the purpose of a particular map the cartographer chooses a technique, a method of cartographic representation, that best displays the desired information. For example, a common type of map used by geographers is one that shows the outline or shape of areas, each of which exhibits a common characteristic within its boundaries. Adjacent areas are differentiated from one another by the use of different colors, patterns of lines, or depths of shading.

Such maps and methods of representation are used to show the distribution of varieties or types of soil, climate, and vegetation (refer again to the color plates throughout the book). They are also used to show different political entities. A map of the United States that shows the separate states in different colors (at least no neighboring states would be the same color) is a good example.

Another type of map in everyday use is one that shows how certain kinds of things are located or arranged on the earth's surface. Maps of this type help to reveal patterns of arrangement of such diverse phenomena as streams, highways, railroads, buildings, types of land use, and deposits of natural resources like coal, petroleum, and iron ore.

A particularly valuable map to the geographer is one that shows the distribution of amounts or quantities of things across the earth's surface. Numerous techniques and a variety of symbols have been developed for the preparation of this type of map. A familiar example is the dot map, which uses each dot to represent the distribution of a specific quantity of a particular thing. Sometimes squares, cubes, or circles of different sizes are used instead of dots to represent quantities of differing magnitudes. Dot maps are often employed to show population density, but they can just as well be used to show the density of any number of things such as earthquakes, oil wells, automobiles, or registered voters (Fig. 2.20).

Another method used to show the distribution of quantities of things on a map is to employ **isarithms.** An isarithm is a line on a map that connects all points with the same numerical value. We have already run across one example of an isarithm, the isogonic line, which is drawn to connect points with the same magnetic declination (Fig. 2.19). Other isarithms we will be using later on include **isotherms,** lines connecting points of equal temperature; **isobars,** lines connecting points of equal barometric pressure; **isobaths,** lines connecting points with equal depths of water; and **isohyets,** lines connecting points having the same annual average precipitation.

CONTOUR MAPS

Of special interest to the physical geographer is the **contour line,** an isarithm connecting points on a map that are at the same elevation above mean sea level. If we should walk around a hill along the 200-foot contour line, we would be 200 feet above sea level at all times and we would walk on the level, not uphill or downhill.

Contour lines are important because they are the best method yet devised for depicting on a map the configuration of the land. By noting the arrangement, spacing, and changing shapes of the lines on a **contour map,** the reader can obtain a remarkably accurate idea of what the lay of the land or the topography of the area covered by the map looks like (Plate 5).

Figure 2.21 gives us a brief introduction to how contour lines can be used to show the character of the land surface. The bottom portion of the diagram is a simple contour map of an asymmetrical (lopsided) volcanic island. Note that the difference in elevation between adjacent contour lines on the map is a constant interval, 20 feet. A constant difference in elevation between contour lines is found on all contour maps and is called the **contour interval.**

What kind of terrain would we cover if we were to walk across our volcanic island from point A to point B? We start from point A at sea level but immediately begin to climb. We soon cross the 20-foot level (contour line), then the 40-foot, the 60-foot, and, near the top of our island, the 80-foot level. After walking over a relatively broad summit that is above 80 feet but not as high as 100 feet (or we would cross another contour line), we once again cross the 80-foot level (contour line) so we must be starting down. During our descent, we cross each lower level in turn until we arrive back at sea level (point B).

In the top portion of Figure 2.21 a **profile** or side view has been constructed to help us visualize the ground we covered in our walk. Now we can see why the trip up the mountain was so much more difficult than the trip down.

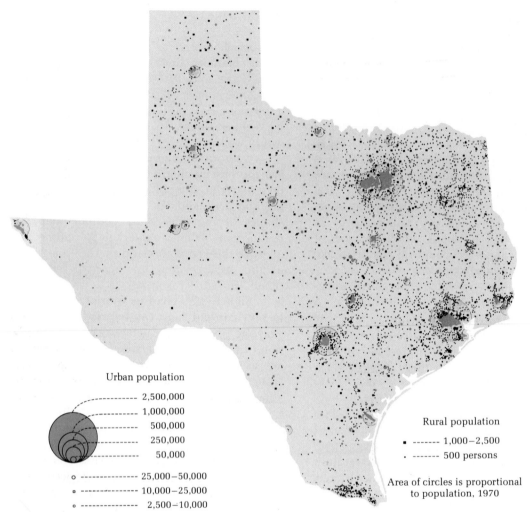

Urban population

- - - - - - - - - - 2,500,000
- - - - - - - - - - 1,000,000
- - - - - - - - - - 500,000
- - - - - - - - - - 250,000
- - - - - - - - - - 50,000

o - - - - - - - - - - - - 25,000–50,000
□ - - - - - - - - - - - - 10,000–25,000
o - - - - - - - - - - - - 2,500–10,000

Rural population

■ - - - - - - - 1,000–2,500
· - - - - - - - 500 persons

Area of circles is proportional
to population, 1970

FIGURE 2.20   This map of Texas population in 1970 shows how graduated circles may be used to represent quantities of differing magnitudes.

The closely spaced contour lines near point A represent a steeper slope than the more widely spaced contour lines near point B. As a matter of fact, we have discovered something that is true of all maps that display isarithms. The closer together the lines are on the map, the steeper the gradient (the greater the rate of change per unit of horizontal distance covered).

One other thing we should understand when studying a contour map is that the contour of the earth itself almost always changes gradually. We should remember that the land

indicated between two lines of a contour map probably slopes gradually toward the line with the lower value; it is unlikely that the land drops in steps down a hill as the contour lines might suggest.

Contour maps often show other features of the earth besides elevation (see Plate 5). For instance, a contour map may show different water bodies such as streams, lakes, rivers, and oceans. It may also show certain man-made features (cultural features) such as towns, cities, bridges, and railroads. When a contour map provides such a variety of information about

FIGURE 2.21 Topographic contour lines connect points of equal elevation in relation to mean sea level. The top portion of the figure shows the vertical profile of an island. The horizontal lines mark 20-foot intervals of elevation above sea level. The lower portion of the figure shows how these lines look from directly overhead. When contour lines are close together, slopes are steep; when far apart, slopes are gentle.

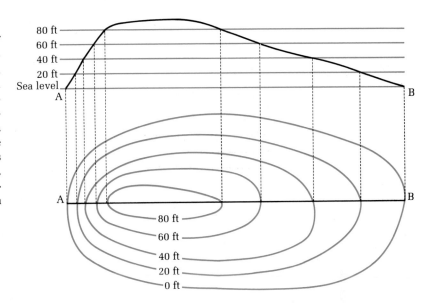

the surface features of a particular area, it is called a *topographic map*. The U.S. Geological Survey produces topograpic maps of the United States at several different scales (Plate 3). Some scales, 1:24,000, 1:52,500, 1:63,360 (1 inch equals one mile), and 1:250,000 are in conventional units of feet and miles. The newer maps at scales of 1:25,000 and 1:100,000 are in metric units of meters and kilometers. Contour maps that show undersea topography are called **bathymetric charts.** In the United States they are produced by the National Ocean Service.

COMPUTER MAPPING

Advanced modern computer technology has revolutionized cartography from a slow and tedious manual process to an automated science and a fascinating art. For most mapping the new automated computer system is faster, more precise, more efficient, and less expensive than the manual cartographic techniques.

Map data can be stored in digital form in data banks until a map is ready to be produced or revised. Hundreds of millions of bits of mapping data can be stored, such as elevations, depths, temperatures, or populations. For instance, a typical U.S. Geological Survey topographic map has over 100 million bits of information to be stored and thousands of bits of data to be plotted on the finished map. A data base for a map may include information on coastlines, political boundaries, city locations, river systems, map projections, latitude and longitude, and other coordinate systems. Systems are now being developed where field data can be entered immediately into the data bank while the field survey is being conducted. The data may even be recorded by voice recognition systems, that is, the data are called in by the field survey persons and directly entered into the map data system.

Computer mapping is particularly suited to map revision without having to redraw the map. Existing map data and new survey data can be displayed on a cathode ray tube (CRT), and the display map can be changed, corrected, or experimented with until the cartographer chooses the final map to be plotted and produced in hardcopy (on paper or plastic sheets) by the computer printer.

The cartographer may "stitch" together separate maps to cover a large area or may "zoom" in on a particular area of the map to give a more detailed larger-scale view. Besides scale changes, computer map revision also allows for metric conversion and rapid changes in projections, contour interval, symbols, colors, and direction of view (orientation). Computer map revision is essential for rapidly changing phenomena, such as weather systems, air pollution, ocean currents, volcanic eruptions, and forest fires. Weather maps, in particular, must be constantly revised, and computer mapping allows rapidly changing atmospheric conditions to be monitored and plotted on a continuous basis.

Of particular interest to physical geographers, geologists, and engineers is the digital terrain model, a three-dimensional (3-D) view of a particular area (Plate 4, top). Such a model is particularly useful for making calculations of volume of rock material, oil, or water resources. The digital terrain model may be modified to show more vertical exaggeration to enhance the relief of an area or to show a cross section side view at a particular location. Different types of terrain displays and maps can be produced from the terrain models (Plate 4, bottom). Some of these are color-scaled contour maps (areas between assigned contours are a certain color, for instance dark green for elevations between sea level and 100 meters), conventional contour maps, color-coded shaded relief maps, and 3-D stereographic shaded relief maps.

Though computer mapping is more electronic and mathematical than the mapmaking of the past, automated cartography maintains the same basic purpose of all mapping, to communicate geographic and spatial knowledge to the user in visual form.

REMOTE SENSING OF THE ENVIRONMENT

The making of maps has changed enormously with changes in technology during the last century. The earliest maps were pieced together from measurements made on the ground. In the 1800s experiments were made with taking pictures from balloons, and in the twentieth century much use has been made by cartographers of photographs taken from airplanes, called **aerial photographs.** These aerial photographs were merely the first beginnings in the rapidly developing technology of remote sensing.

Remote sensing is the mechanical collection of information about objects or the environment, obtained at a distance. A **remote sensor** is the mechanical device by which the information is collected. One common remote sensor, the camera, records data from visible light energy. Other remote sensors, such as radar and thermal infrared, use different forms of energy. Hence, remote sensing is commonly divided into photographic and nonphotographic types.

AERIAL PHOTOGRAPHY

Aerial photographs give us a new perspective on our environment (Plate 6). Mapmakers can use aerial photographs to examine and describe the relationships among objects on the earth's surface. A device called a *stereoscope* allows overlapping pairs of photographs, taken from different angles, to show the earth's surface three-dimensionally. Advanced stereographic equipment has helped in the production of relief or contour maps, and this, along with the use of computers for mapping, has greatly changed the science.

Not all aerial photographs utilize energy in the visible portion of the electromagnetic spectrum. (This spectrum, which includes the various wavelengths of electromagnetic radiation emitted by the sun, is discussed in Chapter 3.) There are types of electromagnetic waves that cannot be seen by the human eye or in normal photographs, and new types of photographic film have been invented in order to allow us to record these waves and to "see" new information about our planetary environment. Infrared photography has extended the sensitivity of film beyond the red end of the spectrum. But because most images produced by infrared photography utilize part of the visible spectrum, they are considered photographic—that is, produced by light. They are sometimes called *camouflage detection films,* since they can find military targets covered by cut vegetation, or *false-color films,* since healthy green plants appear red on the images. The major advantages of infrared photography are its ability to detect crop and plant diseases and its haze penetration capabilities.

NONPHOTOGRAPHIC REMOTE SENSING

Nonphotographic remote sensors may utilize the ultraviolet, infrared, microwave, radio, and X-ray wavelengths of the electromagnetic spectrum. Two of these, radar and thermal infrared (IR), have had wide applicability in environmental remote sensing. The nonphotographic sensors are either "passive" sensors, which sense natural radiation, or "active" sensors, which transmit an energy signal.

Radar (*RAdio Detection And Ranging*) is an active sensor that transmits energy, and the reading of the return signal is the basis for the image produced. It was used at first for military purposes of detecting aircraft and ships. Today its monitoring ability allows us to accurately track thunderstorms, hurricanes, and tornadoes.

A special type of radar called **Side-Looking Airborne Radar (SLAR)** was developed to look down from one side of an aircraft or spacecraft to acquire imagery. Radar has the great advantage of being able to penetrate darkness, cloud cover, and vegetation. It is particularly useful for terrain mapping, ocean wave monitoring, and locating solid objects such as cultural features (buildings, bridges, military equipment, etc.). SLAR produces an image that closely resembles a photographic image (Fig. 2.22).

FIGURE 2.22 An example of imagery produced by Side-Looking Airborne Radar (SLAR). The image is of a large open pit copper mine in Utah. (U.S. Geological Survey)

A **thermal infrared** (IR) sensor, called a thermal scanner or radiometer, records natural heat radiation emitted by clouds, water, land, and man-made features. It "sees" relative temperature differences. Its best applications have been for detecting clouds, storms, ocean currents, forest fires, thermal pollution, and energy (heat) loss by man-made objects. Because darkness does not restrict IR detection, it has obvious military applications. IR sensors may be used on aircraft or spacecraft (Fig. 2.23).

Multispectral scanning (MSS) is now used, especially on satellites. This allows several remote sensors to scan the same area at selected wavelengths. Combinations of visible and infrared bands are most common.

SONAR (Sound Navigation and Ranging) allows us to probe the depths of the oceans and to map undersea topography. The use of seismic waves (earthquake and man-made) has given us knowledge of the interior of the planet and helps us find resources such as gas and oil hidden in the solid earth. Each remote

sensor allows us to "see" our planet better and to gather data to map and plot its complex ever-changing systems.

SATELLITE IMAGERY

Perhaps the most important development for gaining information about our environment is the use of remote-sensing techniques that depend on earth-orbiting satellites in space.

Weather satellites have been in operation since the first TIROS series was launched by the United States in 1960. They have played an important role in weather prediction since that time. The ESSA weather satellite series was the successful follower of the TIROS program. A series of satellites known as NIMBUS was periodically launched to test new instruments and remote sensors on an experimental basis.

Today two NOAA (National Oceanic and Atmospheric Administration) weather satellites cover the globe in a polar orbit 854 kilometers (530 mi) above the earth. They produce visible

FIGURE 2.23 A nighttime infrared image of the central United States from the GOES weather satellite, on September 1, 1985. In the lower right is hurricane "Elena" off the Gulf Coast of Florida. (NOAA/NESDIS)

FIGURE 2.24   A photo image centered over the eastern United States from the GOES weather satellite, on August 22, 1978. The GOES satellites have an orbit altitude of 36,000 kilometers (22,300 mi) and produce new imagery every 30 minutes. They are operated by the National Oceanic and Atmospheric Administration. (NOAA)

and IR coverage of the whole earth's daily weather and cloud patterns (Fig. 2.23). The GOES (Geostationary Operational Environmental Satellite) satellites watch major storm systems from a geostationary orbit (over the

same earth position) approximately 36,000 kilometers (22,300 mi) from earth (Fig. 2.24). GOES-East is centered on a longitude over the United States East Coast, while GOES-West is centered on a longitude over the West Coast and watches for Pacific storm patterns. GOES images are seen on our television weather shows and are reproduced in daily newspapers.

Weather satellites have saved many lives and millions of dollars by increasing our storm and weather forecasting capabilities. Besides NOAA and GOES satellites, the U.S. Air Force operates the Defense Meteorological Satellite Program (DMSP) (see Figs. 16.15 and 16.16). The European Space Agency, the Soviet Union, and Japan also have weather satellites in operation.

Since 1972, LANDSAT satellites (formerly ERTS, Earth Resources Technology Satellites) have been recording geologic, hydrologic, forestry, crop, pollution, and land use activity from their 705 kilometers (438 mi) high orbit (Plate 7). The satellites are equipped with a MSS system covering two visible bands and two photo IR bands. The data from the MSS sensors are often computer-enhanced to produce the beautiful and striking "false-color" images so commonly seen today in books and magazines. The LANDSAT program has been particularly useful for continuous monitoring of environmental data, since the satellites repeat the exact same orbit each 18 days. The satellite data allow for continuous updating of environmental mapping programs. LANDSAT images have also become important in the classroom, where students can see striking visual examples of our natural environment and the continuing impact of human activity.

Spectacular earth photographs were also taken by the astronauts who manned spacecraft in such programs as Gemini, Apollo, and Skylab. The Space Shuttle astronauts are producing even more spectacular hand-held imagery.

The great advantage of these new remote-sensing techniques lies in the fact that they can provide rapid and worldwide coverage of environmental conditions for geographers, cartographers, and other scientists. Satellite coverage is especially important in the less accessible parts of the earth's surface such as ice caps, mountains, tropical forests, and deserts. Important resources like mineral ores, water, and potential energy sources can now be found more rapidly and with greater accuracy. By repeatedly using remote-sensing devices to monitor the same areas of the earth's surface, changes that occur can be detected and mapped.

Though maps remain the major tool of the geographer and other earth scientists, the ability to gather and present data has greatly changed in the last few decades. The use of remote sensors, earth-orbiting satellites, and computers that can record, store, and plot data on a completed map has revolutionized the study of the earth.

QUESTIONS FOR DISCUSSION AND REVIEW

1. Why is the concept of a great circle so useful to navigators?

2. What great circle has been chosen to start latitude? What half of a great circle was chosen to start longitude?

3. What is the latitude and longitude of your city?

4. What time zone are you in? How many hours fast or slow is your time considered to be?

5. If it is 2 A.M. Tuesday in New York (EST), what time and day is it in California (PST)? What time is it in London (GMT)?

6. If you fly across the Pacific Ocean from the United States to Japan, how will the International Date Line affect you?

7. How has the use of the Public Lands Survey System affected the landscape of the United States? Has your immediate locality been affected by its use? How?

8. Why can't maps give an accurate projection of the earth's surface? What is the difference between a conformal map and an equal-area map?

9. How does the Mercator projection differ from the gnomonic projection? What advantages does the Mercator offer the navigator?

10. What is the difference between an R.F. and a verbal map scale?

11. Why is the date of publication of a map important? List several reasons. Relate this to the earth's magnetic forces.

12. What new remote-sensing techniques have revolutionized our mapmaking techniques?

13. Describe the difference between imagery from a weather satellite and LANDSAT. Do the monitoring purposes of these satellites bear a relationship to the different imagery produced by each?

3

SOLAR ENERGY AND TEMPERATURE

All living things on our planet need water and oxygen to survive. Plants need carbon dioxide as well. Most living things we know cannot survive extreme temperatures, nor can they live long while exposed to large doses of the sun's ultraviolet radiation. It is the atmosphere, the layer of air that surrounds the earth, that supplies most of the oxygen and carbon dioxide and that helps maintain a constant level of water and radiation in the earth system.

Sometimes described as a *blanket* of air, the atmosphere serves as an insulator, maintaining the livable temperatures we find on earth. Without the atmosphere, the earth would experience great temperature extremes of as much as 260°C (500°F) between day and night. The atmosphere also serves as a shield, blocking out much of the sun's ultraviolet radiation, as well as protecting us from showers of meteors. At other times the atmosphere is described as an *ocean* of air surrounding the earth. This description reminds us of the currents and circulation of the atmosphere—its dynamics—that create the changing conditions on earth that we know as weather.

For purposes of contrast, we can look at our moon, a celestial body with no atmosphere, in order to see the importance of our own atmosphere. Most obviously, a man standing on the moon without his space suit would not have any oxygen to breathe. Also, astronauts have recorded high temperatures of up to 204°C (400°F) on the hot sunlit side of the moon. On the dark side, the cold, which approaches −121°C (−250°F) would kill an unprotected man.

The next thing our astronaut on the moon might notice is the "unearthly" silence. On earth we can hear because sound waves travel by moving the molecules of the atmosphere. Since the moon has no atmosphere and no molecules to vibrate and carry the sound waves, the lunar visitor would not be able to hear any sounds. Also he could not fly aircraft or helicopters, and it would be fatal to try to use a parachute. In addition, there is no atmosphere to protect him from the bombardment of meteors that fly through space and collide with the moon. In the case of the earth, most meteors burn up before reaching the surface because of friction with the atmosphere. And

Sunshine and shadow, Hale Observatory, Mt. Palomar, California (S. Brazier)

without an atmosphere to protect him, a visitor might also be burned by the ultraviolet rays of the sun. On earth we are protected to a large degree from ultraviolet radiation because the ozone layer of the upper atmosphere absorbs the major portion of it.

We can see that, in contrast to the stark lifelessness of the moon, the earth presents a hospitable environment for life, almost solely because of its atmosphere. All living things are adapted to its presence. For example, many plants reproduce by pollen carried by winds. Birds can fly only because of the air, and the water cycle of the earth is maintained through the atmosphere, as is the heat budget. The atmosphere diffuses sunlight as well, giving us our blue skies and the fantastic reds, pinks, oranges, and purples of sunrise and sunset. Without this diffusion the sky would appear black, as it does from the moon.

Further, the atmosphere provides a means by which the systems of the earth attempt to reach equilibrium. Changes in weather are ultimately the result of this process: attempts to equalize temperature and pressure cause the transfer of heat, energy, and moisture in the atmosphere and maintain the moderate climate of the earth.

CHARACTERISTICS OF THE ATMOSPHERE

The atmosphere extends as far as 9600 kilometers (6000 mi) above the earth's surface. Its density decreases rapidly with altitude, and in fact, 97 percent of the air is concentrated in the first 29 kilometers (18 mi) or so. Since air has mass, the atmosphere exerts pressure on the earth's surface and on all living things and other objects there. At sea level this pressure is about 1034 grams per square centimeter (14.7 lb per sq in), but the higher the elevation, the lower is the atmospheric pressure.

COMPOSITION OF THE ATMOSPHERE

The atmosphere is composed of numerous gases (Table 3.1). Most of these gases remain in the same proportions regardless of the density of the atmosphere. About 78 percent of the atmosphere's volume is made up of nitrogen, and nearly 21 percent consists of oxygen. Argon comprises most of the remaining 1 percent. The percentage of carbon dioxide in the atmosphere varies but is about 0.03 percent by volume. There are traces of other gases as well: ozone, hydrogen, neon, xenon, helium, methane, nitrous oxide ("laughing gas"), and krypton.

NITROGEN, OXYGEN, AND CARBON DIOXIDE Nitrogen makes up the largest proportion of air and is needed by plants. In addition, some of the other gases in the atmosphere serve functions separate from those of the atmosphere as a whole and are vital to the development and maintenance of life on earth. One of the most important of the atmospheric gases is, of course, oxygen, which all animals, including humans, use to oxidize (burn) the food they eat. Oxidation, which is technically the chemical combination of oxygen with other materials to create new products, occurs in situations outside animal life as well. Rapid oxidation takes place, for instance, when we burn fossil fuels or wood and thus release tremendous amounts of heat and light energy. The decay of certain rocks or organic debris and the development of rust are examples of slow oxidation and are processes that are dependent upon the existence of oxygen in the atmosphere.

Carbon dioxide is important to the climate, as it absorbs heat from both the upper atmosphere and the earth and then emits about half of that heat energy back to the earth. This process helps maintain the warmth of the earth and is a factor in the earth's heat energy budget.

Carbon dioxide is also involved in the *system* known as the carbon cycle, whereby plants,

TABLE 3.1   Composition of the Earth's Atmosphere

NAME OF GAS	CHEMICAL SYMBOL	PERCENT OF MASS
Nitrogen	N	78.09
Oxygen	O	20.95
Argon	Ar	0.93
Carbon dioxide	CO_2	0.03 (variable)
Ozone	O_3	Trace (variable)
Water vapor	H_2O	Trace (variable)
Hydrogen	H	Trace
Inert gases:		
Neon, krypton, helium	Ne, Kr, He	Traces

through a process known as **photosynthesis,** use carbon dioxide and water to make carbohydrates (sugars and starches) in which are stored amounts of solar energy (Fig. 3.1). Oxygen is given off as a by-product. Animals then use the oxygen to oxidize the carbohydrates, releasing the stored solar energy. A by-product of this process in animals is the release of carbon dioxide, which completes the cycle when it is in turn used by plants in photosynthesis.

OZONE   Another vital gas in the earth's atmosphere is ozone. The ozone molecule (O_3) is a cousin of the oxygen molecule (O_2), as it is made up of three atoms of oxygen while regular oxygen is made up of only two. Ozone is formed in the upper atmosphere when an oxygen molecule is split into two atoms by shortwave solar radiation and the free unstable atoms join with two other oxygen molecules to form two molecules of ozone consisting of three oxygen atoms each.

Ozone is important to climate because it is capable of absorbing large amounts of the sun's

ultraviolet radiation. Without the ozone of the upper atmosphere, the ultraviolet radiation reaching the earth would severely burn human skin, increase the incidence of skin cancer, destroy certain microscopic forms of life, and damage plants. There is, therefore, increasing concern that emissions from high-flying jet aircraft or fluorocarbons from aerosols may damage this fragile ozone layer.

The small proportion of ultraviolet radiation that ozone allows to reach the earth does serve useful purposes. For instance, it is important in the production of certain vitamins, and it helps in the growth of some viruses and bacteria. It also has a function in the process of photosynthesis. Least important but most visible, ultraviolet radiation produces painful sunburns or beautiful tans, depending on individual skin tolerance and exposure time.

WATER VAPOR, LIQUIDS, AND SOLIDS
Water vapor is always mixed in some proportion with the dry air of the lower part of the atmosphere, although it varies from 0.02 per-

FIGURE 3.1   The equation of photosynthesis shows how solar energy (light) is used by plants to manufacture sugars and starches from atmospheric carbon dioxide and water, liberating oxygen in the process. The stored food energy is then eaten by animals, which also breathe the oxygen released by photosynthesis.

cent by volume in a cold, dry climate to a high of nearly 5 percent in the humid tropics. The percentage of water vapor in the air will be discussed later under the broad topic of humidity, but it is important to note here that the variations in this percentage over time and place are an important consideration in the examination and comparison of climates. In addition, water vapor absorbs heat in the lower atmosphere and so prevents its rapid escape from the earth. Thus, like carbon dioxide, water vapor plays a large role in the insulating action of the atmosphere. In addition to gaseous water vapor, liquid water also exists in the atmosphere as rain and as fine droplets in clouds, mist, and fog. Solid water is found in the atmosphere in the form of ice crystals, snow, and hail. Suspended in the atmosphere are many other solids such as dust, soil particles, pollen, microscopic animals, bacteria, smoke particles, seeds, spores, and salts from ocean spray, all of which can play an important role in absorbing energy and in the formation of rain drops.

VERTICAL LAYERING OF THE ATMOSPHERE

Though for the most part people function in the lowest levels of the atmosphere, there are times, as when we fly in aircraft or climb a mountain, when we leave our normal altitude. The thinness of the atmosphere at higher altitudes then may affect us if we are not accustomed to it. Visitors to Inca ruins in the Andes or high-altitude Himalayan climbers may experience "altitude sickness," and even skiers in the Rockies near "mile-high" Denver may take time to adjust. The air at these levels is much thinner than most of us are used to. In other words, there is more empty space between air molecules, and thus there is less oxygen and other gases in each breath of air inhaled.

The atmosphere can be divided into several layers according to differences in temperature and rates of temperature change (Fig. 3.2). The first of these layers, lying closest to the earth's surface, is the **troposphere** (from Greek: *tropo*, turn, i.e., the turning or mixing zone), which extends about 10 to 16 kilometers (6 to 10 mi) above the earth. Its thickness is least at the poles, greatest at the equator, and tends to vary seasonally. It is within the troposphere that people live and work, plants grow, and virtually all the earth's weather takes place.

The troposphere has two distinct characteristics that differentiate it from other parts of the atmosphere. One is that the water vapor and dust particles of the atmosphere are concentrated in this one layer; they are virtually nonexistent in the layers of the atmosphere above the troposphere. The other characteristic of this layer is that temperature decreases at a nearly uniform rate with increased altitude.

The altitude at which the temperature ceases to fall with increased altitude is called the **tropopause,** which separates the troposphere from the **stratosphere,** the second layer of the atmosphere. The temperature of the lower part of the stratosphere remains fairly constant (about $-57°C$ or $-70°F$) to an altitude of about 32 kilometers (20 mi). It is in the stratosphere that we find the concentration of ozone that does so much to protect life on earth from the sun's ultraviolet radiation. Because of this ozone layer, however, temperatures increase in the upper parts of the stratosphere as the ozone absorbs ultraviolet radiation. Temperatures at the **stratopause,** which is about 56 kilometers (35 mi) above the earth, are about the same as temperatures found on the earth's surface, although little of that heat can be conducted because the air is so thin.

Above the stratopause are the **mesosphere,** in which temperatures tend to drop with increased altitude, and the **thermosphere,** where temperatures increase until they approach something like $1090°C$ ($2000°F$) at noon. However, the air is so thin at this altitude that there is practically a vacuum and little heat can be conducted.

The thermosphere was once called the ionosphere because of the ionization of molecules and atoms that occurs in this layer, mostly as a result of ultraviolet rays but also because

FIGURE 3.2 Vertical temperature changes in the earth's atmosphere are the basis for its subdivision into the troposphere, stratosphere, mesosphere, and thermosphere.

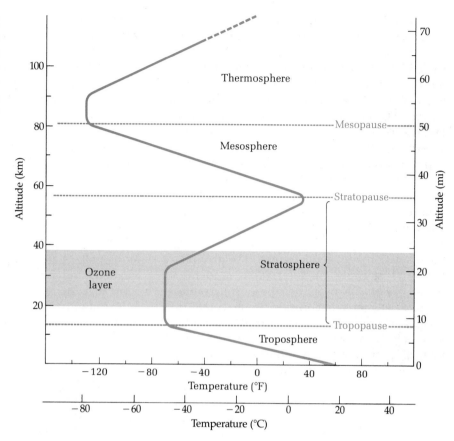

of X-ray and gamma radiation. Ionization refers to the process whereby atoms are changed to ions through the removal or addition of electrons, giving them an electrical charge. The thermosphere merges gradually into the exosphere, the zone where the earth's atmosphere gives way to interplanetary space.

DEFINITION OF WEATHER AND CLIMATE

Weather refers to the condition of atmospheric elements at a given time and for a specific area. That area could be as large as the New York metropolitan area or a spot as small and specific as a weather observation station. While the high-level atmospheric zones are important in such fields as space research, remote sensing, and telecommunications, it is the lowest layer, the troposphere, which is of the greatest interest to physical geographers and weather forecasters who survey the changing conditions of the atmosphere in a study known as **meteorology.**

Many observations of the weather of a place over a period of years provide us with a description of its climate. **Climate** describes an area's average weather, but it also includes those common deviations from the norm or average that are likely to occur, as well as extreme situations, which can be very significant. Thus we could describe the climate of the southeastern United States in terms of average temperatures and precipitation through a year, but we would also have to include mention of the likelihood of hurricanes during certain periods of the year. **Climatology** is the study of the varieties of climates both past and present found on our planet and their distribution over its surface.

Weather and climate are of prime interest to the physical geographer because they affect and are interrelated with other parts of the earth system. The changing conditions of atmospheric elements, like temperature, rainfall, wind, etc., affect soils and vegetation, erode landforms, and cause flooding of towns and farms.

ATMOSPHERIC ELEMENTS

There are four basic characteristics of the atmosphere that serve as the "ingredients" of weather and climate. They are (1) solar energy (or insolation), (2) temperature, (3) precipitation (and moisture), and (4) winds (and pressure). We must examine these four **atmospheric elements** in order to be able to understand and categorize weather and climate. Thus a weather forecast will generally include the probable temperature range that can be expected, the present temperature, a description of the cloud cover, the chance of precipitation, the speed and direction of the winds, and sometimes the barometric trend (the barometer is an instrument that measures air pressure).

In Chapter 1 we noted that the amount of solar energy received at one place on the earth's surface varies during a day and throughout the year. The amount of insolation a place receives is the most important weather element, as the other three are in part dependent upon the intensity and duration of solar energy.

The temperature of the atmosphere at a given place on or near the surface of the earth is largely a function of the insolation received at that location. It also varies with many other factors such as land and water distribution and altitude. Unless there is some form of precipitation occurring, the temperature of the air may be the first element of weather we describe when someone asks us what it is like outside.

However, if it is raining, or the fog is in, or it is snowing, we will probably notice and mention that condition first. We are less aware

of the amount of water vapor or moisture in the air (except at the extreme, as in very humid areas). However, moisture in the air is a vital weather element in the atmosphere, and its variations play an important role in the likelihood of precipitation.

We are probably least aware of variations in air pressure, although the fluctuations in air pressure are basic to the development of winds and storms. However, there are some people who say they can feel a change in the weather "in their bones" because they have arthritis and can probably sense the movement of fluids under pressure in their joints.

We all know that weather varies. It is the momentary state of the atmosphere at a given location, and it varies from time to time and from place to place. There are even variations in the amount that weather varies. In some places or at some times of year the weather changes almost daily from rain to sunshine to clouds to rain to snow. And in other places there may be weeks of uninterrupted sunshine, blue skies, and moderate temperatures and then weeks of persistent rain. At the extreme, there are a few places where there are no great differences in the weather throughout the year. The language of the original people of Hawaii is said to have no word for weather because conditions there varied so little.

ATMOSPHERIC CONTROLS

The variations in the atmospheric elements are caused by **atmospheric controls.** These controls are (1) latitude, (2) land and water relationships, (3) ocean currents, (4) altitude, (5) landform barriers, and (6) human activities.

LATITUDE Latitude is an atmospheric control, that is, a factor that causes variations in the elements of weather and climate, primarily because of the latitudinal changes in insolation, both daily and throughout the year. Because insolation is a prime factor in an explanation of the temperature of the air, we would expect that temperatures would vary with latitude in much the same way that insolation does. That is, we can expect to see lower mean (average) annual temperatures as we move poleward from the equator. Table 3.2 shows the average annual temperatures for several locations in the Northern Hemisphere. We can see that, with one exception, this poleward decrease in temperature is true for these locations, all of which are within a hundred meters or so of sea level. The exception is near the equator itself. Due to the heavy cloud cover in the equatorial regions, average annual temperatures there tend to be lower than at places slightly to the north or south where skies are clearer.

Another very simple way to see this general trend of decreasing temperatures as we move toward the poles is to think what clothes we would take along in one month, say January, if we were to visit Ciudad Bolivar, Venezuela, Raleigh, North Carolina, or Point Barrow, Alaska.

LAND AND WATER CONTRASTS Not only do the oceans and seas of the earth serve as storehouses of water for the whole system, but in their relationship with land and their distribution over the earth's surface the bodies of water act as atmospheric controls that do much to modify the atmospheric elements.

All things heat at different rates. On the earth's surface, bodies of water heat and cool more slowly than do land surfaces. Water is less dense, and heat passes through it to the layers below. In more dense materials like soil or rock, the heat is concentrated on the surface. Thus, a given unit of heat energy that is warming the land only works on the surface layer, while the same amount of energy striking water affects a greater volume of water. Since liquid water flows, it is able to transfer heat to other layers and portions. In addition, being essentially transparent, water allows heat and light energy to penetrate to deeper levels than do various opaque land surfaces. So a given unit of heat energy will be spread through a greater volume of water than of land. Finally, the specific heat of water is greater than that of land, which means that water surfaces must absorb more heat energy than land surfaces in order to be raised the same number of degrees in temperature. In fact, the specific heat of water is over four times that of earth or rock.

For these same reasons water cools off more slowly than does land. The result is that as summer changes to winter, the land cools more rapidly than bodies of water, and as winter becomes summer, the land heats more rapidly. Since the air that is above the land or water gets much of its heat from the surface with which it is in contact and which it overlies, the differential heating of land and water surfaces sets up inequalities and variations in the temperature of the atmosphere.

TABLE 3.2 Typical Temperatures in the Northern Hemisphere

LOCATION	LATITUDE	AVERAGE ANNUAL TEMPERATURE (°C)	(°F)
Libreville, Gabon	0°23′N	26.5	80
Ciudad Bolivar, Venezuela	8°19′N	27.5	82
Bombay, India	18°58′N	26.5	80
Amoy, China	24°26′N	22.0	72
Raleigh, North Carolina	35°50′N	18.0	66
Bordeaux, France	44°50′N	12.5	55
Goose Bay, Labrador, Canada	53°19′N	− 1.0	31
Markuva, USSR	64°45′N	− 9.0	15
Point Barrow, Alaska	71°18′N	− 12.0	10
Mould Bay, NWT, Canada	76°17′N	− 17.5	0

The mean temperature in Seattle, Washington, in July is 18°C (64°F), while the mean temperature during the same month in Minneapolis, Minnesota, is 21°C (70°F) although the two cities have similar latitude. Much of this difference in temperature can be attributed to the fact that Seattle is near the Pacific Coast, while Minneapolis is in the heart of a large continent and far from the moderating influence of an ocean. Consequently, Seattle stays cooler than Minneapolis in the summer because the surrounding water warms up slowly, keeping the air relatively cool. Minneapolis, on the other hand, is in the center of a large landmass that warms very quickly and in turn warms the layer of air above it. In the winter, the opposite is true. Seattle is warmed by the water while Minneapolis is not. The mean temperature in January is 4.5°C (40°F) at Seattle and −15.5°C (4°F) at Minneapolis.

Not only do water and land heat and cool at different rates, but so do various land surfaces. Soil, forest, grass, and rock surfaces all heat differentially and set up variations in the overlying temperatures of the air, which in turn can affect the other climatic elements.

OCEAN CURRENTS   Surface ocean currents are large movements of water pushed by the winds. They may flow from a place of warm temperatures to one of cooler temperatures, or vice versa. These movements result, as we saw in Chapter 1, from the attempt of earth systems to reach a balance, in this instance of temperature and density.

The rotation of the earth affects the movements of both the winds and ocean currents. It causes the currents to move generally in a clockwise direction in the Northern Hemisphere and in a counterclockwise direction in the Southern Hemisphere (Fig. 3.3). The reason for this movement, known as the *Coriolis effect*, will be discussed in Chapter 4.

Since the temperature of a mass of water affects the temperature of the air above it, an ocean current that moves from a warm equatorial area into a colder one or from a polar region toward the equator affects the tempera-

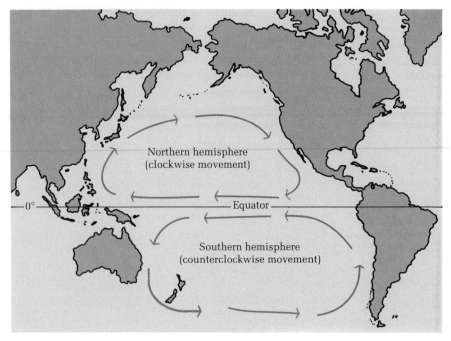

Northern hemisphere
(clockwise movement)

0°

Equator

Southern hemisphere
(counterclockwise movement)

FIGURE 3.3 A highly simplified map of currents in the Pacific Ocean to show their basic rotary pattern. The major currents move clockwise in the Northern Hemisphere and counterclockwise in the Southern Hemisphere because of the Coriolis effect.

FIGURE 3.4 The Gulf Stream (the North Atlantic drift farther eastward) is a warm current that greatly moderates the climate of northern Europe.

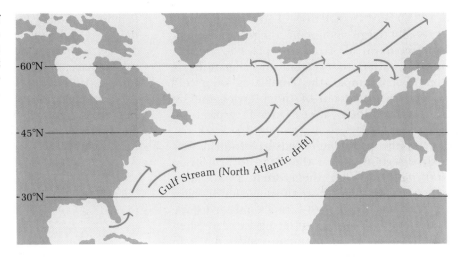

tures of the air in the areas into which it moves. An ocean current can affect coastal climates if it comes close enough to a landmass and if there are consistent onshore winds.

The Gulf Stream, with its extension, the North Atlantic Drift, is an example of a warm ocean current that moves warm water northward, keeping the coasts of Great Britain and Norway ice-free in wintertime and moderating the climates of nearby land areas (Fig. 3.4). We can see the effects of the Gulf Stream if we compare the winter conditions of the British Isles with those of Labrador in northeastern Canada. Though both are at the same latitude, the climate of the British Isles is moderated by the effects of the Gulf Stream (North Atlantic Drift). For example, the average temperature in Glasgow, Scotland, in January is 4°C (39°F), while during the same month it is −21.5°C (−7°F) in Nain, Labrador.

The California Current is a current off the west coast of the United States that helps moderate the climate of the coast as it brings cold water south to relatively warm areas. As the current swings southwest from the coast of central California, additional cold bottom water is brought to the surface, causing further chilling of the air masses above. San Francisco's cool summers (July average, 14°C or 58°F) reflect the effect of this current.

ALTITUDE AS A CONTROL As we have seen, temperatures within the troposphere decrease with increasing altitude. In Southern California you can find snow for skiing if you go to an altitude of 2400 to 3000 meters (8000 to 10,000 ft). Mount Kenya, 5199 meters (17,058 ft) high and located at the equator, is cold enough to have glaciers. Anyone who has hiked upward 500, 1000, or 1500 meters in midsummer has experienced a decline in temperature with increasing altitude. Even if it is hot on the valley floor, you may need a sweater if you climb a few thousand meters. Quito, Ecuador, only 1° south of the equator, has an average temperature of only 13°C (55°F) because it is located at an altitude of about 2900 meters (9500 ft).

Change in altitude also affects air pressure, another element of weather and climate. Air pressure, like temperature, decreases with increasing altitude. As a result, the air pressure on top of a 4000-meter (13,000-ft) mountain will be less than in the plain far below, and it will affect many everyday things. Water boils at 85°C (185°F) instead of 100°C (212°F), making it nearly impossible to make a good cup of tea. Automobile carburetors do not work effectively, and people traveling rapidly up or down the mountain have a popping sensation in their ears because of the change in pressure.

LANDFORM BARRIERS   Landform barriers, especially large mountain ranges, can block movements of air from one place to another and thus affect the weather and climate of an area. For example, the Himalayas keep cold winter Asiatic air out of India, giving the Indian subcontinent a year-round tropical climate.

If the prevailing winds are from a westerly direction, and if they tend to bring rain and moisture with them, then a mountain range that runs north-south will generally have a wet climate on its west-facing windward slope and a dry one on its east-facing leeward (sheltered) slope. While mountain ranges that run north-south, like the Rockies, Cascades, or Sierra Nevada in North America, block the movement of moisture-carrying air from the western oceans to the interior of the continent, thus helping create and maintain desert areas on their eastern sides, they do little to block the movement of cold polar air toward the equator. Therefore, because they are not protected by an east-west mountain range to the north, areas in the southern United States can be subjected to unusual cold spells from the invasion of polar air.

HUMAN ACTIVITIES   Human beings, too, may be considered "controls" of weather and climate. Such activities as the building of cities, burning or clearing forests, draining swamps, or creating large reservoirs may significantly affect local climatic patterns. The exploding of nuclear weapons, air pollution from heavy industry, and automobile emissions in the atmosphere have all had their effect on weather and climate. In addition, people have tried to modify weather almost since the beginning of time. Though we have had only slight success, our potential for influencing weather and climate is considerable.

SOLAR ENERGY AND ATMOSPHERIC DYNAMICS

The sun, like all the other stars in the universe, is a self-luminous mass of gases that emit radiant energy. A slightly less than average-sized star, our sun is the major source of energy, either directly or indirectly, for the entire earth system. The earth does receive very small proportions of energy from other stars and from the interior of the earth itself (volcanoes and geysers provide certain amounts of heat energy). However, when compared with the amount received from the sun, these other sources are insignificant.

ENERGY FROM THE SUN

The energy emitted by the sun comes from nuclear reactions that take place in its interior. There, under high pressure, hydrogen is

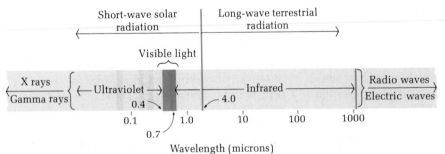

FIGURE 3.5 Radiation from the sun travels toward the earth in a wide spectrum of wavelengths, which are measured in microns (1 micron equals one ten-thousandth of a centimeter). Visible light occurs at wavelengths of approximately 0.4 to 0.7 micron. Solar radiation is short-wave radiation (less than 4.0 microns), whereas terrestrial (earth) radiation is all of long wavelengths (more than 4.0 microns).

changed into helium through nuclear fusion in a process similar to that in a hydrogen bomb. This nuclear reaction releases tremendous amounts of energy that radiate out from the sun in all directions at the speed of light.

Energy emitted by the sun is in the form of **electromagnetic radiation.** The electromagnetic radiation travels in a spectrum of waves of varying lengths that move away from the sun in all directions at the speed of light (Fig. 3.5). About 41 percent of this spectrum of waves is in the form of visible light rays, but much of the sun's radiation cannot be seen by the human eye. About half of the sun's radiant energy is in waves that are longer than visible light rays, and these include heat waves and **infrared waves.** While these cannot be seen, they can sometimes be sensed by the human skin. The remaining 9 percent of solar radiation is made up of X-rays, gamma rays, and ultraviolet rays, all of which are shorter in length than those of visible light. These also cannot be seen but can affect other tissues of the human body, thus the danger in absorbing too many X-rays. Collectively, visible light, ultraviolet rays, X-rays, and gamma rays are known as **short waves.**

Man has learned to harness some of these energy waves for communications (radio, microwave transmission, television), health (X-rays), and use in the field of remote sensing (photography, radar, infrared imagery).

Energy is radiated into space by the sun at a steady rate. The earth's atmosphere intercepts an amount of energy equivalent to 1.94 calories (a **calorie** is the amount of heat required to raise the temperature of 1 g of water 1°C) per square centimeter per minute. (This may also be stated as 1.94 **langleys** per minute, as a langley is 1 calorie per square centimeter.) This rate of emission is known as the **solar constant** and has been measured with great precision outside the earth's atmosphere by orbiting satellites. The atmosphere affects the amount of solar radiation received on the surface of the earth because some energy is absorbed by clouds, some is reflected, and some is refracted. If we could remove the atmosphere from the earth, we would find that the solar energy striking the surface would also be constant.

Of course, the measured value of the solar constant varies with distance from the sun as the same amount of energy radiates out into larger and larger areas. Because of this, if we measured the solar constant for the planet Mercury, it would be much higher than that for the earth. And, when the earth is closest in its orbit to the sun, its solar constant is slightly higher than the yearly average, and when it is farthest away, the solar constant is slightly lower than average. However, this difference does not have a significant effect on the earth's temperatures. When the earth is at aphelion in July and the solar constant is lowest because of the distance from the sun, the Northern Hemisphere is in the midst of a summer with temperatures that are not significantly different from those in the Southern Hemisphere 6 months later. The solar constant also varies slightly with changes in activity on the sun (during intense sunspot or sunstorm activity, for example, the solar constant will be slightly higher than usual). These variations, however, are not even as great as those caused by the earth's elliptical orbit.

THE ROLE OF WATER

Some of this incoming solar radiation, as it penetrates our atmosphere, is involved in energy exchanges as water in our earth system is altered from one state to another. Water is the only material that can exist in all three states of matter—as a solid, as a liquid, and as a gas—within the normal temperature range of the earth's surface. In the atmosphere, water exists as a clear, odorless gas called **water vapor.** It is a liquid in the atmosphere, in the oceans and other water bodies of the earth, in vegetation and animals, and underground. Water is found as solid snow and ice in the atmosphere as well as on and under the surface of the colder parts of the earth.

Not only is water stored in all three states of matter, but it can change from one state to

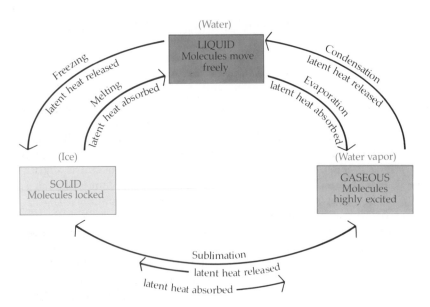

FIGURE 3.6 The three physical states of water and energy exchanges between them.

another, as illustrated in Figure 3.6, and in doing so it is involved in the heat energy exchange of the earth system. The molecules of a gas move faster than do those of a liquid. Thus, when a gas changes to a liquid, its molecules must slow down and some of their energy is released. The molecules of a solid move even more slowly than those of a liquid, so if the liquid is changed to a solid, additional energy is released. When the process is reversed and a solid melts to become a liquid, heat is needed. And when a liquid evaporates and becomes a gas, additional heat is needed to separate the molecules from one another and to put them into free-flowing rapid activity (Brownian movement). This energy is stored as **latent** (or hidden) **heat.**

We can have a direct acquaintance with some of these energy exchanges. For example, when you hold an ice cube in your hand, your hand feels cold because it is giving off the heat needed to melt the ice. We are cooled by perspiration evaporating from our skin, since heat must be absorbed both from our skin and from the remaining perspiration, thereby lowering the temperature of both.

Scientists have become increasingly interested in the energy exchange between the at-

mosphere and the hydrosphere, especially at the ocean's surface. Three fourths of the earth's surface is covered by water. Because water heats up more slowly than land, it retains its heat longer (i.e., it has a higher specific heat). Thus, the oceans act as huge reservoirs of heat energy to power the atmosphere, which directly influences our weather and climate on land. The more scientists learn about this energy exchange between the atmosphere and hydrosphere, the more important the role of water appears to be.

EFFECTS OF THE ATMOSPHERE ON SOLAR RADIATION

In addition to its involvement in these energy exchanges, the sun's energy, as it passes through the earth's atmosphere, loses over half its intensity through various processes. In fact, the amount of insolation actually received at a particular location depends not only on the latitude, the time of day, and the time of year (all of which are related to the angle at which the sun's rays strike the earth), but also upon the transparency of the atmosphere (or the amount of cloud cover, moisture, carbon dioxide, and solid particles in the air).

THE HEAT ENERGY BUDGET

BUDGET AT THE EARTH'S SURFACE

Now that we know the various means of heat transferral, we are in a position to examine what happens to the 47 percent of solar energy that reaches the earth's surface (look again at Fig. 3.7). Approximately 14 percent of this amount is sent back to the atmosphere in the form of long-wave radiation. This 14 percent includes a net loss of 6 percent (of the total originally received by the atmosphere) directly to outer space and 8 percent to the atmosphere. In addition, there is a net transfer back to the atmosphere by conduction and convection of 10 of the 47 percent that reached the earth. The remaining 23 percent returns to the atmosphere through the release of latent heat of condensation. Thus, the 47 percent of the sun's original insolation that reached the surface through the atmosphere is all returned. There has been no long-range gain or loss. Therefore, at the earth's surface the heat energy budget is in balance.

Examination of the heat energy budget of the earth's surface helps us to understand the *open energy system* that is involved in the heating of the atmosphere. The *input* in the system is that of the incoming short-wave solar radiation that reaches the earth's surface, and this is balanced by the *output* of long-wave terrestrial radiation back to the atmosphere. Since these are in balance, we may say that the overall temperature of the earth's surface is in a state of *dynamic equilibrium.*

Of course it should be noted that the percentages mentioned above represent an oversimplification in that they represent *net* losses that occur over a long period of time. In the shorter term, heat may be passed from the earth to the atmosphere and then back to the earth in a chain of cycles before it is finally released into space. In fact, it is the transfer of heat and energy back and forth between earth and atmosphere that produces unusually high atmospheric temperatures over short periods.

We are all familiar with what happens to the inside of a car on a sunny day if all the windows are left closed. Short-wave radiation from the sun is able to penetrate the glass windows (Fig. 3.8). When the insolation strikes the interior of the car and heats up the exposed surfaces, energy is emitted from them as long-wave radiation but cannot escape through the glass as freely. The result is that the interior of the vehicle gets hotter and hotter throughout the day. In extreme cases, windows in some cars have cracked due to differential expansion, or, more seriously, temperatures have become so great in automobiles that pets or babies inside that were unable to open a door or window have died.

A similar phenomenon also occurs in the atmosphere. Like glass, carbon dioxide, water vapor, and dust can block the escape of long-wave radiation by absorbing it and then radiating it back to the earth. Until the moisture, dust, or excessive carbon dioxide are dissipated, heat energy builds up and the earth is kept unusually warm. This has been called the greenhouse effect, although it might better be described as the **atmospheric effect,** since the former phrase is now used in another context, which is discussed in the next section.

BUDGET IN THE ATMOSPHERE   At one time or another, about 60 percent of the solar energy intercepted by the earth system is temporarily retained by the atmosphere. This includes 19 percent of *direct solar radiation* ab-

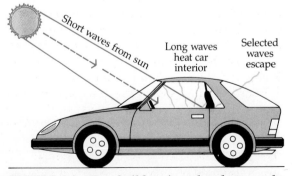

FIGURE 3.8   Heat build-up in a closed car results from penetration of short-wave radiation through the car windows and inability of long-wave reradiation to return outward through the glass.

VIEWPOINT: SOLAR ENERGY

Since humans beings have inhabited this earth, the sun has been an object of curiosity and reverence. From the builders of Stonehenge, who marked the rising of the sun at the summer solstice, to the Hindus bathing in the sacred river Ganges at sunrise or the native American Indians with their sun dances, the sun was a symbol of life-giving power. And now in the late twentieth century we may be rediscovering that fact. As our conventional energy sources dwindle, we are turning to solar energy to provide us with new power sources to heat and cool our homes and factories.

It is perhaps worth remembering that our earlier fuel sources are themselves *locked up* solar energy from past millennia, whether in wood, coal, oil, or gas. But will solar energy be as simple to harness as these more familiar sources were? It has the advantages of being a renewable, indeed infinite, resource, and clean, but is solar energy technologically and economically feasible? Optimists suggest that it can be, and that by the year 2000 we should be able to produce 10 percent and by 2020 up to 25 percent of our energy needs from solar sources.

Small-scale use of solar energy, where cost is not a factor, is already with us. For centuries simple farmers and fisherman have used the power of the sun to evaporate their salt pans, dry fish, and preserve crops. In the space program and in sea buoys, road signs, and offshore oil rigs, solar power lights and heats small-scale operations. On the domestic scene, solar space heating and hot water heating is a reality. One approach, the use of so-called *passive* systems, employs good architectural design and directional siting to warm interiors in winter and prevent overheating in summer.

From the beginning, Indian adobe structures of the Southwest were well adapted to the desert sun, with their thick walls, small windows, and south-facing exposures set in overhangs on canyon walls to shelter inhabitants from the near-vertical summer rays, thus making maximum use of solar heating and cooling. The Chinese, very early on, placed their village homes with doors and windows facing south and thick adobe windowless walls on the north side facing into the cold winter winds. Arizona and California architects, as well as others in Israel and elsewhere, are experimenting with similar passive solar design, utilizing greenhouse-like attachments, ultrathick stone or concrete walls, double-glazed windows, and careful site placement. *Active systems* include flat-plate and collector panels that heat water to 70°C (158°F). The water is then circulated and/or stored for domestic heating arrangements. Obviously, initial installation costs are expensive, but estimates suggest that after about a year and a half, there are savings of 60 percent on the cost of heating water.

Larger-scale solar energy operations are mainly in the experimental stage and, for reasons of cost, are generally not feasible at present. Two types of solar technology are being actively developed in various parts of the world, especially in the United States, France, Israel, and Australia. Photovoltaic cells that convert sunlight directly into electrical power (not unlike a camera's light meter) are already in use in Arizona and California and in remote communications systems. Prices for these photovoltaic cells have dropped dramatically over the last 30 years, since they are now mass-produced.

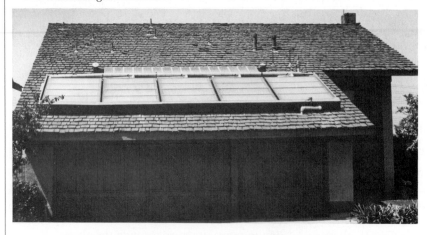

Active domestic solar energy system with flat-plate water heaters below and photovoltaic cells above. (Southern California Edison Co.)

The other major type of solar technology involves solar thermal towers, where racks of tracking mirrors (a heliostat field) follow the sun and focus its heat on a steam boiler perched on a high tower. Temperatures in the boiler may be raised to over 500°C (900°F). In principle no different from our youthful experiments with a magnifying glass to set fire to paper, this device already is operative at experimental sites in France and elsewhere. In California, Solar One in the Mojave Desert is the world's largest solar thermal electric power plant of this kind. However, development is slow because additional solar thermal towers must be considered in terms of current economic feasibility. Only as the costs of other energy sources rise and those of solar devices fall significantly will these become fully competitive.

When sites for possible future plant locations are considered, obviously the "sunbelt states" of the American Southwest are favored, where high percentages of possible annual sunshine amounts are assured. Flat ground is also necessary if large heliostat fields are to be constructed, and, as is the case with all new energy development, the impact of the new sites on the local environment must be carefully considered. Whether it be from earth-based solar energy stations or from insolation-collecting satellites beaming electricity to earth via microwave transmissions or from some other as yet undeveloped technology, the sun will become increasingly important as humanity enters the twenty-first century. Although it may seem a weak pun, many scientists believe that the future of solar energy appears bright . . . and sunny!

"Solar One," a solar electric power plant in the Mojave Desert of California. Over 1800 heliostats (mirrors) focus the sun's heat on a boiler atop a central tower. (Southern California Edison Co.)

sorbed by the clouds and the ozone layer, 8 percent that is emitted by *long-wave radiation* from the earth's surface, 10 percent that is transferred from the surface by *conduction* and *convection*, and 23 percent released by the *latent heat of condensation*. As in the atmospheric effect, some of this energy is recycled to the surface for short periods of time, but eventually all of it is lost into outer space after being replaced by other solar energy. Hence, just as was the case at the earth's surface, the heat energy budget in the atmosphere is in balance over long periods of time— a dynamically stable system.

Now some scientists believe that an imbalance with possible negative effects may be developing. Since the Industrial Revolution, human beings have been adding more and more carbon dioxide to the atmosphere through their burning of fossil (carbon) fuels. Since carbon dioxide absorbs the long-wave radiation from the earth's surface, restricting its escape to space, an increase of heat retention may be occurring in the earth's atmosphere. Nowadays this is referred to as the *greenhouse effect*. Such rising temperatures would have significant effects on other earth features such as the extent of polar ice caps and world sea level.

VARIATIONS IN THE HEAT ENERGY BUDGET

Remember that the figures for heat energy that we have seen are averages for the *whole* earth over a year's time. For any particular location, the factors we have discussed may not be balanced, and adjustments must be made within the entire earth system. Some places have a surplus of incoming solar energy over outgoing energy loss in their budget, while others have a deficit. The main causes of these variations are: (1) differences in latitude and (2) seasonal fluctuation.

As we have previously noted, the amount of insolation received is directly related to latitude. As indicated in Figure 3.9, in the tropical zones where insolation is high throughout the year, more solar energy is received at the earth's surface and in the atmosphere than can be emitted back into space. In the Arctic and Antarctic zones, on the other hand, there is so little insolation during the winter, when the earth is still emitting long-wave radiation, that there is a large deficit for the year. Places in the midlatitude zones have lower deficits or surpluses, but only at about latitude 38° is the budget balanced. If it were not for the heat transfers within the atmosphere and the oceans, the tropical zones would get hotter and hotter, and the polar zones would get colder and colder.

At any location, the heat budget varies throughout the year according to the seasons, with a tendency toward a surplus in the summer or high sun season and a tendency toward a deficit 6 months later. While seasonal differences may be small near the equator, they are very great in the midlatitude and polar zones.

AIR TEMPERATURE

TEMPERATURE AND HEAT

The amount of heat in an object or system is related to the energy within it. Some heat energy can be measured, but some cannot. **Tem-**

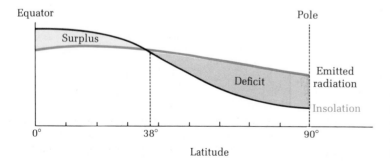

FIGURE 3.9 Latitudinal variation in the energy budget. Low latitudes receive more insolation than they lose by reradiation and have an energy surplus. High latitudes receive less energy than they lose outward and therefore have an energy deficit.

perature is the measurement of available or sensible heat energy in a system. Your body has stored within it a lot of heat energy, much in the form of fats, yet this energy is not reflected in body temperature. Body temperature is an indication of the level of energy that is being used to support body functions and therefore is in a measurable form.

In the same way, the temperature of the earth system does not account for all the heat within the system, such as that stored as latent heat energy in green plants after the process of photosynthesis or in water vapor that has undergone evaporation. Temperature does, however, record the decrease in available energy when, for example, heat has been used to evaporate water from a bowl, a lake, or the ocean. Therefore, since temperature can be used to measure changes in heat energy level, it is a useful indicator of changes in the heat energy budget of a system.

SCALES

Three different scales are used in measuring temperature. One, the Kelvin scale, is used primarily by chemists and physicists and is of minor importance to physical geographers. The one with which Americans are most familiar is the **Fahrenheit** scale, devised in 1714 and included in the English system of measurements. By this scale the temperature at which water boils at sea level is 212°F, while the temperature at which water freezes is 32°F.

The **Celsius** scale (also called the **centigrade** scale) was devised in 1742 by Anders Celsius, a Swedish astronomer. It is considered part of the metric system. The temperature at which water freezes at sea level by this scale was arbitrarily set at 0°C, while the temperature at which water boils was identified as 100°C.

The Celsius scale is used nearly everywhere except in the United States. Even in the United States the Celsius scale is the one used by the majority of the scientific community. By this time you have undoubtedly noted that throughout this book comparable figures in both the Celsius (centigrade) and Fahrenheit

scales are given side by side for all important temperatures. Similarly, whenever important figures for distance, area, weight, or speed are given, we use the metric system followed by the English system. Our interest in the Celsius and metric systems is more than just an educational exercise. The United States is slowly moving toward an acceptance of both in an attempt to increase uniformity of statistical information throughout the world.

Figure 3.10 can assist you in comparing the Fahrenheit and Celsius systems as you encounter temperature figures outside this book. In addition, the following formulas can be used for conversion from Fahrenheit to Celsius or vice versa:

$$C = \tfrac{5}{9}(F - 32) \quad \text{or} \quad F = \tfrac{9}{5}C + 32$$

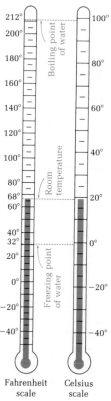

FIGURE 3.10 The Fahrenheit and Celsius temperature scales. The scales are aligned to permit direct conversion of readings from one to the other.

SHORT-TERM VARIATIONS IN TEMPERATURE

Local changes in atmospheric temperature can have a number of causes. These are related to the mechanics of the receipt and dissipation of energy from the sun and to various properties of the earth's surface and the atmosphere.

THE DAILY EFFECTS OF INSOLATION

As we noted earlier, at any particular location the amount of insolation varies both throughout the year (annually) and throughout the day (diurnally). Annual fluctuations are associated with the sun's changing declination and hence, with the seasons. Diurnal changes are related to the rotation of the earth about its axis. Each day insolation begins at sunrise, reaches its maximum at noon (local solar time), and returns to zero at sunset.

Although insolation is greatest at noon, you may have noticed that temperatures usually do not reach their maximum until two or three o'clock in the afternoon (Fig. 3.11). This is because, from shortly after sunrise until the afternoon hours, the insolation received by the earth exceeds the energy being lost through earth radiation. Hence, during that period, as the earth and atmosphere continue to gain energy, temperatures normally show a gradual increase. Sometime around three o'clock, when outgoing earth radiation begins to exceed insolation, temperatures start to fall. The daily lag of earth radiation and temperature behind insolation is accounted for by the time it takes for the earth's surface to be heated to its maximum and for this energy to be transferred to the atmosphere.

Insolation ends with sunset, but on into the night energy that has been stored in the earth's surface layer during the day continues to be lost, and there is a decreasing ability to heat the atmosphere. The lowest temperatures occur just after dawn, when the maximum amount of energy has been emitted and before replenishment from the sun can occur. Thus, if we disregard other factors for the moment, we can see that there is a predictable hourly change in temperature, called the **daily march of temperature.** There is a gentle decline from midafternoon until dawn, but temperature increases rapidly in the 8 hours or so from dawn until the next maximum is reached.

CLOUD COVER   The extent of cloud cover is another factor that affects the temperature of the earth's surface and the atmosphere (Fig. 3.12). Weather satellites have shown that at any time about 50 percent of the earth is covered by clouds. This cover is important, because a

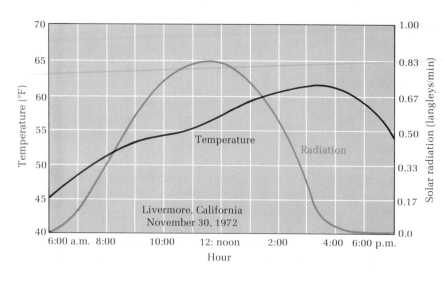

FIGURE 3.11 Diurnal changes in insolation and temperature for Livermore, California (November 30, 1972).

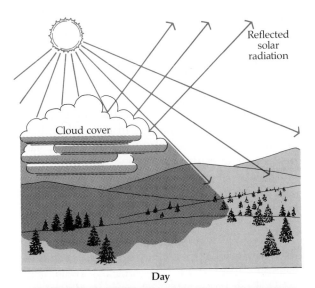

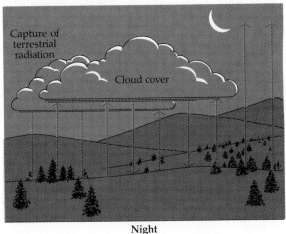

FIGURE 3.12 The effect of cloud cover on temperatures. By intercepting insolation, clouds produce lower air temperatures during the day. By trapping long-wave reradiation from the earth, clouds increase air temperatures at night. The overall effect is a great reduction in the diurnal temperature range.

heavy cloud cover can reduce the amount of insolation a place receives, thereby causing daytime temperatures to be lower than if the sky were clear. On the other hand, we have already noted the atmospheric effect, in which clouds, which are composed in large part of water droplets, are capable of absorbing heat energy

radiating from the earth, thereby keeping temperatures near the surface warmer than they would otherwise be, especially at night. The general effect of cloud cover, then, is to moderate temperature by lowering the potential maximum and raising the potential minimum temperatures.

DIFFERENTIAL HEATING OF LAND AND WATER   Earlier we saw that bodies of water heat more slowly than do land surfaces and that, likewise, they cool more slowly than the land. The air above the earth's surface is heated or cooled in part by what is beneath it. Therefore, temperatures over bodies of water or on land receiving ocean winds tend to be more moderate than those of land-bound places at the same latitude. Thus, the distance of a place to large bodies of water plays a role in its daily temperature pattern. This distance is usually referred to as **continentality.**

REFLECTION   The capacity of a surface to reflect the sun's energy is called its **albedo.** The more solar energy reflected back into space by an earth surface, the less will be available for radiating to the atmosphere and for heating. Temperatures are thus higher for a place with a low albedo than they would be if its surface had a high albedo.

As you may know from experience, snow and ice are good reflectors: they have an albedo of 90 to 95 percent. This is one reason why glaciers on high mountains do not melt away in the summer or why there may still be snow on the ground on a warm day in the spring: solar energy is reflected away. A forest, on the other hand, has an albedo of only 10 to 12 percent, which is good for the trees because they need solar energy for photosynthesis. The albedo of cloud cover varies according to the thickness of the clouds, and it can vary from 40 percent to 80 percent. The high albedo of many clouds is why much solar radiation is reflected directly back into space by the atmosphere. Cities have an albedo of only about 10 percent. This is one reason why hot summer days can be so miserable in the city and the surrounding countryside will be several degrees cooler. The albedo of water varies greatly, depending on the depth of the water body and the angle of the sun's rays. If the angle of the sun's rays is high, smooth water will reflect little: In fact, if the sun is vertical over a calm ocean, the albedo will be about 2 percent. Yet, a low-angle sun, such as just before sunset, causes an albedo of over 90 percent from the same ocean surface. Likewise, a snow surface in winter, when solar angles are lower, can reflect up to 95 percent of the energy striking it, and skiers must constantly be aware of the danger of severe burns from reflected solar radiation.

HORIZONTAL AIR MOVEMENT We have already seen that advection is the major cause of horizontal transfer of heat and energy over the earth's surface. Any movement of air due to the wind, whether on a large or small scale, can have a significant short-term effect on the temperatures of a given location. Thus, wind blowing from an ocean to land will generally bring cooler temperatures in summer and warmer temperatures in winter. Large quantities of air moving from polar regions into the midlatitudes can cause sharp drops in temperature, while air moving in a poleward direction will usually bring with it warmer temperatures.

VERTICAL DISTRIBUTION OF TEMPERATURE

NORMAL LAPSE RATES We have learned that the earth's atmosphere is primarily heated from ground level upwards as a result of long-wave terrestrial radiation, conduction, and convection. Thus temperatures in the troposphere are highest at ground level and decrease with increasing altitude. For every 1000 meters of altitude, the temperature drops at an average of 6°C (3.6°F/1000 ft). This rate, in the free air, is known as the **normal** or **environmental lapse rate.**

The lapse rate at a particular place can vary because of atmospheric conditions and elevation differences with nearby areas (Fig. 3.13). Low lapse rates can exist if denser and colder air is drained into a valley from a higher elevation or if advectional winds bring air in from a cooler region at the same altitude. In each case, the surface is cooled so that its temperature is closer to that at higher elevations directly above it. On the other hand, if the surface is heated strongly by the sun's rays on a hot summer afternoon, the air near the earth will be disproportionately warm, and the lapse rate will be steep. Fluctuations in lapse rates due to abnormal temperature conditions at var-

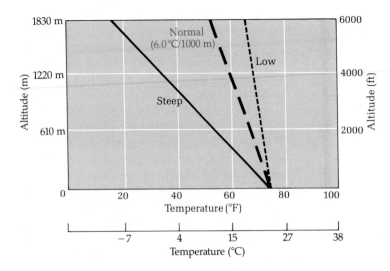

FIGURE 3.13 Steep, normal, and low atmospheric lapse rates.

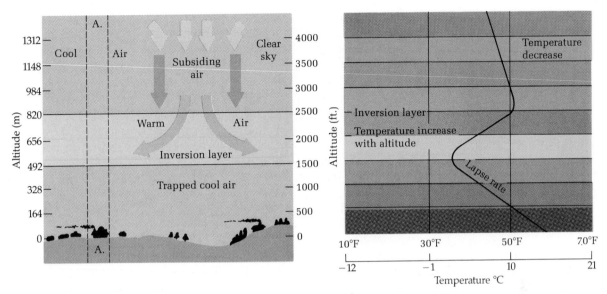

FIGURE 3.14 (Left) Temperature inversion caused by subsidence of air and the resulting adiabatic heating. (Right) Lapse rate associated with the column of air (A) in left-hand drawing.

ious altitudes can play an important role in the weather a place may have on a given day.

INVERSIONS   Under certain circumstances a reversal of the normal pattern of temperature *decrease* with increased altitude may occur in the troposphere; temperature may actually *increase* for several hundred meters. This is called a **temperature inversion.**

Some inversions take place a thousand or two meters above the surface of the earth where a layer of warmer air interrupts the normal decrease in temperature with altitude (Fig. 3.14). Such inversions tend to stabilize the air, causing less turbulence and discouraging both precipitation and the development of storms. Above-surface inversions may occur when air settles slowly from the upper atmosphere. Such air is compressed as it sinks and rises in temperature, becoming more stable and less buoyant. These descending air inversions are common at about 30° to 35°N and S latitudes.

The most noticeable temperature inversions are those that occur near the surface when the earth cools off the lowest layer of air

through conduction and radiation (Fig. 3.15). In this situation the coldest air is nearest the surface and the temperature rises with altitude. Inversions near the surface most often occur on clear nights in midlatitudes and are encouraged by snow cover and the recent importation of cool, dry air into an area. Such conditions produce extremely rapid cooling of the earth's surface at night as it loses the day's insolation through radiation. Then the layers of the atmosphere that are closest to the earth are cooled to temperatures below those at higher altitudes by radiation and by conduction where the air comes in direct contact with the cooled surface. Calm air conditions near the surface both help produce and are a partial result of these temperature inversions.

Another example of surface inversion is common in the coastal area of California, where the cool marine air blowing in from the Pacific Ocean underlies stable, warmer, and lighter air aloft created by the subsidence and compression of air mentioned previously. Such an inversion layer tends to maintain itself. That is, the cold, underlying air is heavier and can-

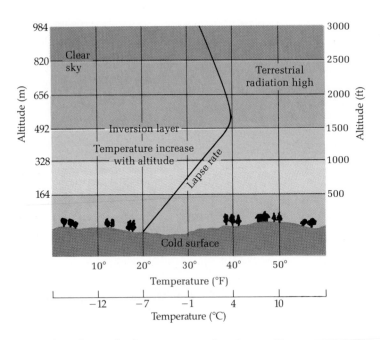

FIGURE 3.15 Temperature inversion caused by the rapid cooling of the air above the cold surface of the earth at night.

not rise through the warmer air above. Not only does the cold air resist rising or moving, but pollutants, such as smoke, dust particles, and automobile exhaust, which are created at the earth's surface, also fail to rise and spread out (see Fig. 1.8). They therefore accumulate, filling the lower atmosphere with pollutants. This situation is particularly acute in the Los Angeles area, which is a basin surrounded by higher mountainous areas (Fig. 3.16). Cooler air blows into the basin from the ocean and then cannot escape either horizontally because of the landform barriers or vertically because of the inversion.

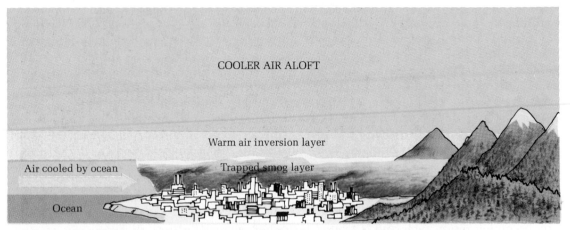

FIGURE 3.16 Conditions producing smog-trapping inversion in the Los Angeles area. Air moving onshore is cooled by the ocean surface and cannot rise because of the resulting temperature inversion. The air pools against the surrounding mountains and absorbs pollutants from the heavily urbanized area below. Often the polluted layer grows high enough to spill out through mountain passes, bringing smog to the interior desert area.

SURFACE INVERSIONS AND FROST

Frost often occurs as the result of a surface inversion. Especially where the earth's surface is hilly, dense cold surface air will tend to slide or flow down the sides of the hills and accumulate in the lower valleys. This air drainage causes colder air to build up in the valleys. Temperatures will decrease there, sometimes resulting in a killing frost while temperatures on the hillsides remain above freezing.

Farmers use a variety of methods to prevent such frosts from destroying their crops. For example, fruit trees in California that can be destroyed by a frost during the growing season are often planted on the warmer hillsides instead of in the valleys. Farmers may also put blankets of straw, cloth, or some other poor conductor over their plants. These take the place of the missing water vapor in the clear atmosphere, preventing the escape of the earth's radiation to outer space, thereby keeping the plants warmer.

Fans are sometimes used to stir up the air in an effort to mix the layers and disturb the inversion. Another device used to prevent frost is huge orchard heaters that heat the air, disturbing the temperature layers. Smudge pots, an older method of preventing frost, have declined in favor because they are major air polluters. The smoke they pour into the air provides an insulation blanket much like the straw or blankets mentioned above, preventing the escape of terrestrial radiation (Fig. 3.17).

ADIABATIC HEATING AND COOLING

We have seen that temperature varies with altitude at the normal lapse rate in the free air. It is also true that air changes its temperature if it is either compressed or allowed to expand. This is known as **adiabatic heating** (with compression) or **adiabatic cooling** (with expansion). In neither case is there any addition or subtraction of energy.

If air in the atmosphere is forced to rise (e.g., over a mountain), it will expand as it gains altitude because there is less air above to exert pressure on it. If the air is "dry," that is, if there is less water vapor in it than it is capable of holding, the temperature of the rising and expanding air will *decrease* at the rate of 10°C/1000 meters (5.6°F/1000 ft). This is known as the **dry abiabatic rate.** It should be emphasized that the dry adiabatic rate and the

FIGURE 3.17 Smudge pots used to protect California orange groves from frost. (S. Brazier)

normal lapse rate should not be confused. The dry adiabatic rate occurs within the *same* rising air parcel and is due to changes in pressure, while the normal lapse rate is associated with static air that is not changing in altitude and is due to differences in distances from the surface of the earth (Fig. 3.18).

When rising and expanding air contains water vapor that is condensing, heat of condensation is released, modifying the dry adiabatic rate of 10°C/1000 meters. The **wet adiabatic rate,** as this is called, is different from the dry rate and varies according to the amount of water vapor that condenses out of the air. It can be as low as 3.6°C/1000 meters, although it averages 5°C/1000 meters (3.2°F/1000 ft).

Air that is descending within the atmosphere has more and more pressure on it due to the increasing weight of the air above it.

This pressure causes its temperature to rise at the dry adiabatic rate of 10°C/1000 meters (5.6°F/1000 ft). This is one reason why winds blowing down slopes into valleys usually bring warmer temperatures.

Since the increasing temperatures of descending, compressed air allow it to hold greater quantities of water vapor, condensation will not occur and, therefore, the heat of condensation will not affect the rate of rise in temperature. Thus the temperature of air that is descending and being compressed always *increases* at the dry adiabatic rate.

TEMPERATURE DISTRIBUTION AT THE EARTH'S SURFACE

Isotherms (from Greek: *isos,* equal; *therm,* heat) are defined as lines that connect places of

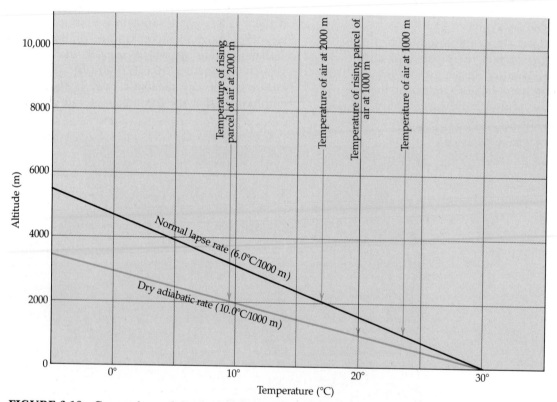

FIGURE 3.18 Comparison of the dry adiabatic lapse rate and the normal lapse rate. The normal lapse rate is the average vertical change in temperature. Air displaced upwards will cool (at the dry adiabatic rate) because of expansion.

FIGURE 3.19 Average sea level temperatures in January (°F).

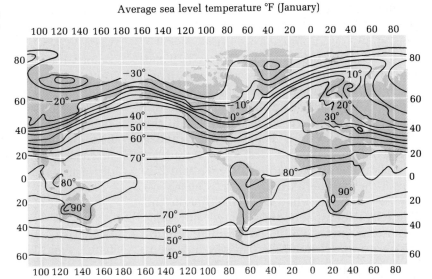

Average sea level temperature °F (January)

equal temperature. When constructing isothermal maps, which show temperature distribution over the earth's surface, elevation has to be accounted for by reducing temperature readings to sea level. This adjustment means adding the normal lapse rate of 6°C for every 1000 meters of elevation. The rate of temperature change on an isothermal map is called the **temperature gradient.** Closely spaced isotherms indicate a steep temperature gradient (or rapid temperature change over distance), and widely spaced lines indicate a weak one (or slight temperature change over distance).

Figures 3.19 and 3.20 show the horizontal distribution of temperatures for the earth at two critical times, during January and July, when the seasonal extremes of high and low temperatures are most obvious in both Northern and Southern Hemispheres. The easiest feature to recognize on both maps is the gen-

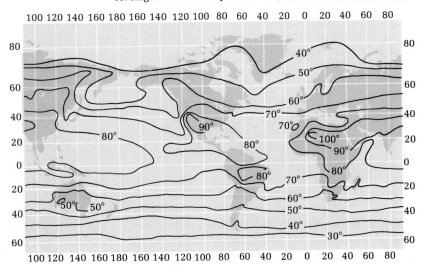

Average sea level temperature °F (July)

FIGURE 3.20 Average sea level temperatures in July (°F).

eral behavior of the isotherms; they run nearly east-west around the earth as do the parallels of latitude.

A more detailed study of Figures 3.19 and 3.20 and a comparison between the two maps reveal some additional important features. The highest temperatures in January are in the Southern Hemisphere; in July they are in the Northern Hemisphere. Look up the latitudes of Lisbon, Portugal, and Melbourne, Australia, in an atlas. Now note on the July map that Lisbon in the Northern Hemisphere is nearly on the 70° isotherm while at Melbourne in the Southern Hemisphere the average July temperature is less than 50°F, even though the two cities are approximately the same distance from the equator. The temperature differences between the two hemispheres are again a product of insolation, this time changing insolation as the sun shifts north and south across the equator between its positions at the two solstices.

Note the greatest deviation from the east-west trend of isotherms occurs where the isotherms leave large landmasses to cross the oceans. As the isotherms leave the land they usually bend rather sharply toward the poles in the hemisphere experiencing winter and toward the equator in the summer hemisphere. This behavior of the isotherms is a direct reaction to the differential heating and cooling of landmasses. The continents are hotter in the summer and colder in the winter than the oceans.

Other interesting features on the January and July maps can be mentioned briefly. Note that the isotherms poleward of 40° latitude are much more regular in their east-west orientation in the Southern than in the Northern Hemisphere. This is because in the Southern Hemisphere (often called the "water hemisphere") there is little land south of 40° latitude to produce land and water contrasts. Note also that the temperature gradients are much steeper in winter than in summer in both hemispheres. The reason for this can be understood when you recall that the tropical zones have high temperatures throughout the year, whereas the polar zones have large seasonal

differences. Hence, the difference in temperature between tropical and polar zones is much greater in winter than in summer. As a final point, observe the especially sharp swing of the isotherms off the coasts of eastern North America, southwestern South America, and southwestern Africa in January, and southern California in July. In these locations the normal bending of the isotherms due to land-water differences is strongly reinforced by the presence of warm or cool ocean currents.

ANNUAL MARCH OF TEMPERATURE

Isothermal maps are commonly plotted for January and July because there is a lag of about 30 to 40 days from the solstices when the amount of insolation is at a minimum or maximum (depending on the hemisphere) to the time of minimum or maximum temperature. This **annual lag of temperature** behind insolation is similar to the daily lag of temperature explained previously. It is a result of the changing relationship between incoming insolation and outgoing earth radiation. Temperatures continue to rise for a month or more after the summer solstice because insolation continues to exceed radiation. Temperatures continue to fall after the winter solstice until the increase in insolation finally matches earth radiation. In short, the lag exists because it takes time for the earth to heat or cool and for those temperature changes to be transferred to the atmosphere.

The annual changes of temperature for a location may be plotted in a graph (Fig. 3.21). The mean temperature for each month at a place like Peoria, Illinois, is recorded and a line drawn connecting the 12 temperatures. The mean monthly temperature is the average of the daily mean temperature recorded at a weather station during a month. The daily mean temperature is the average of the maximum and minimum temperatures for a 24-hour period.

Such a temperature graph, depicting the **annual march of temperature,** is able to show the decrease in solar radiation, as reflected by a decrease in temperature, from midsummer to

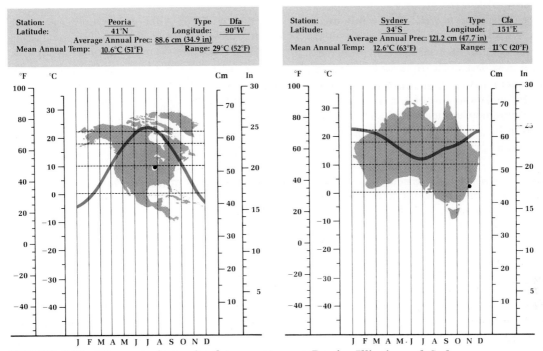

Station:	Peoria	Type	Dfa
Latitude:	41°N	Longitude:	90°W
Average Annual Prec: 88.6 cm (34.9 in)			
Mean Annual Temp: 10.6°C (51°F)		Range: 29°C (52°F)	

Station:	Sydney	Type	Cfa
Latitude:	34°S	Longitude:	151°E
Average Annual Prec: 121.2 cm (47.7 in)			
Mean Annual Temp: 12.6°C (63°F)		Range: 11°C (20°F)	

FIGURE 3.21 The annual march of temperature at Peoria, Illinois, and Sydney, Australia.

midwinter and then the increase in solar radiation by an increase in temperature from midwinter to midsummer.

It is these seasonal fluctuations that impose annual rhythms on our agricultural activities, our recreational pursuits, our clothing styles, and our heating bills. Human activities are constantly influenced by temperature changes, which reflect the input-output patterns of the earth's energy systems.

QUESTIONS FOR DISCUSSION AND REVIEW

1. Why is it useful to think of the atmosphere as a blanket of air? As an ocean of air?

2. Name the gases of which the atmosphere is composed and give the percentage of the total that each supplies.

3. What function does ozone play in the support of life on earth? Where and how is ozone formed?

4. How is the atmosphere subdivided? What level do you live in? Have you been in any of the other levels?

5. What is the difference between a meteorologist and a climatologist?

6. What are the basic characteristics of weather and climate that we call *atmospheric elements*?

7. What factors cause variation in the atmospheric elements?

8. Would you expect an area like Seattle to have a milder or a harsher winter than Grand Forks, North Dakota? Why?

9. What is the solar constant? What would happen if there were a significant change in the solar constant?

10. Discuss the role of water in energy exchange. What characteristics of water make it so important?

11. What is meant by the earth's energy budget? List and define the important energy exchanges that keep it in balance.

12. What is the *greenhouse effect*? How will it affect other earth systems?

13. What temperature is it in Fahrenheit degrees today in your area? In Celsius degrees?

14. At what time of day does insolation reach its maximum? Its minimum? Compare this to the daily temperature maximum and minimum.

15. How is albedo a factor in your selection of outdoor clothes on a hot, sunny day? On a cold, clear, winter day?

16. What is a temperature inversion? Give several reasons why temperature inversions occur.

17. Why do citrus growers use wind machines and heaters? Describe any techniques you are familiar with to prevent frost damage to plants in your area.

18. What is the difference between the normal lapse rate and the dry adiabatic rate? When does the wet adiabatic rate apply?

19. Describe the behavior of the isotherms in Figures 3.19 and 3.20. What factors cause the greatest deviation from an east-west trend? What factors cause the greatest differences between the January and July maps?

4

ATMOSPHERIC PRESSURE AND WINDS

MEASUREMENT OF AIR PRESSURE

Together the many gases of the atmosphere exert a pressure of 1034 grams per square centimeter (14.7 lb/sq in) at sea level. In other words, this is the amount of pressure the atmosphere exerts per square centimeter on all parts of the earth's surface at sea level. The reason that people do not collapse from the weight of the atmosphere is that we have air inside us, in our blood, tissues, and cells, exerting an equal outward pressure that balances the inward pressure of the atmosphere. Atmospheric pressure is important also because it is related to winds and thus helps to determine our weather and climate.

In 1693 Evangelista Torricelli, a student of Galileo, performed an experiment that formed the basis for the development of the **mercury barometer,** an instrument used to measure atmospheric pressure. Torricelli took a tube filled with mercury and inverted it into an open pan of mercury. The mercury in the tube fell until it was at a height of about 76 centimeters above the mercury in the pan, leaving a vacuum bubble at the closed end of the tube (Fig. 4.1). At this point, the pressure exerted by the atmosphere on the open pan of mercury was equal to the pressure from the mercury in the tube. Torricelli observed that as the air pressure increased, it pushed the mercury up into the tube, increasing the height of

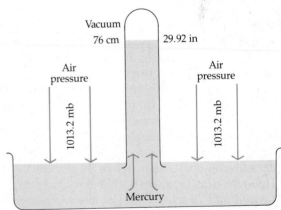

FIGURE 4.1 A simple mercury barometer. Standard sea-level pressure of 1013.2 millibars will cause the mercury to rise 76 centimeters (29.92 in) in the tube.

Windswept landscape. Pennsylvania Mountain, Colorado Rockies. (R. Gabler)

the mercury until the pressure exerted by the mercury would equal the pressure of the air. On the other hand, as the air pressure decreased, the mercury column fell.

In the strictest sense, a mercurial barometer does not actually measure the pressure exerted by the atmosphere on the earth's surface, but instead measures the *response* to that pressure. That is, when the atmosphere exerts a specific pressure, the mercury will respond by rising to a specific height. Meteorologists usually prefer to work with actual pressure units. The unit most often used is the millibar (mb). Standard sea level pressure of 1013.2 mb will cause the mercury to rise 76 centimeters (29.92 in).

VARIATIONS IN ATMOSPHERIC PRESSURE

VERTICAL VARIATIONS IN PRESSURE

Imagine a pile-up of football players during a game. The player on the bottom gets squeezed the most since he has the weight of all the others on top of him, but a player near the top will not get squashed. Similarly, the air pressure decreases with elevation, for the higher we go, the thinner the atmosphere becomes. The molecules are most diffused and there is less pressure because intermolecular space is greater. In fact, at the top of Mount Everest (elevation 8708 meters or 29,028 ft), the air pressure is about two thirds less than it is at sea level.

We humans are usually not sensitive to the small variations in air pressure that occur in our normal routine. However, when we climb or fly to altitudes significantly above sea level, we become aware of the effects of air pressure on our system. When jet aircraft fly at 10,000 meters (33,000 ft), they have to be pressurized and nearly airtight so that a near–sea-level pressure can be maintained. Even then, the pressurization does not work perfectly, so that our ears may pop as they adjust to a rapid change in pressure when ascending or descending. Hiking or skiing at heights that are a few thousand meters in elevation will affect us if we are used to the air pressure at sea level.

We sometimes find that we get out of breath far more easily at high elevations until our bodies adjust to the reduced air pressure.

Air pressure also changes through time for a particular location, as it is not solely related to altitude. At sea level, air pressure is intimately related to the intensity of radiation, the general movement of global circulation, and local humidity and precipitation. A change in air pressure for a given locality is nearly always a sign of a change in the weather.

HORIZONTAL VARIATIONS IN PRESSURE

The causes for horizontal variation in air pressure are grouped into two types: (1) thermal (determined by temperature) and (2) dynamic (related to motion).

We will look at the more simple thermal type first. In Chapter 3 we saw that the earth is heated unevenly because of unequal distribution of insolation, differential heating of land and water surfaces, and different albedos of surfaces. One of the basic laws of gases is that the pressure and density of a given gas vary inversely with temperature. Thus, as the atmosphere (gas) above is heated by the surface below, the air near the earth's surface will become less dense and expand in volume. Such air will have a tendency to rise as its density decreases. When the warmed air rises, there will be less air near the surface, with a consequent decrease in surface pressure. The equator is an area where such low pressure occurs.

In an area with cold air, there will be an increase in density and a decrease in volume. This will cause the air to sink and pressure to increase. The poles are an area where such high pressure occurs. Thus the constant low pressure in the equatorial zone and the high pressure at the poles are thermally induced.

From this we might expect a gradual increase in pressure from the equator to the poles to accompany the gradual decrease in average annual temperature. However, actual readings taken at the earth's surface indicate that pressure does not increase latitudinally in a regular fashion poleward from the equator.

Instead, there are regions of high pressure in the subtropics and regions of low pressure in the subpolar regions. The dynamic causes of these zones, or *belts* of high and low pressure are more complex than the thermal causes.

The dynamic causes are related to the rotation of the earth and the broad patterns of circulation. As air rises steadily and moves away from the equator, it constantly moves toward the poles and descends there owing to the differing temperature conditions, thus creating the thermally induced pressure areas that we just described. However, some of the air drifting from the equator toward the poles in the upper atmosphere is turned aside by the earth's rotation and other dynamic causes and settles in the subtropics. This subsidence creates high pressure in the subtropics.

The relatively low pressures in the subpolar regions can best be explained as a result of air converging and rising in these regions. The converging air comes from the high-pressure belts on both sides (polar and subtropical) of these regions. The rising air is accompanied by a decrease in pressure. Thus, both the subtropical and subpolar pressure regions are dynamically induced.

IDEALIZED WORLD PRESSURE BELTS

On the basis of what we have just learned about pressure on the earth's surface, it is possible to construct a theoretical model of the pressure belts of the world (Fig. 4.2). Later, we will see how real conditions depart from our model and examine why these differences occur.

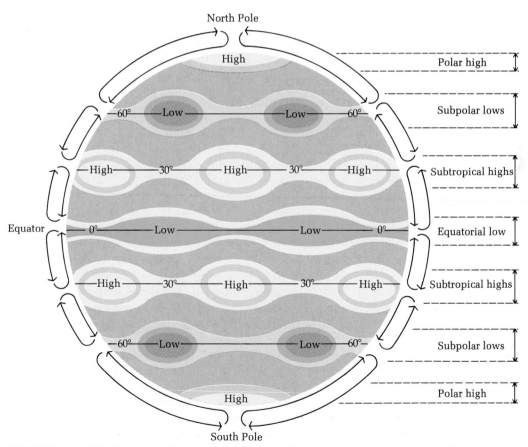

FIGURE 4.2 Idealized world pressure belts. Note the arrows on the perimeter of the globe that illustrate the cross-sectional flow associated with the surface pressure belts.

Centered approximately over the equator in our model is a belt of low pressure or a **trough.** Because this is the region on the globe of greatest annual heating, we can conclude that the low pressure of this area, the **equatorial low,** is determined primarily by thermal factors.

North and south of the equatorial low and centered on the so-called "horse latitudes," which are about 30°N and 30°S, are cells of relatively high pressure. These are the **subtropical highs,** which are the results of dynamic factors involved with the sinking of convectional cells initiated at the equatorial low.

Poleward of the subtropical highs in both the Northern and Southern Hemispheres are large belts of low pressure that extend through the upper middle latitudes. Pressure decreases through these **subpolar lows** until about 65° latitude. Again, dynamic factors play a role in the existence of subpolar lows.

In the polar regions are high-pressure systems, the **polar highs.** These are determined by the cold temperatures and consequent sinking of the dense air in those regions.

This system of pressure belts that we have just developed is a generalized picture. Just as temperatures change from month to month, day to day, and hour to hour, so do pressures vary from time to time at any one place. Our model disguises these changes, but it does give an idea of broad pressure patterns on the surface of the earth.

MAPPING PRESSURE DISTRIBUTION

We will turn to maps that show the actual distribution of barometric pressure on the earth's surface in order to see how well our idealized model of world pressure belts represents the "real world." We have noted that atmospheric pressure varies vertically with altitude as well as horizontally with temperature and other factors. Although we should be aware of the decrease in pressure that takes place as we move vertically through the atmosphere, it is the variations from place to place and time to time over the earth's surface that are basic to an understanding of weather and climate. Therefore, when we map air pressure, we reduce all pressures to what they would be at sea level, just as we changed temperature to sea level in order to eliminate altitude as a factor. This is especially important for atmospheric pressure because the variations due to altitude are far greater than those due to atmospheric dynamics and would tend to mask the more meteorologically important regional differences.

Figures 4.3 and 4.4, which show the average sea level pressure patterns for January and July, are isobaric maps. **Isobars** (from Greek *isos,* equal; *baros,* weight) are lines drawn on these maps to connect places of equal pressure. When the isobars appear close together they portray a significant difference in pressure between places, hence a strong **pressure gradient.** When the isobars are far apart, a weak pressure gradient is depicted.

The atmosphere tends to form cells of high and low pressure. Depicted on a map, these cells are outlined by concentric isobars that form a closed system around centers of low pressure or high pressure. The cells of low pressure are commonly referred to as **lows,** or **cyclones;** the cells of high pressure are called **highs,** or **anticyclones.**

THE GENERAL PATTERN OF ATMOSPHERIC PRESSURE

As our idealized model suggests, the atmosphere tends to form belts of high and low pressure along east-west axes in areas where there are no large bodies of land. These belts are latitudinally arranged and generally maintain their bandlike pattern. However, where there are continental landmasses, belts of pressure are broken and tend to form cellular pressure systems. The landmasses affect the development of belts of atmospheric pressure in several ways. Most influential is the effect of the differential heating of land and water sur-

FIGURE 4.3 Average sea level pressure (mb) in January.

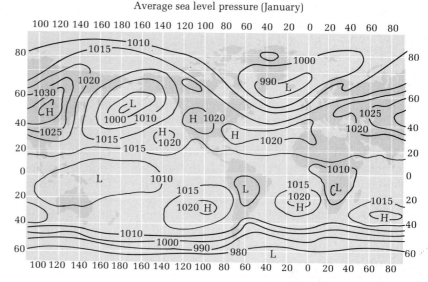

Average sea level pressure (January)

faces. In addition, landmasses affect the movement of air and consequently the development of pressure systems through friction with their surfaces. Landform barriers like mountain ranges also block the movement of air and thereby affect atmospheric pressure.

SEASONAL VARIATIONS IN THE PATTERN

In general, the atmospheric pressure belts shift northward in July and southward in January,

following the changing position of the sun's direct rays as they migrate between the Tropics of Cancer and Capricorn. Thus, there are thermally induced seasonal variations in the pressure patterns as seen in Figures 4.3 and 4.4. These seasonal variations tend to be small in the low latitudes because the low latitudes experience little temperature variation, and larger in high latitudes where there is an increasing contrast in length of daylight and angle of the sun's rays. Furthermore, landmasses tend to alter the general pattern of seasonal

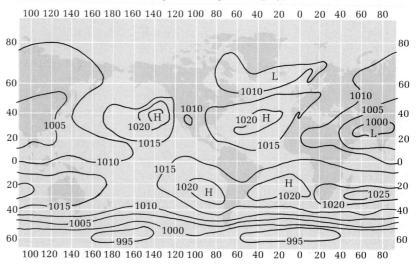

Average sea level pressure (July)

FIGURE 4.4 Average sea level pressure (mb) in July.

variation. This is an especially important factor in the Northern Hemisphere, where land accounts for 40 percent of the total surface as opposed to less than 20 percent in the Southern Hemisphere.

JANUARY Because land cools more quickly than the oceans, its temperatures will be lower in winter than the surrounding seas. Figure 4.3 shows that in the middle latitudes of the Northern Hemisphere this variation leads to the development of cells of high pressure over the land areas in contrast to the subpolar lows of the oceans. Over eastern Asia there is a strongly developed anticyclone during the winter months that is known as the **Siberian High.** Its equivalent in North America, known as the **Canadian High,** is not nearly so well developed.

In addition to the Canadian High and the Siberian High, two low-pressure centers develop: one in the North Atlantic, the **Icelandic Low,** and the other in the North Pacific, the **Aleutian Low.** These regions have not been affected by the seasonal continental temperature changes. The air in them has relatively lower pressure than either the subtropical or the polar high systems. Consequently, air moves toward these low-pressure areas from both north and south. Such low-pressure regions are associated with cloudy, unstable weather and are a major source of winter storms, while high-pressure areas are associated with clear, blue-sky days, calm, starry nights, and cold, stable weather. Therefore, during the winter months cloudy weather tends to be associated with the two oceanic lows and clear weather with the continental highs.

We can also see that the polar high in the Northern Hemisphere is well developed. This development is primarily due to thermal factors, since it is now the coldest time of the year. The subpolar lows have developed into the Aleutian and Icelandic cells described above. At the same time, the subtropical highs of the Northern Hemisphere appear slightly south of their average annual position because of the migration of the sun toward the Tropic of Capricorn. The equatorial trough also appears cen-

tered south of its average annual position over the geographic equator.

In January in the Southern Hemisphere the subtropical belt of high pressure appears as three cells centered over the oceans, for the belt of high pressure has been interrupted by the continental landmasses where temperatures are much higher and pressure tends to be lower than over the oceans. Because there is virtually no land between 45°S and 70°S latitude, the subpolar low circles the earth as a real belt of low pressure and is not divided into cells by interrupting landmasses. Seasonally there is little change in this belt of low pressure other than that in January (or summer in the Southern Hemisphere), where it lies a few degrees north of its July position.

JULY The anticyclone over the North Pole is greatly weakened during the summer months in the Northern Hemisphere, primarily because of the heating of the oceans and landmasses in that hemisphere (Fig. 4.4). The Aleutian and Icelandic Lows nearly disappear from the oceans, while the landmasses, which developed high-pressure cells during the cold winter months, have extensive low-pressure cells slightly to the south during the summer. In Asia, a low-pressure system develops, but it is divided into two separate cells by the Himalayas (see Fig. 4.15). The low-pressure cell over northwest India is so strong that it combines with the equatorial trough, which has moved north of its position 6 months earlier. The subtropical highs of the Northern Hemisphere are more highly developed over the oceans than over the landmasses. In addition, they migrate northward and are highly influential factors in the climate of landmasses nearby. In the Pacific, this subtropical high is termed the **Pacific High;** it is this system of pressure that plays an important role in moderating the temperatures of the west coast of the United States. In the Atlantic Ocean the corresponding cell of high pressure is known as the **Bermuda High** to Americans and the **Azores High** to Europeans. As we have already mentioned, the equatorial trough of low pressure moves north in July, following the migration of the sun's vertical rays, and the subtropical highs of the Southern Hemisphere lie slightly north of their January locations.

In examining the earth's pressure systems at the earth's surface, we have seen that there are essentially seven belts of pressure (two polar highs, two subpolar lows, two subtropical highs, and one tropical low) that are broken into cells of pressure in some places, primarily because of the influence of certain large landmasses. We have also seen that these belts and cells vary in size, intensity, and location with the seasons and with the migration of the sun's vertical rays over the earth's surface.

WIND

Wind is the horizontal movement of air in response to differences in pressure. Winds are the means by which the atmosphere attempts to balance the uneven distribution of pressure over the earth's surface. The movements of the wind also play a large role in compensating for uneven distribution in the atmosphere of incoming solar energy and the loss of earth radiation. Without winds, the equatorial regions would be hotter and the polar areas would be colder. Besides serving a vital function in the advectional transport of heat energy over the earth's surface, winds also transport water vapor from the air above bodies of water, where it has evaporated, to land surfaces. This allows for greater precipitation over land surfaces than could otherwise occur. In addition, winds exert influence on the rate of evaporation itself.

PRESSURE GRADIENTS AND WINDS

Winds vary in speed, intensity, duration, and direction. Much of their strength depends upon the size or strength of the **pressure gradient** to which they are responding. Pressure gradient is the term applied to the rate of change of atmospheric pressure between two points (at the same elevation). The greater this change—that is, the steeper the pressure gradient—the greater will be the wind response (Fig. 4.5). Winds tend to flow down a pressure gradient from high pressure to low pressure, just as water flows down a slope from a high point to a low one. The steeper the pressure gradients involved, the faster and stronger the winds will blow.

Yet wind does not flow directly from high to low as we might expect, for there are other factors involved that affect the direction of wind.

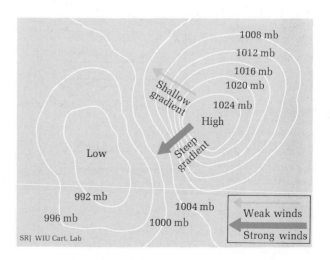

FIGURE 4.5 The relationship of wind to the pressure gradient. The steeper the pressure gradient, the stronger the resulting wind.

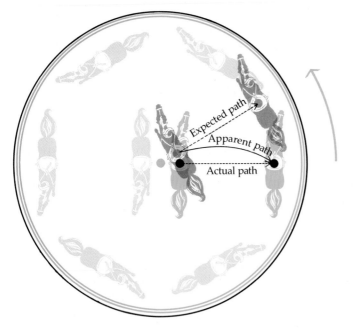

FIGURE 4.6 A ball thrown from the center of a counterclockwise-rotating carousel appears to curve to the right, missing the aiming point. This phenomenon occurs because the aiming point has moved in a counterclockwise arc to a new position by the time the ball has arrived at the original aiming point. However, people on the carousel feel that they are stationary and that the world around them, including the ball, is in motion. This effect is similar to the Coriolis effect.

THE CORIOLIS EFFECT AND WIND

We must remember that the speed of rotation on the earth's surface increases as we move equatorward and decreases as we move toward the poles (see again Fig. 1.13). Thus, to use our previous example, someone living in Leningrad, where the distance around a parallel of latitude is about half that at the equator, travels at about 840 kilometers per hour (525 mph) as the earth rotates, while someone living in Kampala, Uganda, near the equator, moves at about 1680 kilometers per hour (1050 mph).

This situation is analogous to a carousel. If someone rides a horse on the outside of the carousel, he will be going faster than you, if you ride a horse near the center, simply because he must travel a greater distance (a larger circle) in the same amount of time in order to keep up with you. Now, imagine that you are sitting on one of the horses near the center of the carousel, your friend is on a horse near the outside edge, and the carousel is turning counterclockwise (Fig. 4.6). You want to throw him the ball you just won at the shooting gallery. When you do, however, even though you are

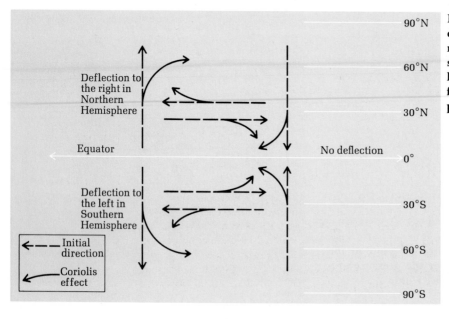

FIGURE 4.7 The apparent deflection of the paths of moving objects on a rotating sphere caused by the Coriolis effect. The Coriolis deflection increases toward the poles.

aiming straight at him (remember you are both still riding around on the carousel), you will miss him, and the ball will apparently be deflected to your right. This is because your friend is traveling at a speed faster than you, and by the time the ball gets to where he had been when you threw it, he will have moved further around the circle. It is true that you are moving also, but you are traveling at a slower speed and covering a shorter distance.

This is what happens to anything moving horizontally over the earth's surface. Because we are unaware of our own rotation with the earth around its axis, we see anything that is moving horizontally as being deflected to the right of the direction in which it is traveling in the Northern Hemisphere and to the left in the Southern Hemisphere.

This apparent deflection is termed the **Coriolis effect** (Fig. 4.7). The degree of deflection, or curvature, is a function of the speed of the object in motion and the latitudinal location of the object. The higher the latitude, the greater will be the Coriolis effect. In fact, not only does the Coriolis effect decrease at lower latitudes, but it does not exist at the equator. Also, the faster the object is moving, the greater will be the apparent deflection. In ad-

dition, the greater the distance something must travel, the greater will be the Coriolis effect.

As we have said, anything that moves horizontally over the earth's surface exhibits the Coriolis effect. Thus, both the atmosphere and the oceans are deflected in their movements. Winds in the Northern Hemisphere moving across a gradient from high to low pressure are apparently deflected to the right of the direction in which they originally blow (and to the left in the Southern Hemisphere). In addition, when considering winds at the earth's surface, we must take into account another force. This force, **friction,** interacts with the pressure gradient and the Coriolis effect.

FRICTION AND WIND

Above the earth's surface frictional drag is of little consequence to wind development. At this level the wind starts down the pressure gradient and is turned 90 degrees in response to the Coriolis effect. At this point the pressure gradient is balanced by the Coriolis effect and the wind, termed a **geostrophic wind,** flows parallel to the isobars (Fig. 4.8a).

However, at or near the earth's surface (up to about 1000 meters above the surface)

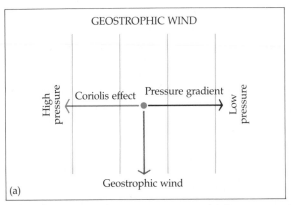
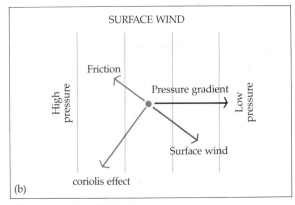

FIGURE 4.8 These Northern Hemisphere examples illustrate that (a) in a geostrophic wind, as a parcel of air starts to flow down the pressure gradient, the Coriolis effect causes it to veer to the right until pressure gradient and Coriolis effect reach an equilibrium and the wind flows between isobars. (b) In a surface wind, this equilibrium is upset by friction, which reduces the wind speed. Since the Coriolis effect is a function of wind speed, it also is reduced. With Coriolis effect reduced, pressure gradient dominates and the wind now flows across isobars in the direction of low pressure.

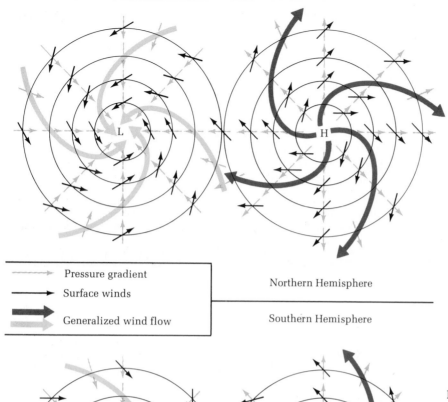

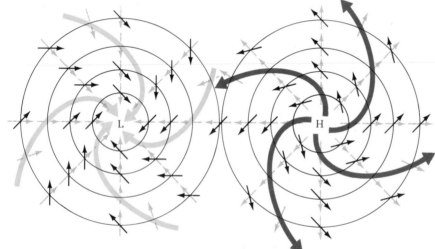

Pressure gradient

Surface winds

Generalized wind flow

Northern Hemisphere

Southern Hemisphere

FIGURE 4.9 Movement of surface winds associated with low-pressure centers (cyclones) and high-pressure centers (anticyclones) in the Northern and Southern Hemispheres. Note that the surface winds are to the right of the pressure gradient in the Northern Hemisphere and to the left of the pressure gradient in the Southern Hemisphere.

frictional drag is important since it reduces the wind speed. A reduced wind speed in turn reduces the Coriolis effect, while the pressure gradient is not affected. With pressure gradient and Coriolis effect no longer in balance, the resultant surface wind does not flow between the isobars like its upper-level counterpart. Instead, a surface wind flows obliquely across the isobars toward the low-pressure area (Fig. 4.8b).

CYCLONES AND ANTICYCLONES

Imagine a high-pressure cell (anticyclone) in the Northern Hemisphere in which the air is moving from the center in all directions down pressure gradients. As it moves, the air will be deflected to the right, no matter which direction it was originally going. Therefore, the wind moving out of an anticyclone in the Northern Hemisphere will move from the center of high pressure in a clockwise spiral (Fig. 4.9).

Air tends to move down pressure gradients from all directions toward the center of a low-pressure area (cyclone). However, since the air is apparently deflected to the right in the Northern Hemisphere, the winds move into the cyclone in a counterclockwise spiral. Because all objects including air and water are

FIGURE 4.10 Winds converge in cyclones (low-pressure centers) and diverge from anticyclones (high-pressure centers).

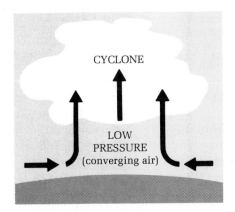

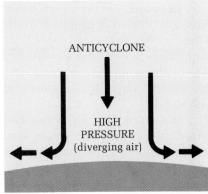

apparently deflected to the left in the Southern Hemisphere, spirals there are reversed. Thus, in the Southern Hemisphere, winds moving away from an anticyclone do so in a counterclockwise spiral and winds moving into a cyclone move in a clockwise spiral.

CONVERGENT AND DIVERGENT CIRCULATION

As we have just seen, winds blow toward the center of a cyclone and can be said to *converge* toward it. Hence, a cyclone is a closed system of isobars whose center serves as the focus for **convergent** wind circulation (Fig. 4.10). The winds of an anticyclone blow away from the center of high pressure and are said to be *diverging*. In the case of an anticyclone, the center of the system serves as the source for **divergent** wind circulation.

Cyclonic circulation is said to be converging whether it is in a counterclockwise spiral as in the Northern Hemisphere or in a clockwise

spiral as in the Southern Hemisphere. Anticyclonic circulation is said to be diverging whether it is doing so in the counterclockwise outward spiral of the Southern Hemisphere or the clockwise outward spiral of the Northern Hemisphere.

WIND TERMINOLOGY

Winds are named after their source. That is, a wind that comes out of the northeast is called a *northeast wind*. One coming from the south, even though going toward the north, is called a *south wind*.

Windward refers to the direction from which the wind blows. The side of something that faces the direction from which the wind is coming is called the *windward* side. Thus, a windward shore is the one that receives onshore winds from the ocean, and a windward slope is the side of a mountain against which the wind blows (Fig. 4.11).

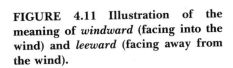

FIGURE 4.11 Illustration of the meaning of *windward* (facing into the wind) and *leeward* (facing away from the wind).

Leeward, on the other hand, means the direction toward which the wind is blowing. Thus, a leeward shore would have offshore winds since it faces the way winds are blowing. Where the winds are coming out of the west the leeward slope of a mountain would be the east slope.

We know that winds can blow from any direction. Yet, in some places winds may tend to blow more from one direction than any other. We speak of these as the **prevailing winds.**

THE EARTH'S SURFACE WIND SYSTEMS

There is a planetary or global wind system that is a response to the worldwide pressures we have already examined. In addition, the global wind system plays a role in the maintenance of those same pressures. This wind system, which is the major means of transport for energy and moisture through the earth's atmosphere, can be examined in an idealized state. To do so, however, we must ignore the influences of landmasses and seasonal variations in solar energy. By assuming, for the sake of discussion, that the earth has a homogeneous surface and that there are no seasonal variations in solar energy received at different latitudes, we will be able to examine a theoretical model of the atmosphere's planetary circulation. Such an understanding will help to explain specific features of climate like the rain and snow of the Sierras and Cascades and the existence of arid regions farther to the east. It will also account for the movement of great surface currents in our oceans that are driven by this atmospheric engine (see Chapter 17).

THE IDEALIZED MODEL OF ATMOSPHERIC CIRCULATION

A system of global winds can be associated with the model of pressures that we previously de-veloped (Fig. 4.2). Winds are related to pressure, and various types of winds are associated with different kinds of pressure cells.

The characteristics of convergence and divergence are very important to our understanding of global wind patterns. We know that because of the pressure gradient, surface winds blow away from high-pressure cells and toward low-pressure cells. In other words, surface air diverges from zones of high pressure and converges on areas of low pressure.

Knowing that surface winds originate in areas of high pressure and taking into account the global system of pressure cells, we can develop our model of the wind systems of the world (Fig. 4.12). This model takes into account differential heating, earth rotation, and atmospheric dynamics.

Note that the winds do not blow in a straight north-south line. The variation is due, of course, to the Coriolis effect, which causes an apparent deflection to the right in the Northern Hemisphere and to the left in the Southern Hemisphere.

Our idealized model of global atmospheric circulation includes six wind belts or zones in addition to the seven pressure zones we have previously identified. Two wind belts, one in each hemisphere, are located where winds move out of the polar highs and move down the pressure gradients toward the subpolar lows. As these winds are deflected to the right in the Northern Hemisphere and to the left in the Southern, they become the **polar easterlies.**

The remaining four wind belts are closely associated with the divergent winds of the subtropical highs. In each hemisphere winds flow out of the poleward portions of these highs toward the subpolar lows. Because of their general movement from the west, the winds of the upper middle latitudes are labeled the **westerlies.** The winds blowing from the highs toward the equator have been called the **trade winds.** Because of the Coriolis effect they are the **northeast trades** in the Northern Hemisphere and the **southeast trades** south of the equator.

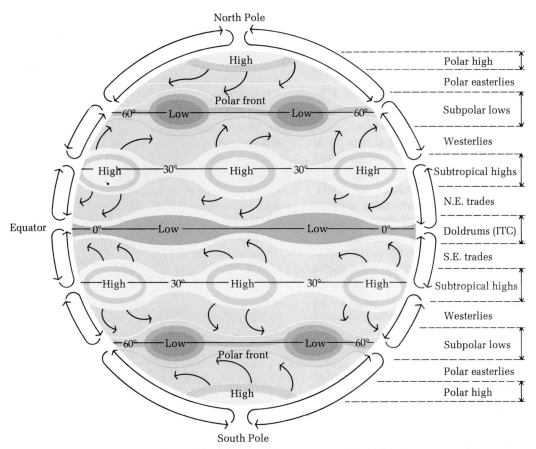

FIGURE 4.12   The idealized model of the earth's pressure and wind systems.

Our model does not conform exactly to actual conditions. First, the vertical sun does not stay precisely over the equator but moves north and south during the year as far as the two tropics. Therefore the pressure systems, and consequently the winds, must move in order to adjust to the change in position of the sun. Then, as we have already discovered, the existence of the continents, especially in the Northern Hemisphere, causes longitudinal pressure differentials that affect the zones of high and low pressure. However, it should prove interesting to compare our model of the planetary wind and pressure system with con-

ditions as they actually exist on the earth's surface.

CONDITIONS WITHIN LATITUDINAL ZONES

TRADE WINDS   A good place to begin our examination of wind and associated weather patterns as they actually occur is in the vicinity of the subtropical highs. On the earth's surface, the trade winds, which blow out of the subtropical highs toward the equatorial trough in both the Northern and Southern Hemispheres, may be identified between latitudes 5° and 25°. Because of the Coriolis effect, the northern trades move away from the subtropical high in a clockwise direction out of the northeast. In the Southern Hemisphere the trades diverge out of the subtropical high toward the tropical low from the southeast, as their movement is counterclockwise. Because the trades tend to blow out of the east, they are also known as the **tropical easterlies.**

The trade winds tend to be constant, steady winds, consistent in their direction. This is most true when they cross the eastern sides of the oceans (near the eastern portions of the subtropical high). The area of the trades varies somewhat during the solar year, moving north and south a few degrees of latitude with the sun. Near their source in the subtropical highs the weather of the trades is clear and dry, but after crossing large expanses of ocean the trades have a high potential for stormy weather.

Early Spanish sea captains depended upon the northeast trade winds to drive their galleons to destinations in Latin America in search of gold, spices, and new lands. Coming eastward toward home, their navigators usually tried to plot a course using the westerlies to the north. The trade winds are one of the reasons that the Hawaiian Islands are so popular with tourists. The steady winds help to keep temperatures pleasant, even though Hawaii is located south of the Tropic of Cancer.

DOLDRUMS Where the trade winds meet in the equatorial trough (or tropical low) there is a zone of calms and weak winds of no prevailing direction. Here the air, which is very moist and heated by the sun, tends to expand and rise, maintaining the low pressure of the area. This zone where the trades meet, which is roughly between 5°N and 5°S, is generally known as the "doldrums." It is also called the **Intertropical Convergence** zone (ITC) and the "equatorial belt of variable winds and calms."

Because of the converging moist air and high potential for rainfall in the doldrums, this region coincides with the world's latitudinal belt of heaviest precipitation and most persistent cloud cover.

Old sailing ships often remained calmed in the doldrums for days at a time. A description of a ship becalmed in the doldrums occurs in *The Rime of the Ancient Mariner* by Samuel Taylor Coleridge (lines 103–108). The ship is sailing northward from the tropical southeasterlies (trades) when it gets to the doldrums.

> The fair breeze blew, the white foam flew,
> The furrow followed free;
> We were the first that ever burst
> Into that silent sea.

> Down dropt the breeze, the sails dropt down,
> 'Twas sad as sad could be;
> And we did speak only to break
> The silence of the sea!

> All in a hot and copper sky,
> The blood Sun, at noon,
> Right up above the mast did stand,
> No bigger than the Moon.

> Day after day, day after day,
> We stuck, nor breath nor motion;
> As idle as a painted ship
> Upon a painted ocean.

SUBTROPICAL HIGHS The areas of subtropical high pressure, generally located between latitudes 25° and 35° N and S, and from which winds blow equatorward as the trades, are often called the subtropical belts of variable winds, or the "horse latitudes." This latter name comes from the occasional necessity for the Spanish conquistadores of throwing their horses overboard in order to conserve drinking water when their ships were becalmed in these latitudes. The subtropical highs are areas, like the doldrums, in which there are no strong prevailing winds. However, unlike the doldrums, which are characterized by convergence, rising air, and heavy rainfall, the subtropical highs are areas of sinking and settling air from higher altitudes, which tend to build up the atmospheric pressure. Weather conditions are typically clear, sunny, and rainless, especially over the eastern portions of the oceans where the high-pressure cells are strongest.

WESTERLIES   The winds that move poleward out of the subtropical high-pressure cells in the Northern Hemisphere are deflected to the right and thus blow from the southwest. Those in the Southern Hemisphere are deflected to the left and blow out of the northwest. Thus, these winds have correctly been labeled the **westerlies.** They tend to be less consistent in direction than the trades, but they usually are stronger winds and are often associated with stormy weather. The westerlies occur between about 35° and 65° N and S latitudes. In the Southern Hemisphere, where there is less land than in the Northern Hemisphere to affect the development of winds, the westerlies attain their greatest consistency and strength. Most of the United States, except Florida, Hawaii, and Alaska, is under the influence of the westerlies.

POLAR WINDS   Specific, accurate records to describe the movement of the atmosphere in the two polar regions are lacking. Our best estimate is that pressures are consistently high throughout the year at the poles and that prevailing easterly winds blow from the polar regions to the subpolar low-pressure systems.

POLAR FRONT   Despite the fact that we cannot fully describe the wind systems of the polar regions, we do know that the winds can be highly variable, blowing at times with great speed and intensity. When the cold air flowing out of the polar regions and the warmer air moving in the path of the westerlies meet, they do so like two warring armies: One does not absorb the other. Instead, the cold air pushes the warm air out of the way, forcing it to rise rapidly. The line along which these two great wind systems battle is appropriately known as the **polar front.** The weather that results from the meeting of the cold polar air and the warmer air from the subtropics can be very stormy. In fact, many of the storms that move slowly through the middle latitudes in the path of the prevailing westerlies are born at the polar front.

THE EFFECTS OF SEASONAL MIGRATION

Just as insolation, temperature, and pressure systems migrate north and south as the earth revolves around the sun, the earth's wind systems also migrate with the seasons. During the summer months in the Northern Hemisphere, maximum insolation is received north of the equator. This condition causes the pressure belts to move north as well, and the wind belts of both hemispheres shift accordingly. Six months later, when maximum heating is taking place south of the equator, the various wind systems have migrated south in response to the migration of the pressure systems. Thus, seasonal variation in wind and pressure conditions is one important way in which actual atmospheric circulation differs from our idealized model.

The seasonal migration will most affect those regions near the boundary zone between two wind or pressure systems. During the winter months such a region will be subject to the moods of one system. Then, as summer approaches, that system will migrate poleward and the next equatorward system will move in to influence the region. Two such zones in each hemisphere have a major effect on climate. The first lies between latitudes 5° and 15°, where the wet equatorial low of the high-sun season (summer) alternates with the dry subtropical high and trade winds of the low-sun season (winter). The second occurs between 30° and 40°, where the subtropical high dominates in summer but is replaced by the wetter westerlies in winter.

(Text continues on page 108.)

VIEWPOINT: WHITHER THE WIND

While winds are important in moving large and small bodies of air across the earth's surface, creating significant climatic and weather patterns, they also affect us in ways of which we may be less aware. Let us consider a few examples. Insect pests are sometimes wind-borne. Locusts migrate with prevailing winds, and their movements show a marked correlation with the location of the intertropical convergence zone. Research into outbreaks of the highly infectious hoof-and-mouth disease in cattle in England in 1967 and 1968 has shown that airborne transmission was responsible for spreading it from the continent of Europe. Crop dusting with insecticides is most efficiently carried out in very light wind conditions; otherwise the applications may be excessively dispersed and cause ecological harm to the environment. The distribution of pollutants may be affected by prevailing winds: the interior valleys of the Los Angeles basin are under siege from severe smog conditions when the air flow is onshore, while offshore air flows with the Santa Ana wind clear the basin of smog.

Early sailing vessels, of course, were highly dependent on the world's prevailing winds. Ships sailing the Atlantic trade triangle carried slaves from West Africa to the West Indies and to sugar and cotton plantations of the southern United States by following the trades. They then followed the westerlies from New England for the return journey to Europe. Today, airport locations are chosen with the main runways normally aligned in the direction of the prevailing wind in order to reduce the danger of crosswinds on landings. Inflight weather forecasts stress wind conditions, as these help to determine the fastest and most economical flight paths. Long-range flights may follow different routes in summer and winter owing to the seasonal shift of the pressure and wind belts. Turbulence, often at low levels due to thermal convection, creates discomfort for passengers and may ultimately contribute to structural fatigue in aircraft.

The glider pilot, of course, seeks thermal wind currents for soaring, and he or she is very sensitive to vertical movements of air. Sailing is another sport in which 'reading the wind' and a knowledge of sea-breeze patterns enable competitors to make the best tactical use of wind flows and wind gusts. Likewise, competitive skiing and especially ski-jumping may be adversely affected by high winds or gusty conditions, and ski lift and chairs may be closed if winds reach high velocities.

The wind speed has a critical effect on what is known as the wind-chill factor. Wind makes us feel cooler or affects our **sensible temperature** by increasing our loss of body heat. Thus at high temperatures, wind will make us more comfortable, while at lower temperatures wind can make us very uncomfortable. The faster the wind speed and the lower the temperature, the greater will be the chilling effects caused by the wind.

Wind chill can be transposed into equivalent temperatures. These temperatures represent the effective air temperature resulting from the combination of wind at a certain speed and actual air temperature. That is, if the air temperature is 20°F and the wind is blowing at 15 mph, the effective temperature is −6°F according to the wind-chill table. Or, if the wind is blowing twice as fast, at 30 mph, while the air temperature is still 20°F, the result is a temperature that feels like −18°F.

In designing our urban environments, architects must take wind into account. The built-up surface of an urban area typically presents more frictional drag on the air moving across it than a rural area, and thus overall wind speeds are generally lower in cities than in the country. However, high-rise towers or slab buildings have sometimes resulted in turbulent air flows between them; in one such instance in an English city, a shopping area between a low building and a new high-rise structure had to be glassed in so that shoppers could avoid the gusty air flows that resulted. Midwesterners are all too familiar with the effects of a tornado, and the collapse of the Tacoma suspension bridge in Washington in 1940 was attributed to severe wind stress.

Texel, the Netherlands. (R. Gabler)

Wind-Chill Factors*

MPH	Dry-Bulb Temperature (°F)																
	35	30	25	20	15	10	5	0	−5	−10	−15	−20	−25	−30	−35	−40	−45
	Equivalent Temperature† of Wind Chill Index (°F)																
calm	35	30	25	20	15	10	5	0	−5	−10	−15	−20	−25	−30	−35	−40	−45
5	33	27	21	16	12	7	1	−6	−11	−15	−20	−26	−31	−35	−41	−47	−54
10	21	16	9	2	−2	−9	−15	−22	−27	−31	−38	−45	−52	−58	−64	−70	−77
15	16	11	1	−6	−11	−18	−25	−33	−40	−45	−51	−60	−65	−70	−78	−85	−90
20	12	3	−4	−9	−17	−24	−32	−40	−46	−52	−60	−68	−76	−81	−88	−96	−103
25	7	0	−7	−15	−22	−29	−37	−45	−52	−58	−67	−75	−83	−89	−96	−104	−112
30	5	−2	−11	−18	−26	−33	−41	−49	−56	−63	−70	−78	−87	−94	−101	−109	−117
35	3	−4	−13	−20	−27	−35	−43	−52	−60	−67	−72	−83	−90	−98	−105	−113	−123
40	1	−4	−15	−22	−29	−36	−45	−54	−62	−69	−76	−87	−94	−101	−107	−116	−128
45	1	−6	−17	−24	−31	−38	−46	−54	−63	−70	−78	−87	−94	−101	−108	−118	−128
50	0	−7	−17	−24	−31	−38	−47	−56	−63	−70	−79	−88	−96	−103	−110	−120	−128

*Wind speeds greater than 40 mph have little additional chilling effect.
† Equivalent in cooling power on exposed flesh under calm condition.
Source: Mimeographed report, ESSA, Washington, D.C.

Can wind ever be harnessed as an energy source? Historically it has been used for grinding grain and pumping water; in the seventeenth century the Dutch used windmills to lift water into canals to drain their marshlands, and by 1850 windmill power was used in much of the Midwest and Great Plains to draw water to the surface. Today, as other fuels show signs of depletion, there is renewed interest in generating power by using wind energy, a free, nonpolluting, and inexhaustible power source. Unfortunately, wind power generators need consistent 10–30 mph winds to operate efficiently, and there is the visual impact of such wind towers—blades and power lines in the landscape. A Lockheed study suggests that wind will be an unlikely substitute for fossil or nuclear power but would be a good supplemental energy source in certain areas and might account for an estimated 1 to 2 percent of energy production in the United States by the year 2000. Southern California Edison has funded 2 million dollars for an aerogenerator in San Gorgonio Pass at the southern entrance to the Los Angeles Basin; with 165-foot blades, it is planned to produce enough energy to electrify 1000 homes. Whether it will be a prototype for the future remains to be seen.

A commercial wind turbine generator. A modern adaptation to an ancient energy source. (Natural Power and Light Company, Macomb, Illinois)

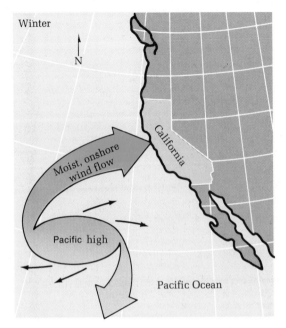
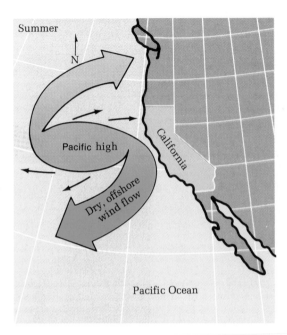

FIGURE 4.13   Summer and winter positions of the Pacific anticyclone in relation to California. The influence of the anticyclone dominates during the summer. The high pressure blocks out cyclonic storms and produces warm, sunny, and dry conditions. In the winter the anticyclone lies further south and feeds the westerlies that bring occasional cyclonic storms and rain from the North Pacific.

California is an example of a region located within a zone of transition between two wind or pressure systems (Fig. 4.13). During the winter months this region is under the influence of the westerlies blowing out of the Pacific High. These winds, being turbulent and full of moisture from the ocean, bring winter rains and storms to "sunny" California. As summer approaches, however, the subtropical high moves north and the westerlies along with it. And as California comes under the influence of the calm and steady high-pressure system, it experiences again the climate for which it is famous: day after day of warm, clear, blue, cloudless skies. This alternation of moist winters and dry summers is typical of the western sides of all landmasses between 30° and 40° latitude.

LONGITUDINAL DIFFERENCES IN WINDS

We have seen that there are sizable latitudinal differences in pressure and winds. In addition, there are significant longitudinal variations, especially in the zone of the subtropical highs.

As we have previously mentioned, the subtropical high-pressure cells, which are generally centered over the oceans, are much stronger on their eastern than on their western sides. Thus over the eastern portions of the oceans (west coasts of the continents) in the subtropics, subsidence and divergence are especially noticeable. The above-surface temperature inversions, so typical of anticyclonic circulation, are close to the surface and the air is calm and clear. The air moving equatorward from this portion of the high produces the classic picture of the steady trade winds with clear, dry weather.

Over the western portions of the oceans (eastern sides of the continents) conditions are markedly different. In its passage over the ocean, the diverging air is gradually warmed and moistened, the above-surface inversion occurs at higher elevations, and turbulent, stormy weather conditions are likely to develop. As indicated in Figure 4.14, wind movement in the

FIGURE 4.14 The circulation pattern in a subtropical anticyclone. Subsidence of air is strongest in the eastern part of the anticyclone, producing calm air and arid conditions over adjacent land areas. The southern margin of the anticyclone feeds the persistent northeast trade winds. (After Trewartha)

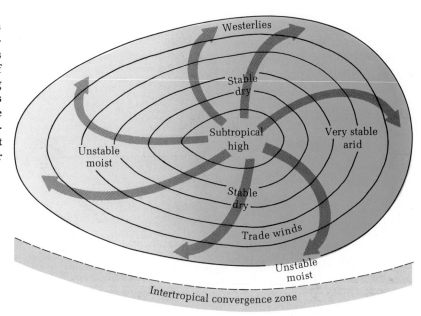

western portions of the anticyclones may actually be poleward and directed toward landmasses. Hence the trade winds in these areas will be especially weak or nonexistent much of the year.

Farther toward the poles, especially in the Northern Hemisphere, as we have pointed out in discussing Figures 4.3 and 4.4, there are great land-sea contrasts in temperature and pressure throughout the year. In the cold continental winters, the land will be associated with pressures that are higher than the oceans, and thus there will be strong, cold winds from the land to the sea. In the summer, the situation changes, with relatively low pressure existing over the continents because of higher temperatures. Wind directions are thus greatly affected, and the pattern is reversed so that winds flow from the sea to the land.

MONSOON WINDS

The term *monsoon* comes from the Arabic word *mausim*, meaning season. This word has been used by Arab seamen for many centuries to describe seasonal changes in wind direction across the Arabian Sea between Arabia and India. As a meteorological term, **monsoon** refers to the directional shifting of winds from one season to the next. Usually, the monsoonal shifting is from a humid wind blowing from the ocean toward the land in the summer to a dry, cooler wind blowing seaward off the land in the winter, and it involves a full 180° change in the wind.

The development of the monsoon is most characteristic of southern Asia, although it occurs on other continents as well. As the large landmass of Asia cools more quickly than the surrounding oceans, the continent develops a strong center of high pressure from which there must be an outflow of air in winter (Fig. 4.15). This outflow blows across much land toward the tropical low before reaching the oceans. It brings cold, dry air south.

In summer the Asian continent heats quickly and develops a large low-pressure center. This development is reinforced by a poleward shift of the Intertropical Convergence to a position over southern Asia. Warm, moist air from the oceans is attracted into this low. Though full of water vapor, this air does not in itself cause the wet summers with which the monsoon is associated. However, any turbulence or landform barrier that makes this moist air rise and, as a result, cool off, will bring

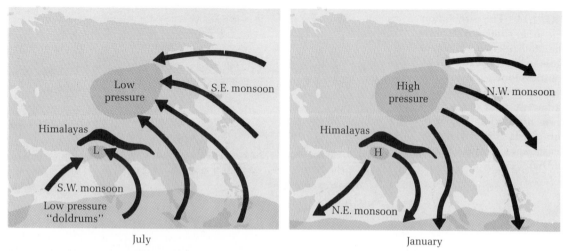

FIGURE 4.15 Seasonal changes in surface wind direction that create the Asiatic monsoon system. The burst of the "wet monsoon," or sudden onshore flow of tropical humid air in July, is apparently triggered by changes in the upper-air circulation, resulting in heavy precipitation. The offshore flow of dry continental air in winter creates the "dry monsoon" and drought conditions in southern Asia.

about precipitation. This precipitation is particularly noticeable in the foothills of the Himalayas, the western Ghats of India, and the Annamese Highlands of Vietnam. This is the time of year when the rice crop is planted in many parts of Asia.

In the lower latitudes, a monsoonal shift in winds can come about simply through the seasonal shifting of wind belts. For example, the winds of the equatorial zone migrate during the summer months northward toward the southern coast of Asia, bringing with them warm, moist, turbulent air. The tropical southeasterlies also migrate north with the sun, some crossing the equator. As they do so, their direction will be changed by the Coriolis effect, and they will become southwesterly trades in the Northern Hemisphere. They also bring warm, moist air (from their travels over the ocean) to the southern (and especially southeastern) coast of India. In the winter months the equatorial winds and the southern trades migrate south, leaving southern Asia under the influence of the dry, calm trades of the Northern Hemisphere. Asia and northern Australia are true monsoon areas, with a full 180° wind shift.

Other regions like the southern United States and West Africa have "monsoonal tendencies."

The phenomenon of monsoon winds and their characteristic seasonal shifting cannot, however, be fully explained by the differential heating of land and water, nor by the seasonal shifting of tropical and subtropical wind belts. There are aspects of the monsoon system, for example, its "burst," or sudden transition between dry and wet in southern Asia, that must have other causes. Meteorologists looking for a more complete explanation of the monsoon are examining the role played by the jet stream (described in the following section) and other wind movements of the upper atmosphere.

UPPER AIR WINDS

Thus far we have closely examined the wind patterns near the earth's surface. Of equal, or perhaps even greater, importance is the flow of air *above* the earth's surface—in particular, the flow of air at altitudes above 5000 meters (16,500 ft), in the upper troposphere. The formation, movement, and

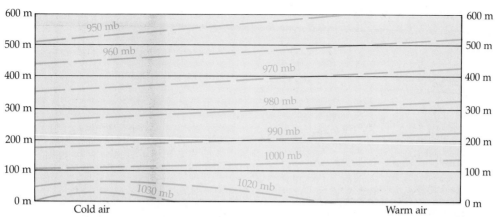

FIGURE 4.16 Variation of pressure surfaces with height. Note that the horizontal pressure gradient is from cold to warm air at the surface and in the opposite direction at higher elevations (e.g., 400 meters).

decay of surface cyclones and anticyclones depends to a great extent on the flow of air above the earth's surface.

The circulation of the upper air winds is far less complex than the surface wind circulation. In the upper troposphere an average westerly flow, the *upper air westerlies,* is maintained poleward of about 15–20° latitude in both hemispheres. Because of the reduced frictional drag, the upper air westerlies move much more rapidly than their surface counterparts. Between 15–20° north and south latitudes are the *upper air easterlies,* which can be considered the upper air extension of the trade winds. The flow of the upper air winds became very apparent during World War II, when high-altitude bombers moving in an eastward direction were found to cover similar distances faster than those flying westward. American pilots had encountered the upper air westerlies, or perhaps even **jet streams,** very strong air currents imbedded within the upper air westerlies.

The upper air westerlies form as a response to the temperature difference between warm tropical air and cold polar air. The air in the equatorial latitudes is warmed; rises convectively to high altitudes; and then flows toward the polar regions. At first this seems to contradict our previous statement, relative to surface winds, that air flows from cold areas (high pressure) toward warm areas (low pressure). This apparent discrepancy disappears, however, if you recall that the pressure gradient, down which the flow takes place, must be assessed between two points *at the same elevation.* A column

of cold air will exert a higher pressure at the earth's surface than a column of warm air. Consequently, the pressure gradient established at the earth's surface will result in a flow from the colder air toward the warmer air. However, cold air is denser and more compact than warm air. Thus pressure decreases with height more rapidly in cold air than in warm air. As a result, at a specific height above the earth's surface a lower pressure will be encountered in cold air than in warm air. This will result in a flow (pressure gradient) from the warmer air toward the colder air at that height. Figure 4.16 illustrates this concept.

Returning to our real-world situation, as the upper air winds flow from the equator toward the poles (down the pressure gradient), they are turned eastward because of the Coriolis effect. The net result is a broad circumpolar flow of westerly winds throughout most of the upper atmosphere (Fig. 4.17). Since the up-

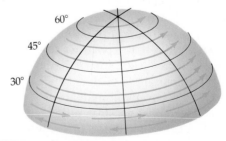

FIGURE 4.17 The upper air westerlies form a broad circumpolar flow throughout most of the upper atmosphere.

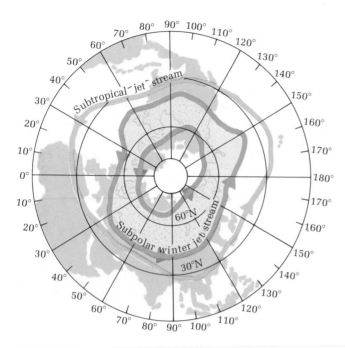

FIGURE 4.18 Approximate location of the subtropical jet stream and area of activity of the polar jet stream (shaded) in the Northern Hemisphere Winter. (After Riehl)

per air westerlies form in response to the thermal gradient between tropical and polar areas, it is not surprising that they are strongest in winter (low-sun season) when the thermal contrast is greatest. On the other hand, during the summer (high-sun season) when the contrast in temperature over the hemisphere is much reduced, the upper air westerlies move more slowly.

The temperature gradient between tropical and polar air, especially in winter, is not uniform but is instead concentrated where the warm tropical air meets cold polar air. This boundary, or front, with its stronger pressure gradient, marks the location of the **polar jet stream.** Ranging from 40 to 160 kilometers (25 to 100 mi) in width and up to 2 or 3 kilometers (1 to 2 mi) in depth, the polar jet stream can be thought of as a faster, internal ribbon of air within the upper air westerlies. While the polar jet stream flows over the midlatitudes, another westerly **subtropical jet stream** flows above the sinking air of the subtropical highs in the lower midlatitudes. Like the upper air westerlies, both jets are best developed in winter when hemispherical temperatures exhibit their steepest gradient (Fig. 4.18). During the summer

both jets weaken in intensity, and the subtropical jet stream frequently disappears completely.

We can now go one step further and combine our knowledge of the upper air circulation and the surface circulation to yield a more realistic portrayal of the vertical circulation pattern of our atmosphere (Fig. 4.19). In general

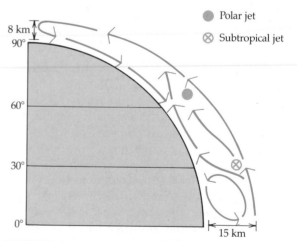

FIGURE 4.19 A more realistic schematic cross section of the average circulation in the atmosphere.

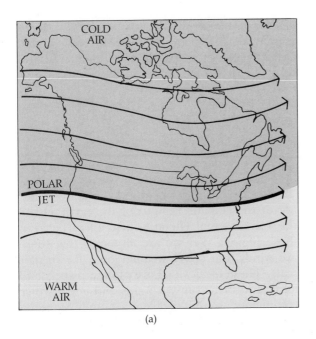

(a)

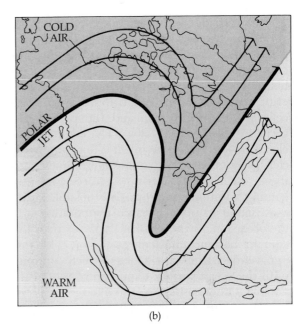

(b)

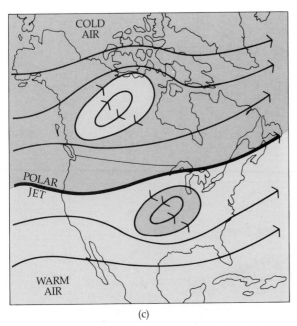

(c)

FIGURE 4.20 Development and dissipation of Rossby waves in the upper air westerlies. (a) A fairly smooth flow prevails. (b) Rossby waves form, with a ridge of warm air extending into Canada and a trough of cold air extending down to Texas. (c) The trough and ridge are cut off and will soon dissipate. The flow will then return to a pattern similar to (a).

the upper air westerlies, and associated polar jet stream, flow in a fairly smooth pattern (Fig. 4.20a). At times, however, the upper air westerlies develop oscillations, termed *long* waves or **Rossby waves,** after the meteorologist who proposed and proved their existence (Fig. 4.20b). Rossby waves result in cold polar air pushing into the lower latitudes and forming *troughs* of low pressure, while warm tropical air moves into higher latitudes, forming *ridges* of high pressure. It is when the upper air circulation is in this configuration that surface weather is most influenced. In Chapter 6 we will examine this influence in more detail.

Eventually the upper air oscillations become so extreme that the "tongues" of displaced air are cut off, forming upper air cells of warm and cold air (Fig. 4.20c). This process helps to maintain a net poleward flow of energy from equatorial and tropical areas. The cells eventually dissipate and the pattern returns to normal (Fig. 4.20a). The complete cycle takes 4 to 8 weeks. While it is not completely clear why the upper atmosphere goes into these oscillating patterns, we are currently gaining additional insights. One possible cause is variation in ocean surface temperatures. If the oceans in, say, the North Pacific or near the equator become unusually warm or cold, this apparently triggers oscillations, which continue until the ocean surface temperature returns to normal. Other causes are also likely.

In addition to this influence on weather, jet streams are important for other reasons. They can carry pollutants, such as radioactive wastes or volcanic dust, over great distances and at relatively rapid rates. It was the polar jet stream that carried the ash from Mount St. Helens eastward into Idaho and Montana. Nuclear fallout from experiments in the USSR and China can be monitored in succeeding days as it crosses the Pacific and later the United States in the jet streams. Pilots flying eastward, for example from America to Europe, take advantage of the jet stream, so that the flying times in this direction may be significantly shorter than those in the reverse direction.

LOCAL WINDS

In this chapter, we have been discussing the major circulation patterns of the earth's atmosphere, which are vital to an understanding of the climatic regions of the earth and to the fundamental climatic differences between those regions. Yet we are all aware that there are winds that affect weather on a far smaller scale. These local winds are often a response to local landform configurations and add further complexity to the problem of understanding the dynamics of weather.

LAND BREEZE—SEA BREEZE

The **land breeze—sea breeze cycle** is a diurnal (daily) one in which the differential heating of land and water again plays a role (Fig. 4.21). During the day, when the land, and consequently the air above it, is heated more quickly and to a higher temperature than the nearby

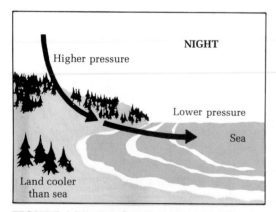
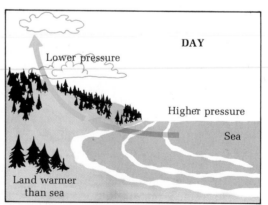

FIGURE 4.21 Land and sea breezes. This day-to-night reversal of winds is a consequence of the different rates of heating and cooling of land and water areas. The land becomes warmer than the sea during the day and colder than the sea at night; the air flows from the cooler to the warmer area.

ocean (or large lake or sea), the air above the land expands and rises. This process creates a local area of low pressure, and the rising air tends to be replaced by the denser, cooler air from over the ocean. Thus there is a sea breeze of cool, moist air blowing in over the land during the day. This sea breeze helps explain why seashores are so popular in summer, since cooling winds help alleviate the heat. The winds can mean a 5°C to 9°C (9°F to 16°F) reduction in temperature along the coast, as well as a lesser influence on land perhaps as far from the sea as 15 to 50 kilometers (9 to 30 mi). During hot summer days such winds cool cities like Chicago, Milwaukee, and Los Angeles.

At night, the land and the air above it cool off more quickly and to a cooler temperature than the nearby water body and the air above it. Consequently, the pressure builds up over the land and air flows out toward the lower pressure over the water, creating a land breeze.

For thousand of years, fishermen in sailboats have left their coasts at dawn, when there is still a land breeze, and have returned with the sea breeze of the late afternoon.

MOUNTAIN BREEZE—VALLEY BREEZE

Under the calming influence of a high-pressure system, there is a daily **mountain breeze—valley breeze** that is somewhat similar in mechanism to the land breeze—sea breeze cycle just discussed (Fig. 4.22). During the day, when the valley and slopes of mountains are heated by the sun, the air expands and rises up the sides of the mountains. This warm daytime breeze is the valley breeze, named for its place of origin. Clouds, which can often be seen hiding mountain peaks, are actually the visible evidence of condensation in warm air rising from the valleys. At night, when the valley and slopes are cooled because the earth is giving off more radiation than it is receiving, the air cools and sinks once again into the valley as a cool mountain breeze.

DRAINAGE WINDS

Also known as **katabatic winds, drainage winds** are local to mountainous regions and can only occur under calm, clear conditions. Cold, dense air will accumulate in a high valley, plateau, or snowfield within a mountainous area. Because the cold air is very dense, it tends to flow downward, escaping through passes and pouring out onto the land below. Drainage winds can be extremely cold and strong, especially when they result from cold air accumulating over ice caps such as Greenland and Antarctica. These winds are known by many local names; for example, in Yugoslavia they are called the *bora,* in France the *mistral,* and in Alaska the *Taku.*

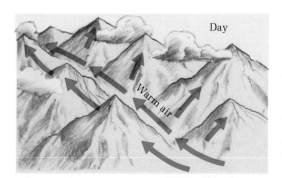 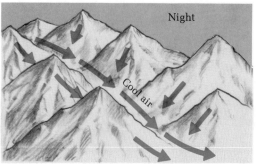

FIGURE 4.22 Mountain and valley breezes. This daily reversal of winds results from heating of mountain slopes during the day and their cooling at night. Warm air is drawn up slopes during the day, and cold air drains slopes at night.

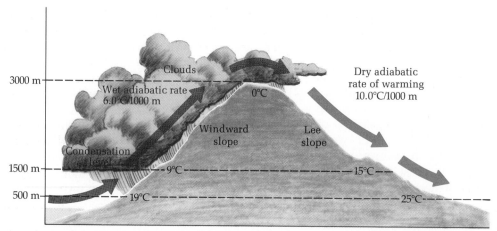

FIGURE 4.23   Foehn winds are warm, dry winds that result from the passage of air over a mountain range. Descending the lee side of the range, the air is warmed at the dry adiabatic rate.

FOEHN WINDS

A fourth type of local wind is also known by several names in different parts of the world (e.g., **Chinook** in the Rocky Mountain area and **foehn** (pronounced "fern") in the Alps). Foehn-type winds occur when air originating elsewhere must pass over a mountain range. As these winds flow down the leeward slope after crossing the mountains, the air is compressed and heated at the dry adiabatic lapse rate (Fig. 4.23). Thus they enter the valley below as warm, dry winds. The rapid temperature rise brought about by such winds has been known to damage crops, increase forest fire hazard, and set off avalanches. In addition, some research supports the notion that foehn-type winds affect people's personalities. The research indicates that suicides and crimes of passion increase when these winds are blowing.

An especially dry foehn-type wind is the **Santa Ana** of southern California. It forms when the eastern portion of the subtropical high strengthens and moves over the western United States. The clockwise circulation of the high drives the already warm, dry air of the desert southwest over the mountains of eastern California, accentuating the already oppressive conditions as the air moves down the western slopes.

There is no question that winds, both local and global, are effective elements of atmospheric dynamics. We all know that a hot but windy day is not nearly as unpleasant as a hot day without the wind. This difference exists because winds increase the rate of evaporation, and heat must be removed from our bodies, the air, animals, and plants in order for evaporation to take place. For this same reason, the wind on a cold day increases our discomfort.

QUESTIONS FOR DISCUSSION AND REVIEW

1. What is atmospheric pressure at sea level? How do you suppose the earth's gravity is related to atmospheric pressure?

2. How does incoming insolation affect pressure in the atmosphere? Give an example of an area where incoming insolation would create a pressure system. Would high or low pressure occur?

3. What is the difference between a cell and a belt of pressure?

4. What kind of pressure, high or low, would you

expect to find in the center of an anticyclone? Describe and diagram the wind pattern of an anticyclone in the Northern and the Southern Hemispheres.

5. Explain how water and land surfaces affect the pressure overhead, by seasons. How does this relate to the afternoon sea breeze?

6. How does the Pacific High affect the temperature of the West Coast?

7. What kinds of winds are characteristic of the area in which you live? What causes these winds? How is plant life affected by them? How are your outdoor activities affected?

8. Map the trade winds of the Atlantic Ocean and compare your map with one of trade routes in the nineteenth century or earlier.

9. What are the horse latitudes? the doldrums?

10. What are monsoons? Have you ever experienced one? What causes them? Name some nations that are vitally concerned with the arrival of the "wet monsoon."

11. Why are meteorologists concerned with upper air observations? What methods do they use to make these observations?

12. Describe the movements of the upper air. How have pilots applied their experience of the upper air to their flying patterns?

13. What effect on valley farms could a strong drainage wind have?

14. What effect would foehn-type winds have on agriculture, forestry, and ski resorts?

15. How can you apply your knowledge of pressure and winds in your everyday life?

5

MOISTURE, CONDENSATION, AND PRECIPITATION

THE SIGNIFICANCE OF WATER

Water is vital to all life on earth. Although some living things can survive without air, nothing can survive without water. Water is necessary for photosynthesis, cell growth, protein formation, soil formation, and the absorption of nutrients by plants and animals.

Water affects the earth's surface in innumerable ways. Because of the structure of the water molecule, water is able to dissolve an enormous number of substances—so many, in fact, that is has been called *the universal solvent*. Because water acts as a solvent for so many substances, it is almost never found in a pure state. Even rain is filled with impurities picked up in the atmosphere. Indeed, without these impurities to condense around, neither clouds nor precipitation could occur. In addition, rainwater usually contains some dissolved carbon dioxide from the air. Therefore, rain is a very weak form of carbonic acid. We shall see later (Chapter 14) that this fact affects the way in which water shapes certain landforms. The weak "material" acidity of rainwater should not

be confused with the environmentally damaging *acid rain*, which is at least ten times more acidic.

Not only can water dissolve and transport many minerals, it can also transport solid particles in suspension. These characteristics make water a unique transportation system for the earth. Water is able to make nutrients available to plants that would not otherwise be available. Water carries minerals and nutrients down streams, through the soil, through the openings in subsurface rocks, and through plants and animals. It deposits solid matter on stream floodplains, in river deltas, and on the ocean floor.

The surface tension of water and the behavior of water molecules make possible **capillary action**—the ability of water to pull itself upward through small openings against gravity. Capillary action also permits transport of dissolved material in an upward direction. Capillary action moves water into the stems and leaves of plants—even to the topmost needles of the great California redwoods and the top leaves of the rain-forest trees. Capillary action is also important in the movement of blood

Storm over the South Pacific. (R. Sager)

through our bodies. Without it, many of our cells could not receive the necessary nutrients carried by the blood.

Another important and highly unusual property of water is that it expands when it freezes. Most substances contract when cooled and expand when heated. Water follows the rules until it is cooled below 4°C (39°F). Then it begins to expand instead of contracting. Ice is therefore less dense than water and consequently will float on water as do ice floes and icebergs.

Lastly, compared to most other liquids and solids, water is very slow to heat up and very slow to cool off. Therefore, as we saw in Chapter 3, large bodies of water on the earth act as reservoirs of heat during winter and have a cooling effect in the summer. This moderating effect on temperature can be seen in the vicinity of lakes as well as on seacoasts.

The earth's water, or hydrosphere (from the Latin: *hydros,* water) is found in all three states: as a liquid in rivers, lakes, oceans and rain; as a solid in the form of snow and ice; and as a gas in our atmosphere. Even the water

temporarily stored in living things can be considered part of the hydrosphere. About 73 percent of the earth's surface is covered by water, with the largest proportion within the world's oceans (Fig. 5.1). In all, the total water content in the earth's system, whether liquid, solid, or vapor, is about 1.33 billion cubic kilometers (326 million cubic mi). Although, as we shall see next, water cycles in and out of the atmosphere, lithosphere, and biosphere, this total amount of water in the hydrosphere remains constant.

THE HYDROLOGIC CYCLE

The circulation of water from one part of the general earth system to another is known as the **hydrologic cycle.** The air contains water vapor that has entered the atmosphere through evaporation from the earth's surface. When water vapor condenses and falls as precipitation, several things may happen to it. First, it may go directly into a body of water, where it is immediately available

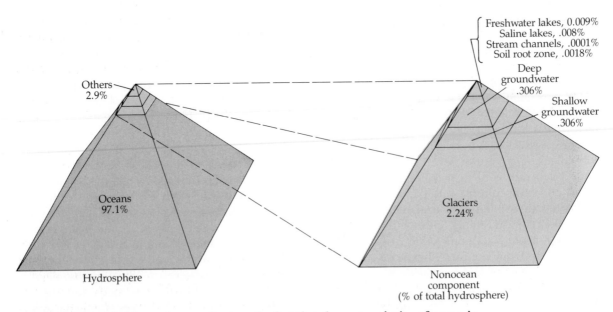

FIGURE 5.1 This illustration emphasizes the fact that the vast majority of water in the hydrosphere is salt water, stored in the world's oceans. The bulk of the supply of fresh water is relatively unavailable because it is stored in polar ice caps.

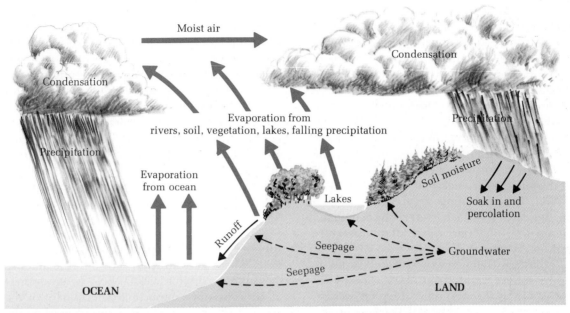

FIGURE 5.2 The hydrologic cycle. Water is cycled endlessly between the atmosphere, the soil, subsurface storage, lakes and streams, plants and animals, glacial ice, and the principal reservoir—the oceans.

for evaporation back into the atmosphere. Alternatively, it may fall onto the land. Then, it may run off the surface to form streams, ponds, or lakes. Or it may be absorbed into the ground, where it either can be contained by the soil or flow through open spaces, called *interstices,* that exist in loose sand, gravel, clay, and voids in solid rock. Ultimately much of the water in or on the earth's surface reaches the oceans. Some water that reaches the surface as snow becomes a part of the large amounts of ice over Greenland and Antarctica, as well as in high mountain glaciers in other parts of the world. Other water is used by plants and animals and temporarily becomes a part of living things. Thus there are six storage areas for water in the hydrologic cycle: the atmosphere, the oceans, bodies of fresh water on the surface, plants and animals, open spaces beneath the earth's surface, and glacial ice.

Liquid water is returned to the atmosphere as a gas through evaporation. Water evaporates from all bodies of water on the earth, from plants and animals, from soils, and

it can even evaporate from falling precipitation. Once the water is an atmospheric gas again, the cycle can be repeated.

Basically, then, the hydrologic cycle is one of condensation, precipitation, and evaporation. It is an ongoing cycle, without any known beginning or end, that occurs throughout the earth system. The diagram in Figure 5.2 provides a schematic illustration of the circulation of water in the hydrologic cycle.

WATER IN THE ATMOSPHERE

THE WATER BUDGET AND ITS RELATION TO THE HEAT BUDGET

Water exists in the atmosphere in all three states. As a solid it can be found as snow. As a liquid, water exists as rain and in fine droplets within clouds or fog. Most commonly, however, water exists as a tasteless, odorless, transparent gas known as water vapor, which is mixed with the other gases of the atmosphere in varying proportions. Water vapor is found within approximately the first 5500 meters (18,000 ft) of the troposphere and makes up a small, but highly variable, percentage of the atmosphere by volume. Atmospheric water is the source of all condensation and precipitation. Through these processes, as well as through evaporation, water plays a significant role as the earth's temperature regulator and modifier. In addition, as we noted in Chapter 3, water vapor in the atmosphere absorbs a significant portion of both incoming solar energy and outgoing earth radiation. By preventing great losses of heat from the earth's surface, water vapor helps to maintain the moderate range of temperature found on this planet.

Although the quantity of water vapor in the atmosphere varies from one place to another as well as through time, the total amount of atmospheric water remains nearly constant. In fact, the earth is a closed system with respect to water (that is, water is neither received from outside the earth system nor given off from it). Thus, an increase in water within one subsystem must be accounted for by a loss in another. Put another way, we say that the earth system operates with a water budget in which the total quantity of water remains the same and in which the deficits must balance the gains throughout the entire system.

We know that the atmosphere gives up a great deal of water, most obviously in clouds and through several forms of precipitation (rain, snow, hail, sleet) and condensation (fog, dew). If the quantity of water in the atmosphere remains at the same level through time, the atmosphere must be absorbing from other parts of the system an amount of water equal to that which it is giving up. During 1 minute, as much as 1 billion tons of water are given up by the atmosphere through some form of precipitation or condensation while another billion tons of water are evaporated and absorbed as water vapor by the atmosphere.

If we look again at our discussion in Chapter 3 of the heat energy budget, we can see that a part of that budget is the latent heat of condensation. Of course, this energy is orig-

inally derived from the sun. The sun's energy is used in evaporation and is then stored in the molecules of water vapor to be released only during condensation. Although the heat transfers involved in evaporation and condensation within the total heat energy budget are proportionately small, the actual energy is significant. Imagine the amount of energy released every minute when 1 billion tons of water condense and fall as precipitation. It is this vast storehouse of energy, the latent heat of condensation, which provides the major source of power for the storms of the earth: hurricanes, tornadoes, and thunderstorms.

The quantity of water vapor in the atmosphere varies from place to place and over time. There are, however, limits to the amount of water vapor that can be held by any parcel of air. The most important determinant of the amount of water vapor that can be held by the air is temperature. The warmer air is, the greater the quantity of water vapor it can hold. Therefore, we can make a generalization that air in the polar regions holds far less water vapor (approximately 0.2 percent by volume) than the hot air of the tropics and equatorial regions of the earth, where the air can contain as much as 5 percent by volume.

SATURATION AND DEW POINT

When air of a given temperature holds all of the water in vapor form that it possibly can, it is said to be in a state of **saturation** and has reached its **capacity.** If a constant temperature is maintained in a quantity of air, there will come a point, if more water is added, when the air will be saturated and unable to hold any more water vapor. For example, when you take a shower, the air in the room becomes increasingly humid, and the point is reached where the air cannot contain more water. Then, excess water vapor condenses onto mirrors and walls and even onto you when you try to dry off. We know that the capacity of air to hold water vapor varies with temperature. In fact, as we can see in Figure 5.3, this capacity of air to contain moisture increases with rising temper-

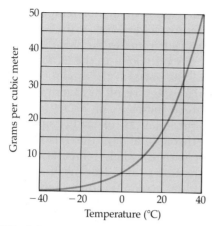

FIGURE 5.3 The graph shows the maximum amount of water vapor that can be contained in a cubic meter of air over a wide range of temperatures. Warm air is able to contain more water vapor per unit volume than cool air.

atures. Some examples will help illustrate the relationship between temperature and water vapor capacity. If we assume that a parcel of air at 30°C is saturated, then it will contain 30 grams of water vapor in each cubic meter of air (30 g/m^3). Now suppose we increase the temperature of the air to 40°C *without* increasing the water vapor content. The parcel is no longer saturated, since air at 40°C can hold more than 30 g/m^3 of water vapor (actually 50 g/m^3). Conversely, if we decrease the temperature of saturated air from 30°C (which contains 30 g/m^3 of water vapor) to 20°C (which only has a water vapor capacity of 17 g/m^3), some 13 grams of the water vapor will be forced to condense out of the air because of the reduced capacity.

It is also evident that if an unsaturated parcel of air is cooled, it will eventually reach a temperature where the air will become saturated. This critical temperature is known as the **dew point.** For example, if a parcel of air at 30°C contains 20 g/m^3 of water vapor, it is not saturated since it could hold 30 g/m^3. However, if we cool that parcel of air to 21°C, it would become saturated since the capacity of air at 21°C is 20 g/m^3. Thus, that parcel of air at 30°C has a dew point temperature of 21°C. It is the

cooling of air to below its dew point temperature that brings about the condensation that must precede precipitation.

Because the capacity of air to hold water vapor increases with rising temperatures, air in the equatorial regions has a higher dew point than does air in polar regions. Thus, because the atmosphere can hold more water in these equatorial regions, there is greater potential for large quantities of precipitation than in polar regions. Likewise, in the middle latitudes, summer months, because of their higher temperatures, have more potential for large-scale precipitation than do winter months.

HUMIDITY

The amount of water vapor in the air at any one time and place is called **humidity.** There are three common ways to express the humidity content of the air, and each method provides information that contributes to our discussion of weather and climate.

ABSOLUTE AND SPECIFIC HUMIDITY

Absolute humidity is the measure of the mass of water vapor that exists within a given *volume* of air. It is expressed either in the metric system as the number of grams per cubic meter (g/m^3), or the English system as grains per cubic foot (gr/ft^3). **Specific humidity** is the mass of water vapor (given in grams) per mass of air (given in kilograms). Obviously, both are measures of the actual amount of water vapor in the air. Since most water vapor gets into the air through the evaporation of water from the earth's surface, it stands to reason that absolute and specific humidity will decrease with *vertical* distance from the earth.

We have also learned that air is compressed as it sinks and expands as it rises. Thus, a given parcel of air changes its volume as it moves vertically, although its weight remains the same and there may be no change in the amount of water vapor in that quantity of air. We can see, then, that absolute humidity, although it measures the amount of water vapor, can vary simply as a result of the vertical movement of a parcel of air. Specific humidity, on the other hand, changes *only* as the quantity of the water vapor changes. For this reason, when assessing the changes of water vapor content in large masses of air, which often have vertical movement, specific humidity is the preferred measurement.

RELATIVE HUMIDITY **Relative humidity,** which is commonly given on television and radio weather reports, is probably the best known means of describing the content of water vapor in the atmosphere. It is simply the ratio between the amount of water vapor in air of a given temperature and the maximum amount of vapor that the air could hold at that temperature; it expresses how close the air is to saturation.

If the temperature and absolute humidity of an air parcel are known, its relative humidity can be determined by using Figure 5.3. For instance, if we know that a parcel of air has a temperature of 30°C and an absolute humidity of 20 g/m^3, we can look at the graph and determine that if it were saturated, its absolute humidity would be 30 g/m^3. Then to determine relative humidity all we do is divide 20 grams (actual content) by 30 grams (potential content) and multiply by 100 (to get an answer in percentage):

$$(20 \text{ grams} \div 30 \text{ grams}) \times 100 = 67\%$$

The relative humidity in this case is 67 percent. In other words, the air is holding only two thirds of the water vapor it could contain at 30°C; it is only at 67 percent of its capacity.

There are two important factors in the *horizontal* distribution and variation of relative humidity. One of these is the availability of moisture. For example, air above bodies of water is apt to contain more moisture than similar air over land surfaces because there is simply more water available for evaporation. Conversely, the air overlying a region like the central Sahara Desert is usually very dry because it is far from the oceans and there is little water to be evaporated. The second factor in the horizontal variation of relative humidity is temper-

ature. In regions of higher temperature, relative humidity will be lower if the actual amount of water vapor remains the same.

At any one point in the atmosphere, relative humidity varies if the amount of water vapor increases as a result of evaporation *or* if the temperature increases or decreases. Thus, although the quantity of water vapor may not change through a day, the relative humidity will vary with the daily temperature cycle. As air temperature increases from shortly after sunrise to its maximum in the early afternoon, the relative humidity decreases as the air becomes capable of holding greater and greater quantities of water vapor. Then, as the air becomes cooler, decreasing toward its minimum temperature right after sunrise, the relative humidity increases.

Relative humidity affects our comfort through its relationship to the rate of evaporation. Perspiration evaporates into the air, leaving behind a salty residue, which you can taste if you lick your lips after sweating a great deal. Evaporation is a cooling process, since the heat used to change the perspiration to water vapor (and which becomes locked in the water vapor as latent heat) is subtracted from your skin. This is the reason on a hot August day, when the temperature approaches 35°C (95°F), you will be far more uncomfortable in Atlanta, Georgia, if the relative humidity is 90 percent than in Tucson, Arizona, where it may be only 15 percent at the same temperature. At 15 percent your perspiration will be evaporated at a faster rate than with a higher relative humidity, and you will benefit from the resultant cooling effects. When the relative humidity is 90 percent, the air is nearly saturated and far less evaporation can take place.

SOURCES OF ATMOSPHERIC MOISTURE

In our earlier discussion of the hydrologic cycle, we saw that the atmosphere receives water vapor through the process of evaporation. Water evaporates into the atmosphere from many different places, most important of which are the surfaces of the earth's bodies of water. Water also evaporates from wet ground surfaces and soils, from droplets of moisture on vegetation, from city pavements and other man-made surfaces, and even from falling precipitation.

Vegetation provides another source of water vapor. Plants give up water in a complex process known as **transpiration,** which can be a significant source of atmospheric moisture. A mature oak tree, for instance, can give off 400 liters (105 gal) of water per day, and a cornfield may add 11,000–15,000 liters (2900–4000 gal) of water to the atmosphere per day for each acre under cultivation. In some parts of the world, notably tropical rain forests of heavy, lush vegetation, transpiration accounts for a significant amount of atmospheric humidity. Together, evaporation and transpiration, or **evapotranspiration,** account for the water vapor in the air that is available for precipitation.

RATE OF EVAPORATION

The rate of evaporation is affected by several factors. First it is affected by the amount of accessible water. Thus, Table 5.1 shows that the rate of evaporation tends to be greater over the oceans than over the continents. The only time this generalization is not true is in equatorial regions between 0° and 10° N and S, where the vegetation is so lush on the land that transpir-

ation provides a large amount of water for the air.

Temperature also affects the rate of evaporation. First, as the temperature of the air is increased, its capacity to contain moisture also increases, providing room for additional water in the atmosphere. Also, as air temperature increases, so too does the temperature of the water at the evaporation source. Such increases in temperature mean that more energy is available to the water molecules for their escape from a liquid state to a gaseous one. Consequently, more molecules can make the transition.

Mentioned previously, but deserving more attention, is another factor affecting the rate of evaporation. That is the degree to which the air is saturated with water vapor. The drier the air and the lower the relative humidity, the greater the rate of evaporation can be. Some of us have had direct experience with this principle: Compare the length of time it takes your bathing suit to dry as the water in it evaporates on a hot, humid day to how long it takes on a day when the air is dry.

Wind, too, affects the rate of evaporation. If there is no wind, the air that overlies a water surface will approach saturation as more and more molecules of liquid water change to water vapor. This evaporation will cease once saturation is reached. However, if there is a wind, it will blow that saturated or nearly saturated air away from the evaporating surface, replacing it

TABLE 5.1   Distribution of Actual Mean Evaporation

ZONE	LATITUDE					
	60°–50°	50°–40°	40°–30°	30°–20°	20°–10°	10°–0°
Northern Hemisphere						
Continents	36.0 cm(14.2 in)	33.0(13.0)	38.0(15.0)	50.0(19.7)	79.0(31.1)	115.0(45.3)
Oceans	40.0(15.7)	70.0(27.6)	96.0(37.8)	115.0(45.3)	120.0(47.2)	100.0(39.4)
Mean	38.0(15.0)	51.0(20.1)	71.0(28.0)	91.0(35.8)	109.0(42.9)	103.0(40.6)
Southern Hemisphere						
Continents	20.0 cm(7.9 in)	NA	51.0(20.1)	41.0(16.1)	90.0(35.4)	122.0(48.0)
Oceans	23.0(9.1)	58.0(22.8)	89.0(35.0)	112.0(44.1)	119.0(47.2)	114.0(44.9)
Mean	22.5(8.8)	NA	NA	99.0(39.0)	113.0(44.5)	116.0(45.7)

VIEWPOINT: WATER AND THE QUALITY OF LIFE

We literally cannot live without water. Seventy-five percent of our body weight is water, and it is so important to our health and well-being that the loss of only 25 percent of it by dehydration can bring death. A seventeenth-century mystic, Thomas Traherne, once wrote: "You can never enjoy the world aright, till the sea itself floweth in your veins," and he had more in mind than a purely physiological interpretation. He surely was aware of the refreshment and enjoyment that water in nature can bring us. "Nothing is lovelier than moving water," says a poet, and we must agree if we've ever sat by a falling cascade in the mountains or watched the snowflakes swirl from a wintry sky.

And how we take these things for granted! Too many of us turn on the tap and fill our glasses without a second thought. Do we realize that each of us in the United States uses an average of 600 liters (160 gal) of water a day and in summer up to 2500 liters (660 gal)? Simple figures for domestic use are as follows:

Drinking	2 liters (½ gal)/day
Laundry	60–150 liters (15–40 gal)/load
Automatic dishwasher	60 liters (15 gal)/load
Toilet	20 liters (5 gal)/flush
Bath	100 liters (25 gal)
Shower	20 liters (5 gal)/minute
Washing, brushing teeth, shaving, etc.	20 liters (5 gal)/day

Although individuals consume a great deal of water, the total amount used at home is small compared with the rapidly increasing amount of water used by city governments, agriculture, and industry. The total for the whole country amounts to more than 1500 billion liters (400 billion gal) daily. Examples of municipal use are sewage disposal, fire extinguishing, heating and air-conditioning, street cleaning, fountains, and swimming pools.

Today's industry could not survive without an abundant supply of water. It takes 40 liters (10 gal) of water to make 4 liters (1 gal) of gasoline, 1000 liters (260 gal) to make a pound of paper, and 1200 liters (320 gal) to make a barrel of beer or a pound of steel. In industry, water is used as an ingredient in products, as an agent for cooling, and as a vehicle for removing impurities and for diluting and carrying off waste. The nation's steel mills alone use five times the amount of water consumed directly by the population of New York City, and one good-sized paper mill requires enough water to support a town of 50,000 people. Without water, industry would be at a standstill.

Water is also valuable as a power source. The energy of falling water has long been used to power mills producing lumber, flour, woolens, cotton textiles, and paper. Today it is used to supply power in hydroelectric plants. In the United States alone there are over 1500 hydroelectric plants that have harnessed water power to supply heat, light, and energy to homes and industry. These hydroelectric plants supply about 20 percent of the nation's electricity, while steam plants powered by fossil fuels and nuclear energy produce 70 percent. The total amount of water used annually in electric power generation is some 1500 trillion liters (400 trillion gal).

In agriculture, water is used for crop irrigation via sprinkler systems, ditches, or drip irrigation. It has been estimated that an average of 4000 liters (1000 gal) of water are needed to produce each pound of food we consume. A pound of bread takes 500 liters (136 gal) to grow the wheat, a pound of tomatoes about 475 liters (125 gal), and each pound of steak that we eat has taken 9500 liters (2500 gal) of water to produce feed and water for the livestock.

Irrigation has always been important for food production since the days of the Pharoahs in the Nile valley and throughout Chinese and Indian history for rice and other crops. Indeed, many historians feel that the strength of these so-called "hydraulic civilizations" grew from the political control necessary to manipulate large amounts of water for food production. Historically in the United States,

Hoover Dam. (R. Gabler)

controversy has often developed over water rights along river banks. What were known as *riparian rights,* a concept developed in the wetter climates of Western Europe, specified that water diverted from a stream must be returned to that stream undiminished. These traditional laws were challenged in water-short California by a new *doctrine of appropriation,* which allows diversion of water away from the river. The irrigation of desert and semidesert areas like the Great Valley and Imperial Valley of California has engendered constant struggle and litigation over water use, and the long debate between Arizona and California over Colorado River water is well known.

Essentially every drop of water that we use has been recycled by the earth's atmosphere and hydrosphere an infinite number of times. We use it from our faucets and hoses as if it is an unlimited resource by right; perhaps if we still had to draw it by hand from wells or share a standpipe in the street with other families, as do many inhabitants of the Third World, we would use it more sparingly and with more care.

Pollution is a world problem, as this photograph of a canal in Leeuwarden in the Netherlands indicates. The once spotless reputation of the Dutch for preservation of the environment has been badly damaged in recent years. (R. Gabler)

We need to be aware that the world's hydrosphere will not automatically purify itself forever in the light of the expanding demands of an evergrowing population. Already we see signs of damage. Rain in some areas of the industrialized world shows increasing acidity, buildings like the Parthenon in Athens are crumbling under the attack of sulfuric moisture in the atmosphere from industrial pollution, and the lakes and streams in the northeastern United States are losing their fish populations as the runoff into them becomes more toxic from *acid rains.* Wastes are dumped into lakes, rivers, and oceans and into underground water to levels beyond those that can be successfully absorbed, decomposed, and dispersed by the water body. Some wastes pollute waters with chemicals such as nitrates and phosphates from agricultural runoff. These chemicals increase the growth of oxygen-consuming algae, sometimes to such an extent that the water body becomes depleted of free oxygen. This process is known as **eutrophication.** Eutrophication, along with the addition of toxic wastes, drastically disturbs the biological system of the water body. Similarly, water used as a coolant by industry can raise the temperature of the lake or stream into which it is dumped. Sometimes a temperature increase of even a few degrees (*thermal pollution*) can affect aquatic plant and animal communities adversely.

In some areas we already see the danger of overdraft, or overpumping of ground water. Water tables below the surface are falling: in the Great Valley of California, for example, or the Ganges Valley in India, as deeper *tube wells* are drilled. Thus, still more energy is required for pumping the water from deeper wells, the quality of water may diminish, and along coastal areas, saltwater intrusion may occur. Further, with inadequate drainage in irrigated areas, *salinization* of soils may appear, in which mineral salts accumulate in the soil as the irrigation water evaporates.

Can we stop the danger before it is too late and once again respect the purifying hydrologic cycle? There are good signs: The clean-up of Lake Erie in recent years and the return of fish to Los Angeles harbor and to the River Thames in London all indicate that we can reverse the damage—if we only recognize that the hydrosphere of our earth is indeed a closed system, that it cannot replenish itself from outside our atmosphere. The water on our *Spaceship Earth* is all we have, and we need to cherish it with care.

with air of a lower humidity. This allows evaporation to continue as long as the wind keeps blowing saturated air away and bringing in drier air. Anyone who has gone swimming on a windy day has experienced the chilling effects of rapid evaporation.

POTENTIAL EVAPOTRANSPIRATION

So far, we have discussed actual evaporation and transpiration (evapotranspiration). Geographers and meteorologists also concern themselves with **potential evapotranspiration** (Fig. 5.4). This term refers to idealized conditions in which there would be enough precipitation to provide sufficient moisture for all possible evapotranspiration in an area. Various formulas have been derived for estimating the potential evapotranspiration at a location since it cannot be measured directly. These formulas commonly employ temperature, latitude, vegetation, and soil character (permeability, water retention ability) as factors that could affect the potential.

In places where precipitation exceeds potential evapotranspiration, there is a surplus of water for storage in the ground and bodies of water, and water can even be exported to other places if canal construction is feasible. When potential evapotranspiration exceeds precipitation, as it does during the dry summer months in California, then there is no water available for storage, and in fact, the water stored during previous rainy months evaporates quickly into the warm dry air (Fig. 5.5). Soil becomes dry and vegetation turns brown as any available water is soaked up by the atmosphere. For this reason, fires become a potential hazard during the late summer months in California.

Knowledge of potential evapotranspiration is used by irrigation engineers to learn how much water will be lost through evaporation so that they can determine whether the water that is left is enough to justify a canal. Farmers, by assessing the daily or weekly relationship between potential evapotranspiration and precipitation, can determine when to irrigate, and how much to irrigate, their crops.

CONDENSATION

Condensation is the process by which a gas is changed to a liquid. In our present discussion of atmospheric moisture and precipitation, condensation refers to the change of water vapor to liquid water.

Condensation occurs when air saturated with water vapor is cooled. Viewed in another way, we can say that if we lower the temperature of air until it has a relative humidity of 100 percent (the air has reached the dew point), condensation will occur with any more cooling. Such condensation can occur without the addition of any further water vapor.

It follows, then, that condensation is dependent upon (1) the relative humidity of the air and (2) the degree of cooling. In the arid air of Death Valley, a huge amount of cooling must take place before the dew point is reached. In contrast, on a humid summer afternoon in Mississippi, a minimal amount of cooling will bring on condensation.

This is the principle behind the formation of droplets of water on the side of a can of cold cola on a warm afternoon: The temperature of the air is lowered when it comes in contact with the cold can. Consequently, the air's capacity to hold water vapor is diminished. If air touching the can is cooled sufficiently, its relative humidity will reach 100 percent. Any cooling beyond that point will result in condensation in the form of water droplets on the can.

CONDENSATION NUCLEI

In order for condensation to occur, one other factor is necessary: **condensation nuclei.** These are minute particles in the atmosphere that provide a surface upon which condensation can take place. Theoretically, if all such particles were removed from a volume of air, we could cool that air below its dew point without condensation occurring. Conversely, if there is a superabundance of such particles, condensation may take place at relative humidities just below 100 percent. For example, ocean fogs, which are an accumulation of condensation

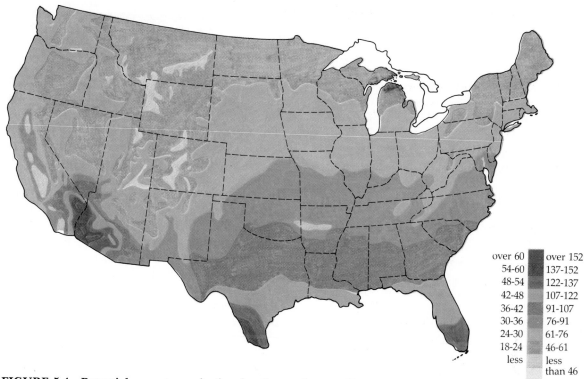

FIGURE 5.4 Potential evapotranspiration for the contiguous 48 states. (After C. W. Thornthwaite, *Geographical Review*)

over 60	over 152
54-60	137-152
48-54	122-137
42-48	107-122
36-42	91-107
30-36	76-91
24-30	61-76
18-24	46-61
less	less than 46

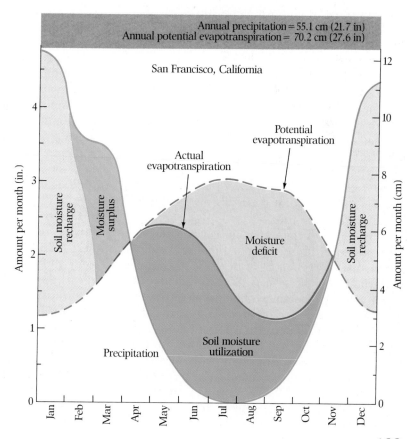

FIGURE 5.5 This is an example of the Thornthwaite water budget system, which "keeps score" of the balance between water input by precipitation and water loss to evaporation and transpiration, permitting month-by-month estimates of both runoff and soil moisture.

129

droplets formed on sea-salt particles in the air from the evaporation of sea water, can form when the relative humidity is as low as 92 percent.

Condensation nuclei are most often the sea-salt particles mentioned above. They can also be particles of dust, smoke, dirt, pollen, or volcanic material. More and more commonly, they are chemical particles that are the by-products of industrialization. The condensation that takes place on such chemical nuclei is often corrosive and dangerous to human health; when it is, we know it as smog.

In nature, condensation appears in a number of forms. Fog, clouds, frost, and dew are all the results of condensation of water vapor in the atmosphere. The type of condensation produced is dependent upon a number of factors, including the cooling process itself. The cooling that produces condensation in one form or another can occur as a result of radiation cooling, through advection, through convection, or a combination of these processes.

FOGS

Fogs and clouds appear when water vapor condenses on nuclei and a large number of these droplets form a mass. Not being transparent to light in the way that water vapor is, these masses of condensed water droplets appear to us as fog or clouds, in any of a number of shapes and forms, and usually in shades of white or gray.

In terms of the water budget and the hydrologic cycle, fog is a minor form of condensation. Yet, in certain areas of the world, it has important climatic effects. The so-called drip factor helps to sustain vegetation along desert coastlines where fog occurs. Fog also plays havoc with our modern transportation systems. Navigation on the seas is made more difficult by fog, and air travel can be greatly impeded. In fact, fog sometimes causes major airports to shut down until visibility improves. Highway travel is also greatly hampered by heavy fogs that can lead to huge chain reaction pile-ups of cars.

RADIATION FOG Radiation cooling can lead to **ground-inversion** or **radiation fog.** This kind of fog is likely to occur on a cold, clear, calm winter night, usually in the middle latitudes. These conditions allow for maximum outgoing radiation from the ground with no incoming radiation. As the ground gets colder and colder during the night and gives up more and more of the heat that it has received during the day, the air directly above is cooled by conduction through contact with the ground surface. Since the cold ground can only cool the lower few meters of the atmosphere, an inversion is created in which cold air at the surface is overlain by warmer air above. If this cold layer of air at the surface is cooled to a temperature below its dew point, then condensation will occur, usually in the form of a low-lying fog. However, wind strong enough to disturb this inversion layer can prevent the formation of the fog by not allowing the air to stay at the surface long enough to become cooled below its dew point.

The chances for a ground-inversion fog occurring are increased by certain types of land formation. In valleys and depressions, cold air accumulates through air drainage. During a cold night, this air can be cooled below its dew point, and a fog will be formed, like a lake, in the valley. It is common in mountainous areas to see an early morning radiation fog on the valley floor, while snow-capped mountain tops shine against a clear blue sky. Radiation fogs have a diurnal cycle: They form during the night and are usually the most dense right after sunrise, when temperatures are lowest. They then "burn off" during the day, when the heat from the sun slowly penetrates the fog and warms the earth. The earth in turn warms the air directly above it, increasing its temperature and consequently its capacity. With a greater capacity the fog evaporates into the air. As the earth's heat penetrates to higher and higher layers of air, the fog continues to burn off—from the ground up!

Radiation fog often forms even more densely in industrial areas where the high concentration of chemical particles in the air provides an abundance of condensation nuclei. Such a fog is usually thicker and denser than conventional "natural" radiation fogs and less easily dissipated by wind or sun.

ADVECTION FOG The most common type of fog, the **advection fog,** occurs through the movement of warm, moist air over a colder surface, either land or water. When the warm air is cooled below its dew point through heat loss by radiation and conduction from the colder surface below, condensation occurs in the form of fog. Advection fog is usually less localized in nature than radiation fog. It is also less likely to have a diurnal cycle, though if not too thick it can be burned off early in the day, to return again in the afternoon or early evening. More common, however, is the persistent advection fog that spreads itself over a large area for days at a time.

Advection fog forms over land during the winter months in middle latitudes. It forms, for example, in the United States when warm moist air from the Gulf of Mexico flows northward over the cold, frozen, and sometimes snow-covered upper Mississippi Valley.

During the summer months, advection fog may form over large lakes or over the oceans. Formation over lakes occurs when warm continental air flows over a colder water surface, as when a warm air mass passes over the cool surface of Lake Michigan. An advection fog also can be formed when a warm air mass moves over a cold ocean current, and the air is cooled sufficiently to bring about condensation. This variety of advection fog is known as a *sea fog*. Such a situation accounts for the fogs along the west coast of the United States. During the summer months the Pacific subtropical high moves north with the sun, and winds flow out of the high toward the west coast where they pass over the cold California Current. When condensation occurs, fogs form that flow in over the shore, pushed from behind by the eastward movement of air, and pulled by the low pressure of the warmer land (Fig. 5.6). Advection fogs also occur in New England, especially along the coasts of Maine and the Canadian Maritime Provinces, when warm, moist air from above the Gulf Stream flows north over the colder waters of the Labrador Current. Advection fog over the Grand

FIGURE 5.6 An advection fog caused by warm, moist air passing over colder water. (R. Sager)

Banks off Newfoundland has long been a hazard for cod fishermen there.

OTHER MINOR FORMS OF CONDENSATION

Dew, which is made up of tiny droplets of water, is formed by the condensation of water vapor at or near the surface of the earth. Dew collects on surfaces that are good radiators of heat (your car, blades of grass, the bike that stayed outside last night). These good radiators give up large amounts of heat during the night hours, when there is no incoming radiation. When the air comes in contact with these cold surfaces, it cools, and if cooled sufficiently, droplets of water will form as beads on the surface. When the temperature of the air is below 0°C (32°F), **white frost** forms. It is important to note that frost is not frozen dew but instead represents a sublimation process—water vapor changing directly to the frozen state.

Sometimes, under very still conditions, though air temperatures may be below 0°C (32°F), the liquid droplets that make up clouds or fogs are not frozen into solid particles because of low air pressure. When such *supercooled* water droplets come in contact with a surface, like the edge of an airplane wing or a tree branch or a window, ice crystals are created on that surface in a formation known as **rime.**

CLOUDS

Clouds are the most important form of suspended water droplets caused by condensation. We have already seen that fog, dew, and other phenomena are the result of the cooling of air to a temperature below its dew point. The cooling that produces fog, dew, and the other forms we have already examined can be either radiational cooling or cooling through advection.

Cloud formations usually develop from a cooling process that is the result of the upward movement of air. In Chapter 3 we saw that as air rises, it expands and cools at the dry adiabatic rate. If this cooling process lowers the temperature of the moving air to a temperature below its dew point, then condensation will occur, provided there are condensation nuclei available.

It is this cooling process resulting from the upward movement of air and its consequent expansion that allows for the condensation of water vapor within large air masses above the surface of the earth. As more and more water vapor is condensed within an air mass, water droplets multiply, some of the sun's light is blocked, and the visible result is a cloud.

Clouds are important for several reasons. First, they are the source for all precipitation. **Precipitation** is made up of condensed water particles, either liquid or solid, which fall to earth. Obviously not all clouds result in precipitation, but we cannot have precipitation without the formation of a cloud first. Also, clouds serve an important function in the heat energy budget. We have already noted that clouds absorb some of the incoming solar energy. They also reflect some of that energy back to space and scatter and diffuse other parts of the incoming energy before it strikes the earth as diffuse radiation. In addition, clouds absorb some of the earth's radiation, so that it is not lost to space, and then they reradiate it back to the surface. Finally, clouds are a beautiful and ever-changing aspect of our environment. The colors of the sky, the variations in the shapes and hues of clouds, have provided us all with a beautiful backdrop to the natural scenery here on earth.

Clouds appear white or in shades of gray or even deep gray approaching black. They differ in color depending upon how thick or dense they are and if the sun is shining on the surface that we see. The thicker a cloud, the darker it will appear, for the more light it is able to absorb and thus block from our view. Clouds also seem dark when we are seeing their shaded side instead of their sunlit side.

There are three basic forms of clouds: cirrus, stratus, and cumulus. (See Plate 8.) Classi-

FIGURE 5.7   Cirrus clouds are wisps of fine ice crystals in the upper air. They often indicate an approaching storm. (Courtesy of Kenneth R. Martin Collection)

fication systems categorize these cloud formations into many subtypes; however, most such subtypes are overlaps of the three basic forms. (See Plate 10.).

Cirrus (from Latin: *cirrus,* a lock or wisp of hair) clouds form at very high altitudes, normally 6000 to 11,000 meters (19,800 to 36,300 ft), and are made up of ice crystals rather than droplets of water. They are thin, wispy, white clouds that trail like feathers across the sky. When associated with fair weather, cirrus clouds are scattered white patches in a clear blue sky (Fig. 5.7).

Stratus (from Latin: *stratus,* layer) clouds can appear anywhere from near the surface of the earth to almost 6000 meters (19,800 ft). The variations of stratus clouds are based in part upon their altitude. The basic characteristic of stratus clouds is their horizontal sheetlike appearance lying in layers (Fig. 5.8). In contrast to the vertical structure of the cumulus clouds, stratus clouds have a strong horizontal development with a fairly uniform thickness throughout. Often, stratus clouds cover the entire sky with a gray cloud layer. It is stratus clouds that make up the dull gray overcast sky common to winter days in much of the midwestern and eastern United States. Often, the

stratus cloud formation will overlie an area for days and any precipitation will be steady and persistent.

Cumulus, or **cumuliform** (from Latin: *cumulus,* heap or pile), clouds develop vertically rather than forming the more horizontal

FIGURE 5.8   Stratus clouds are extensive horizontal sheets of water droplets. They may cover thousands of square kilometers. Stratus indicate stable air. Rainfall from stratus is usually a long-lasting drizzle. (S. Brazier)

FIGURE 5.9 Cumulus clouds form where there are vertical updrafts of unstable air. The base indicates the level at which condensation occurs in upward-moving air. (Courtesy of Kenneth R. Martin Collection)

structures of the cirrus and stratus forms. Cumulus are massive piles of clouds, usually with a flat base that can be anywhere from 500 to 12,000 meters (1650 to 39,600 ft) above sea level (Fig. 5.9). From this base, they pile up into great rounded structures, often with tops like cauliflowers. The cumulus cloud is the visible evidence of an upward movement of air; its base is the point where condensation has begun in a column of air as it moves upward.

Another term used in describing clouds is **nimbus,** meaning precipitation (i.e., rain is falling). Thus there is the nimbostratus cloud, which may bring a long-lasting drizzle, and the cumulonimbus or the thunderhead. This latter cloud has a flat top called an anvil head, as well as a flat base, and it becomes darker as condensation within it increases and blocks the sun (Fig. 5.10). The cumulonimbus is the source of the gusty winds, torrential rain, and lightning common on hot afternoons in humid regions.

FIGURE 5.10 Cumulonimbus clouds as seen from the air. Though rainfall from these enormous clouds may be torrential, it usually does not last very long. A cumulonimbus cloud is often associated with lightning and thunder and occasionally hail. (NOAA)

STABILITY AND INSTABILITY AS CONDITIONS OF AIR

The various forms of clouds are related to differing degrees of vertical air movement. Some clouds are associated with little air movement, while others form in buoyant, rising currents of air.

An air parcel will rise of its own accord as long as it is warmer than the surrounding atmospheric air. When it reaches a layer of the atmosphere that is the same temperature as itself, it will stop rising. Thus, an air parcel warmer than the surrounding atmospheric air will rise and is said to be **unstable.** On the other hand, an air parcel that is colder than the surrounding atmospheric air will resist any upward movement and is said to be **stable.**

To fully comprehend the concept of stability and instability, it is imperative to differentiate between *environmental lapse rate* and *adiabatic lapse rate*. In Chapter 3 we found that the temperature of our atmosphere, in general, decreases with increasing height above the earth's surface—the so-called normal lapse rate or environmental lapse rate. Thus, the environmental lapse rate reflects nothing more than the vertical temperature structure of the atmosphere. Although it averages 6°C/1000 meters (3.6°F/1000 ft), it is quite variable and must be measured through the use of meteorologic instruments sent aloft. Also, recall that when an air parcel is *lifted* through the atmosphere it expands and cools at a fixed rate of 10°C/1000 meters (5.6°F/1000 ft)—the *dry adiabatic lapse rate*. If condensation takes place within this rising air parcel, latent energy is released, thus retarding the cooling to approximately 5°C/1000 meters (3.2°F/1000 ft)—the *wet adiabatic lapse rate*. A rising air parcel *will* cool at one of these two adiabatic rates. Which rate is in operation depends upon whether condensation *is* (wet adiabatic rate) or *is not* (dry adiabatic rate) occurring.

Determining the stability or instability of an air parcel involves nothing more than asking the question: If an air parcel were lifted to a specific elevation (cooling at an adiabatic lapse rate), would it be warmer, colder, or the same temperature as the atmospheric air (determined by the environmental lapse rate at that time) at that same elevation?

If the air parcel is warmer than the atmospheric air at the selected elevation, then the parcel would be unstable and continue to rise since warmer air is less dense and, therefore, buoyant. Thus, under conditions of **instability** the environmental lapse rate must be *greater than* the adiabatic lapse rate in operation. For example, if the environmental lapse rate is 12°C/1000 meters and the ground temperature is 30°C, then the atmospheric air temperature at 2000 meters would be 6°C. On the other hand, an air parcel (assuming no condensation occurs) lifted to 2000 meters would have a temperature of 10°C. Since the air parcel is warmer than the atmospheric air around it, it is unstable and will continue to rise (Fig. 5.11). However, let us assume that it is another day and all the conditions are the same except that measurements indicate the environmental lapse rate on this day is 2°C/1000 meters. Consequently, while our air parcel lifted to 2000 meters would still have a temperature of 10°C, the temperature of the atmosphere at 2000 meters would now be 26°C. Thus, the air parcel would be colder and would sink back toward the earth as a result of its greater density (Fig. 5.11). As you can see, under conditions of **stability,** the environmental lapse rate is *less than* the adiabatic lapse rate in operation. If an air parcel, upon being lifted to a specific elevation, has the same temperature as the atmospheric air surrounding it, then it is neither stable nor unstable. Instead, it is considered **neutral** and will neither rise nor sink but will remain at that elevation.

Whether an air parcel will be stable or unstable is related to the amount of cooling and heating of air at the earth's surface. With chilling of the air through radiation and conduction on a cool, clear night, air near the surface will be relatively close in temperature to that aloft, and the environmental lapse rate will be low, thus enhancing stability. With the rapid heating of the surface on a hot summer day, there will be a very steep environmental lapse rate because the air near the surface is so much warmer than that above, and instability is enhanced.

Pressure zones can also be related to atmospheric stability. In areas of high pressure, stability is maintained by the slow subsiding of relatively cool air from aloft. In low-pressure regions, on the other hand, instability is promoted by the tendency for air to converge and then rise.

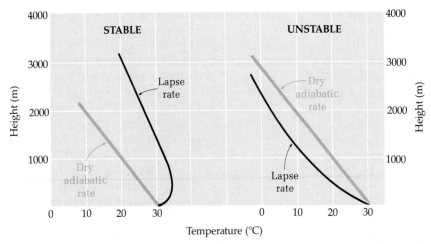

FIGURE 5.11 Relationship between lapse rates and air-mass stability. When air is forced to rise, it cools adiabatically. Whether it continues to rise or resists vertical motion depends on whether adiabatic cooling is less rapid or more rapid than the prevailing vertical temperature lapse rate. If the adiabatic cooling rate exceeds the lapse rate, the lifted air will be colder than its surroundings and will tend to sink when the lifting force is removed. If the adiabatic cooling rate is less than the lapse rate, the lifted air will be warmer than its surroundings and will therefore be buoyant, continuing to rise even after the original lifting force is removed.

PRECIPITATION

Condensed droplets within cloud formations stay in the air and do not fall to the earth because of their general buoyancy and the upward movement of the air. These droplets of condensation are so minute that they are kept floating in the cloud formation, their mass and the consequent pull of gravity being insufficient to overcome the buoyant effects of air and the vertical currents, or updrafts, within the clouds.

Precipitation occurs when the droplets of water, ice, or frozen water vapor coalesce and develop masses too great to be held above the earth. They then fall to the earth as rain, snow, hail, or sleet. The form that precipitation takes is largely dependent upon the method of formation and temperature during formation.

FORMS OF PRECIPITATION

Rain, consisting of droplets of liquid water, is by far the most common form of precipitation. The precise ways in which condensed droplets

of water within clouds develop into raindrops that fall to the earth as precipitation are not clearly understood. There is evidence, however, that in many cases raindrops develop through the grouping of water droplets around ice crystals formed at high altitude within the same cloud. That much precipitation results from the cumulus clouds that reach into altitudes with subfreezing temperatures tends to lend support to this theory of raindrop development. Precipitation is also produced, however, by clouds that do not reach such high altitudes. To explain the development of raindrops within these clouds, meteorologists suggest the fusion of small droplets into bigger drops, which in beginning their fall attract even more droplets. It has been estimated that one raindrop is due to the coalescence of about 1 million cloud droplets. Raindrops vary in size but are generally about 2.5 to 6 millimeters (0.1 to 0.25 in) in diameter. As we all know, rain can come in many ways: as a brief afternoon shower, a steady three-day rainfall, or the deluge of a tropical rainstorm. When the temperature of an air mass is only slightly below dew point, the raindrops may be very small (about 0.5 millimeters or less in diameter) and close together. The result is a fine mist or haze called **drizzle.** Drizzle is so light that it is greatly affected by the direction of air currents and the variability of winds. Consequently, drizzle seldom falls vertically.

Snow is the second most common form of precipitation. When water vapor is frozen directly into a solid without passing first through a stage as liquid water, it forms minute ice crystals around certain types of nuclei, and these crystals are usually in six-sided symmetrical shapes. Combinations of these ice crystal shapes make up the intricate and delicate patterns of snowflakes.

Sleet is frozen rain, formed when rain, in falling to the earth, passes through a cold layer of air and freezes. The result is the creation of solid particles of clear ice. In English-speaking countries outside the United States, sleet refers not to this phenomenon of frozen rain but rather to a mixture of rain and snow.

Hail is a less common form of precipitation than the three just described. It occurs most often during summer months and is the result of certain phenomena that are especially peculiar to the cumulonimbus cloud form. Hail appears as rounded lumps of ice, called **hailstones,** which can vary in size from 5 millimeters (0.2 in) in diameter up to the size of a baseball. The world record is a hailstone 30 centimeters (12 in) in diameter that fell in Australia. Dropping from the sky, hailstones can be highly destructive to crops and other vegetation, as well as to cars and buildings. Hailstones have even been known to kill animals and humans. Children think it is strange that they must leave a pool or lake where they are swimming simply because hailstones are falling, but this is a sensible precaution since the atmospheric conditions that produce hailstones also produce thunder and lightning. In fact, these phenomena often occur in conjunction with one another.

Hail forms when ice crystals are lifted by strong updrafts in a cloud. Then as these ice crystals fall through the cloud, supercooled water droplets attach themselves and are frozen in a layer. Sometimes these pellets are lifted up into the cold layer of air and then dropped again and again. The resulting hailstone, made up of concentric layers of ice, has a frosty, opaque appearance when it finally breaks out of the strong updrafts of the cloud formation and falls to earth.

On occasion, a raindrop can form and have a temperature below 0°C (32°F). These supercooled droplets will change to ice the instant they fall onto a cold surface. The icy covering over trees, plants, wires, etc. that results is known as **glaze.** People usually call the rain and its resultant blanket of ice an "ice storm." Because of the weight of ice, glazing can break off large branches of trees, bringing down telephone and power lines. It can also make roads practically impassable. A small counterbalance against the negative effects of glazing is the beauty of the natural landscape after an ice storm. Against the background of a clear blue sky, sunlight catches on the ice, reflecting and making diamonds out of the most ordinary weeds and tree branches.

CONDITIONS CAUSING PRECIPITATION

Adiabatic cooling is the only process that can lower the temperature of a large air mass to its dew point in order to produce sufficient condensation for precipitation, and adiabatic cooling can take place only in an air mass that rises and expands.

There are three major ways in which a parcel of air may be forced to rise, and each of these produces its own characteristic types of precipitation (Fig. 5.12). There is **convectional precipitation,** which results from the displacement of warm air upward in a convectional system. In **orographic precipitation,** an air mass encounters a land barrier, usually a mountain, and must rise above it in order to pass. **Cyclonic or frontal precipitation** takes place when a warm air mass rises after encountering a colder and denser air mass.

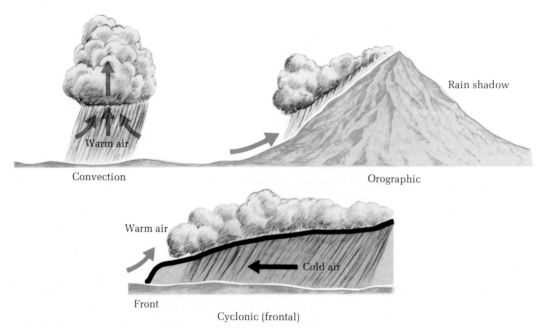

Rain shadow

Warm air

Convection

Orographic

Warm air

Cold air

Front

Cyclonic (frontal)

FIGURE 5.12 The principal causes of precipitation are upward movement of moist air resulting from convection, orographic lifting, and frontal activity.

CONVECTIONAL PRECIPITATION The simple explanation of convection is that when air is heated near the surface, it expands, becomes lighter, and rises. It is displaced by the cooler, denser air around it to complete the convection cycle. The important factor in convection for our discussion of precipitation is that the heated air rises and thus fulfills the one essential criterion for significant condensation and ultimately precipitation.

To enlarge our understanding of convectional precipitation, let us apply what we have learned about instability and stability. Figure 5.13 illustrates two cases where air is heated at the surface, instability occurs, and air rises due to convection. But case (a) is quite different from case (b). In both, the lapse rate in the free atmosphere is the same, and it is especially high during the first few thousand meters but slows after that (as on a hot summer day). In case (a) the air parcel is not very humid, and thus the dry adiabatic rate applies throughout its ascent. By the time the air reaches 3000 meters (9900 ft), its temperature and density are the same as the surrounding atmospheric air. At this point convectional lifting ceases.

In case (b) we have introduced the latent heat of condensation. As in case (a), the as yet unsaturated rising column of air cools at the dry adiabatic rate of 10°C/1000 meters (5.6°F/1000 ft) for the first 1000 meters (3300 ft). But since the air parcel is humid, the dew point is soon reached in the rising air column, condensation takes place, and cumulus clouds begin to form. As condensation occurs, the heat locked up in the water vapor is released and heats the moving parcel of air, retarding the adiabatic rate of cooling so that the rising air is now cooling at the wet adiabatic rate (5°C/1000 meters). Hence, the temperature of the rising air parcel remains warmer than the atmospheric air it is passing through, and the air parcel will continue to rise and rise and rise. It is obvious in case (b), the case that incorporates the latent heat of condensation, that we have massive condensation, towering cumulus clouds, and potentially heavy precipitation.

Convectional precipitation is most common in the humid equatorial and tropical areas that receive much of the sun's energy, and during summers in middle latitudes. Though differential heating of land surfaces surely plays

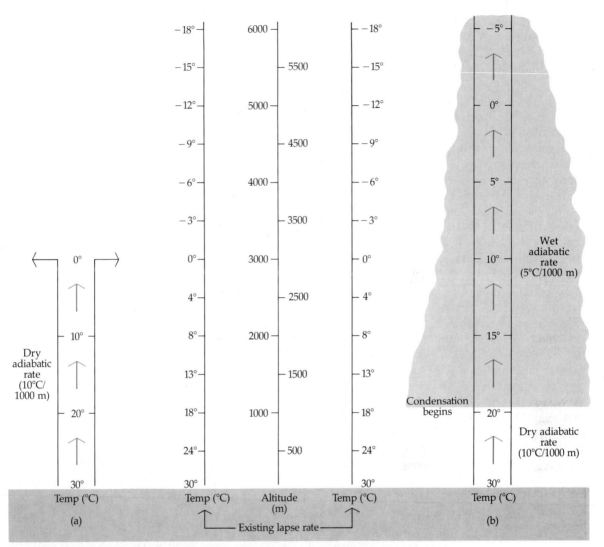

FIGURE 5.13   Effect of humidity on air-mass stability. In (a) warm, *dry* air rises and cools at the dry adiabatic rate, soon becoming the same temperature as the surrounding air, at which point convectional uplift terminates. Since the rising dry air did not cool to its dew point temperature by the time convectional lifting ended, no cloud formed. In (b) rising warm, *moist* air soon cools to its dew point temperature. The upward moving air subsequently cools at an assumed average wet adiabatic rate, which keeps the air warmer than the surrounding atmosphere so that the uplift continues. Only when all moisture is removed by condensation will the air cool rapidly enough at the dry adiabatic rate to become stable.

an important role in convectional precipitation, many meteorologists do not feel that it is the sole factor. Vertical air currents and turbulence, as well as surface obstructions such as hills, mountains, and even skyscrapers may provide the initial upward push for air that is

potentially unstable. Once condensation begins in a convectional column, there is additional energy available from the latent heat of condensation for further lifting.

It is such convectional lifting that can result in the heavy precipitation, thunder, lightning, and tornadoes of summer afternoon thunderstorms. When the convectional currents are strong in the characteristic cumulus clouds, hail can result.

 OROGRAPHIC PRECIPITATION As was the case with convectional rainfall, there is a simple definition of orographic rainfall and a somewhat more complex explanation. When land barriers such as mountain ranges, hilly regions, or even the escarpments (steep edges) of plateaus or tablelands lie in the path of prevailing winds, large portions of the atmosphere are forced to rise above these barriers (Fig. 5.12). This fills the one main criterion for significant precipitation—that large masses of air are cooled by ascent and expansion until large-scale condensation takes place. The resultant precipitation is termed orographic (from Greek: *oros,* mountain). As long as the air parcel rising up the mountainside remains stable (i.e., cooling at a greater rate than the environmental lapse rate), any resulting cloud cover and precipitation will be stratiform in nature. However, the story is not really complete until one realizes that the situation can be complicated by the same circumstances described in case (b) of Figure 5.13. A potentially unstable air parcel may only need the initial lift provided by the orographic barrier to set it in motion. In this case it will continue to rise of its own accord (no longer forced), as it seeks air of its own temperature and density. The land barrier only served to provide the initial thrust; it has performed its function as a lifting mechanism.

Because the air has deposited most of its moisture on the windward side of the mountain, there will normally be a great deal less precipitation on the leeward side, since on this side the air will be much drier and the dew point consequently much lower. The leeward side of the mountain is thus said to be in the **rain shadow.** Just as being in the shade, or in shadow, means that you are not receiving any direct sun, so being in the rain shadow means that you do not receive much rain. If you live near a mountain range, you can see the effects of orographic precipitation and the rain shadow in the pattern of vegetation (Fig. 5.14). The windward side of the mountains (say, the Cascades in Oregon and Washington) will be heavily forested and thick with vegetation. The opposite slopes in the rain shadow will usually be drier and the cover of vegetation more sparse. This is why eastern Oregon is desertlike, while the part of that state west of the Cascades is covered with green forests.

FIGURE 5.14 (a) Orographic uplift over the windward (western) slope of the Sierra produces condensation, cloud formation, precipitation, and the resulting dense stands of forest. (b) Semiarid or rainshadow conditions occur on the leeward (eastern) slope of the Sierras. (R. Gabler)

FIGURE 5.15 Average monthly precipitation at San Francisco, California, is represented by colored bars along bottom of graph. Such a graph of monthly precipitation figures gives a much more accurate picture than the annual precipitation total, which does not tell us that nearly all the precipitation occurs in only half of the year.

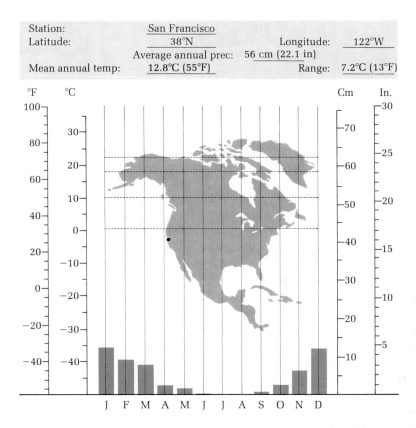

Station:	San Francisco		
Latitude:	38°N	Longitude:	122°W
	Average annual prec:	56 cm (22.1 in)	
Mean annual temp:	12.8°C (55°F)	Range:	7.2°C (13°F)

FRONTAL PRECIPITATION

The zones of contact between relatively warm and relatively cold air masses are known as **fronts.** When two large air masses of different densities and temperatures meet, the warmer one is lifted above the colder. When this happens, the major criterion for large-scale condensation and precipitation is once again met. Frontal precipitation thus occurs as the moisture-laden warm air rises above the front caused by contact with the cold air.

For fronts to be fully understood, we must examine air masses, their characteristics, how they come together, and what happens when they do. This discussion of air masses will be taken up in Chapter 6, and with it we will have a more detailed look at frontal disturbances and precipitation.

DISTRIBUTION OF PRECIPITATION

There are different ways to describe the precipitation a region receives. We can look at its average annual precipitation to get an overall picture of the amount of moisture it gets during a year. We could also look at its number of raindays [in a rainday, 0.25 millimeters (0.01 in) or more of rain is received during 24 hours]. If we were to divide the number of raindays in a year or month by the total number of days in that period, we would have a percentage that would be the probability of rain. Such a measure is important to farmers and to ski or summer resort owners whose incomes may depend on precipitation or the lack of it.

We can also look at the average monthly precipitation. This provides a picture of the seasonal variations in precipitation (Fig. 5.15). For instance, in describing the climate of the west coast of California we would not be giving the full story were we just to give the average annual precipitation, for this figure would not show the distinct wet and dry seasons that characterize this region.

HORIZONTAL DISTRIBUTION OF PRECIPITATION

Plate 9 shows average annual precipitation for the world's continents. We can see that there is great variability in the distribution of precipitation over the earth's surface. Although there is a zonal distribution of precipitation related to latitude, this distribution is obviously not the only factor involved in the amount of precipitation an area receives.

The likelihood and amount of precipitation is based on two factors. First, precipitation depends upon the degree of lifting that occurs in air of a particular region. This lifting, as we have already seen, may be due to the convergence of different air masses, or to differential heating of the earth's surface, or to the lifting that results when an air mass encounters a rise in the earth's surface, or to a combination of these factors. Also, precipitation is dependent upon internal characteristics of the air itself, which include its degree of instability, its temperature, and its humidity.

Since higher temperatures, as we have seen, allow air masses to hold greater amounts of water vapor, and conversely, cold air masses can hold less water vapor, we can expect a general decrease of precipitation from the equator to the poles that is related to the unequal zonal distribution of incoming solar energy, discussed in Chapter 3.

However, if we look again at Plate 9, we see that there is a great deal of variability in average annual precipitation beyond the general pattern of a decrease with increased latitude. In the following discussion, we will examine some of these variations and give the reasons for them. Basically, we will be applying what we have already learned about temperature, pressure systems, wind belts, and precipitation.

DISTRIBUTION WITHIN LATITUDINAL ZONES

The equatorial zone is generally an area of high precipitation (over 200 centimeters annually). This high level is largely due to the zone's high temperatures, high humidity, and the instability of its air. High temperatures and instability lead to a general pattern of rising air, which in turn allows for precipitation. This tendency is strongly reinforced by the convergence of the trades as they move toward the equator from opposite hemispheres. In fact, the Intertropical Convergence Zone is one of the two great zones where frontal precipitation occurs. (The other is the polar front within the westerlies zone.)

In general, the air of the trade wind zones is stable compared to the instability of the equatorial zone. Under the control of these steady winds, there is little in the way of atmospheric disturbances to lead to convergent or convectional lifting. However, since the trade winds are basically easterlies, when they move onshore along east coasts or high islands they bring moisture from the oceans with them. Thus, within the trade wind belt continental east coasts tend to be wetter than continental west coasts.

In fact, where the air of the equatorial and trade wind regions with its high temperatures and vast amounts of moisture moves onshore from the ocean and meets a landform barrier, record rainfalls can be measured. The windward slope of Mount Waialeale on Kauai, Hawaii, at approximately 22°N latitude, holds the world's record for greatest average annual rainfall—1160 centimeters (460 in).

Moving poleward from the trade wind belts, we enter the zones of subtropical high pressure where the air is subsiding. As it sinks lower, it is warmed adiabatically, increasing its moisture-holding capacity and consequently reducing the amount of precipitation in this area. In fact, if we look at Figure 5.16, which shows average annual precipitation on a latitudinal basis, we can see that there is a dip in precipitation level corresponding to the latitude of the subtropical high-pressure belts and cells. These areas of subtropical high-pressure are, in fact, where we find most of the great deserts of the world: in northern and southern Africa, in Arabia, North America, and Australia. The exceptions to this subtropical aridity occur along the eastern sides of the landmasses, where, as we have already noted, the subtropical high-pressure cells are weak and wind direction is

FIGURE 5.16 Latitudinal distribution of average annual precipitation. This figure illustrates the four distinctive precipitation zones: high precipitation caused by convergence of air in the tropics and in the middle latitudes along the polar front and low precipitation caused by subsidence and divergence of air in the subtropical and polar regions.

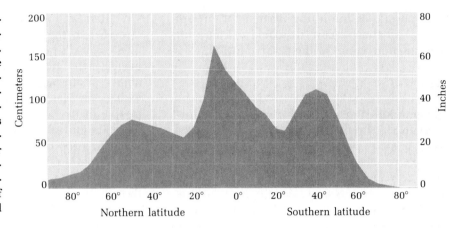

often onshore. This exception is especially true of regions affected by the monsoons.

In the zones of the westerlies from about 35° to 65° N and S latitude, precipitation occurs largely as a result of the meeting of cold, dry polar air masses and warm, humid subtropical air masses along the polar front. Thus, there is much cyclonic or frontal precipitation in this zone.

Naturally, the continental interiors of the middle latitudes are drier than the coasts since they are farther away from the oceans. Furthermore, where air in the prevailing westerlies is forced to rise, as it is when it crosses the Cascades and Sierra Nevada of the Pacific Northwest and California, especially during the winter months, there is heavy orographic precipitation. Thus, in the middle latitudes, continental west coasts tend to be wet, and precipitation decreases with movement eastward toward continental interiors. Along eastern coasts within the westerlies precipitation usually increases once again because of proximity to humid air from the oceans.

In the United States, the interior lowlands are not as dry as we might expect within the prevailing westerlies. This is because of the great amount of frontal activity resulting from the conflicting northward and southward movements of polar and subtropical air. If there were a high east-west mountain range ex-

tending from central Texas to northern Florida, the lowlands of the continental United States north of that range would be much drier than they actually are because they would be cut off from moist air originating in the Gulf of Mexico.

Also characteristic of the belt of the westerlies are desert areas that occur in the rain shadows of prevailing winds that are forced to rise over mountain ranges. This effect is in part responsible for the development and maintenance of Death Valley as well as the desert zone of eastern California and Nevada in the United States, the mountain-ringed deserts of eastern Asia, and Argentina's Patagonian Desert, which is in the rain shadow of the Andes. Note in Figure 5.16 that there is greater precipitation in the middle latitudes of the Southern Hemisphere than there is in the Northern. This occurs largely because there is a lot more ocean and less landmass in the Southern Hemisphere westerlies than in the corresponding zone of the Northern Hemisphere.

Moving poleward, we find that temperatures decrease, and along with them the moisture-holding capacity of the air diminishes as well. The low temperatures also lead to low evaporation rates and, in addition, the air in the polar regions shows a general pattern of subsidence that yields areas of high pressure. This settling of the air in the polar regions is the opposite of the lifting needed for precipitation. All these factors combine to cause low precipitation values in the polar zones.

VARIABILITY OF PRECIPITATION

The rainfall depicted in Plate 9 is an annual average. It should be remembered, however, that for many parts of the world there are significant variations in precipitation, both within any one year and between years. For example, areas like the Mediterranean region, California, Chile, South Africa, and Australia that are on the west sides of the continents and roughly between 30° and 40° latitude get much more rain in the winter than in the summer. And there are areas between 10° and 20° latitude that get much more of their precipitation in the summer (high sun season) than in the winter (low sun season).

Rainfall totals can change markedly from one year to the next, and, tragically for many of the world's people, the drier a place is on the average, the greater will be the statistical variability in its precipitation (compare Fig. 5.17 with Color Plate 9). To make matters worse for people in dry areas, a year with a particularly high amount of rainfall may be balanced with several years of below-average precipitation. This situation has occurred recently in West Africa's Sahel, the Soviet steppe, and the American Great Plains.

Thus, there are years of drought and years of flood, each bringing its own kind of disasters upon the land. Farmers, resort owners, construction workers, and others whose economic well-being depends in one way or another on precipitation or the lack of it can determine only a probability of rainfall on an annual, monthly, or even a seasonal basis.

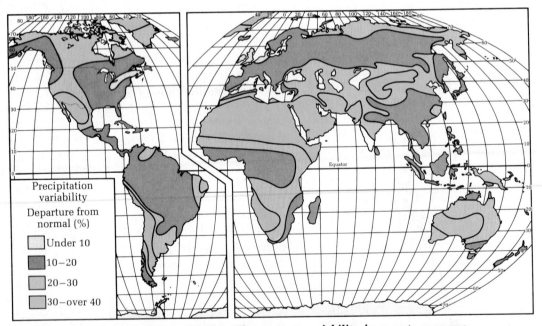

FIGURE 5.17 Precipitation variability. The greatest variability in year-to-year precipitation totals is in the dry regions, accentuating the critical problem of moisture supply in those parts of the world. (After William Van Royen, 1954, *Atlas of the World's Agricultural Resources*, Prentice-Hall, Englewood Cliffs, New Jersey)

Plate 8　Major Cloud Forms

Cirrus clouds. These are wisps of fine ice crystals in the upper air. *(NOAA)*

Stratus clouds. These are extensive horizontal sheets of water droplets, indicating stable air. Rainfall from stratus is often a long-lasting drizzle. *(NOAA)*

Cumulus clouds. They indicate convectional uplift of heated, unstable air. *(Courtesy Kenneth R. Martin Collection)*

Plate 9 World Map of Average Annual Precipitation

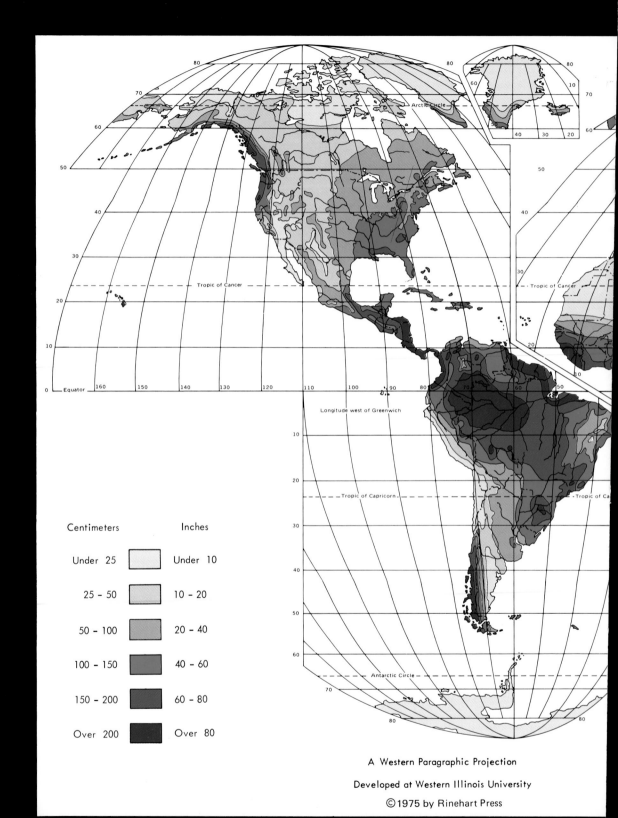

Centimeters	Inches
Under 25	Under 10
25 – 50	10 – 20
50 – 100	20 – 40
100 – 150	40 – 60
150 – 200	60 – 80
Over 200	Over 80

A Western Paragraphic Projection

Developed at Western Illinois University

©1975 by Rinehart Press

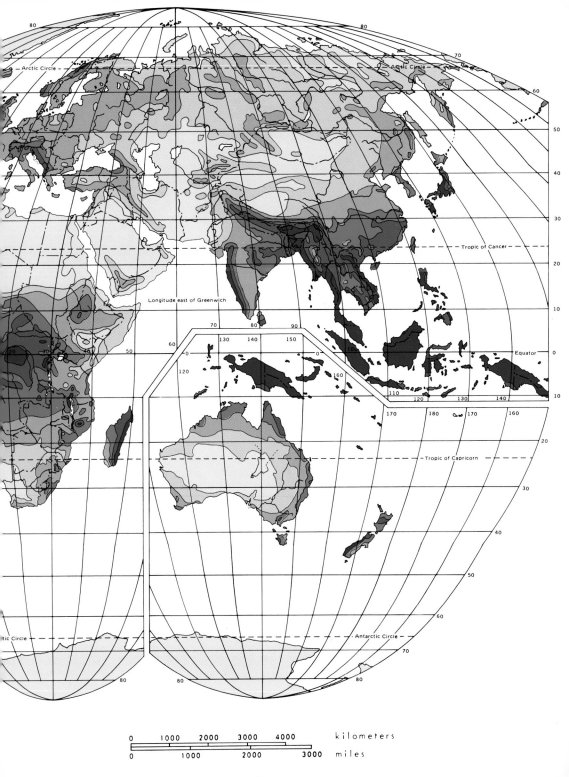

Arctic Circle

Arctic Circle

Tropic of Cancer

Longitude east of Greenwich

Equator

Tropic of Capricorn

Antarctic Circle

kilometers

miles

0   1000   2000   3000   4000

0   1000   2000   3000

clouds. "Alto" indicates a middle-level
e a horizontal stratiform layer spread
sky. *(S. Brazier)*

**Altocumulus clouds. A middle layer of globular,
and therefore cumuliform clouds.** *(NOAA)*

s clouds. The sun and moon can often
rough them producing a halo. *(NOAA)*

**Stratocumulus clouds seen from above. The lay-
ered horizontal sheet (stratiform) with its globular
(cumuliform) patches is clearly visible.** *(S. Brazier)*

Meteorologists are not yet able to predict rainfall with 100 percent accuracy. This failure is due to the many factors involved in causing precipitation (temperature, available moisture, atmospheric disturbances, landform barriers, frontal activity, air mass movement, and differential surface heating, among others). In addition, the interaction of these factors in the development of precipitation is very complex and not completely understood.

QUESTIONS FOR DISCUSSION AND REVIEW

1. How is the hydrologic cycle related to the earth's water budget?

2. Is the saturation or dew point in your area generally higher or lower than in Korea? Honduras?

3. What is the difference between absolute and specific humidity? What is relative humidity?

4. Follow the weather report in your area for a week by watching TV, listening to the radio, or reading the newspapers. What figures were given for temperature and humidity? What did each day feel like in terms of your own comfort with the weather?

5. Why is the inside of a greenhouse generally more humid than an ordinary room?

6. Why is wearing wet clothes sometimes bad for your health? Under what conditions might wearing wet clothes be good for your health?

7. Imagine that you are deciding when, in your daily schedule, to water the garden. What time of day would be best in terms of conserving water? Why?

8. What factors affect the formation of ground-inversion fogs?

9. What kinds of fogs occur in the area you live in?

10. Describe the most recent cloud formations in your area. What kinds of weather (temperature, humidity, precipitation) were the cloud types associated with?

11. How is atmospheric stability related to the adiabatic rates?

12. What atmospheric conditions are necessary for precipitation to occur?

13. Find out how many inches of precipitation have fallen in your area this year. Is that average or unusually high or low?

14. Compare and contrast convectional, orographic, and frontal precipitation.

15. How is rainfall variability related to total annual rainfall? How might this relationship be considered a double problem for people?

6

AIR MASSES AND ATMOSPHERIC DISTURBANCES

AIR MASSES

In the previous three chapters, we have looked at the elements of the atmosphere and investigated some of the controls that act upon those elements, causing them to vary from place to place and through time. There is, however, even more involved in the examination of weather. We have not yet looked at storms (atmospheric disturbances)—their types and characteristics, their origin, and their development. Storms are an important part of the weather story. As background for an understanding of these atmospheric disturbances, we need first to look at the subject of air masses.

During World War I, Norwegian meteorologists Vilhelm and Jacob Bjerknes (a father and son) developed what has become known as **air-mass analysis.** This is a way of looking at the actions and interactions of parts of the atmosphere. Air-mass analysis is useful in explaining storms and many of the day-to-day weather changes common in middle latitudes. It is not unusual in some parts of the United States to go to bed at the end of a beautiful warm day in early spring and wake up to falling snow the next morning. Such changeability

is a common middle-latitude weather characteristic; especially during certain times of the year, the weather changes from a period of cold, clear, dry days to a period of snow, only to be followed by one or two more moderate but humid days.

An **air mass** is a large body of air, sometimes subcontinental in size, which may move over the earth's surface as a distinct entity. An air mass is relatively homogeneous in regard to temperature and humidity. That is, at approximately the same altitude within the air mass, the temperature and humidity will be similar. Of course, since an air mass may extend over 20° or 30° of latitude, we can expect some slight modifications due to variations in insolation, which are significant over that distance, and to changes caused by contact with differing land surfaces. As a result of this temperature and moisture uniformity, the density of air will be much the same throughout any one level within an air mass.

The similar characteristics of temperature and humidity within an air mass are determined by the nature of its **source region**—that is, the place where the air mass is originally formed. Only a few areas on the earth make

Thunderstorm with lightning discharge. (NOAA)

good source regions. In order for the air mass to have similar characteristics throughout, the source region must have a nearly homogeneous surface. In addition, the air mass must have sufficient time to acquire the characteristics of the source region. Hence, gently settling, slowly diverging air will accompany a source region, while converging, rising air will not.

There are five terms used to describe air masses, and each reflects a property of a source region. They are: (1) tropical (warm, symbolized by capital *T*), (2) polar (cold, *P*), (3) arctic (very cold, *A*), (4) continental (dry, *c*), and (5) maritime (wet, *m*). These five can be combined to give us the classic classification of air masses first described by Bergeron (another Norwegian) in 1928: Maritime Tropical *(mT)*, Continental Tropical *(cT)*, Continental Polar *(cP)*, Continental Arctic *(cA)*, and Maritime Polar *(mP)*. These five types are described more fully in Table 6.1. From now on, we will use the symbols rather than the full names as we discuss each type of air mass.

STABILITY OF AIR MASSES

As a result of the general circulation patterns within the atmosphere, air masses do not re-

TABLE 6.1   Types of Air Masses

	SOURCE REGION	USUAL CHARACTERISTICS AT SOURCE	ACCOMPANYING WEATHER
Maritime Tropical (*mT*)	Tropical and sub-tropical oceans	Subsiding air; fairly stable, but some instability on western side of oceans; warm and humid	High temperatures and humidity, cumulus clouds, convectional rain in summer; mild temperatures, overcast skies, fog, drizzle, and occasional snowfall in winter; heavy precipitation along *mT/cP* fronts in all seasons
Continental Tropical (*cT*)	Deserts and dry plateaus of subtropical latitudes	Subsiding air aloft; generally stable, but some local instability at surface; hot and very dry	High temperatures, low humidity, clear skies, rare precipitation
Maritime Polar (*mP*)	Oceans between 40° and 60° latitude	Ascending air and general instability, especially in winter; mild and moist	Mild temperatures, high humidity; overcast skies and frequent fogs and precipitation, especially during winter; clear skies and fair weather common in summer; heavy orographic precipitation, including snow, in mountainous areas
Continental Polar (*cP*)	Plains and plateaus of subpolar and polar latitudes	Subsiding and stable air, especially in winter; cold and dry	Cool (summer) to very cold (winter) temperatures, low humidity; clear skies except along fronts; heavy precipitation, including winter snow, along *cP/mT* fronts
Continental Arctic (*cA*)	Arctic Ocean, Greenland, and Antarctica	Subsiding very stable air; very cold and very dry	Seldom reaches United States, but when it does: bitter cold, subzero temperatures, clear skies, often calm conditions

main stationary over their source regions indefinitely. When an air mass begins to move over the earth's surface along a path known as a **trajectory,** it retains its distinct and homogeneous characteristics to a large extent. However, modification does occur as the air mass gains or loses some of its heat energy and moisture content to the ground below. Although this modification is generally slight, the gain or loss of heat energy can make an air mass more stable or unstable.

Because it has a bearing on its stability, an air mass is further classified as to whether it is warmer or colder than the surface over which it is in contact. If an air mass is colder than the surface over which it passes, then the surface will heat the air mass from below. This will, in turn, increase the environmental lapse rate, enhancing the prospect of instability. To describe such a situation the letter *k* (from German: *kalt,* cold) is added to the other letters that symbolize the air mass. For example, an *mT* air mass originating over the Gulf of Mexico in summer that moves onshore over the warm land would be denoted *mTk.* Such an air mass is often unstable with copious, convective precipitation likely. On the other hand, this same *mT* air mass moving onshore during the winter would now be warmer than the land surface. Consequently, the air mass would be cooled from below, decreasing its environmental lapse rate, which enhances the prospect of stability. We describe this situation with the letter *w* (from German: *warm,* warm) and the air mass would be denoted *mTw.* In this case stratiform, not convective, precipitation is most likely.

NORTH AMERICAN AIR MASSES

Because most of us are familiar with the weather in at least one region of the United States, we will concentrate in this chapter on the air masses of North America and their effects on weather. What we learn will be applicable to the rest of the world, and, as we examine climatic regions in some of the following chapters, we will be able to understand that weather everywhere is most often affected by the movements of air masses. Especially in middle-latitude regions, the majority of atmospheric disturbances result from the confrontations of different air masses.

All five types of air masses (*cA, cP, mP, cT,* and *mT*) influence the weather of North America, some more than others. Air masses assume characteristics of their source regions (Fig. 6.1). Consequently, as the source regions change with the seasons, primarily because of changing insolation, the air masses also will vary.

CONTINENTAL ARCTIC AIR MASSES

The frigid, *frozen* surface of the Arctic Ocean serves as the source region for this air mass. It is extremely cold, very dry, and very stable. Even during the winter, when this air mass is best developed, it seldom travels far enough south to affect the United States. However, on those few occasions when it does extend down into the midwestern and eastern United States, its impact is awesome. Below zero, record-setting cold temperatures often result. If the *cA* air mass remains in the Midwest for an extended period, vegetation—not accustomed to the extreme cold—can be severely damaged or killed.

CONTINENTAL POLAR AIR MASSES

At its source, a continental polar air mass is cold, dry, and stable, since it is warmer than the surface beneath it; the weather of a *cP* air mass is cold, crisp, clear, even sparkling. Because there are no east-west landform barriers in North America, *cP* air can migrate south across Canada and the United States. A tongue of *cP* air can sometimes reach as far south as the Gulf of Mexico or Florida. When winter *cP* air extends into the United States, its temperature and humidity are raised only slightly, and the movement of such an air mass into the Midwest and South brings with it a cold wave characterized by colder than average temperatures and clear, dry air, causing freezing temperatures as far south as Florida and Texas.

Because of the general westerly direction of atmospheric circulation in the middle lati-

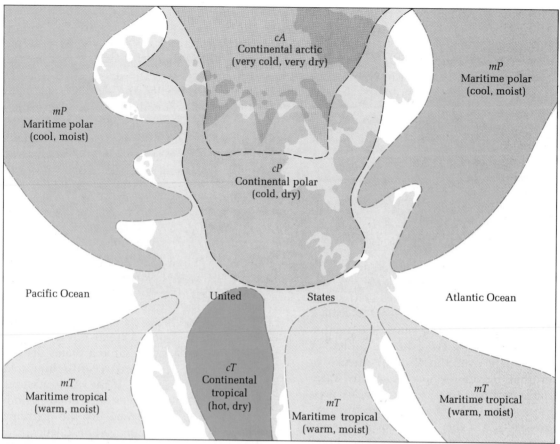

FIGURE 6.1   Source regions of North American air masses. Air-mass movements import the temperature and moisture characteristics of these source regions into far distant areas. (After Trewartha)

tudes, a *cP* air mass is rarely able to break through the great western mountain ranges to the West Coast of the United States. When such an air mass does reach the California, Washington, and Oregon coasts, however, it brings with it unusual freezing temperatures that do great damage to agriculture.

MARITIME POLAR AIR MASSES   During winter months, because the oceans tend to be warmer than the land, an *mP* air mass tends to be warmer than its cousin on land, the *cP* air mass. Much *mP* air was originally cold, dry, *cP* air that has moved to a position over the ocean. There it is heated by the warmer water and collects moisture. Thus, at its source, *mP* air is

cold (although not nearly so cold as *cP* air), damp air with a tendency toward instability. The northern Pacific Ocean serves as the source region for maritime polar air masses, which, because of the general westerly circulation of the atmosphere in the middle latitudes, affect the weather of the United States. As such an air mass moves south and east, it gains moisture and warmth until it reaches land. When this *mP* air meets any lifting agent (such as a mass of colder, denser air or coastal mountain ranges), the result is usually very cloudy weather with a great deal of precipitation. An *mP* air mass is still the source of many midwestern snowstorms even after crossing the western mountains.

Generally, an *mP* air mass that develops over the northern Atlantic Ocean does not affect the weather of the United States since such an air mass tends to flow instead toward Europe. However, on occasions there may be a reversal of wind direction accompanying the migration of a low-pressure system, and New England can be made miserable by the cool, damp winds, rain, and snow of a "northeaster."

MARITIME TROPICAL AIR MASSES   The Gulf of Mexico and subtropical Atlantic Ocean serve as a source region for maritime tropical air masses that have a great influence on the weather of the United States. During winter the waters are warm, and the air above is warm and moist. As the warm, moist air moves northward up the Mississippi Lowlands, it travels over increasingly cooler land surfaces. The lower layers of air are chilled, and dense advection fog often results. And when it reaches the *cP* air migrating southward from Canada, the warm air is forced to rise, and major precipitation can occur.

The longer days and more intense insolation of summer months modify a Gulf and Atlantic air mass at its source region by increasing its temperature and moisture content. However, during summer the land is warmer than the nearby waters, and as the *mT* air mass moves onto the land, the instability of the air mass is increased. This air mass is the source of great thunderstorms and convective precipitation on hot, humid days, and it is also responsible for much of the hot, humid weather of the southeastern and eastern United States.

Other *mT* air masses form over the Pacific Ocean in the subtropical latitudes. These air masses tend to be slightly cooler than those that form over the Gulf and the Atlantic, partly because of their passage over the cooler California Current. A Pacific *mT* air mass is also more stable because of the strong subsidence associated with the eastern portion of the Pacific subtropical high. This air mass contributes to the dry summers of southern California and occasionally brings moisture in winter as it rises over the mountains of the Pacific Coast.

CONTINENTAL TROPICAL AIR MASSES   There is a fifth type of air mass, but it is the least important to the weather of the United States. This is the continental tropical air mass that develops over large, homogeneous land surfaces in the subtropics. The Sahara Desert of North Africa is a prime example of a source region for this type of air mass. The weather typical of the *cT* air mass is usually very hot and dry, with clear skies and major heating from the sun during daytime.

In North America, there is little land in the correct latitudes to serve as a source region for a *cT* air mass of any significant proportion. A small *cT* air mass can form over the deserts of the southwestern United States and northwestern Mexico in the summer. In its source region, a *cT* air mass provides hot, dry, clear weather. When it moves eastward, however, it is usually greatly modified as it comes in contact with larger and stronger air masses of different temperatures, humidities, and densities.

FRONTS

We have seen that air masses migrate with the general circulation of the atmosphere. Over the United States there is thus a general eastward flow of the air masses with the direction of the westerly wind belt. In addition, air masses tend to diverge from areas of high pressure and converge toward areas of low pressure. This tendency means that the tropical and polar air masses, formed within systems of divergence, tend to flow toward the areas of continental low pressure within the United States. An important feature of an air mass is that it maintains the primary characteristics first imparted to it by its source region, although some slight modification may occur during its migration.

Air masses differ primarily in their temperature and in their moisture content, which in turn affect the air masses' density and atmospheric pressure. Air masses with different properties do not mix easily, but instead come in contact along sloping boundaries called

fronts. Although usually depicted on maps as a two-dimensional boundary line separating two different air masses, a front is actually a three-dimensional surface with length, width, and height. To emphasize this concept, a front is sometimes referred to as a **surface of discontinuity.**

The area where the surface of discontinuity between two contrasting air masses intersects the earth's surface is represented on weather maps by a line. However, this intersection is not actually a line but a zone that can cover an area 2 or 3 kilometers (1–2 mi) wide to one as wide as 150 kilometers (90 mi). It is more accurate, then, to speak of a frontal zone rather than a frontal line.

The sloping surface of a front is created as the warmer and lighter of the two contrasting air masses is lifted or rises above the cooler and denser air mass. Such rising is known as **frontal lifting,** and it is a major source of precipitation, especially in middle-latitude countries like the United States, for the major convergence of contrasting air masses occurs in the middle latitudes.

The steepness of the frontal surface is governed primarily by the degree of difference between the two converging air masses. When there is a sharp difference between the two air masses, as when an *mT* air mass of high temperature and moisture content meets a *cP* air mass with its cold, dry characteristics, the slope of the frontal surface will be steep. With a steep slope, there will be greater frontal lifting. Provided other conditions (e.g., temperature and moisture content) are equal, a steep slope, with its greater frontal lifting will produce heavier precipitation than will a gentler slope.

Fronts are differentiated by determining whether the colder air mass is moving on the warmer one or vice versa. The weather that occurs along a front is also dependent on which air mass is the "aggressor."

COLD FRONT

A **cold front** occurs when a cold air mass actively moves upon a warmer air mass and pushes it upward. The colder air, being denser and heavier than the warm air it is displacing, stays at the surface while forcing the warmer air to rise. As we can see in Figure 6.2, a cold front usually results in a relatively steep slope in which the warm air may rise 1 meter in the vertical for every 40 to 80 meters of horizontal distance. If the warm air mass is unstable and has a high moisture content, heavy precipitation can result, sometimes in the form of violent thunderstorms. In any case, cold fronts are usually associated with strong weather disturbances or sharp changes in temperature, air pressure, and wind.

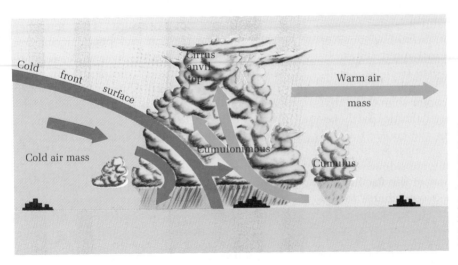

FIGURE 6.2 Cross section of a cold front. Cold fronts generally move rapidly, with a blunt forward edge that drives adjacent warmer air upward, producing violent precipitation from the warmer air.

FIGURE 6.3 Cross section of a warm front. Warm fronts advance more slowly than cold fronts and replace (rather than displace) cold air by sliding upward over it. The gentle rise of the warm air produces stratus clouds and gentle drizzles.

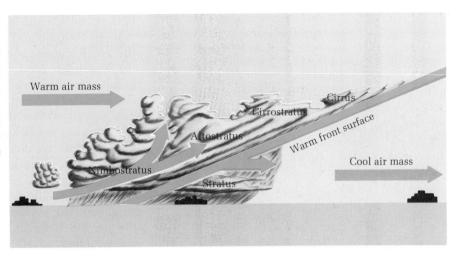

Warm air mass

Cirrus

Cirrostratus

Altostratus

Warm front surface

Nimbostratus

Stratus

Cool air mass

WARM FRONT

When a warmer air mass is the aggressor and invades a region occupied by a colder air mass, a **warm front** is the result. In a warm front, the warmer air, as it slowly pushes the cold air back, rises over the colder, denser air mass, which again stays in contact with the earth's surface. The slope of the surface of discontinuity that results is usually far gentler than that occurring in a cold front. In fact, the warm air may rise only 1 meter in the vertical for every 100 or even 200 meters of horizontal distance. Thus, the frontal lifting that develops will not be as great as that occurring along a cold front. The result is that the weather associated with the passage of a warm front *tends* to be less violent and the changes less abrupt than those associated with cold fronts. If we look at Figure 6.3, we can see why the advancing warm front affects the weather of areas ahead of the actual surface location of the frontal zone. Changes in the weather resulting from fronts that have not yet reached us can be indicated by the cloud forms that precede them.

STATIONARY AND OCCLUDED FRONTS

When two air masses have converged and formed a frontal boundary, but neither is moving, then we have a situation known as a **stationary front.** The weather that results from a stationary front is apt to affect an area with clouds, drizzle, and rain for several days. In fact, a stationary front and its accompanying weather will remain until the contrasts between the two air masses are reduced or the circulation of the atmosphere finally causes one or the other or both air masses to move on.

An **occluded front** occurs when a faster moving cold front overtakes a warm front and all of the warm air is pushed aloft. This frontal situation usually occurs in the latter stages of a midlatitude cyclonic storm, which will be discussed next. (Map symbols of the four frontal types are shown in Fig. 6.4.)

ATMOSPHERIC DISTURBANCES

In addition to the general circulation of the atmosphere within the wind belts described in Chapter 4, there are secondary circulations. These are made up of storms and other atmospheric disturbances.

Our examination of air masses and fronts will serve as a background for our look at atmospheric disturbances. We use the term **atmospheric disturbance** because it is more general than **storm,** and it includes variations in the secondary circulation of the atmosphere that cannot correctly be classified as storms.

Partly because our primary interest is in the weather of North America, we will concen-

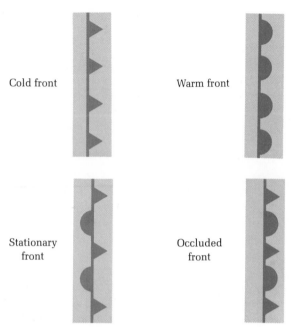

FIGURE 6.4 The four major frontal symbols used on weather maps.

trate on an examination of **extratropical disturbances,** as the atmospheric disturbances of the middle latitudes are sometimes called. Furthermore, the middle latitudes are an area of convergence, and it is here that unlike air masses such as the cold polar air and the warm subtropical air commonly meet. The **polar front** marks a zone along which these air masses most often meet and where many middle-latitude storms develop. The polar front is not a permanent line; in fact, it may sometimes be depicted on a surface weather map as more than one line. Furthermore, the zone of the polar front tends to move north and south with the seasons and is apt to be stronger in winter than in summer.

CYCLONES AND ANTICYCLONES

NATURE, SIZE, AND APPEARANCE ON MAPS We have previously distinguished cyclones and anticyclones according to differences in pressure and wind direction. Also, when studying maps of world pressure distri-

bution, we identified large areas of semipermanent cyclonic and anticyclonic circulation in the earth's atmosphere (the subtropical high, for example). Now, when examining middle-latitude atmospheric disturbances, we will use the terms *cyclone* and *anticyclone* to describe the moving cells of low and high pressure that drift with varying regularity in the path of the prevailing westerly winds. As systems of higher pressure the anticyclones are usually characterized by clear skies, gentle winds, and a general lack of precipitation. As centers for converging, rising air, the cyclones are the true storms of the middle latitudes with associated fronts of various types.

As we know from experience, no middle-latitude cyclonic storm is ever exactly like any other. They vary in their intensity, in the number of hours or days they last, in the speed with which they pass, in the strength of their winds, in the amount and type of cloud cover, in the quantity and kind of precipitation they deposit, and in the surface area they affect.

Because of the seemingly endless variety of cyclones and because of our limited knowledge about them, we will be describing *model* cyclones in the following discussions. Not every storm will act in the way we describe, but there are generalizations that will be helpful in understanding the general operation of middle-latitude storms.

Because a cyclone has a low-pressure center, winds tend to converge toward that center as an attempt to equalize pressure. If we visualize this air moving in toward the center of the low-pressure system, we can see that air that is already at the center must be displaced upward. The lifting that occurs in a cyclone results in clouds and precipitation, and the warm air rises like a corkscrew.

Anticyclones are high-pressure systems where atmospheric pressure decreases toward the outer limits of the system. Visualizing an anticyclone, or high, we can see that air in the center of the system must be subsiding, which in turn displaces surface air outward, away from the center of the system. Hence, an anticyclone has diverging winds. In addition, an anticyclone tends to be a fair-weather system, as the subsiding air in its center increases in temperature and stability, reducing the opportunity for condensation.

We should note here that the pressures we are referring to in these two systems are relative. What is important is that in a cyclone pressure *decreases* toward the center and in an anticyclone pressure *increases* toward the center. Furthermore, the intensities of the winds involved in these systems will depend on the steepness of the pressure gradients involved. Thus, if there is a steep pressure gradient in a cyclone, with the pressure much lower at the center than at the outer portions of the system, the winds will converge toward the center with considerable velocity.

The situation is easier to visualize if we imagine these pressure systems as landforms. A cyclone is shaped like a basin (Fig. 6.5). If we are filling the basin with water, we know that the water will flow in faster the steeper the sides and the deeper the depression. If we visualize an anticyclone as a hill or mountain, then we can also see that just as water flowing down the sides of such landforms will flow

faster with increased height and steepness, so will air blowing out of an area of very high pressure move rapidly.

On a surface weather map, cyclones and anticyclones are depicted by concentric isobars of increasing pressure toward the center of a high and of decreasing pressure toward the center of a low. Usually, a high will cover a larger area than a low. Both pressure systems are capable of covering and affecting extensive areas. There are times when nearly the entire midwestern United States is under the influence of the same system. The average diameter of an anticyclone is about 1500 kilometers (900 mi), while that of a cyclone is about 1000 kilometers (600 mi).

GENERAL MOVEMENT The cyclones and anticyclones of the middle latitudes are steered, or guided, along a path reflecting the configuration and speed of the upper air westerlies. The upper air flow can be quite variable with wild oscillations. However, a general west-to-east pattern does prevail. Consequently, people in the eastern United States look at the weather occurring to the west to see what they might

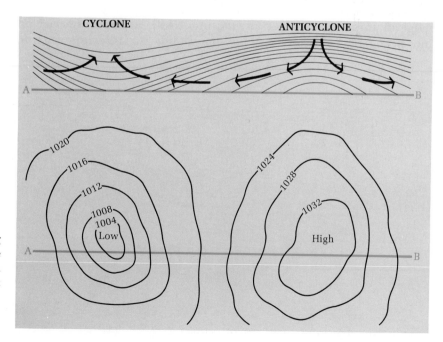

FIGURE 6.5 Close spacing of isobars around a cyclone or anticyclone indicates a steep pressure gradient that will produce strong winds. Wide spacing of isobars indicates a weaker system.

expect in the next few days. Most storms that develop in the Great Plains or Far West move across the United States during a period of a few days at an average speed of about 10 meters per second (23 mph), and then travel on into the North Atlantic before occluding.

Although neither cyclones nor anticyclones develop in exactly the same places at the same times each year, they do tend to develop in certain areas or regions more frequently than in others, and they do follow the same general paths, which are known as **storm tracks** (Fig. 6.6). These storm tracks vary with the seasons. In addition, because the temperature variations between the air masses are stronger during the winter months, the atmospheric disturbances that develop in the middle latitudes during those months are greater in number and intensity.

CYCLONES Now let us look more closely at cyclones—their origin, development, and characteristics. Warm and cold air masses meet at the polar front, where most cyclones develop. The air of these two contrasting masses does not merge but may move in opposite directions along the frontal zone. Although there may be some slight uplift of the warmer air along the edge of the denser, colder air, the uplift will not be significant. There may be some cloudi-

ness and precipitation along such a frontal zone, though not of storm caliber.

For reasons not completely understood, but certainly related to the wind flow in the upper troposphere, warmer air may start to push in a northerly direction into the colder air. This is the beginning of the wave formation that will lead to the fully grown cyclone (Fig. 6.7). As the warmer air pushes into the cold air, forming a warm front, portions of the cold air may begin to move in a southerly direction, pushing on the warm air and forming a cold front. This process continues, resulting in a fully developed cyclone.

As the contrasting air masses jockey for position, the clouds and precipitation that exist along the fronts are greatly intensified and the area affected by the storm is much greater. Along the warm front, precipitation will be more widespread but less intense than along the cold front. One factor that can vary the kind of precipitation occurring at the warm front is the stability of the warm air mass. If it is stable, then its upgliding over the cold air mass may cause only a fine drizzle or a gentle, powdery snow if the temperatures are low enough. On the other hand, if the warm air mass is moist and unstable, the upgliding may set off heavier precipitation. As you can see by referring again to Figure 6.3, the precipitation

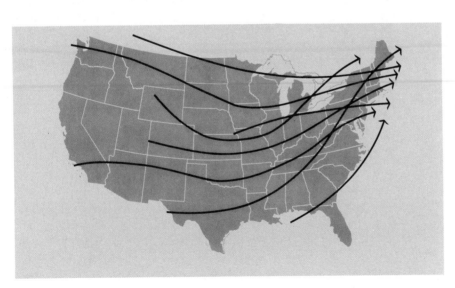

FIGURE 6.6 Common storm tracks for the United States. Virtually all cyclonic storms move from west to east in the prevailing westerlies and swing northeastward across the Atlantic Coast. Storm tracks originating in the Gulf of Mexico represent tropical hurricanes.

FIGURE 6.7 Stages in the development of a midlatitude cyclone. Each view represents the development somewhat eastward of the preceding view as the cyclone travels along its storm track. Note the occlusion in (e).

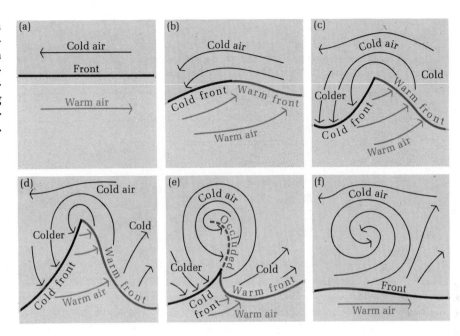

that falls at the warm front may *appear* to be coming from the colder air, and the weather may feel cold and damp. Yet the precipitation is actually coming from the warmer air mass above, though it must fall through the colder air mass to reach the earth's surface.

Because the cold front moves faster, it will eventually overtake the warm front. This produces the situation we previously identified as an occluded front. Because additional warm, moist air will not be lifted after occlusion, condensation and the release of latent energy will diminish and the storm will soon die. Occlusions are usually accompanied by rain and are the major process by which midlatitude cyclones are destroyed.

CYCLONES AND LOCAL WEATHER Different portions of a wave cyclone exhibit different weather. Therefore, the weather that a location experiences at a *particular* time depends upon which portion of the wave cyclone is over the location. Also, since the entire cyclonic system tends to travel from west to east, a specific *sequence* of weather can be expected at a given location as the cyclonic system, with its "mixed bag of weather," passes over that location.

Let us assume that it is late spring and a cyclonic storm has originated in the southeast corner of Nebraska and is following a track (refer again to Fig. 6.6) across northern Illinois, northern Indiana, northern Ohio, through Pennsylvania, and finally out over the Atlantic Ocean. A bird's-eye view of this storm, at one point in its journey, is presented in Figure 6.8a. A cross-sectional view *north* of the center of the cyclone is presented in Figure 6.8b, while a cross-sectional view *south* of the center of the cyclone is presented in Figure 6.8c. As the storm continues eastward, at 9–13 meters per second (20–30 mph), the sequence of weather will be different for Detroit, where the warm and cold fronts will pass just to the south, than for Pittsburgh, where both fronts will pass over. To illustrate this point, let us examine, element by element, the variation in weather that will occur in Pittsburgh, with reference, where appropriate, to the differences that occur in Detroit as the cyclonic system moves east.

Temperature and pressure are highly interrelated. As temperature increases, pressure decreases. Therefore, these two elements will be discussed together. Since a cyclonic storm is composed of two dissimilar air masses, there

(a)

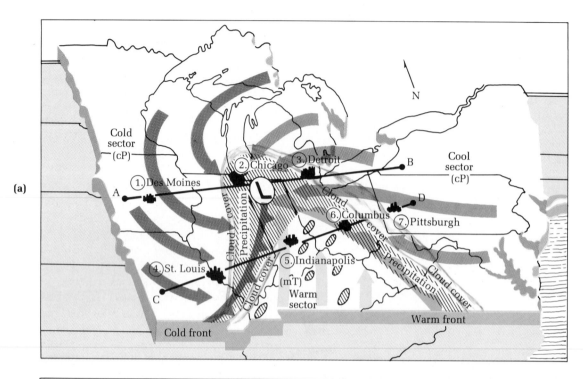

(b)

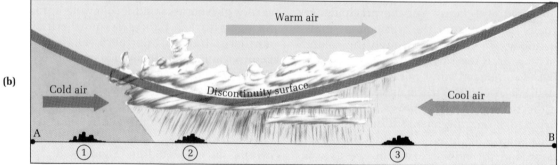

(c)

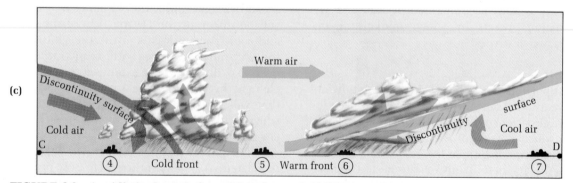

FIGURE 6.8   A midlatitude cyclone positioned over the Midwest on its movement to the east. (a) A bird's-eye view of the cyclonic system. (b) A cross section along line AB to the north of the center of low pressure. (c) A cross section along line CD to the south of the center of low pressure.

are usually important temperature contrasts. The sector of warm, humid *mT* air in the southern and southeastern part of the cyclone is usually considerably warmer than the cold *cP* air surrounding it. The temperature contrast is accentuated in the winter, when the source region for *cP* air is the cold cell of high pressure normally found in Canada at that time of year. During the summer the contrast between these air masses is greatly reduced.

As a consequence of the temperature difference, the atmospheric pressure in the warm sector is considerably lower than the atmospheric pressure in the cold air behind the cold front. Far in advance of the warm front the pressure is also high, but as the warm front (Fig. 6.3) approaches, increasingly more cold air is replaced by overriding warm air, thus steadily reducing the pressure.

Therefore, as the warm front of this late-spring cyclonic storm approaches Pittsburgh, the pressure will decrease. After the warm front passes through Pittsburgh (where the temperature may have been 8°C/46°F or more), the pressure will stop falling and the temperature could easily rise to 18° to 20°C (64° to 69°F) as *mT* air invades the area. Indianapolis has already experienced the passage of the warm front. After the cold front passes, the pressure will rise rapidly, while the temperature will drop. In this late-spring storm the *cP* air temperature behind the cold front might be 2°C to 5°C (35° to 40°F). Detroit, which is to the north of the center of the cyclone, will miss the warm air sector entirely and therefore will experience a slight increase in pressure and a temperature change from cool to cold, as the cyclone moves to the east.

Changes in wind direction are one signal of the approach and passing of a cyclonic storm. Since a cyclone is a center of low pressure, winds usually flow toward its center. Also, winds are caused by differences in pressure. Therefore, the winds associated with a cyclonic storm are stronger in winter when the pressure (and temperature) variations between air masses are greatest.

In our example, Pittsburgh is located to the south and east of the center of low pressure, ahead of the warm front, and is experiencing winds from the southeast. As the entire cyclonic system moves east, the winds in Pittsburgh will shift to the south-southwest after the warm front passes. Indianapolis is currently in this position. After the cold front has passed, the winds in Pittsburgh will be out of the north-northwest. St. Louis has recently experienced the passage of the cold front and currently has winds from the northwest. The changing direction of wind, clockwise around the compass, from east to southeast to south to southwest to west and northwest, is called a **veering wind shift** and indicates that the center of a low has passed to the north of your position. On the other hand, Detroit, which is also experiencing winds from the southeast, will undergo a completely different sequence of wind directional changes as the cyclonic storm moves eastward. Detroit's winds will shift to the northeast as the center of the storm passes to the south. Chicago has just undergone this shift. Finally, after the storm has passed, the winds blow from the northwest. Des Moines, to the west of the storm, currently has northwest winds. Such a change of wind direction, from east to northeast to north to northwest, is called a **backing wind shift,** as the wind "backs" counterclockwise around the compass. A backing wind shift indicates that you are north of the cyclone's center.

The type and intensity of precipitation and cloud cover also vary as a cyclonic disturbance moves through a location. In Pittsburgh, the first sign of the approaching warm front will be high cirrus clouds. As the warm front continues to approach, the clouds will thicken and lower. When the warm front is within 150–300 kilometers (90–180 mi) of Pittsburgh, light rain and drizzle will begin and stratus clouds will blanket the sky.

After the warm front has passed, precipitation will stop and the skies will clear. However, if the warm, moist *mT* air is unstable, convective shower activity may result.

As the cold front passes, warm air in its path is forced to move aloft rapidly. This may mean that there will be a cold, hard rain, but the band of precipitation normally will not be very wide because of the steep angle of the sur-face of discontinuity along a cold front. In our example, the cold front and the band of pre-cipitation has just passed St. Louis.

Thus, Pittsburgh can expect three zones of precipitation as the cyclonic system passes over its location: a broad area of cold showers and drizzle in front of the warm front; a zone within the moist, subtropical air from the south where there can be scattered convectional showers; and a narrow band of hard rainfall associated with the cold front (Fig. 6.8c). How-ever, locations to the north of the center of the cyclonic storm, such as Detroit, will usually ex-perience a single, broad band of light rains re-sulting from the lifting of warm air above cold air from the north (Fig. 6.8b). In winter, the precipitation is likely to be snow, especially in locations to the north of the center of the storm, where the humid *mT* air overlies ex-tremely cold northern air.

As you can see, different portions of a wave cyclone are accompanied by different weather. If we know where the cyclone will pass relative to our location, we can make a fairly accurate forecast of what our weather will be like as the storm moves east along its track.

CYCLONES AND THE UPPER AIR FLOW

The upper air wind flow greatly influences our surface weather. We have already discussed the role of these upper air winds in the steering of surface storm systems. Another less obvious in-fluence of the upper air flow is related to the undulating, wavelike flow so often exhibited by the upper air. As the air moves its way through these waves, it undergoes divergence or con-vergence because of the rather complex dy-namics associated with curved flow. This upper air convergence and divergence greatly influ-ences the surface storms below.

The region between a ridge and the next downwind trough (A–B in Fig. 6.9) is an area of upper level convergence. In our atmosphere an action taken in one part of the atmosphere is compensated for by an opposite reaction somewhere else. In this case the upper air con-

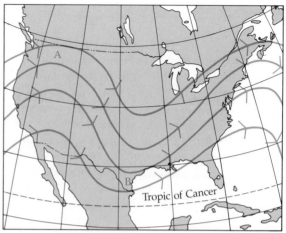

FIGURE 6.9 The upper air wind pattern, such as that depicted above, can have a significant influ-ence on temperatures and precipitation at the earth's surface.

vergence is compensated for by *divergence at the surface,* which will inhibit the formation of a midlatitude storm or cause an existing storm to weaken or even dissipate. On the other hand, the region between a trough and the next downwind ridge (B–C in Fig. 6.9) is an area of upper level divergence, which in turn is com-pensated for by surface convergence. Conver-gence at the surface will enhance the prospects of storm development or strengthen an already existing storm.

In addition to storm development or dissi-pation, upper air flow (Fig. 6.9) will have an impact on temperatures as well. If we assume that our "average" upper air flow is from west to east, then any deviation from that pattern will cause either colder air from the north or warmer air from the south to be advected into an area. Thus, after the atmosphere has been in a wavelike pattern for a few days, the areas in the vicinity of a trough (B in Fig. 6.9) will be colder than normal as polar air from higher latitudes is brought into that area. Just the op-posite occurs at locations near a ridge (C in Fig. 6.9). In this case, warmer air, from more south-erly latitudes than would be the case with west-to-east flow, is advected into the area near the ridge.

WEATHER FORECASTING Weather forecasting, at least in principle, is fairly straightforward. Meteorologic observations are made, collected, and mapped to depict the current state of the atmosphere. From this information, the probable movement, as well as any anticipated growth or decay, of the current weather systems is projected for a specific amount of time into the future.

When a forecast is wrong—which we all know occurs—it is usually because of either limited or erroneous information collected and processed in the first place, or more likely, because of errors in the anticipated path or growth of the storm systems. Little errors will compound themselves over time. For example, a few degrees' shift in a storm's path may result in an error of a few miles in the projected location of that storm in a two-hour forecast. But this same few-degree error may result in a projected locational error of hundreds of miles in a 48-hour forecast. Consequently, the further into the future one tries to forecast, the greater the chance of error.

While forecasts are not perfect, they are much better today than in the past. Much of this improvement can be attributed to our current sophisticated technology and equipment. Increased knowledge and surveillance of the upper atmosphere have improved the accuracy of weather prediction. Weather satellites have helped tremendously by providing meteorologists with a better understanding of weather and weather systems. They have been of particular value to forecasters on the west coast of the United States. Before the advent of weather satellites these forecasters had to rely on information relayed from ships, leaving enormous areas of the Pacific unobserved. Thus forecasters were often caught off guard by unexpected weather events.

In addition, high-speed computers allow for the rapid processing and mapping of observed weather conditions. Computers also allow for the processing of numerical forecasts, which are based upon the solution of physical equations that govern our atmosphere. Numerical forecasts, and long-term forecasts based upon statistical relationships, would not be possible without computers. In fact, computers now play such an important role in forecasting that some of the world's largest and fastest computers are used to forecast the weather.

Though forecasters now possess a great deal of knowledge and a variety of highly sophisticated devices that were previously unavailable, none of these devices are foolproof. Understanding some of the problems the weather forecaster faces may make us more understanding when a forecast fails. No one can promise a sunny day. Nor can anyone say that it will definitely rain tomorrow, for no one knows the future. The weather forecaster combines science and art, fact and interpretation, data and intuition, to come up with some probabilities about future weather conditions.

ANTICYCLONES Just as cyclones are centers of low pressure that are typified by the convergence of air, so anticyclones are cells of high pressure in which air diverges. The substance of air in the center of an anticyclone encourages stability, as the air is warmed adiabatically while sinking toward the surface. Consequently, the air is able to hold additional moisture as its capacity increases with increasing temperatures. The weather resulting from the influence of an anticyclone is often clear, with no rainfall. There are, however, certain conditions under which there can be some precipitation within a high-pressure system. When such a system passes near or crosses a large body of water, the resulting evaporation can cause variations in humidity significant enough to result in precipitation.

There are two sources for the relatively high pressures that are associated with anticyclones in the midlatitudes of North America. Some anticyclones move into the middle latitudes from northern Canada and the Arctic

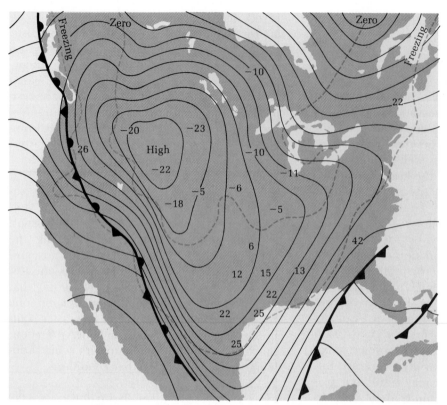

FIGURE 6.10 Weather map showing a massive outbreak of cold polar air during the winter, which produces anticyclonic conditions over most of the United States, with a stationary front to the west and a cold front still advancing southeastward into Florida and the Caribbean Sea. The Arctic source region of the air, combined with clear skies permitting surface cooling by radiation, dropped temperatures 10° to 20°F below zero from Montana southward to New Mexico and eastward to Illinois. (Adapted from *Weatherwise*, June, 1962)

Ocean, in what are called outbreaks of cold polar air (Fig. 6.10). These outbreaks can be quite extensive, covering much of the midwestern and eastern United States. The temperatures in an anticyclone that has developed in a *cA* air mass can be markedly lower than those expected for any given time of year. They may be far below freezing in the winter and are often associated with the first frost in the fall and the last frost in spring.

Other anticyclones are generated in zones of high pressure in the subtropics. When they move across the United States toward the north and northeast, they bring waves of hot, clear weather in summer and unseasonably warm days in the winter months.

The actual movement of anticyclones and cyclones across the United States during the fall can serve as a useful illustration of the relationships between these systems of high and low pressure. Such movement can be seen in the four consecutive weather maps of Figure 6.11. Note the alternation of cyclones and anticyclones and their steady movement eastward from the Pacific Northwest to the central plains to the eastern seaboard. Precipitation in the form of rain and snow (in the mountains) is associated with the fronts and cyclonic centers; there is clear weather in those portions of the United States dominated by the anticyclones. Winds from the alternating highs carry air in the general direction of the intervening lows.

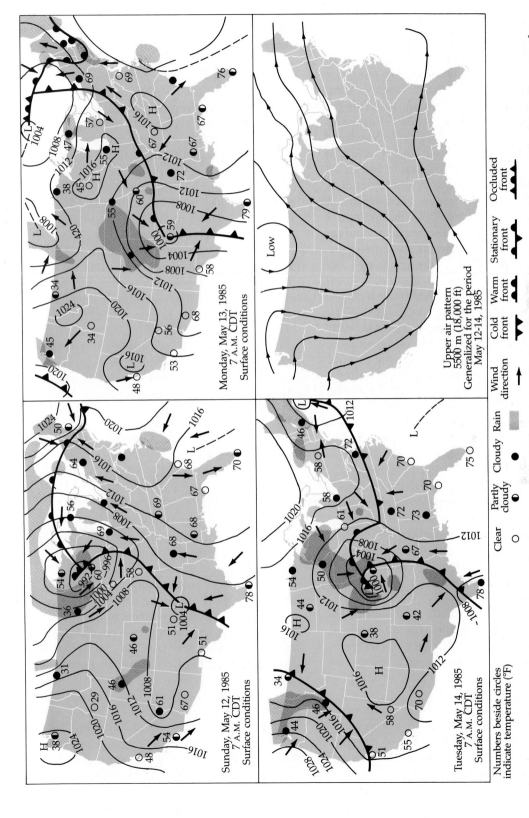

FIGURE 6.11 Sequence of daily weather maps, and prevailing upper air flow, illustrating movements of weather systems across the United States.

THUNDERSTORMS

Thunderstorms are common local storms of the middle and lower latitudes. Very simply, a **thunderstorm** is a convectional storm accompanied by thunder and lightning. **Lightning** is an intense discharge of electricity. For lightning to occur, positive and negative electrical charges must be generated within the cloud. It is believed that the intense friction of the air on moving ice particles within a cumulonimbus cloud generates these charges. A clustering of positive charges tends to occur in the upper portion of the cloud, with negative charges clustering in the lower portion. When the potential difference between these charges becomes large enough to overcome the natural insulating effect of the air, a lightning flash, or discharge, takes place. These discharges, which often involve over a million volts, can occur within the cloud, between two clouds, or from cloud to ground. The air immediately around the discharge is momentarily heated to temperatures in excess of 25,000°C (45,000°F)! The heated air expands explosively, creating the shock wave we call **thunder.**

Thunderstorms usually cover a small area of a few miles, although there may be a series of related thunderstorms covering a larger region. The intensity of a thunderstorm depends upon the degree of instability of the air and the amount of water vapor it holds. A thunderstorm will die out when most of its water vapor has condensed, for there will no longer be energy available for continued vertical movement. In fact, most thunderstorms last less than an hour.

As an intense form of convectional precipitation, thunderstorms result from the uplift of moist air. As is the case for other convectional precipitation, the trigger actions causing that uplift can be thermal convection, orographic lifting, or frontal lifting.

Thunderstorms are most common in lower latitudes, during the warmer months of the year, and during the warmer hours of the day. It is apparent, then, that the amount of solar heating affects the development of thunderstorms. This is true because the intense heating of the surface steepens the environmental lapse rate. This in turn leads to increased instability of the air, allowing for greater moisture-holding capacity and adding to the buoyancy of the air.

Orographic thunderstorms occur when air is forced to rise over land barriers, providing the necessary initial trigger action leading to the development of convectional cells. Thunderstorms of such orographic origin play a large role in the tremendous precipitation of the monsoons of Southeast Asia. In North America, they occur over the mountains in the West (the Rockies and the Sierra Nevada), especially during summer afternoons when heating of the land surfaces increases the air's instability. For this reason, pilots of small planes try to avoid flying in the mountains during the afternoon in summer for fear of getting caught in the turbulence of a thunderstorm.

Thunderstorms are often associated with cold fronts where a cooler air mass forces a warmer air mass to rise. This action can bring about the strong, vertical updrafts necessary for convectional precipitation. In fact, a cold front may be preceded by a line of thunderstorms, part of a **squall line** resulting from such frontal lifting.

As we mentioned in the discussion of precipitation types in Chapter 5, hail can be a product of thunderstorms when the vertical updrafts of the convection cells are sufficiently intense to carry water droplets repeatedly into a freezing layer of air. Fortunately, since thunderstorms are primarily associated with warm-weather areas, only a very small percentage around the world produce hail. In fact, hail seldom occurs in thunderstorms in the lower latitudes. In the United States, there is relatively little hail along the Gulf of Mexico, where thunderstorms themselves are most common.

TORNADOES

Tornadoes are the smallest and most intense storms of the atmosphere. They can occur almost anywhere but are far more common in the interior of North America than anywhere else in the world (Fig. 6.12). In fact, Oklahoma and Kansas lie in the path of so many "twisters" that they are sometimes referred to as tornado alley.

A **tornado** is actually a small, intense cyclonic storm of very low pressure, violent updrafts, and converging winds of enormous contrast. Until fairly recently the wind speed within the tornado vortex could only be estimated since conventional instrumentation in the path of a tornado was literally blown away by the strong winds. Remotely sensed observations, primarily with Doppler radar, now allow

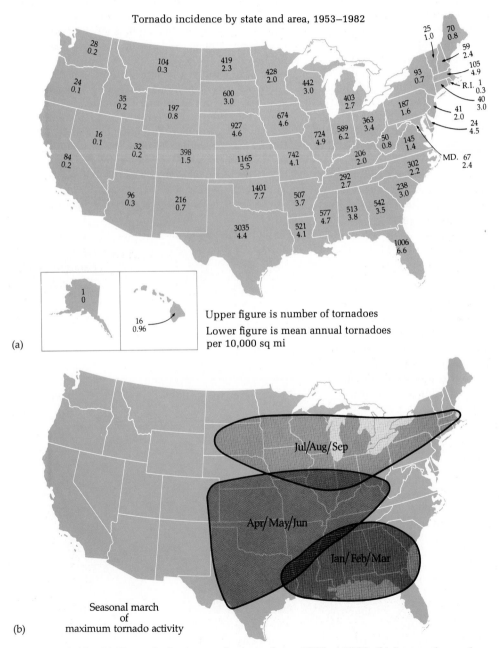

(a) Tornado incidence by state and area, 1953–1982

Upper figure is number of tornadoes
Lower figure is mean annual tornadoes per 10,000 sq mi

(b) Seasonal march of maximum tornado activity

FIGURE 6.12 **(a) Tornado frequency by state from 1953 to 1982. (b) Seasonal march of peak tornado activity.**

VIEWPOINT: THE TWISTER by Walter Sullivan

The following article appeared in The New York Times *in 1965. Since that time tornado detection and warning systems have been improved. In addition, many communities like Macomb, Illinois, have shifted their civil defense efforts almost completely from an air raid emphasis to a tornado watch. Though these improvements have helped to reduce loss of life and injury, there is still a great deal to be learned about tornadoes. Perhaps someday scientists will be able to divert tornadoes into uninhabited areas or into ways of dissipating their energy safely.*

Since midday, radio and television announcers had reported a danger of tornadoes in various Midwest areas. Along Route 32, east of Shannondale, Ind., heavy rain had been falling on the fields of corn stubble, but the family of Wayne Rose felt snug in his spacious farmhouse. The great trees on both sides of the driveway swished and groaned in the wind.

Toward evening it became very dark. Farmer Rose stepped out and found the air oppressively warm and humid. Suddenly he noticed that the wind had died. "It was as quiet as a morgue," he said later. "Then I heard an express train—no, 10 express trains—and there it came, not an upright funnel, but lying almost on its side as it snaked across the fields.

"I shoved the family into the Buick, but it outran us." For a few moments there was a chaos of noise and bombardment. Then it was over. House, barn, silos, outbuildings all were gone. A heavy washing machine, torn from its plumbing in the heart of the house, had been dropped in the middle of a debris-strewn field. Distant barbed wire fences had caught upright within the trees, each a nightmare of twisted trunks and stripped branches.

From the other houses along Route 32 came groans and screams, but the Rose family was lucky. Their car, battered and dripping mud, had saved them. Only a few homes along the road had storm cellars, for tornadoes are not common that far north, "but you can bet everyone around here is going to build one now," Mr. Rose said.

Over and over, last Sunday, deadly funnels dropped from the black sky, devastating a patch of Chicago suburb, an Indiana trailer park, a community near Toledo, Ohio. Close to 2,500 were injured and at least 237 killed. Not since 1925 had a series of tornadoes taken such a toll. The casualties were high in part because the storms swept heavily settled sections instead of the more sparsely inhabited region, centered on Oklahoma, where tornadoes are common and the residents better prepared to take shelter. Most important of all: there is at present no effective way to warn those in the immediate path of a tornado. The best that can be done is to announce that the occurrence of such storms is likely in a certain area.

In fact, little is known about what initiates the wild vortex of a tornado. A hurricane is a spiralling wind system hundreds of miles wide, and it can be explored by aircraft and ground observations. A tornado is so compact, violent and short-lived that it defies analysis. Its lifetime may be measured in minutes, even though the squall line of which it is the child endures many hours, marching across the land and spawning a succession of tornadoes.

While there were reports of 37 tornadoes last Sunday, the number has little meaning. A single funnel may drop repeatedly to the ground, or it may cut a swath close to 100 miles in length.

Tornadoes seem to have some points of similarity with the whirlpools that appear along the front where two moving bodies of water meet one another.

The encounter that produces a tornado-breeding squall line is typically between moist, warm air from the Gulf of Mexico and cold air that has crossed the Rockies, from the West, and been drained of moisture. In such an encounter the cold air, like a wedge moves in under the warm air, which rises. The warm air then cools and this squeezes out its moisture in heavy rain.

The tremendous force of a tornado is able to level entire city blocks. (NOAA)

The energy originally required to evaporate the rain reappears as latent heat when the raindrops form and this further warms the humid air, making for greater turbulence, more upward motion and more rain—a sort of wet chain reaction.

This effect is typical of cold fronts the world over, but there is something unusual about these phenomena when they occur over the Midwest in spring. Tornadoes are then a peculiar feature of this region. They occur nowhere else in the world with such frequency, although they have been reported from as placid places as the British Isles.

Dr. Edwin Kessler, head of the Weather Bureau's storm laboratory at Norman, Okla., has suggested that in spring the Gulf air may be cool enough to hug the ground as it moves inland. In that season the strong winter westerlies of the upper air, whose core is known as the jet stream, still retain considerable strength. Finally, the smooth terrain offers little resistance to the surface winds.

These factors conspire to produce violence, be it in thunderstorms, hail or tornadoes. The latent heat released when the rain begins to pour down is enormous. A two-inch downpour on a region measuring 10 by 10 miles, according to Dr. Kessler, releases energy equivalent to 350 atomic bombs of the Hiroshima variety. Only about one per cent of this goes into air motion, but that is still a great deal. . . .

The eastward advance of the squall line from the Mississippi Valley was monitored almost to the East Coast minute by minute by giant Weather Bureau radar stations. There are more than 30 of these from coast to coast, each with a range of about 250 miles. Their observations are compiled and plotted on a map of the country at RADU—the Radar Analysis and Detection Unit at the weather station in Kansas City. Every hour or two RADU transmits facsimiles of this map to some 200 weather stations throughout the country.

Also in Kansas City is SELS, the Severe Local Storm Center, which examines the reports and issues warnings.

The chief difficulty is that a radar operator can only rarely identify a tornado from the pattern on his scope. The squall line stands out sharply because of its heavy rains, hail and dense clouds. Once in a while there is, within it, a suggestion of spiral structure—a hook-like form that is the earmark of a tornado—but tornadoes are often indistinguishable from heavy thunderstorms.

At the Oklahoma laboratory efforts are being made to develop radar systems that will overcome this weakness. One exploits the so-called Doppler effect. If a musical note is echoed off a stationary wall, the pitch of the echo is identical to that of the outgoing sound. However, if the wall were moving toward the source, each reflected sound wave would be shortened by this motion and hence the pitch of the echo would be raised. If the wall were receding, the pitch would be lowered. The extent of the change in pitch could be used to assess the speed of this motion.

Doppler radar works on the same principle. It is hoped that it will be able to detect the violent motions of dust, raindrops, and debris within a tornado vortex—movements that probably reach many hundreds of miles an hour.

Another line of research seeks to devise a system for computer analysis of radar data, plotting target densities from the echo strength with an accuracy that cannot be displayed on an ordinary radar scope. Some believe the ultimate solution will be to combine these various tools into an automated system that will spot incipient tornadoes, several at a time, and warn each community in their probable paths.

Tornado damage in southern Illinois. Note the water tower standing in the background—evidence of the narrow path followed by the storm. (R. Gabler)

SOURCE: *The New York Times*, April 18, 1965, p. 12E. © 1965 by The New York Times Company. Reprinted by permission. The author was a science writer for *The New York Times*.

us to observe the windfield of a tornado. Most tornadoes (62 percent) are fairly *weak* and have wind speeds of 45 meters per second (100 mph) or less. About 35 percent of tornadoes can be classified as *strong,* with wind speeds reaching 90 meters per second (200 mph). Nearly 70 percent of all tornado fatalities result from *violent* tornadoes. Although very rare (only 2 percent of all tornadoes reach this stage), they may last for hours and have wind speeds approaching 135 meters per second (300 mph).

A tornado appears as a swirling, twisting funnel that moves across the landscape at 10 to 15 meters per second (22 to 32 mph). Its narrow end may be only 100 meters (330 ft) across. When this narrow end is in contact with the ground, the greatest damage is done. Above the ground, the end can swirl and twist, but little or nothing is done to the ground below. The color of a tornado can be milky white to black depending upon the amount and direction of sunlight and the type of debris being picked up by the storm as it travels across the land. While most tornado damage is caused by the violent winds, most tornado injuries and deaths result from flying debris.

Tornadoes often occur as a part of the turbulence and eddies that are part of a larger cyclonic system. Most often, they occur within the squall line that precedes the arrival of a rapidly advancing cold front, or within the cold front itself, and they are usually associated with the cumulonimbus clouds so characteristic of squall lines and cold fronts.

Tornadoes are so small and usually so short-lived that they are virtually impossible to forecast except in the most general sense—in the sense that conditions are right in an area for *possible* tornado development. However, *Doppler radar,* with its ability to observe wind speeds as well as rainfall intensity, greatly assists in the detection and tracking of a tornado once it does form. The National Weather Service hopes to have a Doppler Radar network across the United States within a decade. Such a network would result in more accurate and faster warning, which in turn will certainly help reduce tornado fatalities.

WEAK TROPICAL DISTURBANCES

Until World War II, the weather of the tropical regions had been described as hot and humid, generally fair, but basically pretty monotonous. The only tropical disturbance given any attention was the tropical cyclone (also called a *hurricane* or *typhoon*), a spectacular though relatively uncommon storm that affects only islands, coastal lands, and ships at sea.

Even though more attention has been focused recently upon tropical weather, an aura of mystery remains about the weather of this region. One reason for this lack of information is that the few weather stations in the tropical areas are widely scattered and often poorly equipped. As a result, it is difficult to understand completely the passing weather disturbances in the tropics.

After some examination of the weather in the tropics, it is now agreed that there are a variety of weak atmospheric disturbances that affect the weather and relieve the monotony, although it is likely that the full number of these disturbances has not yet been recognized. The primary result of these weak tropical disturbances upon the weather of the tropical region is not on the temperature but rather on the cloud cover and on the amount of precipitation. Temperatures in the tropics are largely unaffected during the passage of a tropical storm, except that as the cloud cover is increased, temperature extremes are consequently reduced.

EASTERLY WAVE

The **easterly wave** is the best known of the weak tropical disturbances. It shows up in Figure 6.13 as a trough-shaped, weak, low-pressure cell that is generally aligned on an approximate north-south axis. Traveling slowly in the trade-wind belt from east to west, it is preceded by fair, dry weather and followed by cloudy, showery weather. This occurs because air tends to converge into the low from its rear, or the east, causing lifting and convectional showers. The resulting divergence and subsidence to the west accounts for the fair weather. It is possible, although not usual, for this type of disturbance to intensify into a tropical hurricane.

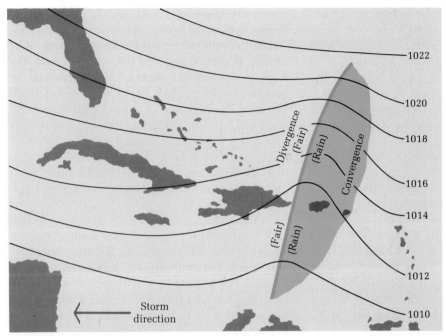

FIGURE 6.13 **A typical easterly wave in the tropics. Note that the isobars (and resulting winds) do not close in a circle but merely make a poleward "kink," indicating a low-pressure trough rather than a closed cell. The resulting weather is a consequence of convergence of air coming into the trough, producing rain, and divergence of air coming out of the trough, producing clear skies. (After Riehl)**

POLAR OUTBREAK   Occasionally, there may be an outbreak of polar air that follows a low into the subtropics and tropics. Such an outbreak would, of course, be preceded by the squalls and clouds and rain of a cold front. Following, however, would be a period of cool, clear, fair weather as the modified polar air influences are felt. On rare occasions near the equator in the Brazilian Amazon, such an Antarctic outburst, known locally as a *friagem*, can bring freezing temperatures and widespread damage to vegetation. Farther to the south near São Paulo, the coffee crop can be ruined, causing coffee prices in North America to rise.

HURRICANES

Hurricanes, or typhoons, are severe tropical cyclones, which, though not nearly so frequent as midlatitude cyclones, receive a great deal of attention from scientists and lay people alike, pri-marily because of their awesome intensity and strength and their great destructive powers. Abundant, even torrential rains and winds of great speed, often exceeding 45 meters per second (100 mph), characterize hurricanes. Though these storms develop on the western side of oceans and often spend their entire lives there, at times their paths do take them over islands and coastal lands. The results can be devastating destruction of property and sometimes of life. It is not just the rains and winds, however, that can produce such damage to people and their surroundings, for accompanying the hurricane are unusually high seas, called **storm surges,** which can flood entire coastal communities.

A **hurricane** is a circular, cyclonic system with wind speeds of at least 33.5 meters per second (75 mph). It has a diameter anywhere from 160 to 640 kilometers (100 to 400 mi). Extending upward to heights of 12 to 14 kilometers (40,000 to 45,000 ft), the hurricane is a great, towering column of spiraling air (Fig. 6.14). At its base, air is sucked in by the very low pressure at the center and then spirals inward. Once within the hurricane structure, air rises rapidly to the top and spirals outward. It is this rapid upward movement of great quantities of moisture-laden air that produces the enormous amounts of rain during a hurricane. Furthermore, the resulting latent heat of condensation that is released provides the power to drive the storm.

At the center of the hurricane is the eye of the storm, an area of calm, clear, usually warm and humid, but rainless air. People on ships who have traveled through a hurricane and looked up into the eye have been surprised to see great numbers of birds flying there. Caught in the eye and unable to escape because of the great winds whirling around the center, these birds will often alight on the passing ship as a resting spot.

Hurricanes have very strong pressure gradients with isobars that decrease in amount toward a center of unusually low pressure. The strong pressure gradients are the reason behind the powerful winds of the hurricane. In contrast to the midlatitude cyclone, this tropical storm is formed out of a single air mass and does not have the different temperature sectors that are a part of the frontal system. Rather, a hurricane has a fairly even, circular temperature distribution, which we might have been led to expect from its circular isobars and spiraling winds.

Although a great deal of time and effort, to say nothing of money, has been spent on studying the development, growth, maturity, and paths of hurricanes, much is still not known. For example, it is not possible to predict the path of a hurricane, even though it can be tracked with radar and studied through the use of planes and weather satellites. In addi-

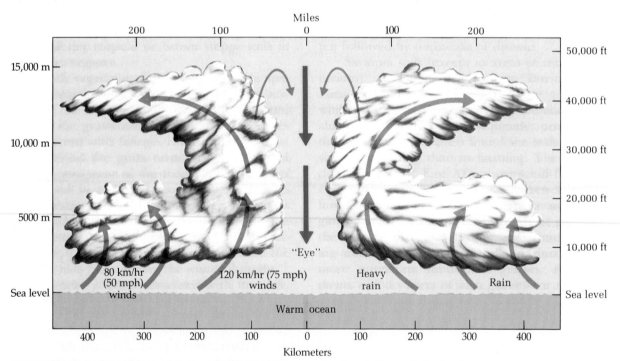

FIGURE 6.14   Circulation pattern within a hurricane, showing inflow of air in the spiraling arms of the cyclonic system, rising air in the towering circular wall cloud, and outflow in the upper atmosphere. Subsidence of air in the storm's center produces the distinctive calm, cloudless "eye" of the hurricane.

tion, meteorologists can list factors that are favorable for the development of a hurricane, but they cannot say that in a certain situation a hurricane will definitely develop and travel along a particular path.

Among the factors that can lead to hurricane development is a warm ocean surface of about 25°C (77°F). Along with this, there must be warm, moist, overlying air. These preconditions are probably the reasons hurricanes occur most often in the late summer and early fall, when the air masses have their maximum humidity. Also, the Coriolis effect must be sufficient to support the rapid spiraling of the hurricane. Because of this, hurricanes neither develop nor survive in the equatorial zone from about 8°S to 8°N, for there the Coriolis effect is far too weak. Most meteorologists believe that hurricanes are born out of weak tropical disturbances such as the easterly wave and, in fact, will not develop without the impetus of such a disturbance. It is further speculated that some sort of turbulence in the upper air also plays a part in a hurricane's initial development.

Hurricanes do not last long over land, for their source of moisture (and consequently of energy) is cut off. Also, friction with the land surface produces a drag on the whole system.

North Atlantic hurricanes that move first toward the west with the trades and then north and northeast as they intrude into the westerlies become polar cyclones if they remain over the ocean and eventually die out. Over the land, they will also become simple cyclonic storms, but even when they have lost some of their hurricane power, they can still do great damage.

Hurricanes occur over subtropical and tropical oceans and seas (the South Atlantic is an exception, though it is not known why). In Australia and the South Pacific these violent storms are called *tropical cyclones*. Near the Philippines, they are known as *baguios*, but in most of East Asia they are called *typhoons*. In the Bay of Bengal, east of India and Bangladesh, where over 200,000 people were killed by the flooding resulting from one in 1971, they are referred to as *cyclones*.

It has been proposed by some people that we seek ways to control hurricanes so they are not so destructive, but it has been pointed out by others that hurricanes are a major source of rainfall and an important means of transferring energy within the earth system away from the tropics. Eliminating them might cause unwanted and unforeseen climate changes.

QUESTIONS FOR DISCUSSION AND REVIEW

1. What is an air mass?

2. Do all areas on earth produce air masses? Why or why not?

3. Use Table 6.1 to find out what kinds of air masses are most likely to affect your local area. How do you suppose they affect weather in your area?

4. What forces modify the behavior of air masses? What kinds of weather may be produced when an air mass begins to move?

5. Why do you suppose air masses can be classified by whether they develop over water or over land?

6. Why does *mP* air affect the United States? Are there any deviations from this tendency?

7. What kind of air forms over the southwestern United States in summer? Have you ever experienced weather in such an air mass? What was it like? What kind of weather might you expect to experience if such an air mass met an *mP* air mass?

8. What is a front? How does it occur?

9. In a meeting of two contrasting air masses, how can the aggressor be determined?

10. Compare warm and cold fronts. How do they differ in duration and precipitation characteristics?

11. What kind of weather often results from a stationary front? What kinds of forces do you suppose tend to break up stationary fronts?

12. How does the westerly circulation of winds affect air masses in your area? What kinds of weather result?

13. What are the major differences between middle-latitude cyclones and anticyclones?

14. If a wind changes clockwise, what is the shift called? Where does it locate you in relation to the center of a low-pressure system? Explain why this happens.

15. Describe the sequence of weather events over a 48-hour period if a typical low-pressure system (cyclone) passed 300 kilometers (180 mi) north of your location in the spring.

16. List three major causes of thunderstorms. How might the storms that develop from each of these causes differ?

17. Have you ever experienced a tornado or hurricane? Describe your feelings during it and the events surrounding it. How do the news media prepare us for such natural disasters?

7

CLIMATE:
Low-Latitude and
Arid Regions

THE SPECIAL ROLE OF CLIMATE

A persistent theme thus far in our study of physical geography has been that all parts of the total earth system are interrelated. This interrelationship is well illustrated by the relationship of climate to the earth system. Stop and think for a moment of all the ways in which temperature, precipitation, wind, and the other elements of climate affect your life.

In Chapter 3, weather and climate were defined briefly. It seems advantageous at this point to discuss climate in greater detail. Whereas weather is the state or condition of the atmosphere at a given place and time, climate provides a description of the atmospheric conditions that are likely at a given time of year in a given location, or, as a child once stated, "Climate is what you should have, weather is what you get." Climatic description includes such things as the average annual precipitation and temperature; the patterns of wind direction, temperature, and precipitation as those elements change throughout the year; the anticipated number of days of sunshine or cloud cover; and the ranges between the statistical highs and lows that may be expected of each climatic element.

Devised by scientists in an attempt to predict the weather, climate is a concept based upon averages and statistical probabilities. The concept of climate is a useful tool. The state of the atmosphere is so much a part of our environment and the elements of weather influence so many parts of the earth system that it is helpful to be able to describe the average or typical condition of the atmosphere for a particular region. Scientists have found it useful to devise systems of climate classification by which they can group locations with similar climates and differentiate those locations from others with significantly different climatic conditions.

In studying climate, it has become apparent to climatologists that certain plants, soils, and even animals are associated with each climatic type. For example, what type of climate do you associate with palm trees? polar bears? sand dunes? These associations are not surprising since climate is a composite of variations in the weather elements over an extended period

---

Tropical Savanna, Africa (Donald Marshall collection, WIU International Programs)

of time, and such elements as insolation, temperature, moisture, and winds greatly influence the vegetation, soils, and living conditions in a region. Therefore, we can expect certain climates to be more acceptable to certain varieties of plants, animals, and soils.

MICROCLIMATES

Within a climatic region it is possible to find many small local variations in which the atmospheric elements differ significantly. Some of these variations may be caused by local landform configurations or human-related changes in the environment. Just as the global wind belts vary because of the different sizes and shapes of continents and oceans, so can local variations in terrain create local wind variations that in turn may cause differences in the other weather elements. These local wind variations can affect the weather of a particular locale consistently enough so that a **microclimate** (from Greek: *micro,* small) develops. Thus, land on the coast will have a different microcli-

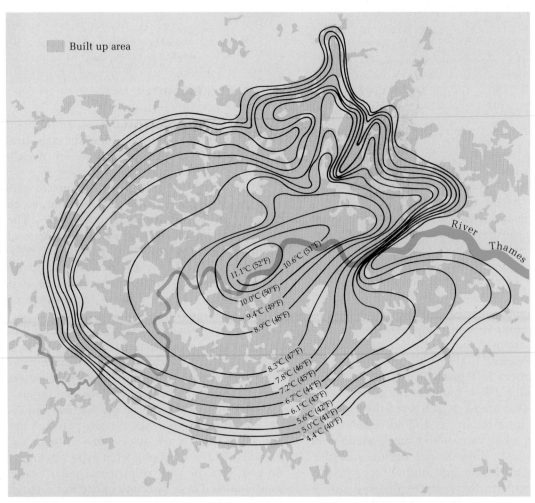

FIGURE 7.1 Urban heat island effect around London, England. Most cities are significantly warmer than surrounding vegetation-covered areas—a phenomenon that is noticeable to the person in the street as well as to heat-sensing instruments. (After Chandler)

mate from land 20 miles inland. The windward side of a mountain usually has a wetter and cooler microclimate than that of the leeward side of the mountain (refer back to Fig. 5.14).

Microclimates can vary greatly in scale (from an acre of farmland to the coast of a continent) and type (e.g., microclimate of plants, animals, forests, and valleys). One example of a microclimate is that associated with large cities. Urban centers tend to be warmer than their outlying rural areas and are sometimes referred to as **heat islands** (Fig. 7.1). Climatologists are attempting to describe the characteristics and influences of the urban microclimate (see Viewpoint, p. 270). Farmers are interested in the microclimate of their acreage in order to determine the crops for which their land is best suited. Even insects respond to microclimates: to the warm, dark, wet soil under a log or to the different microclimates of the top and underside of a leaf.

THE DELICATE BALANCE

The integration and balance of all parts of the earth system are such that a change in one part will affect other parts as well as the system as a whole. This is the basis for the fears of environmentalists and ecologists about our water supply, the quality of our atmosphere, and plant and animal life. These people claim, not without justification, that if we interfere too much with one subsystem of the earth and disturb the delicate balances that presently exist, we may cause irreparable damage to our planet.

The role of our atmosphere relative to our supply of oxygen, fresh water, and solar energy is vital for life and therefore cannot be ignored. The dynamics of the atmosphere are also important to the quality of life that has developed on earth. Life in general depends upon the dynamics of weather and climate and on the variations in these phenomena over space and time.

Over time, short-term variations in the weather elements often balance each other out. For example, a cold winter may be followed by a warm spring, or dry years may be followed by

wet years. Since climate is based upon the averages of the weather elements over a period of years (20 to 30 years for the United States), these short-term variations in weather tend to cancel each other out. As a result, climate does not show the marked variations that weather does. The fact that we are able to describe climates, to classify them and talk about their associated vegetation and soils, and that this information remains useful over the years shows us that there is a continuing thread of consistency throughout the ever-changing dynamics of the atmosphere. This consistency is based on the relatively stable earth-sun relationships, on the fixed composition of the atmosphere, and on the overall balance of the heat energy cycle and the hydrologic cycle. This consistency allows geographers to develop a system of climatic classification and permits an orderly and reliable description of the world's climates.

SCIENTIFIC CLASSIFICATION

Scientists create order out of complexity by organizing similar entities together into groups that are distinguished by particular characteristics that they share to the exclusion of other groups. Scientists refer to this type of organization as **classification.** It is very much like the organization of a grocery store, which separates canned goods, produce, beverages, and meats, but puts all brands of coffees together and likewise associates varieties of rice and noodles. Classification brings order out of apparent chaos while allowing us to analyze difficult subjects. Through classification we can look at a few groups that share important similarities instead of an infinite number of individual instances.

GEOGRAPHIC CLASSIFICATION

Like other scientists, the geographer's chief purpose in classifying is to organize a diversity of facts. Because he studies his facts on the basis of their location or distribution, the geographer is most interested in the likenesses and differences among places and in phenomena as they occur in spatial patterns. His classification must be based on what he considers to be the most *geographically* significant variables. Frequently his problem is how to (1) identify, (2) separate, and (3) characterize regions that are unlike in rather subtle ways. Such regions are rarely clear-cut and generally have gradational boundaries.

Geographers deal with areas of all sizes. Thus their classification schemes vary enormously in scale and degree of sophistication. However, the type of classification encountered in an introduction to physical geography must apply to the world as a whole, since that is the scope of our interest. This can readily be seen by the classifications used on the maps throughout the book. Each map has a limited number of recurring classes. Each classification type is easily distinguished from the other types. But because each map is on a worldwide scale, with phenomena presented in a broad, general, and integrated way, there will always be local exceptions to these broad generalizations.

THE CLASSIFICATION OF CLIMATE

As we have seen in earlier chapters, the climatic elements vary greatly over the earth. But we can reduce the infinite number of worldwide climatic variations to a comprehensible number of groups or varieties by combining elements with *similar* statistics. That is, we can classify climate strictly on the basis of the similarity of weather elements, ignoring the causes of those variations (e.g., air-mass analysis). Such a classification, based on statistical parameters or physical characteristics, is called an **empirical classification.** On the other hand, if we based our classification on the causes or the genesis of climatic variation, we would have a **genetic classification.**

Ordering the vast wealth of available climatic data into descriptions of major climatic groups, either on an empirical or on a genetic basis, enables us to concentrate on the larger-scale causes for climatic differentiation. In ad-

dition, we are also able to examine the exceptions to the general relationships, the causes of which are often one or more of the minor climatic controls. And, finally, differentiating climates helps us explain the distribution of other climate-related natural phenomena of importance to man.

Despite its value, climatic classification is not without its problems. One reason for this dilemma is that climate is a generalization about observed facts based on the averages and probabilities of weather. Therefore, climate does not describe a real weather situation; instead, it presents a composite weather picture. In addition, it is impossible within such a generalization to include the many variations that actually exist. On a global scale, generalizations, simplifications, and compromises are made to distinguish between climatic regions.

On a map of climatic regions, distinct lines separate one region from another (Fig. 7.2). Obviously the lines on our map that separate the climatic regions do not mark the point where there are abrupt changes in temperature or precipitation conditions. Rather, the lines signify **zones of transition** between different climatic regions. Furthermore, these zones or boundaries between climatic regions are based on monthly and annual averages and shift as temperature and moisture statistics change over the years.

The actual transition from one climatic region to another is gradual except in cases where the change is brought on by an unusual climatic control such as a mountain barrier. It would actually be more accurate to depict climatic regions and their zones of transition on a map by showing one color fading into another.

THE KÖPPEN SYSTEM

In the early twentieth century Wladimir Köppen, a German botanist and climatologist, developed a system of climatic classification that today remains the most widely used. Though it is an empirical classification based on average temperature and precipitation sta-

FIGURE 7.2 This map shows the diversity of climates possible in a relatively small area, including portions of Chile, Argentina, Uruguay, and Brazil. The climates range from dry to wet and from hot to cold, with many possible combinations of temperature and moisture characteristics.

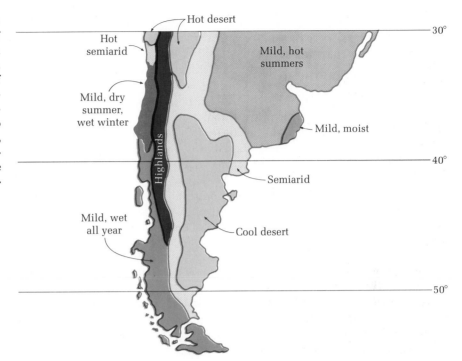

tistics, Köppen's climatic regions were formulated to coincide with well-defined vegetation regions, so it is recognized as a descriptive system. Evidence of the strong influence of Köppen's system is seen by the wide usage of his climatic terminology, even in nonscientific literature (e.g., steppe climate, tundra climate, rain-forest climate).

ADVANTAGES AND LIMITATIONS OF KÖPPEN'S SYSTEM

Temperature and precipitation are not only two of the easiest weather elements to measure, they are also measured more often and in more parts of the world than any other variables. By using temperature and precipitation statistics to define his boundaries, Köppen was able to develop precise definitions for each climatic region, eliminating the imprecision that can develop in verbal and sometimes in genetic classifications.

Moreover, temperature and precipitation are the most important and effective weather elements. Variations caused by the climatic controls will show up most obviously in temperature and precipitation statistics. On the other hand, temperature and precipitation are the weather elements that most directly affect humans, other animals, vegetation, soils, and the form of the landscape. By tying his classification to these two elements, Köppen devised a system that is very much related to the visible aspects of our environment.

Furthermore, by relating temperature and precipitation in the descriptions of his climatic regimes, Köppen included the concept of **effective precipitation,** which takes into account the temperature-controlled potential evapotranspiration rate. This further explains the strong association of climate and vegetation with Köppen's system.

Köppen's climatic boundaries were designed to define the vegetation regions in a classification of world vegetation produced earlier by the Swiss botanist, Alphonse de Candolle. Thus, Köppen's climatic boundaries are "vegetation lines." For example, the 10°C (50°F) monthly isotherm is used by Köppen because it has relevance to the timberline, the line beyond which it is too cold for trees to thrive. Trees need a certain amount of warmth to grow, and unless at least one month has an average temperature exceeding 10°C, trees will ordinarily be unable to survive. For this reason, Köppen defined the treeless polar climates as including those areas where the mean temperature of the warmest month is below 10°C. Clearly, if climates are divided according to associated vegetation types, and if the division is based on the weather elements of temperature and precipitation, the result will be a visible association of vegetation with climatic types. The relationship with the visible world in Köppen's climate classification system is one of its most appealing features to geographers.

There are, of course, limitations to Köppen's system. For example, Köppen considered only average monthly temperature and precipitation in making his climatic differentiations. Monthly temperature and precipitation permit estimates of precipitation effectiveness but do not measure it with enough precision to permit comparison from one locality to another. In addition, for the purposes of generalization and simplification, Köppen ignored winds, cloud cover, intensity of precipitation, humidity, and daily temperature extremes—much, in fact, of what makes local weather and climate distinctive.

Finally, an inherent limitation of Köppen's system is that it is empirical and therefore deals with facts or observations, not with causes. However, genetic classification may not be as stable as empirical ones because our knowledge of causes is apt to change. For example, any climatic classification based on causes known in 1900 would be of limited usefulness today. The empiricism that seems to be one of the limitations of Köppen's climatic classification is therefore at the same time one of its strong points.

SIMPLIFIED KÖPPEN CLASSIFICATION

In this text we will present and use a version of Köppen's climate classification, somewhat modified and simplified. Köppen gave names to his climatic regions. He also used combinations of letters to symbolize them. The letter symbols provide an international shorthand describing climatic regions that are often difficult to characterize in words. In the following discussion we will refer to the various climatic regions by name, including their letter symbols in parentheses. At first glance the classification scheme appears complex; however, close examination of Table 7.1 will reveal that the scheme is based upon a few critical values with which all geography students soon become familiar. In some cases we have combined certain subdivisions of climatic regions described by Köppen and included them within a broader zone. This is for the purpose of simplification, but it also demonstrates that the Köppen system is understandable and useful at various levels of sophistication. Our present objective in using Köppen's scheme is to make clear the systematic arrangement of different climates on the earth's surface.

There are six major climate categories within this simplified version of Köppen's climatic classification system, each symbolized by a capital letter. They are tropical (A climates); arid (B climates); mesothermal, mild winter (C climates); microthermal, severe winter (D climates); polar (E climates); and undifferentiated highland (H climates). Of these six large climate groups, all but the arid and highland climates are defined by temperature parameters (see Plate 11). The arid B climates are differentiated from all the others as being moisture-deficient, since the precipitation they receive during the year is exceeded by their potential evapotranspiration, which is a function of temperature. The sixth climatic group, the undifferentiated highland H climates, are highly complex because of their altitude and exposure variations over short distances. These are simply too intricate to show on a world map and could only be resolved into the appropriate Köppen climatic types on large-scale maps.

The four climatic groups that are related directly to temperature parameters form more or less symmetrical latitudinal bands in both hemispheres, progressing from the equator to

TABLE 7.1   Simplified Köppen Classification of Climates (Courtesy of James C. Reck)

FIRST LETTER	SECOND LETTER	THIRD LETTER	
E Warmest month less than 10°C (50°F) ICE CLIMATES	T Warmest month between 10°C (50°F) and 0°C (32°F) F Warmest month below 0°C (32°F)	NO THIRD LETTER (with Ice Climates) SUMMERLESS	ET EF
B Arid or Semiarid Climates ARID CLIMATES = BS-Steppe, BW-Desert	S Semiarid Climate (see Graph 1 below)* W Arid Climate (see Graph 1)	h Mean annual temperature is greater than 18°C (64.4°F) k Mean annual temperature is less than 18°C (64.4°F)	BSh BSk BWh BWk

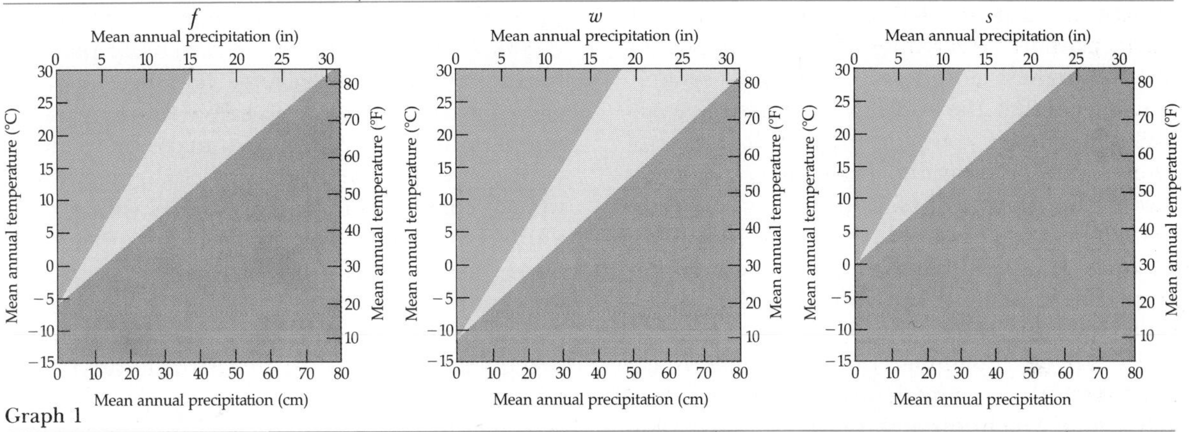

Graph 1

Coolest month greater than 18°C (64.4°F) TROPICAL CLIMATES: Am-Tropical Monsoon; Aw-Tropical savanna; Af-Tropical rain forest	f Driest month has at least 6 cm (2.4 in) m Seasonally, excessively moist (see Graph 2 at right) w Dry winter, wet summer (see Graph 2)	NO THIRD LETTER (with Tropical Climates) WINTERLESS Graph 2	Af Am Aw

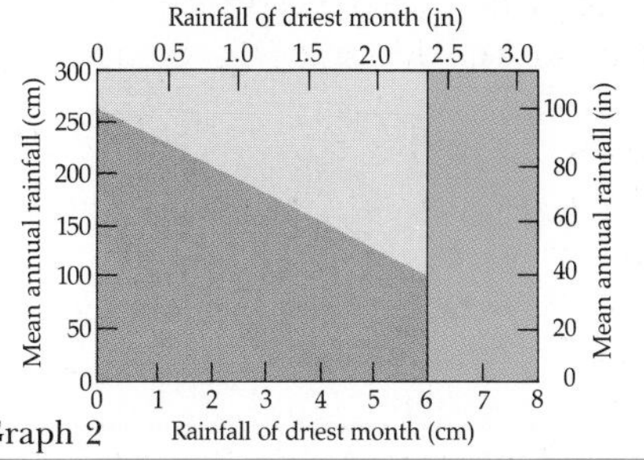

C Coldest month between 18°C (64°F) and 0°C (32°F), at least one month over 10°C (50°F) WARM TEMPERATE CLIMATES	s (DRY SUMMER) Driest month in the summer half of the year, with less than 3 cm (1.2 in) of precipitation and less than ⅓ of the wettest winter month	a Warmest month above 22°C (71.6°F)	Csa Csb Cwa Cwb
	w (DRY WINTER) Driest month in the winter half of the year, with less than ¹⁄₁₀ of the wettest summer month	b Warmest month below 22°C (71.6°F), with at least 4 months above 10°C (50°F)	Cfa Cfb Cfc
D Coldest month less than 0°C (32°F), at least one month over 10°C (50°F) SNOW CLIMATES	f (ALWAYS MOIST) Does not meet conditions for s or w above	c Warmest month below 22°C (71.6°F), with 1 to 3 months above 10°C (50°F)	Dwa Dwb Dwc Dfa Dfb Dfc
		d Same as c, but coldest month is below −38°C (−36.4°F)	Dfd Dwd
H HIGHLAND CLIMATES	NO SECOND LETTER CHARACTERIZED BY VERTICAL ZONATION	NO THIRD LETTER OF CLIMATES GIVEN ABOVE	H

*Boundaries of dry climates.

the polar regions. Tropical *A* climates are located in the equatorial region, mesothermal *C* climates approximately in the lower middle latitudes, microthermal *D* climates in the higher middle latitudes, and polar *E* climates, as their name implies, in the polar regions. The arid *B* climates, which are defined by moisture availability rather than temperature, appear in all but the highest latitudes. Their lack of symmetry reveals that the controls of moisture availability are more complex than the controls of temperature. Naturally, highland *H* climates occur without regard to latitude or general climatic patterns. To simplify matters, just as the climates progress from one type to another from the equator to the poles, so the letters that symbolize those climates advance from *A* to *E*. The exception is, of course, the *B* climates, which are not defined by temperature alone and which impinge upon the latitudinal boundaries of the *A, C,* and *D* climates.

CLIMOGRAPHS

It is possible to summarize the nature of the climate at any point on earth by graphic means as in Figure 7.3. Given information on mean monthly temperature and rainfall, we can express the nature of the changes in these elements throughout the year simply by plotting mean monthly temperature and precipitation values as points above or below (in the case of temperature) a zero line. To make the pattern of the monthly temperature changes clearer, we can connect the monthly values with a continuous line, producing an annual temperature curve. To avoid confusion, monthly precipitation amounts are usually shown as bars reaching to various heights above the line of zero precipitation. Such a display of a location's or station's climate is called a **climograph.** To read it, one must relate the temperature curve to the values given along the left margin of the graph,

Station:	Nashville, Tenn.	Type:	Cfa
Latitude:	36°N	Longitude:	88°W
	Average annual prec.:	119.6 cm(47.1 in)	22.5°C
Mean annual temp:	15.2°C(59.5°F)	Range:	(40.5°F)

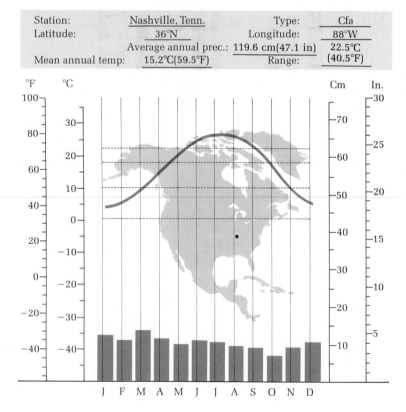

FIGURE 7.3 A standard climograph showing average monthly temperature (curve) and rainfall (bars). The horizontal index lines at 0°C (32°F), 10°C (50°F), 18°C (64.4°F), and 22°C (71.6°F) are the Köppen temperature parameters by which the station is classified.

FIGURE 7.4 A generalized continent showing the characteristic locations of the Köppen climatic regions.

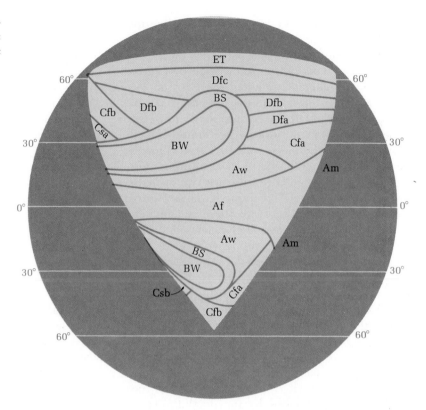

while the precipitation amounts are read on the right margin. Other information may also be displayed, depending upon the type of climograph used. Figure 7.3 represents the type we will use. A glance at it will indicate that the climograph can be used to determine the Köppen classification of the station as well as to show its specific temperature and rainfall regimes.

THE DISTRIBUTION OF CLIMATIC TYPES

The controls of atmospheric phenomena that combine in different ways to produce different climates are well-enough understood that the climates of a landmass of any size or shape may be predicted with considerable accuracy simply on the basis of general principles. Suppose a large, unexplored continent divided the vast Pacific Ocean, stretching almost from pole to pole. On the basis of what

we have learned about the general controls of temperature and precipitation and assuming that this continent does not have a mountainous landscape, it would not be difficult to construct a map of its climatic patterns (see Fig. 7.4). Throughout the following section, frequent reference should be made to this figure (and also to Plate 14, which provides pictorial illustration of each climatic type).

TROPICAL CLIMATES

Near the equator of our hypothetical continent we would find high temperatures year-round. At noon the sun would never be far from the zenith and would be directly overhead twice a year. Humid climates of this type with no winter season are Köppen's tropical *(A)* climates. Köppen chose 18°C (64.4°F) for the average temperature of the coldest month as his boundary for tropical climates because above this temperature vegetative growth and productive cycles are not influenced by cold. Actually, this temperature criterion closely coincides with the geographic limit of certain palms.

On the basis of our knowledge of factors that control climatic conditions, we would expect major differences in the amount and distribution of rainfall among locations within the tropical regions. In addition we would anticipate that the boundaries between *A* climates and *C* climates (which have at least one month averaging below 18°C) would not run straight across our hypothetical continent. Thus Figure 7.4 indicates that there are three *A* climates. These climates are expanded poleward to 30° latitude or more in the continent's interior and contracted near the coasts. This reflects the moderating influence of the oceans on monthly temperatures. We also note that the *A* climates extend poleward over the East Coast but are pinched sharply along the West Coast. This reflects the fact that cool currents move equatorward along tropical west coasts while tropical east coasts are bathed by warmer waters.

We have learned that regions near the equator are influenced by the doldrums and the principal rain-bringing phenomenon in these latitudes, the Intertropical Convergence (ITC). But the ITC is not anchored in one place. Instead it migrates with the seasons. Thus the convergence and rising air that bring rain to the tropics more or less follow the sun. Within 5° to 10° of the equator the doldrums and rainfall occur year-round, since the ITC moves through twice a year and is never far away. Likewise, tropical coasts facing the easterly trade winds rarely experience drought, since there is a steady onshore flow of humid air. But poleward of this zone precipitation becomes seasonal. When the noon sun is nearly overhead (high sun or summer), the ITC brings adequate rainfall. During the low sun period (winter season), the subtropical highs and the trades over the continents bring clear, dry weather conditions.

Table 7.1 indicates that the Köppen classification system takes into account the importance of seasonal precipitation in producing climatic variation. The letter *f* (from German:

feucht, wet) is used to designate year-round precipitation. The letter *s* indicates summer drought, and the letter *w* indicates winter drought. With the *A* climates the letter *m* (monsoon) is also used to indicate seasonal rainfall, but under conditions where the rainy season is moist enough to support forest vegetation. Thus, on our hypothetical continent, we find the tropical rain-forest *(Af)* climate in the equatorial region supporting luxurious rain forests and expanding eastward in the absence of highland barriers. The equatorial rain-forest climate is flanked both north and south by the dry winter tropical savanna *(Aw)* climate. In the latter, unlike the *Af* climate, the vegetation must adapt to several months of drought and is of the savanna type discussed in Chapter 9. Finally, along coasts facing the strong moisture-laden inflow of air associated with the summer monsoon, we find the tropical monsoon *(Am)* climate. This climate owes its existence to the seasonal reversal of atmospheric pressure and winds over both land and water described in Chapter 4.

POLAR CLIMATES

The tropical *(A)* climates are distinguished from all others as lacking winters (cold periods). It seems a happy circumstance that they occupy the largest area of any distinctive climatic type, bracketing almost 50° of latitude (25° north and south of the equator). At the opposite extreme are the polar *(E)* climates that, statistically, have no summers. These were defined by Köppen as areas in which no month has an average temperature exceeding 10°C (50°F). Such climates are found at very high latitudes where the sun's rays, when they strike, always strike at a low angle, spreading their energy over a large area. Poleward of this temperature boundary the climatic stress is such that trees cannot survive at all.

The polar climates are subdivided into tundra *(ET)* and ice-cap *(EF)* climates. Tundra climate occurs where at least one month averages above 0°C (32°F) and the characteristic vegetation is tundra, consisting of low-growing shrubs and herbs. The ground beneath may be frozen solid, only the upper portion thawing during the summer. In the ice-cap climate no month has an average temperature above 0°C and a permanent cover of snow and ice exists. Except for the temporary visits of polar bears, seals, penguins, and a few other bird species, the ice-cap climate is virtually lifeless.

The 10°C isotherm for the warmest month more or less coincides with the Arctic Circle, poleward of which the sun does not rise above the horizon in midwinter. Again the effect of ocean currents prevents this isotherm from being a straight line. It is deflected northward on the west sides of our hypothetical continent, where tropical waters move poleward, and dips to the south on the east sides of the landmass, where cold arctic water leaks into the general oceanic circulation.

MESOTHERMAL AND MICROTHERMAL CLIMATES

Except where arid climates prevail, the lands on our hypothetical continent between the tropical and polar climates are occupied by the transitional midlatitude C and D climates. As they are neither tropical nor polar, the C and D climates must have at least one month averaging below 18°C (64.4°F) and one month averaging above 10°C (50°F). Although both C and D climates have distinct temperature seasons, the D climates have severe winters with at least one month averaging below freezing. Once again vegetation reflects the climatic differences. In the D climates all but needle-leaf trees defoliate naturally during the winter as soil water is temporarily frozen and unavailable. Much of the natural vegetation of the C climates retains its foliage throughout the year since liquid water is always present in the soil. The line separating mild from severe winters usually lies in the vicinity of the 40th parallel.

There are a number of important internal differences with the C and D climates that produce individual climatic types based on precipitation patterns or seasonal contrasts. On our hypothetical continent, the Mediterranean or dry summer mesothermal (Cs) climate appears along west coasts between 30° and 40° latitude where the dry subtropical highs of summer alternate with the humid conditions of the westerly wind belt in winter. On the east coasts in generally the same latitudes the humid subtropical (Cfa) climate is found. The contrast between coasts should be expected when we recall that the subsidence, dry weather, and temperature inversions of the subtropical highs are only in evidence along the eastern sides of the oceans where they come in contact with west coasts of the continents. In contrast, continental east coasts receive both cyclonic and convectional precipitation.

The distinction between the humid subtropical (Cfa) and marine west coast (Cfb) climates illustrates the second important criterion for the internal subdivisions of C and D climates—seasonal contrasts. Small letter symbols are used to illustrate the transitional character and rapid changes in temperature that occur as a consequence of latitude and the proximity of either cold or warm maritime influences. The C and D climates are recognized as seasonally hot, mild, cool, and cold. As the statistical definitions of Table 7.1 indicate, the letter a stands for hot summers, b for mild summers, c for cool summers, and d for cold winters. Thus the Cfb climate has year-round precipitation and a mild summer. Its location along continental west coasts poleward of the Csa climates and often extending beyond 60° latitude is a direct result of mild, moist atmospheric conditions associated with the westerly winds throughout the year. In these latitudes the westerlies move onshore across warm ocean currents.

Under usual conditions the D climates receive year-round precipitation associated with the traveling cyclonic depressions of the polar front. Internal subdivision is associated with seasonal contrasts caused by latitude and increasing continentality. On our generalized continent the D climates are found exclusively in the Northern Hemisphere. As in the real world, there is no land in the Southern Hemisphere latitudes that would normally be occupied by the humid microthermal climates. In the Northern Hemisphere the *D* climates progress poleward through the *Dfa* climate (humid continental, hot summer) and the *Dfb* climate (humid continental, mild summer) to the *Dfc* climate (subarctic). Because of increasing aridity and temperature extremes, the latitudinal belts of the *D* climates either are broader or are replaced entirely by the *B* climates toward the interior of our model continent.

ARID CLIMATES

All of our climatic differentiation thus far has been based on temperature boundaries and the seasonality of precipitation. However, our ideal continent will contain a large area in which all of these variations are less important than the simple fact that the physical environment and all of its subsystems are dominated by year-round moisture deficiency. This area will penetrate deep into the heart of the continent, interrupting the neat zonation of climates that would otherwise exist. This is an area of Köppen's *B* climates.

The definition of climatic aridity is that precipitation received is less than potential evaporation. Aridity itself is a relative concept that does not depend solely on the amount of precipitation received. Evaporation rates and temperature must also be taken into account. In a climate in the low latitudes with relatively high temperatures, the evaporation rate is higher than it is in higher latitudes with lower temperatures. As a result, more rain must fall in the lower latitudes to produce the same effects (especially noticeable on vegetation) as a smaller amount of precipitation in areas with lower temperatures and consequently lower evaporation rates. This fact also explains the differences in the parameters that define *f, s,* and *w* climates when used with tropical climates on the one hand and mesothermal and microthermal climates on the other.

On our hypothetical continent the arid climates are concentrated in a zone from about 15°N and S latitude to about 30°N and S along the western coasts, expanding much further poleward over the heart of the landmass. This eastward expansion is a consequence of re-

moteness from the oceanic moisture supply, since in these latitudes the major flow of air is from west to east. The main factor that causes the deficiency of precipitation is air subsidence in the subtropical high-pressure systems. The correspondence between the arid climates and the belt of subtropical high pressure is quite unmistakable.

The pattern of ocean currents reinforces the effect of the subtropical anticyclones in suppressing precipitation in the subtropical zone. It is on the eastern sides of the oceanic anticyclones that cold water from the polar seas is moving equatorward along the continental west coast. Air moving landward from the west is cooled by these currents, increasing its stability and often producing a temperature inversion. This allows the zone of deficient precipitation to be accentuated on the coast itself. Coastal deserts may have considerable fog and a minor water supply in the form of condensed dew, but they receive the lowest precipitation amounts of any region on earth.

Desert *(BW)* climates are defined as areas where the annual amount of precipitation is less than one-half the potential evaporation rate. The result is that there is very sparse vegetation that is especially adapted to survive for long periods with a minimal water supply.

Bordering the deserts are steppe *(BS)* climates or semiarid climates that are transitional between the extreme aridity of the deserts and the moisture surplus of the humid climates. The steppe climate receives more precipitation than the desert does. As a result there is a more complete cover of vegetation, usually in the form of short grass.

Our definition for the steppe climate is an area where precipitation is less than potential evaporation but more than one-half the potential evaporation. However, the transition in vegetation from the desert climate to the steppe climate is gradual and constantly fluctuating in position.

HIGHLAND CLIMATES

We have been speaking of our hypothetical continent as if it were regular in form and had no significant relief features. Of course the pattern of climates and extent of aridity would be affected by irregularities in configuration such as the presence of deep gulfs, interior seas, or significant highlands. The climatic patterns of Europe and North America are quite different because of such variations.

Highlands can channel air-mass movements and create abrupt climatic divides. Their own microclimates form an intricate pattern related to elevation, cloud cover, and exposure. One significant effect of highlands aligned at right angles to the prevailing wind direction is the creation of rain-shadow conditions. Thus mountains transverse to the atmospheric flow not only have their own climates but generally produce arid regions extending tens to hundreds of kilometers to their lee.

REGIONALIZATION: A GEOGRAPHIC CONCEPT

A region, as the term is used by geographers, refers to an area that has recognizably similar internal characteristics that are distinct from those of other areas. A region may be described on any basis that unifies it and differentiates it from others. Regions can be defined on the basis of surface configuration (Colorado Plateau, Wyoming Basin) or by agricultural use and associated landscape (Corn Belt, Cotton Belt, Spring Wheat Belt). Regions can also be described in terms of vegetation (tropical rain forest, midlatitude grassland) as well as by political boundaries (countries or states).

In defining any one of these regions the geographer examines the features that give the area its unity and that differentiate it from other regions. He may also look at the region as a whole. That is, he may analyze the region as a total environment in which the earth's systems are interacting. Thus, through regionalization, the study of natural, cultural, or economic regions, the geographer is able to examine the integration of the various functioning systems as they work together in a real life situation. It is this *total view* of a working,

integrated region that helps us distinguish one region from another. For instance, our ability to describe a coastal zone as distinct from inland areas is based on descriptions of the interrelationships between climate, vegetation, soils, landforms, and even animals found in that region. So, too, we distinguish the humid continental climate with a long, hot summer (*Dfa*) from that with a warm or moderate summer (*Dfb*) not only on the basis of our climatic parameters but also through the differences expressed in associated natural elements of the environment (especially vegetation and soils). In some cases the geographer may also consider the differences between regions in terms of the cultural life of the people or their response to their physical situation. Thus the integration of the various physical and cultural systems within an environment is an important aspect of the concept of regionalization.

There are several reasons why the study of regionalization appeals to geographers. For one, it is related to their interest in describing the world in which we live. More importantly, it is an extension of their interest in the ways the physical and cultural elements of the environment are integrated and expressed.

Later in this chapter we will be examining climatic regions of the world by the criteria of precipitation and temperature, which we have just defined. But before we examine the individual climatic regions of the earth, let us look again at the distribution of the large climatic zones (Plate 12) and the relationship of that distribution to other physical phenomena.

One thing that is immediately noticeable is the change in climate with latitude. This is especially apparent in North America. Looking at the East Coast of the United States and Canada, we see that the southern tip of Florida has a tropical savanna (*Aw*) climate. This gives way almost immediately to the humid subtropical (*Cfa*) climate of the southern states. Moving poleward from Virginia toward Maryland and southern New Jersey, the humid subtropical climate changes to a humid continental, hot summer (*Dfa*) climate. Further north, New Hampshire, Vermont, and Maine have a humid continental climate with mild summers

(Dfb). Moving still further north, we see that most of Canada has a subarctic *(Dfc)* climate until we reach a latitude of 55° to 60°N, or 65° to 70°N in the western part of the country. Poleward of these latitudes the climate is described as polar tundra *(ET)*.

It is also interesting to discover locations that have similar climates. For instance, how many areas in the world have climates similar to the one with which you are most familiar? The Mediterranean climate is found in many parts of the world: California, southern Europe, the Middle East, central Chile, and small sections of the southwest coasts of Africa and Australia. These places are all in similar latitudes (between about 30° and 40°) and receive winds from oceans to the west. The marine west coast climate often is found poleward of the Mediterranean type: in Washington and Oregon, southern Chile, and in much of northwest Europe, at latitudes between 40° and 60°. This climate is found as well in New Zealand, on the southeast coast of Australia, and in southeast Africa. The subarctic subtype of the humid microthermal climate is found over the vast continental areas of both northern Canada and the Soviet Union. Follow the equator around the world, and you will spend most of your time in a tropical rain-forest climate except for excursions into the mountains and deserts of eastern Africa.

A first look at the world climate map may have given you the impression that climatic regions are scattered about haphazardly. A closer inspection, however, has shown that in many areas there is a latitudinal progression of climate. Furthermore, we have seen that similar climates usually appear in similar latitudes and/or in similar locations with respect to landmasses or topography. These climatic patterns emphasize the close relationship between climate, the weather elements, and the climatic controls. As we predicted when we examined the hypothetical continent, there is an order to the earth's atmospheric conditions and so also to its climatic regions.

Carrying this idea even further, let us compare the world maps of temperature (p. 85), precipitation (Plate 9), and pressure and winds (p. 95) with Plate 12, which shows world climates. To a large degree, the superposition of the maps showing the individual atmospheric elements or the combination of them has produced the map of climatic regions. We know, of course, that ultimately the order and pattern of the climatic regions are based *first* on the patterns produced by the earth-sun relationships. The exceptions to those patterns are the result of the uneven distribution and irregular sizes and shapes of land and water bodies, of the Coriolis effect on wind and ocean currents, and of interruption of the continents by mountain barriers.

A striking variation in these global climatic patterns becomes apparent when we compare the Northern and Southern Hemispheres. Because the Southern Hemisphere lacks the large landmasses of the Northern Hemisphere in higher latitudes, there are no climates (in land regions) that can be classified as humid microthermal *(D)*, and only one small peninsula of Antarctica can be said to have a tundra climate.

CLIMATE, VEGETATION, AND SOILS

Let us turn now to Plate 13, which shows the world distribution of the major vegetation groups. When we compare this map with the one showing the world distribution of climate, the similarities are strong enough to convince us that the quality of climate is of major importance to the character of vegetation. Note the similar locations of tropical rain-forest vegetation and the tropical rain-forest climates or the location of Mediterranean woodlands and Mediterranean climates. As a third example, see how much of the subarctic climate regions in Canada and the Soviet Union are covered with coniferous forests.

Soils are similarly influenced by climate. To illustrate the extent of this influence, compare the map of world soil distribution (Plate 16) with the climate map. Note the correlation between the soils and climates of the subarctic regions of Canada and the Soviet Union. At the same time, recall that both of these subarctic regions are dominated by coniferous forests. As with vegetation, soils demonstrate the influence of climate on their development through their global distribution. Besides their common dependence on climate, there is a strong influential relationship between vegetation and soils. As another example of such interrelationships, consider the world locations of desert soils, desert vegetation, and arid climates. This particular example shows the similarity in patterns from one system to another, stemming in this instance from the effects of precipitation or, more accurately, from the lack of precipitation.

Figure 7.5 shows schematic diagrams of the world distribution of climate, vegetation, and zonal soils. These, like the hypothetical continent, can serve as simplified models of the real situation. The sides of the boxes represent continuums from hot at the bottom to cold at the top and from dry at the left to wet at the right. The result is that each of the four corners is a climatic extreme: cold and dry, cold and wet, hot and dry, hot and wet.

The boxes show the correlations among the systems of climate, soils, and vegetation. Also apparent from a comparison of these boxes is the fact that the correlations are imperfect. That is, there are climates that we would not classify as dry that have a grassland vegetation typical of semiarid climates. There are middle-latitude forests found in microthermal climates, and there are middle-latitude forests that grow in soil, usually found in association with a tropical rain forest and its climate.

As the above discussion indicates, a survey of world climatic regions without inclusion of the associated vegetation types and characteristic soils would be incomplete. Thus, our descriptions of the various climatic regions, cov-

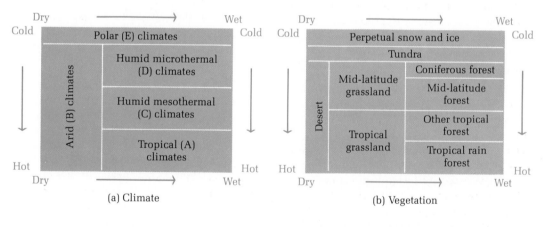

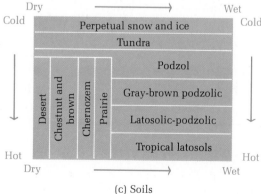

FIGURE 7.5 Diagrammatic representation of climate, vegetation, and soils as functions of temperature and moisture characteristics.

ered in the remainder of this chapter as well as in Chapter 8, will emphasize these associations. Bear in mind, however, that these associations are not absolute—there will be some exceptions. Also, while some soil and vegetation terminology will be introduced in our discussion of climatic regions, detailed information of soils and vegetation will be provided in later chapters. This will allow our descriptions to focus on the *broad*, world-scale associations that exist among climates, vegetation types, and soils.

These descriptions will be useful in forming an orderly knowledge of the diversity of the earth's physical environment. As you become familiar with the total environment of a region, you will be better able to see the interrelationships of the earth's systems in action. And from your knowledge of a particular region, you will eventually be able to generalize about the physical environment of areas with similar climates.

HUMID TROPICAL CLIMATIC REGIONS

We have already learned a good deal about the climatic regions of the humid tropics through our study of a hypothetical continent and our examination of the maps throughout this book. It now remains for us to identify the major characteristics of each humid tropical climatic type in turn, along with its associated world regions.

A careful reading of Table 7.2 provides a preview of the significant facts associated with the humid tropical climates. The table also reminds us that although each of the three *A* climates has high average temperatures throughout the year, there are major differences among them based on the amount and distribution of precipitation.

TROPICAL RAIN-FOREST CLIMATE (*Af*)

Tropical rain-forest regions probably come most readily to mind when someone says the word *tropical*. Hot and wet throughout the year, the tropical rain-forest climate has been the stage for many stories of both fact and fiction.

One cannot easily forget the classic struggle with the elements portrayed by Bogart and Hepburn in *The African Queen*. Even scientists are impressed by what they observe.

The heat increased rapidly towards two o'clock (92° and 93°F), by which time every voice of bird or mammal was hushed; only in the trees was heard at intervals the harsh whirr of a cicada. The leaves, which were so moist and fresh in the early morning, now become lax and drooping; the flowers shed their petals. . . . On most days, a heavy shower would fall some time in the afternoon, producing a most welcome coolness. The approach of the rain-clouds was after a uniform fashion very interesting to observe. First, the cool sea-breeze, which commenced to blow about ten o'clock, and which had increased in force with the increasing power of the sun, would flag and finally die away. The heat and electric tension of the atmosphere would then become almost insupportable. Languor and uneasiness would seize on everyone; even the denizens of the forest betraying it by their motions. White clouds would appear in the east and gather into cumuli, with an increasing blackness along their lower portions. The whole eastern horizon would become almost suddenly black, and this would spread upwards, the sun at length becoming obscured. Then the rush of a mighty wind is heard through the forest, swaying the tree-tops; a vivid flash of lightning bursts forth, then a crash of thunder, and down streams the deluging rain. Such storms soon cease, leaving bluish-black motionless clouds in the sky until night. Meantime all nature is refreshed: but heaps of flower-petals and fallen leaves are seen under the trees. Towards evening life revives again, and the ringing uproar is resumed from bush and tree. The following morning the sun again rises in a cloudless sky, and the cycle is completed; spring, summer and autumn, as it were, in one tropical day. The days are more or less like this throughout the year in this country. . . . With the day and night always of equal length, the atmospheric disturbances of each day neutralizing themselves before each succeeding morn; with the sun in its course proceeding midway across the sky, and the daily temperature the same within two or three degrees throughout the year—how grand in its perfect equilibrium and simplicity is the march of Nature under the equator.

The Naturalist on the River Amazon
H. W. Bates

TABLE 7.2 The *A* Climates

NAME AND DESCRIPTION	CONTROLLING FACTORS	GEOGRAPHIC DISTRIBUTION	DISTINGUISHING CHARACTERISTICS	RELATED FEATURES
Tropical Rain Forest *(Af)* Coolest month above 18°C (64.4°F); driest month with at least 6 cm (2.4 in) of precipitation.	High year-round insolation and precipitation of doldrums (ITC); rising air along trade wind coasts.	Amazon R. Basin, Congo (Zaire) R. Basin, east coast of Central America, east coast of Brazil, east coast of Madagascar, Malaysia, Indonesia, Philippines.	Constant high temperatures; equal length of days and nights; lowest (2°C–3°C/3°F–5°F) annual temperature ranges; evenly distributed heavy precipitation; high amount of cloud cover and humidity.	Tropical rain-forest vegetation (selva); jungle where light penetrates; tropical iron-rich soils, climbing and flying animals, reptiles, and insects; slash-and-burn agriculture.
Tropical Monsoon *(Am)* Coolest month above 18°C (64.4°F); one or more months less than 6 cm (2.4 in) of precipitation; excessively wet during rainy season.	Summer onshore, winter offshore air movement related to shifting ITC and changing pressure conditions over large landmasses; also transitional between *Af* and *Aw*.	Coastal areas of southwest India, Sri Lanka (Ceylon), Bangladesh, Burma, southwestern Africa, Guyana, Surinam, French Guiana, northeast and southeast Brazil.	Heavy high-sun rainfall, (especially with orographic lifting), short low-sun drought; 2°C–6°C (3°F–10°F) annual temperature range, highest temperature just prior to rainy season.	Forest vegetation with fewer species than tropical rain forest grading to jungle and thorn forest in drier margins; iron-rich soils; rain-forest animals with larger leaf-eaters and carnivores near savannas; paddy-rice agriculture.
Tropical Savanna *(Aw)* Coolest month above 18°C (64.4°F); wet during high-sun season, dry during lower-sun season.	Alternation between high-sun doldrums (ITC) and low-sun subtropical highs and trades caused by shifting winds and pressure belts.	Northern and eastern India, interior Burma and Indo-Chinese Peninsula; northern Australia; borderlands of Congo (Zaire) R., south central Africa; llanos of Venezuela, campos of Brazil; western Central America, south Florida, and Caribbean Islands.	Distinct high-sun wet and low-sun dry seasons; rainfall averaging 75–150 cm (30–60 in); highest temperature ranges for *A* climates.	Grasslands with scattered drought-resistant trees, scrub, and thorn bushes; fluctuating water table, leached soils common, iron-rich soils on wetter margins, brown soils near steppes; large herbivores, carnivores, and scavengers.

This account of a day in a tropical rain forest brings its climate vividly to life. We feel the high temperatures, oppressive humidity, and the almost daily rains for which it is known.

CONSTANT HEAT AND HUMIDITY
Most weather stations in the tropical rain-forest climatic regions record average monthly temperatures of 25°C (77°F) or more (Fig. 7.6). Because these regions are usually located within 5° or 10° of the equator, the sun's noon rays are always close to directly overhead. Days and nights are of almost equal length, and the amount of insolation received remains nearly constant throughout the year. Consequently, there are no appreciable temperature variations that can be linked to the sun and therefore considered seasonal.

The **annual range,** or the difference between the average temperatures of the warmest and coolest months of the year, reflects the consistently high angle of the sun's rays. As indicated in Figure 7.6, the annual range is seldom more than 2°C or 3°C (4°F or 5°F). In fact, at Ocean Island in the Central Pacific the annual range is 0°C because of the additional moderating influence of a large water body on the already nearly uniform pattern of insolation.

One of the most interesting features of the tropical rain-forest climate is that the **daily ranges,** or the differences between the highest and lowest temperatures during the day, are usually far greater than the annual range. Highs of 30°C to 35°C and lows of 20°C to 24°C produce daily or diurnal ranges of 10°C to 15°C (18°F to 27°F). However, the drop in temperature at night is small comfort. The high humidity causes even the cooler evenings to seem oppressive.

The climographs of Figure 7.6 illustrate that significant variations in precipitation can occur even within rain-forest regions. Although most rain-forest locations receive more than 200 centimeters (80 in) a year of precipitation and the average is in the neighborhood of 250 centimeters (100 in), there are some that record an annual precipitation of over 500 centimeters. Ocean locations, near the greatest source of moisture, tend to receive the most rain. As a group, climate stations in the humid tropics experience much higher annual totals than typical humid midlatitude stations. Compare, for example, the 365 centimeters at Akassa with the average 112 centimeters received annually at Portland, Oregon, or the 61 centimeters received in London.

We should recall that the heavy precipitation of the tropical rain-forest climate is associated with the warm, humid air of the doldrums

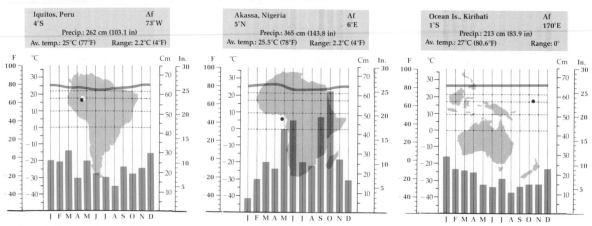

FIGURE 7.6   Climographs for tropical rain-forest stations.

FIGURE 7.7 **The multistoried broad-leaf evergreen forest of the Amazon Basin near Iquitos, Peru. (R. Sager)**

and the unstable conditions along the ITC. Both convection and convergence serve as trigger mechanisms, causing the moist air to rise and resulting in the heavy rains that are characteristic of this climate. There is heavy cloud cover during the warmer daylight hours when convection is at its peak, although the nights and early mornings can be quite clear. Variations in rainfall can usually be traced to the ITC and its weak low-pressure systems. Many tropical rain-forest locations (Akassa, for example) exhibit two maximum precipitation periods during the year, one during each appearance of the ITC as it follows the migration of the sun's direct rays. In addition, although no season can be called dry, during some months it may rain on only 15 or 20 days.

THE BALANCE BETWEEN FOREST AND SOIL The relationship between the soils of the tropical rain forest and the vegetation they support is so close that there exists a nearly perfect ecological balance between the two, threatened only by people's efforts to earn a living from the soil.

The most common vegetative growth is multistoried, broad-leaf evergreen forest made up of many species whose tops form a thick, almost continuous cover that blocks out much of the sun's light (Fig. 7.7). This type of rain forest is called **selva.** Within the selva there is usually little undergrowth on the forest floor because sunlight cannot penetrate. But conditions vary. The following passage describes a situation where a lower layer of growth undergirds the true forest giants.

The thing that astonished me most was the absolute straightness, the perfect symmetry of the tree-trunks, like the pillars of a dark and limitless cathedral. The ground itself was covered with a thick carpet of dead leaves and seedling trees. There was practically no earth visible and certainly no grass or flowers. Up to a height of ten feet or so a dense undergrowth of young trees and palms of all kinds hid the roots of the giants, but out of this wavy green sea of undergrowth a myriad tree-trunks rose straight upwards with no apparent decrease in thickness—that was the most extraordinary thing—for a hundred or a hundred and fifty feet before they burgeoned into a solid canopy of green which almost entirely shut out the sky. The tree-trunks, though similar in that they were all straining straight upward towards the light, were of every colour and texture of bark—smooth and black like Purbeck marble, red and scaly as our own Scots pine, pale grey or ghostly green like the nightmare jungles in *Snow White and the Seven Dwarfs,* or beautifully marbled and dappled like a moth's wing.

Next to the remarkable symmetry of the tree-trunks, the most astonishing thing was the amount of parasitic growth. Many of the boles of the trees were almost hidden by a network of creepers, occasionally breaking out into huge leaves but usually bare and often as thick as a man's thigh. In other places the vines and creepers hung straight down from the branches to the ground, where they had taken root again and looped themselves from tree to tree like the crazy rigging of a thousand wrecked ships. Up in the tree-tops where the great trunks suddenly burst into branches were huge hanging gardens of mosses and ferns, whose rotting foliage seemed to provide its own soil so that the whole mass might be six or eight feet in diameter and was festooned in its own little world of lianas and creepers.

The Jungle is Neutral
F. Spencer Chapman

The trees of the rain forest supply the tropical soils with the nutrients the trees need for growth. As leaves, flowers, and branches fall to the ground, or as roots die, the numerous soil animals and bacteria act on them, transforming the organic matter into usable nutrients. However, if the trees are removed, there is no replenishment of these nutrients and no natural counterbalance to laterization, the process by which large amounts of precipitation remove the soluble minerals from the soil. With all soluble minerals leached from the soil and with rapid deterioration of the remaining humus due to the intense activities of microorganisms, worms, termites, ants, and other insects, all that remains is an infertile mixture of insoluble manganese, aluminum, and iron compounds. After such soil depletion, the once-magnificent rain forest may never again be able to establish itself. If it does, it will take years to achieve the ecological balance it once had with the soil.

Of course the typical situation we have just described does not apply everywhere in the tropical rain-forest climate. Some regions are covered by the true jungle, a term often mis-used when describing the rain forest. **Jungle** is a dense tangle of vines and smaller trees that develops where direct sunlight does reach the ground, as in clearings and along streams. Other regions have soils that remain fertile or have bedrock that is chemically basic and provides the soils above with a constant supply of soluble nutrients through the natural weathering processes. Examples of the former region are found along major river floodplains, and examples of the latter are the volcanic regions of Indonesia and the limestone areas of Malaysia and Vietnam. Only in such regions of continuous soil fertility can agriculture be intensive and continuous enough to support population centers in the tropical rain-forest climate.

HUMAN ACTIVITIES Throughout much of the tropical rain-forest climate humans are far outnumbered by other forms of animal life. A comparison of Plates 2 and 12 suggests that most rain-forest regions are among the least inhabited areas of the world. Though there are few large animals of any kind, a great variety of smaller tree-dwelling and aquatic species live in the rain forest. Birds, monkeys, bats, alligators, crocodiles, snakes, and amphibians such as frogs of many varieties abound. Animals that can fly or climb into the food-rich leaf canopy have become the dominant animals in this world of trees.

Most common of all, though, are the insects: mosquitoes, ants, termites, flies, beetles, grasshoppers, butterflies, and bees live everywhere in the forest. Because they can breed continuously in this climate without danger from cold or drought, insects thrive. The problem this poses for man is best described by someone who has been there.

One of our chief troubles at night was from insects. In the daytime they did not trouble us much, for there are not nearly so many mosquitoes in the great jungle as in the rubber, or near cultivation in the plains; but at night they bit us severely. Worse even than the bite is the shrill humming that seems to be just beside your ear. However much you slap your face, the noise soon breaks out again. Far worse than the mosquitoes were the midges, whose wings made no noise, but whose bite was really a bite and itched like a nettle-sting. They were particularly bad in the early morning and often woke us up long before dawn. As a result of the bites we received in the night, our faces would be so enlarged and distorted that in the morning we were almost unrecognizable. Often our cheeks were so swollen that the eyes were closed and we could not see until we had bathed them in the cool water of a stream.

The Jungle is Neutral
F. Spencer Chapman

In addition to the insects themselves, there are genuine health hazards for tropical rainforest inhabitants. Not only does the oppressive, sultry weather impose physical discomfort and pain, it also allows a variety of human parasites and disease-carrying insects to thrive. Malaria, yellow fever, and sleeping sickness are all insect-borne diseases of the tropics, frequently fatal, and uncommon in more temperate climates. As they do today, native populations during the nineteenth and early twentieth centuries often developed partial immunity to such diseases, but settlers and colonizers from the midlatitudes, where the diseases had been eradicated or had never existed, had no such protection. It is no wonder that tropical areas of Africa became known as "The White Man's Grave."

Whenever native populations have existed in the rain forest, subsistence hunting and gathering of fruits, berries, small animals, and fish have been important. Since the introduction of agriculture, land has been cleared and crops such as manioc, yams, beans, maize (corn), bananas, and sugar cane have been grown. It has been the practice to cut down the smaller trees, burn the resulting debris, and plant the desired crops (Fig. 7.8). With the forest gone, this kind of farming is only possible for a year or two, or perhaps even three, before the soil is completely exhausted of its small supply of nutrients and the surrounding area is depleted of game. At this point the native population moves to another area of forest to begin the practice over again. This kind of subsistence agriculture is variously known as **slash-and-burn, swidden,** or simply **shifting cultivation.** Its impact upon the close ecological balance between soil and forest is obvious in many rain-forest regions. Sometimes the dam-

(Text continues on page 198)

FIGURE 7.8 Getting an area ready for planting in Jamaica; an example of subsistence slash-and-burn (swidden) agriculture. (R. Gabler)

VIEWPOINT: AMAZONIA: THE LAST FRONTIER?

Chamber of Deputies, Brazilia. (Reece Jones)

Brazil is a vast South American country of rich contrasts. On the one hand, there is the modern capital, Brasilia, with its industry, skyscrapers, and the up-to-date lifestyle of its citizens. On the other, there is the immense rain-forest region inhabited largely by primitive tribes of Indians who depend on hunting and gathering for survival.

The Amazon River flows 6500 kilometers (4000 miles) from its headwaters in the Peruvian Andes across Brazil to the Atlantic Ocean. Vast areas of the Amazon basin are virgin rain forest and relatively unexplored jungle. In fact, this area makes up the largest unexplored area left in the world.

The Indians who inhabit this rain forest have lived communally in tribes for centuries, supporting themselves by hunting the animals of the forest and gathering its fruits, roots, and other foods. Until the twentieth century the relatively small number of Brazilian Indians have lived in harmony with their environment, maintaining a balance between themselves and the animals and plants of the forest. Despite appearances to the contrary, the rain forest is not a fertile environment: Its soils are weak and it can support only small animal and human populations. Consequently the rain-forest Indians have lived near subsistence level, without the opportunity to build up the reserves and surpluses that support growth, ruling classes, war, etc.

This situation would no doubt have continued had representatives of twentieth-century civilization not intervened and lured the Indians out of their primitive lifestyle and into the modern economic system—one that the Indians were ill-equipped to handle.

The rubber industry, which in the early part of this century depended on the gathering of wild rubber in the rain forest, was one of the most effective contributors to the demise of many Indian tribes. Indians were hired—sometimes at gunpoint—to collect wild rubber. In return for their efforts they received such products of modern society as axes, knives, pots, and pans. By this apparently simple exchange the primitive Indians left behind their own tribal and communal system and entered today's cash economy. In the process they lost much of their cultural heritage and lifestyle.

Twentieth century society not only hired Indians to gather rubber, they also paid for pelts, hides, and live animals (to be resold as pets or for scientific research). As the Indians began to hunt for profit rather than simply for food, they upset the delicate equilibrium of the rain-forest ecosystem. In fact, they overhunted to such a degree that they seriously threatened their own existence in the rain forest.

Modern man has also brought to the Brazilian Indian, like the North American Indian before him, diseases against which the primitives have little or no defense. Such diseases as measles and the common cold have killed off entire tribes.

Since 1900, 87 of the 230 tribes of Brazilian Indians known at that time have become extinct, and only about 30 can still be said to be isolated. In addition, many other Indian tribes are losing the cultural identification that they had developed over the centuries. Certainly the influence—whether direct or indirect—of civilized man in Brazil has been just as much a factor in the demise of the Indian there as in the United States a century ago.

The Brazilian government is aware of the Indian "problem" and even has an agency, FUNAI (National Foundation for the Indian), to handle Indian issues. Article 198 of Brazil's constitution guarantees that the primitive Indians of the rain forest have permanent possession of their lands and rights to their resources. However, more and more Indians are being moved to special parks and reserves that provide 31.1 square kilometers (12 sq mi) of land per Indian. As many Americans are now becoming aware, one of the biggest losses in the destruction of Indian tribes and the herding of others onto reservations is the passing of a strong, vital cultural heritage. Brazil, too, may regret this loss. However, one of the directors of FUNAI has been quoted as saying, "There is no interest in preserving Indian culture. We want to integrate the Indians into Brazilian society—to make them Brazilians like we are."

The problems of Indians and their cultural clash with modern-day Brazil have been greatly intensified over the last ten years by Brazil's great push to colonize the Amazon basin. This drive is epitomized by the building of the 4800–6400 kilometer (3000–4000 mi) long Trans-Amazon Highway, which stretches across Brazil through the Amazon rain forest. This highway has opened up to development lands and resources that hitherto have been out of civilization's reach. The Brazilian government, through loans and other kinds of support, has greatly encouraged the settlement of these lands. Many Brazilians are caught up in the excitement of this project and the colonization of their country's frontier. One author, Alfonso Henriques, in comparing Brazil's settlement of the Amazon basin to America's settlement of the West, wrote, "It can be expected that . . . the Amazon homestead drive will likewise transform Brazil into one of the great economic powers of the world . . . and it may be that the Trans-Amazon, combined with the extensive network of highways planned for the basin of that great river, will be considered as the greatest undertaking of the twentieth century."

Building this road has brought with it a frontier atmosphere reminiscent of that which surrounded the settling of the American West and the completion of the transcontinental railroad. As one writer has described it, "Barroom disputes are often settled by the gun. Hundreds of women from the countryside have flooded into the towns to work as bar girls and prostitutes, entertaining truckloads of highway workers each week-end at $3 each." Also like the American frontier, there is a problem of hostile Indians whose lands are being invaded and taken over for a project that threatens the only existence they have ever known.

Critics of the Trans-Amazon Highway and the settlement of the Amazon have focused on two issues. One deals with the problem of poor jungle and rain-forest soils. Though slash-and-burn farming is being forbidden and soil experts are providing advice on crops, fertilization, and soils, there is still a fear that the weak soils of the rain forest will be worn out within a couple of years and then will be unable to support the new settlers. The other issue is raised by ecologists who wonder what effect the destruction of the vast Brazilian rain forest will have on the balance of the various earth systems, particularly the supply of oxygen.

As an American resident of Brazil said, "As an economic investment, the Trans-Amazon Highway is quite debatable. But this project is based on larger considerations—tying the whole country together. It's a matter of national pride, something like settling our West."

It is expected that the Trans-Amazon Highway, which is being extended across the Amazonian rain forest, will greatly expand the economy of Brazil. What has been its impact on the forest and the primitive peoples who originally inhabited the region? (Copyright by Manchete)

age done to the system is irreparable, and only jungle, thornbush, or scrub vegetation will return to the cleared areas.

In terms of numbers of people supported, the most important agricultural use of the tropical rain-forest climate is the wet-field (paddy) rice agriculture of southeastern Asia. However, this is best developed in the *monsoon* variant of this climate. Commercial plantation agriculture is also significant. The principle plantation crops are rubber and cacao, both of which originally grew wild in the forests of the Amazon Basin but are now of greatest importance elsewhere—rubber in Malaysia and Indonesia and cacao in West Africa and the Caribbean area.

The true productivity of the tropical rain-forest regions undoubtedly awaits the development of additional technology. Associated with the great rivers of these regions is a vast potential for hydroelectric power. Although the nature of the selva limits commercial lumbering (there are many different trees but no stands of one species), many important tropical plants such as quinine, cola, and coca await further exploitation. In the bedrock and soils beneath the forests are rich deposits of bauxite (aluminum), manganese, and iron ores. It is not unreasonable to suggest that these regions are among the frontiers of the future.

TROPICAL MONSOON CLIMATE (Am)

We associate the monsoon most closely with the peninsula lands of Southeast Asia. Here the alternating circulation of air (from land in winter, from water in summer) is strongly related to the shifting of the ITC. During the summer months the ITC moves suddenly north into the Indian subcontinent and adjoining lands to a latitude of 20° or 25°. This is due in part to the attracting force of the deep low-pressure area of the Asian continent. But, as we have previously noted, the mechanism is complex and involves changes in the upper air flow as well as in surface currents. Several months later the moisture-laden summer monsoon is replaced by an outflow of dry air from the massive Siberian high-pressure area that develops in the

winter season over central Asia. By this time the ITC has shifted far to the south.

However, Figure 7.9 and Plate 12 confirm the fact that there are climatic regions outside of Asia that fit the simplified Köppen classification of tropical monsoon. Although a modified version of the monsoonal wind shift occurs at Freetown, Sierra Leone, in Africa, the climate there might also be described as transitional between the constantly wet rain-forest climate and the sharply seasonal wet and dry conditions of the tropical savanna. Cayenne, French Guiana, on the other hand, receives much of its rainfall during the low-sun (winter) season when the trades move onshore along the northeast coast of South America, bringing abundant moisture from the nearby ocean.

DISTINCTIONS BETWEEN RAIN FOREST AND MONSOON Whatever the factors that produce tropical monsoon climatic regions, there are strong similarities between these regions and those that are classified as tropical rain forest. In fact, although their core regions are distinctly different from one another, the two climates are often intermixed over transition areas. It is for this reason that Plate 12 indicates tropical monsoon and tropical rain forest as one category and distinguishes the core regions by letter symbol. A major reason for the similarity between *Am* and *Af* climates is that an *Am* area has enough precipitation to allow continuous vegetative growth with no dormant period during the year. Rains are so abundant and intense and the dry season is so short that the soils usually do not dry out completely. As a result this climate and its soils support a plant cover much like that of the tropical rain forests.

However, there are clear distinctions between rain-forest and monsoon climatic regions. The most important, of course, concern precipitation, including both distribution and amount. The monsoon climate has a short dry season, and the rain forest does not. Perhaps even more interesting is the fact that average annual rainfall in monsoon regions varies more widely from place to place. It usually totals be-

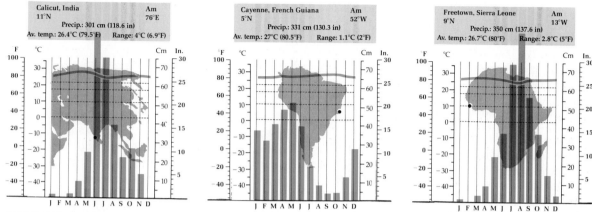

FIGURE 7.9 **Climographs for tropical monsoon stations.**

tween 150 and 400 centimeters (60 and 150 in) and may be massive where the onshore monsoon winds are forced to rise over mountain barriers. Mahabaleshwar (altitude 1362 meters), on the windward side of India's Western Ghats, averages more than 630 centimeters (250 in) of rain during the 5 months of the summer monsoon.

The annual march of temperature of the monsoon climate differs appreciably from the monotony of the rain-forest climate. The heavy cloud cover of the rainy monsoon considerably reduces insolation and temperatures during that time of year. But during the period of clear skies just prior to the onslaught of the rains, higher temperatures are recorded. As a result the annual temperature range in a monsoon climate is from 2°C to 6°C (compared to 2°C to 3°C in the tropical rain forest). This is illustrated by the temperature curves on the climographs of Figure 7.9.

Some additional distinctions between monsoon and rain-forest regions can be found in vegetation and animal life. Toward the wetter margins, the tropical monsoon forest resembles the tropical rain forest, but fewer species are present and certain ones become dominant, as teak does in Burma. The seasonality of rainfall narrows the range of species that will prosper. Toward the drier margins of the monsoon climate, the trees grow farther apart and the monsoon forest often gives way to jungle or a

dwarfed thorn forest. The composition of the animal kingdom also changes. The climbing and flying species that dominate the forest are joined by larger, hoofed leaf eaters and carnivores, such as the famous tigers of Bengal.

EFFECTS OF SEASONAL CHANGE The wet and dry seasonality of the monsoon regions has often been compared to the four temperature seasons of the middle latitudes. Just as some writers have sought to catch the essence of spring, summer, fall, and winter on paper, so have those whose lives have been intertwined with the monsoon sought to capture its unique qualities. Following is such a description by an Indian writer.

What the four seasons of the year mean to the European, the one season of the monsoon means to the Indian. It is preceded by desolation; it brings with it the hopes of Spring; it has the fullness of summer and the fulfillment of autumn all in one.

Those who mean to experience it should come to India some time in March or April. The flowers are on their way out and the trees begin to lose their foliage. The afternoon breeze has occasional whiffs of hot air to warn one of the days to come. For the next three months the sky becomes a flat and colorless gray without a wisp of a cloud anywhere. People suffer great agony. Sweat comes out of every pore and the clothes stick to the body. Prickly heat erupts behind the neck and spreads over the body till it bristles like a porcupine and one is afraid to touch oneself. The thirst is unquenchable, no matter how much one drinks. The nights are spent shadow-boxing in the dark trying to catch mosquitoes and slapping oneself in an attempt to squash those hummings near one's ears. One scratches and curses when bitten; knowing that the mosquitoes are stroking their bloated bellies safely perched in the farthest corners of the nets, that they have gorged themselves on one's blood. When the cool breeze of the morning starts blowing, one dozes off and dreams of a paradise with ice cool streams running through lush green valleys. Just then the sun comes up strong and hot and smacks one in the face. Another day begins with its heat and its glare and its dust.

After living through all this for ninety days or more, one's mind becomes barren and bereft of hope. It is then that the monsoon makes its spectacular entry. Dense masses of dark clouds sweep across the heavens like a celestial army with black banners. The deep roll of thunder sounds like the beating of a billion drums. Crooked shafts of silver zigzag in lightning flashes against the black sky. Then comes the rain itself. First it falls in fat drops; the earth rises to meet them. She laps them up thirstily and is filled with fragrance. Then it comes in torrents which she receives with the supine gratitude of a woman being ravished by her lover. It impregnates her with life which bursts forth in abundance within a few hours. Where there was nothing, there is everything: green grass, snakes, centipedes, worms, and millions of insects.

I Shall Not Hear the Nightingale
Khushwant Singh

The seasonal precipitation of the tropical monsoon climate is also important for economic reasons, especially to the people of Southeast Asia and India (Fig. 7.10). Most of the people living in those areas are farmers, and their major crop is rice, which is the staple food for millions of Asians. Rice is an irrigated crop, so the monsoon rains are very important to its growth. Harvesting, on the other hand, must be done during the dry season.

Each year an adequate food supply for India depends on the arrival and departure of the monsoon rains. The difference between famine and survival for many of India's people is very much involved with the climate.

TROPICAL SAVANNA CLIMATE (*Aw*)

Because of its location well within the tropics (usually between latitudes 5° and 20° on either side of the equator) the tropical savanna cli-

FIGURE 7.10   Seasonal change in the monsoon area of central Java. (U.S. Geological Survey)

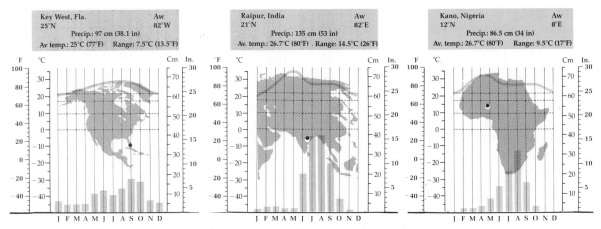

FIGURE 7.11   Climographs for tropical savanna stations.

mate has much in common with the tropical rain forest and monsoon. The sun's vertical rays at noon are never far from overhead and temperatures remain constantly high. Days and nights are of nearly equal length throughout the year, as they are in other tropical regions.

However, as previously noted, its distinct seasonal precipitation pattern identifies the tropical savanna. As the latitudinal wind and pressure belts shift with the climate, regions are under the influence of the rain-producing ITC (doldrums) for part of the year and the rain-suppressing subtropical highs for the other part. In fact, the poleward limits of the savanna climate are approximately the poleward limits of migration of the ITC, while the equatorward limits of this climate are the equatorward limits of movement by the subtropical high-pressure systems.

As you can see in Table 7.2 and on Plate 12, the greatest areas of savanna climate are found peripheral to the rain-forest climates of Latin America and Africa. Lesser but still important savanna regions occur in India, peninsular Southeast Asia, and Australia. In some instances the climate extends poleward of the tropics as it does in the southernmost portion of Florida.

TRANSITIONAL FEATURES OF THE SAVANNA   Of particular interest to the geographer is the transitional nature of the tropical savanna. Often located between the humid rain-forest climate on the one hand and the rain-deficient steppe *(BS)* climate on the other, the savanna experiences some of the characteristics of both. During the rainy, high-sun season atmospheric conditions resemble those of the rain forest, while the low-sun season can be as dry as nearby arid lands are all year. Because of the gradational nature of the climate, precipitation patterns vary considerably (Fig. 7.11). Savanna locations close to the rain forest may have rain during every month, and their total annual precipitation may exceed 180 centimeters (70 in). In contrast, the drier margins of the savanna have longer and more intensive periods of drought and lower annual rainfalls (less than 100 centimeters). The following excerpt illustrates conditions at Kano, Nigeria, one of the drier locations.

Kano can be hot, sizzling hot, in March and April and again about October, when the sun is cruel and the north wind brings the Sahara to choke your throat and crack your lips. From May to September the rain cools the land and luxurious greenness cheers the eye and the spirit. Around Christmas-time, Kano is far enough north to have a winter. The ladies wear fur coats and electric stoves are switched on.

Tremendous dust storms come down from the Sahara with the harmattan wind, which blows from October to March. This wind is intermittent, but when it rises, the belt bordering the Sahara, from Dakar to Kano, gets a dust which may achieve the obscurity of a dense London fog.

West Africa
F. J. Pedler

There are other characteristics of the savanna that help to demonstrate its transitional nature. The higher temperatures just prior to the arrival of the ITC produce annual temperature ranges 3°C to 6°C (5°F to 10°F) wider than those of the rain forest but still not as wide as the steppe and desert. Although the typical savanna vegetation is a mixture of grasslands with trees, scrub, and thorn bushes (known as **llanos** in Venezuela and **campos** in Brazil), there is considerable variation. Near the equatorward margins of these climates grasses are higher, and where there are trees they grow fairly close together. Toward the drier, poleward margins, trees are more widely scattered and smaller, and the grasses are shorter. Soils, too, are affected by the climatic gradation as the iron-rich reddish soils of the wetter sections are replaced by darker-colored wet and dry tropical or brown steppe soils in the drier regions.

Both vegetation and soils have made special adaptations to the alternating wet-dry seasons of the savanna. During the wet (high-sun) period the grasslands are green and the trees are covered with foliage. During the dry (low-sun) period the grass turns dry, brown, and lifeless, and most of the trees lose their leaves as an aid in reducing moisture loss through transpiration. The trees have deep roots that can reach down to water in the soil during the dry season. They are also fire-resistant, an advantage for survival in the savanna, since the grasses may burn during the winter drought.

Laterite is often associated with the soils. This is a zone in the subsoil, as much as 6 meters (20 ft) thick, depleted of all but oxides and hydroxides of iron, aluminum, and manganese. When wet this material may be cut into bricks that become rock-hard when dried out. The peculiar aspect of the soil-forming regime that results in the formation of laterite is a water table that fluctuates widely in level. This is clearly a result of the strong seasonality of precipitation in the savanna climate. The iron and aluminum oxides are concentrated at the top of the zone of water table fluctuation.

SAVANNA POTENTIAL   Conditions within tropical savanna regions are not well suited to agriculture, although many of our domesticated grasses (grains) are presumed to have grown wild there. Rainfall is far more unpredictable than in the rain forest or even the monsoon climate. For example, Nairobi, Kenya, has an average annual rainfall of 86 centimeters. Yet from year to year the amount of rain received may vary anywhere from 50 to 150 centimeters. As in the steppe and desert, the drier the savanna station, the more unreliable the rainfall. However, the rains are essential for human and animal survival in this region. When they are late or deficient, as they have been in West Africa in recent years, severe drought and famine result. On the other hand, when the rains last longer than usual or are excessive, they can cause major floods, often followed by outbreaks of disease.

Savanna soils (except in areas of recent alluvium) also limit productivity. During the rains of the wet season they become gummy, while during the dry season they are hard and almost impenetrable. Consequently, people in the savannas have often found the soils better suited to grazing than to farming. The Masai cattle herders of East Africa are world-famous examples (Fig. 7.12). However, even animal husbandry has its problems. Many savanna regions make poor pasture lands at least part of the year, and large-scale commercial cattle raising must await the successful introduction of more nutritious subtropical grasses. Experiments with a variety of such grasses are already underway in several regions of South America. In addition, soil erosion can be a problem in the savannas during the rainy season, especially in areas that have been overgrazed by both wild and domesticated animals.

The savannas of Africa have exhibited the greatest potential of the world's savanna regions. They have been veritable zoological gar-

FIGURE 7.12 The Masai of Kenya are among the world's best-known cattle herders. Unfortunately their herds frequently bring them into conflict with neighboring non-pastoral peoples, as well as with the dwindling animal population of the savannas. (Courtesy of J. Reck)

dens for the larger tropical animals. These grassland regions support many different herbivores (plant eaters), such as the elephant, rhinoceros, giraffe, zebra, and antelope. The herbivores in turn are eaten by the carnivores (flesh eaters) such as the lion, leopard, cheetah, hyena, jackal, and wild dog. During the dry season the herbivores find grasses and water along stream banks and forest margins and at isolated water holes. The carnivores follow the herbivores to the water, and a few human hunters still follow them both. However, the days of the great herds of game animals in Africa are numbered. Most are now crowded into overgrazed national parks, which are themselves under pressure from those who desire more range for their growing cattle herds. In a few decades the great numbers of wild animals may be reduced to a few zoo specimens.

ARID CLIMATIC REGIONS

Regions which are classified *B* in the simplified Köppen system are widely distributed over the earth's surface. A brief study of Figure 7.13 confirms that they are found from the vicinity of the equator to more than 50° N and S latitude. There are two major concentrations of desert lands, and each illustrates one of the important causes of climatic aridity. The first is centered on the Tropics of Cancer and Capricorn (23½° N and S latitudes) and spreads 10° to 15° poleward and equatorward from there. The second is located poleward of the first and occupies continental interiors, particularly in the Northern Hemisphere.

The concentration of deserts in the vicinity of the tropical sun lines is directly related to the subtropical high-pressure cells. Although the

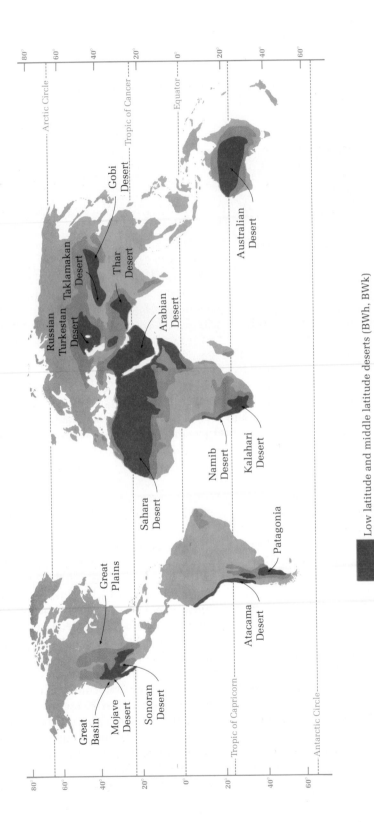

FIGURE 7.13 A map of the world's deserts.

boundaries of the subtropical highs may migrate north and south with the sun, their influence remains constant in these latitudes. We have already learned that the subsidence and divergence of air associated with these cells is strongest along the eastern portions of the oceans (western portions of the continents). Hence the clear weather and dry conditions of the subtropical high pressure extend inland from the western coasts of each landmass in the subtropics. The Atacama, Namib, and Kalahari Deserts and the desert of Baja California are restricted in their development by the small size of the landmass or by landform barriers to the interior. But the western portion of the Afro-Eurasian landmass comprises the greatest stretch of desert in the world and includes the Sahara, Arabian, and Thar Deserts. Similarly, the Australian Desert occupies most of the interior of the Australian continent.

The second concentration of deserts is located within continental interiors remote from moisture-carrying winds. Such arid lands include the vast cold-winter deserts of inner Asia and the Great Basin of interior drainage in the western United States. The dry conditions of the latter region extend northward into the Columbia Plateau and southward into the Colorado Plateau and are increased by the mountain barriers that restrict the movement of rain-bearing air masses from the west. Similar rain-shadow conditions help to explain the Patagonia Desert of Argentina and the arid lands in western China (Sinkiang).

Both wind direction and ocean currents can accentuate aridity in coastal regions. Where prevailing winds blow parallel to a coastline instead of onshore, a situation is created in which desert conditions are likely to occur because little moisture is brought inland. This seems to be the case in eastern Africa and perhaps in northeastern Brazil. Where a cold current flows next to a coastal desert, foggy conditions may develop. Warm moist air from the ocean may be chilled to its dew point as it passes over the cooler current. A temperature inversion is created, increasing stability and preventing the upward movement of air required for precipitation. The unique fog-shrouded coastal deserts in Chile (the Atacama), southwest Africa (the Namib), and Baja California have the lowest precipitation of any regions on earth.

The map of Figure 7.13 indicates that the deserts of the world are core areas of aridity, usually surrounded by the slightly moister steppe regions. Hence our explanations for the location of deserts hold true for the steppes as well. The steppe climates are either subhumid borderlands of the humid tropical (A), mesothermal (C), and microthermal (D) climates or are transitional between these climates and the deserts. We classify both steppe and desert on the basis of the relation between precipitation and potential evaporation. In the desert climate the amount of precipitation received is less than one half the potential evaporation. In the steppe climate the precipitation is more than one half but less than the total potential evaporation.

The criterion for determining whether a climate is desert, steppe, or humid is *precipitation effectiveness*. The amount of precipitation actually available for use by plants and animals is the effective precipitation. Not all precipitation is effective. In desert areas, the air is so hot and dry that falling rain frequently evaporates before it reaches the ground. Such precipitation is obviously ineffective in supplying moisture to the land.

Precipitation effectiveness is related to temperature. At higher temperatures it takes more precipitation to produce the same effects on vegetation and soils than at lower temperatures. The result is that areas with higher temperatures and therefore greater potential for evaporation can receive more precipitation than cooler regions and yet have a more arid climate.

Because of the temperature influence, precipitation effectiveness depends on the season in which an arid region's meager precipitation is concentrated. Obviously, precipitation received during the low-sun period will be more effective than that received when temperatures are higher, for less will be lost through evaporation.

The graphs in Table 7.1 can be used to help determine whether a particular location fits within the parameters for a desert, steppe, or a more humid climate. These graphs are based on the concept of precipitation effectiveness, for they take into consideration not only the average amount of precipitation received in a year and the average annual temperature but also whether precipitation is concentrated in the coldest or warmest part of the year.

DESERT CLIMATES (BWh, BWk)

The deserts of the world extend through such a wide range of latitudes that the simplified Köppen system recognizes two major subdivisions: the low-latitude deserts *(BWh)*, where temperatures are relatively high year-round and frost is absent or infrequent even along poleward margins, and high-latitude deserts *(BWk)*, which have distinct seasons including below freezing temperatures during winter

(Table 7.3). However, the significant fact about all deserts is their aridity. The relative unimportance of temperature is emphasized by the small number of occasions that we distinguish between low-latitude and middle-latitude deserts in the discussion that follows.

LAND OF EXTREMES   By definition deserts are associated with a minimum of precipitation, but they represent the extremes in other atmospheric conditions as well. Because there are few clouds and little water vapor in the air, as much as 90 percent of insolation reaches the earth in desert regions. This is why the highest insolation and the highest temperatures are recorded in low-latitude desert areas and not in the more humid tropical climates that are closer to the equator. Again because of light cloud cover, there is little atmospheric effect and much of the energy received by the earth during the day is radiated back to the atmosphere at night. Consequently, night tempera-

TABLE 7.3   The *B* Climates

NAME AND DESCRIPTION	CONTROLLING FACTORS	GEOGRAPHIC DISTRIBUTION	DISTINGUISHING CHARACTERISTICS	RELATED FEATURES
Desert (BW) Precipitation less than half of potential evaporation; mean annual temperature above 18°C (64.4°F) *(h)*; below *(k)*.	Descending, diverging circulation of subtropical highs; continentality often linked with rainshadow location.	Coastal Chile and Peru, southwest Africa, central Australia, Baja California and interior Mexico, North Africa, Arabia, Iran, Pakistan, and western India *(h)*; inner Asia, and western United States *(k)*.	Aridity; low relative humidity, irregular and unreliable rainfall; highest percentage of sunshine, highest diurnal temperature range, highest daytime temperatures; windy conditions.	Xerophytic vegetation, often barren, rocky, or sandy surface; desert soils, excessive salinity; usually small, nocturnal, or burrowing animals; nomadic herding.
Steppe (BS) Precipitation more than half but less than potential evaporation; mean annual temperature above 18°C (64.4°F) *(h); below (k).*	Same as deserts; usually transitional between deserts and humid climates.	Peripheral to deserts, especially in Argentina, northern and southern Africa, Australia, central and southwest Asia, and western United States.	Semiarid conditions, annual rainfall distribution similar to nearest humid climate; temperatures vary with latitude, elevation, and continentality.	Dry savanna (tropics) or short grass vegetation; highly fertile black and brown soils; grazing animals in vast herds, predators and smaller animals; ranching, dry farming.

tures in the desert drop far below their daytime highs. The excessive heating and cooling give low-latitude deserts the greatest diurnal temperature ranges in the world, and middle-latitude deserts are not far behind. In the spring and fall these ranges may be as great as 40°C (72°F) in a day. More common diurnal temperature ranges in deserts are 22°C to 28°C (40°F to 50°F).

Daytime air temperatures, especially in low-latitude deserts, must be experienced to be appreciated, but the following quotation may convey the idea of desert heat.

The first summer in Kharga was rather a shock; though I had been in Egypt for over four years. . . . At 6 A.M. in the morning the mercury stood at 98°, and this being the cool of the day the house was shut right up until evening, so that coming from the glare and heat at mid-day one got the impression—a totally false one—that the house was cool. At mid-day one saw the most ghastly sight I have ever seen and that was a bright patch of sunlight in the fireplace caused by the sun shining straight down the chimney.* A warm red glow from a fireplace in mid-winter is one of the pleasantest things I know, but a staring yellow patch of sunlight where the glowing coals ought to be, lightening up the gloom of a darkened room that is pretending to be cool, has a most grisly effect.

During the whole of the day the temperature remained at from 110° to 115° with a hot wind. At 6 P.M. the wind dropped and it seemed to get hotter till 11 P.M. when the wind started again, feeling quite as blistering and unpleasant as it had been at mid-day. It was quite impossible to sleep, and I used to walk about on the verandah throwing water on the mosquito-curtains in a vain attempt to bring down the temperature and do something to moisten the intense dryness that caused the tables and chests of drawers in the house to split with loud reports. At about 2 A.M. there was a slight but appreciable cooling off and one could usually get to sleep then till 5:30 A.M. when the heat of the newly-born sun awoke one to another day of hell.

Three Deserts
C. S. Jarvis

*Kharga is in Lat. 25°26′N, and therefore the sun at midsummer would be about 2° from the zenith at noon.

Because of the deficiency of atmospheric moisture in the desert, temperatures in shade are much lower than those a few steps away in direct sunlight. (Keep in mind that all temperatures recorded for statistical purposes, including those in the desert, are shade temperatures.) Khartoum (in the Sahara of the Sudan) has an *average* annual temperature of 29.5°C (85°F), which is a *shade* temperature. Temperatures in the bright desert sun at Khartoum under cloudless skies are often 43°C (110°F) or more. Soil temperatures rise close to 95°C (200°F) in midsummer in the Mojave Desert of southern California.

During low-sun or winter months deserts experience colder temperatures than more humid areas at the same latitude, and in summer they experience hotter temperatures. Just as with the high diurnal ranges in deserts, these high annual temperature ranges can be attributed to the lack of moisture in the air.

Annual temperature ranges are usually greater in middle-latitude deserts like the Gobi in Asia than in low-latitude deserts because of the colder winters experienced at higher latitudes. Compare, for example, the climograph for Aswan in South Central Egypt at 24°N, a low-latitude desert location, with the climograph for Turtkul, Soviet Turkestan, at 46°N, a middle-latitude desert location (Fig. 7.14). The annual range for Aswan is 17°C; in Turtkul it is 34°C.

Precipitation in the desert climate is irregular and unreliable, but when it comes, it may arrive in an enormous cloudburst, bringing more precipitation in a single rainfall than has been recorded in years. This happened in the extreme in the port of Walvis Bay, located on the coast of the Namib Desert, a cold-current coastal desert of southwest Africa. The equivalent of 10 years of rain was received in one night when a freak storm dumped 3.2 centimeters of rain.

Relative humidity in deserts is low (10 to 30 percent) because of the high temperatures. Absolute humidity, or the actual amount of water vapor in the air, is often high compared to colder climates; hence the relative humidity rises as temperatures fall during the night.

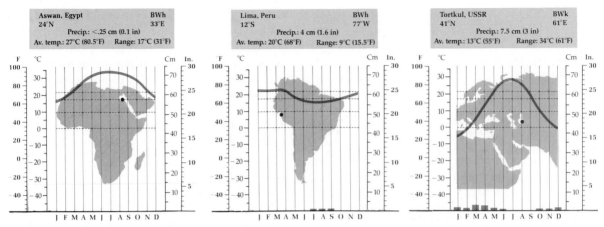

Aswan, Egypt	BWh
24°N	33°E
Precip.: <.25 cm (0.1 in)	
Av. temp.: 27°C (80.5°F)	Range: 17°C (31°F)

Lima, Peru	BWh
12°S	77°W
Precip.: 4 cm (1.6 in)	
Av. temp.: 20°C (68°F)	Range: 9°C (15.5°F)

Tortkul, USSR	BWk
41°N	61°E
Precip.: 7.5 cm (3 in)	
Av. temp.: 13°C (55°F)	Range: 34°C (61°F)

FIGURE 7.14   Climographs for desert stations.

Though precipitation is rare in many desert locations, the formation of dew in the cool hours of early morning is not. Where measurements have been made, the amount of dew formed has sometimes considerably exceeded the annual rainfall for that location. It has been suggested, in fact, that dew may be of great importance to plant and animal life in the desert. Studies are now being carried out on the use of dew as a moisture source for certain crops, thereby minimizing the need for large-scale irrigation.

The convection currents set up by the intense heating of the land during the day help to make the desert a windy place. In addition, the sparseness of vegetation and the frequent absence of topographic interruptions allow winds to sweep across the desert plains unimpeded. Sand and dust are carried by the desert winds, lowering visibility and irritating eyes and throats.

ADAPTATIONS BY PLANTS AND ANI-MALS   Deserts tend to have sparse vegetation, and large tracts may be barren bedrock, sand, or gravel. The plants that do exist are **xerophytic,** or adapted to extreme drought. They may have thick bark, thorns, little foliage, and waxy leaves, all of which reduce loss of water by transpiration (Fig. 7.15). Other characteristic adaptations are the storage of moisture in stem or leaf cells as in the cactus (e.g., barrel cactus,

prickly pear, saguaro). Other vegetation, like creosote bush, mesquite, and acacias, have deep root systems to reach water, while still others, such as the Joshua tree, spread their roots widely near the surface for their moisture supply. Many desert plants react to prevent moisture loss through the surface of their leaves in the process of transpiration. The ocotillo, for example, sheds its leaves in dry periods. Others, like the darning-needle cactus and the palo verde, have no leaves. The acacia tree folds its leaves, protecting and shading the inner surface of each leaf from the desert sun. Many desert plants, such as the cactus, desert dandelion, and desert primrose, produce brightly colored flowers, which encourage pollination and thus ensure survival of the species. Still others, like creosote and brittle bush, secrete poison, serving to keep out competitors. In addition to the xerophytic shrub forms, herbaceous annuals spring up periodically as a result of rains. During early spring and again in fall, California's Mojave Desert may be carpeted with flowers. The richness of the herbaceous cover varies from season to season and year to year, as the rains vary in location and amount.

Desert animals also have adapted to the arid conditions. Many keep cool and preserve body water and energy by burrowing beneath the surface, often hibernating when water is insufficient. Most are nocturnal, sleeping during the day and hunting or searching for food and water at night. Some have protective coloration and most are small, which reduces their requirements for both food and water. Reptiles such as lizards and snakes and a variety of rodents are probably the most common and successful of desert animals.

Survival is the key question of everyday life in the desert for animal, plant, and man, as we can see in the following excerpt about the Mojave Desert.

And there are true secrets in the desert. In the war of sun and dryness against living things, life has its secrets of survival. Life, no matter on what level, must be moist or it will disappear. I find most interesting the conspiracy of life in the desert to circumvent the death rays of the all-conquering sun. The beaten earth appears defeated and dead, but it only appears so. A vast and inventive organization of living matter survives by

seeming to have lost. The gray and dusty sage wears oily armor to protect its inward small moistness. Some plants engorge themselves with water in the rare rainfall and store it for future use. Animal life wears a hard, dry skin or an outer skeleton to defy the desiccation. And every living thing has developed techniques for finding or creating shade. Small reptiles and rodents burrow or slide below the surface or cling to the shaded side of an outcropping. Movement is slow to preserve energy, and it is a rare animal which can or will defy the sun for long. A rattlesnake will die in an hour of full sun. Some insects of bolder inventiveness have devised personal refrigeration systems. Those animals which must drink moisture get it at second hand—a rabbit from a leaf, a coyote from the blood of a rabbit.

One may look in vain for living creatures in the day time, but when the sun goes and the night gives consent, a world of creatures awakens and takes up its intricate pattern. Then the hunted

FIGURE 7.15 This diverse association of vegetation is well adapted to survive the year-round drought conditions prevailing in desert climates. (R. Gabler)

come out and the hunters, and hunters of the hunters. The night awakes to buzzing and to cries and barks.

Travels with Charley
John Steinbeck

Even humans, the most adaptable of animals, find the desert environment a lasting challenge. For the most part people have been hunters and gatherers, nomadic herders, and subsistence farmers wherever there was a water supply from wells or *exotic streams* (streams that bring water from outside the region) like the Nile, Tigris, Euphrates, Indus, and Colorado Rivers. They have learned to adjust their habits to the environment. For example, they wear loose clothing to protect themselves from the burning rays of the sun and to prevent moisture loss by evaporation from the skin. At night, when the temperatures drop, the clothing keeps them warm by insulating and minimizing the loss of body heat.

Given an adequate water supply that is carefully applied, desert regions can be agriculturally productive. Soils are little leached and rich in plant nutrients, but they present numerous problems. These often include excessive salinity, the presence of a dense lime subsoil **(caliche),** low organic content, few microorganisms, and excessive permeability. Normally, only alluvial soils are used for agriculture, other types being thin and often rocky.

Great care must be used in irrigating desert lands. Excessive applications of water may raise the water table, causing the rise of saline capillary water that evaporates in the soil or at the surface, leaving its load of salts there. Vast agricultural areas in Iraq, Iran, Pakistan, and elsewhere have been destroyed and abandoned as a consequence of man-induced salinization. The solution to the problem is adequate drainage and the use of irrigation not only to provide water to the crops but also to flush the soil.

Of course, permanent agriculture has been established in desert regions all around the world, wherever river or well water is available. Every desert has its irrigated oases, although they vary considerably in size. Some produce mainly subsistence crops, but others have be-

come significant producers of commercial crops for export.

STEPPE CLIMATES (*BSh, BSk*)

Further study of Figure 7.13 and Table 7.3 provides a reminder that the distribution of the world's steppe lands is closely related to the location of the deserts. Both of the moisture-deficient climate types share the controlling factors of continentality, rain-shadow location, or the subtropical high-pressure cells, or some combination of the three. The transitional nature of the steppes may make them seem like better-watered deserts at one time and like slightly subhumid versions of their humid climate neighbors at another. Herein lies the major problem of steppe regions. How and to what extent should these variable and unpredictable climatic regimes be used by man?

SIMILARITIES TO DESERTS We are already aware that steppe regions are differentiated from deserts by their greater precipitation. For example, while most low-latitude desert locations receive fewer than 25 centimeters (10 in) of rain annually, low-latitude steppe regions usually receive between 25 and 50 centimeters (10 and 20 in). However, the similarities between deserts and steppes are often greater than the differences. In both climates the potential evaporation exceeds the precipitation. As in the deserts, precipitation in steppe regions is unpredictable and varies widely in total amount from year to year. Annual rainfall differs significantly from place to place within both desert and steppe regions and vegetation varies accordingly.

To be more specific, both the general precipitation pattern and the nature of the vegetation of a steppe region are usually closely related to the more humid climate immediately adjacent. That is, when the steppe is located between the desert and the tropical savanna *(Aw),* the steppe's rains come with the high-sun season. Where next to a Mediterranean *(Cs)* climate, the steppe receives primarily winter precipitation. Similarly, the short, shallow-rooted grasses most commonly associated with the

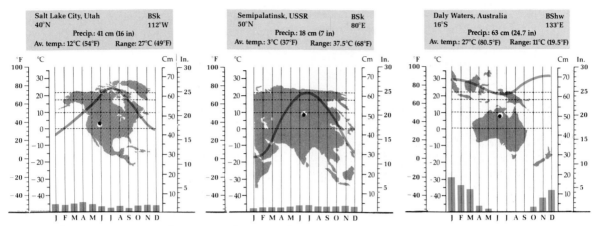

FIGURE 7.16 Climographs for steppe climate stations.

steppe climate occur in the areas of transition from mesothermal *(C)* and microthermal *(D)* climates to the desert. But in the areas of transition from tropical savanna to the desert the vegetation is the dry savanna type, including scrub tree and brush growth, which becomes more stunted and sparse toward the drier margins, until the typical desert shrub vegetation type is dominant.

Due to similar conditions of atmospheric moisture and cloud cover, temperatures in the steppe regions also have much in common with those of the desert. Both low-latitude *(BSh)* and middle-latitude *(BSk)* varieties are identified by mean annual temperature (Table 7.1 and Fig. 7.16). As in the desert, steppe temperatures vary throughout the climate type with latitude, distance from the sea, and elevation. The climographs of Figure 7.16 demonstrate that although summer temperatures are high in all steppe regions, the differences in winter temperatures can produce annual ranges in middle-latitude steppes that are two or three times as great as those in low-latitude steppes.

A DANGEROUS APPEAL Although the surface cover is often incomplete, in more humid regions of the middle-latitude steppes the grasses have been excellent for pasture. In North America this was the realm of the bison and antelope and in Asia the domain of wild horses (Fig. 7.17). Steppe soils are usually high

FIGURE 7.17 It is hard to imagine the magnitude of the animal population on the North American steppes little more than a century ago. One is now fortunate to see a single remnant of the former herds of millions of bison and antelope. The scene pictured here is rare indeed. (R. Sager)

VIEWPOINT: DUST BOWL DANGERS

The end result of plowing land in semiarid regions where periodic droughts are a danger to agriculture. Topsoil that is no longer protected by sod blows away. (U.S. Forest Service)

The Great Plains cut across the United States from Canada south to Texas. The climate of this region is semiarid with aridity increasing toward the west and south. Simply to describe the climate as semiarid, however, does not give an adequate picture of the regime. Precipitation in this area can be highly variable. Total precipitation in a wet year may be as much as three times that in a dry one. Even when the amount of precipitation received is normal, it may be distributed poorly. For example, a good rainfall may be followed by weeks or even months of drought. Then, too, rains received during summer months when temperatures and wind speeds are high and the air relatively dry are far less effective than rains received at other times.

In one sense it is the wet years that cause the greatest problems in the semiarid plains because they encourage settlement and farming in this unpredictable area. Then when drought comes, as inevitably it does, farmers are put under a severe strain.

A look at the Great Plains during the 1920s and 1930s demonstrates what can happen when people gamble with nature. War-time demands for wheat, economic prosperity, and a number of wet years encouraged farmers to push the wheat belt further and further west into the semiarid plains. The increased availability of power equipment further intensified use of this land. The *Yearbook of Agriculture* for 1938 elaborates on the situation of the 1920s and 1930s in the Great Plains.

Natural human optimism, local pride, and commercial interests that profit from new settlement and intensive land development, all exert pressure for an intensive use of the land and an accompanying commercial and social development in the Great Plains that are suitable only to the most productive soils and to climatic conditions prevailing only in the most favorable years. After a succession of years of heavy rainfall in a Great Plains community, the opinion that the climate has changed permanently for the better becomes widespread. Settlement is stimulated. Land prices rise. Range lands are heavily stocked. Large farms and ranches are subdivided to form several small grain farms. The crop acreage is expanded indiscriminately on good land and bad. New business enterprises spring up overnight. New roads and schools are built to serve the growing population, and bonds are issued on the inflated land valuations. . . .

Everything goes well as long as the rains continue. But inevitably the drought years come, as with dismaying regularity they have come since 1930. The crops fail and the pasture and range grasses stop growing. Stockwater reservoirs dry up, and cattle are rushed to market to prevent them from starving or dying of thirst. The dust begins to blow and "black blizzards" lay bare the soil down to the furrow bottoms, pile drifts of dust around the farmsteads and in the fence rows, and make life unbearable for all the people within a radius of several hundred miles. With dwindling farm income, principal and interest payments and taxes become delinquent. Mortgages are foreclosed. The local government finds it difficult to raise sufficient revenue to continue operation, and road and school bond interest payments

in organic matter and soluble minerals. The fertile black soils are most common toward the wetter margins of these regions, while nutrient-rich chestnut and brown soils are found in the drier portions. Attributes such as these have attracted farmers and herdsmen alike to the rich

grasslands but not without penalty to both man and land.

The climate is dangerous for agriculture, and man takes a sizable risk when he attempts to farm. Although dry-farmed wheat and drought-resistant barley and sorghum can be

are defaulted. The Federal Government is called upon to make emergency feed and seed loans, many of which are never repaid. Local business enterprises fail. . . . Families by the score leave the community. Many of those who remain go on relief.

Misuse of the land of the Great Plains has been brought to public attention most forcibly during recent years by the dust storms, which have blanketed the country with great dust clouds as far east at times as the Atlantic seaboard. Wind erosion is serious throughout the Great Plains, except on the eastern margins of the region. In many places 3 or 4 inches of topsoil have been blown away, and in others, dunes 15 to 20 feet high have been formed. In an area covering 20 counties in southwestern Kansas, the Oklahoma Strip, the Texas Panhandle, and southeastern Colorado, a soil-erosion survey by the Soil Conservation Service showed 80 percent of the land more or less affected by wind erosion, 40 percent of it to a serious degree. . . .

Yearbook of Agriculture, 1938
Dept. of Agriculture

Soil disappearing from a plowed field. (USDA)

The Dust Bowl in large part was the result of human interference in the ecosystem. The native prairie and steppe grasses had provided a year-round blanket of protection for the soils. Farmers replaced this natural protection with wheat, which grew from seed in a bare field. Harvesting left the fields bare again, subject to the erosive effects of both wind and rain.

In addition, native plant communities like the prairie grasses can withstand the impact of such natural events as hail, wind, drought, or grasshoppers far better than monocrops such as wheat. This was made painfully clear to the wheat farmers of the 1930s who saw their crops dry up and blow away along with inches of unprotected topsoil. These same farmers had only to look at neighboring unplowed lands to see the grasses, perhaps a little sparse because of the drought, but still standing and still protecting the soil beneath.

The tragedy of the Dust Bowl can be counted in lives wasted, crops destroyed, and in fortunes lost. It must also be counted in the quantities of topsoil removed from the land by wind and rain. Such a loss has effects that reach far into the future. As the land is denuded of its nutrient-rich topsoil, subsequent crops are weaker, less able to withstand hardships, and ultimately worth less in the marketplace.

Largely as a result of what occurred during the Dust Bowl many techniques have been developed to prevent such widespread devastation from happening again. Hardier varieties of wheat have been developed, irrigation is more widely available, and fertilization has increased plant nutrition and therefore stamina. Such techniques have enabled many wheat farmers to make it through the dry years and to reap the benefits of the wet years. However, these dry farmers must always remember the vulnerability of their crop, and that their soil—above all else—must be protected. For once the soil starts to blow, as the farmers of the Dust Bowl discovered, it is difficult to stop. And once it is gone, it takes years to replace.

successfully raised because both farming methods and crops are adapted to the environment, the use of techniques employed in more humid regions can lead to serious problems. Where farmers have attempted to grow crops by normal methods near the wetter margins of the

steppe, they have suffered during the years of drought, which are inevitable and occur at unpredictable intervals. During dry cycles, crops fail year after year, and with the land stripped of its natural sod, the soil is exposed to wind erosion. Even using the grasses for the herding of domesticated animals is not always the answer, for overgrazing can just as quickly create "Dust Bowl" conditions.

The difficulties in making steppe regions more productive point out again the sensitive ecological balance of the earth's systems. The natural rains in the steppe are usually sufficient to support a vegetation cover of short grasses that in turn can feed the roaming herbivores that graze on them. The herbivore population in turn is kept in check by the carnivores who prey on them. But when people enter the scene, sending out more animals to graze, plowing the land, or merely killing off the predators, the ecological balance is tipped and sometimes the results are disastrous.

QUESTIONS FOR DISCUSSION AND REVIEW

1. What kinds of things are included in a description of climate? In your own words describe the climate of the region in which you live by referring to some of these.

2. Describe the microclimate of a covered terrarium, using what you have learned about the interrelationships of plants, air, water, sun, etc. How would the microclimate change if you did not water the plants?

3. Why is scientific classification important? In what ways are the geographer's classification schemes related to regions or areas of the earth's surface?

4. What is the difference between genetic and empirical classification? Give an example of each. Is a statistical classification empirical or genetic?

5. Why is Köppen's climatic classification system considered both statistical and descriptive? What are the advantages and disadvantages of Köppen's system for geographers?

6. Summarize the meaning of the letters Köppen combined to symbolize climatic regions. Using these letter combinations, again describe the climate of the region in which you live. Which description do you prefer, the one in Question 1 or 6? Why?

7. What is a climograph? What is its function?

8. Describe the three tropical (A) climates. Why does each exist?

9. How do ocean currents affect tropical and polar climates?

10. How are the mesothermal (C) and the microthermal (D) climates similar? In what important respect do they differ?

11. In what two important ways do the C and D climates differ internally? What Köppen letters are used to symbolize these differences?

12. What is the major factor producing the dryness of arid climates? What factors produce the variations in highland climates?

13. What is a region as defined by geographers? Give several examples. Why is regionalization such a useful concept?

14. How many areas in the world have climates similar to the one with which you are most familiar? (Plate 12 may help you identify these areas.)

15. What exceptions are there to the general rule that the pattern of climatic regions is produced by earth-sun relationships? How do the Northern and Southern Hemispheres differ in climatic pattern, and why?

16. State in your own words the relationships between climate, vegetation, and soils illustrated by the three diagrams in Figure 7.5.

17. What are the controlling factors that explain the tropical rain-forest climate? Give a brief verbal description of this climatic type.

18. What aspects of tropical rain forests are favorable to human use? Unfavorable? Describe the delicate balance between forest and soil. How might humans affect that balance?

19. What are the differences in climate, vegetation, and human use between the tropical rain forest and the monsoon climate?

20. Explain the seasonal precipitation pattern of the tropical savanna climate. State some of the transitional features of this climate. How have vegetation and soils adapted to the wet-dry seasons?

21. What conditions give rise to desert climates?

22. Describe the characteristics of desert soils. How can they be used for agriculture?

23. What kinds of adaptations have desert animals made to their environment?

24. How do steppes differ from deserts? Why might human use of steppe regions in some ways be more hazardous than use of deserts?

8

CLIMATE:
Middle-Latitude, High-Latitude, and Highland Regions

HUMID MESOTHERMAL CLIMATIC REGIONS

When we use the term *mesothermal* (from the Greek: *mesos,* middle) in describing climates, we are usually referring to the *moderate* temperatures that characterize such regimes. However, we could also be referring to their *middle* position between those climates that have high temperature throughout the year and those that experience severe cold. By definition the mesothermal climates experience seasonality, with distinct summers and winters that distinguish them from the humid tropics. But their summers are long and their winters are mild, and this separates them climatically from the microthermal climates, which lie poleward.

Table 8.1 presents the three distinct mesothermal climates discussed in Chapter 7. In all three the annual precipitation exceeds the annual potential evaporation, but in the Mediterranean (*Cs*) there is a lengthy period of precipitation deficit in the summer season that distinguishes this climate from the humid subtropical (*Cfa*) and the marine west coast

(*Cfb*) climates. The latter two are further differentiated by the fact that the humid subtropical regions have hot summers (*a*) while the marine west coast regions experience mild summers (*b*).

MEDITERRANEAN CLIMATE (*Csa, Csb*)

The Mediterranean climate is one of the best arguments a geographer can use for organizing a study of the environment or developing an understanding of world regions on the basis of climatic classification. Such a climate appears with remarkable regularity in the vicinity of 30° to 40° latitude along the west coasts of each landmass. The alternating controls of subtropical high pressure in summer and westerly wind movement in winter are so predictable that all Mediterranean lands have notably similar and easily recognized temperature and precipitation characteristics (Fig. 8.1). Although soils may differ considerably throughout the climate, this variation is common to all Mediterranean regions. The special appearance, combination, and climatic adaptations of Mediterranean vegetation are not only unusual but are

Fall in New England (R. Gabler)

217

TABLE 8.1   The *C* Climates

NAME AND DESCRIPTION	CONTROLLING FACTORS	GEOGRAPHIC DISTRIBUTION	DISTINGUISHING CHARACTERISTICS	RELATED FEATURES
Mediterranean (*Csa, Csb*) Warmest month above 10°C (50°F); coldest month between 18°C (64.4°F) and 0°C (32°F); summer drought; hot summers (*a*), mild summers (*b*).	West coast location between 30° and 40° N and S latitudes; alternation between subtropical highs in summer and westerlies in winter.	Central California; central Chile; Mediterranean Sea borderlands, Iranian highlands; Capetown area of South Africa; southern and southwestern Australia.	Mild, moist winters, hot, dry summers inland (*Csa*), with cooler, often foggy coasts (*Csb*); high percentage of sunshine; high summer diurnal temperature range; frost danger.	Sclerophyllous vegetation; low, tough brush (chaparral); scrub woodlands; varied soils, erosion in Old World regions; wintersown grains, olives, grapes, vegetables, citrus, irrigation.
Humid Subtropical (*Cfa, Cwa*) Warmest month above 10°C (50°F); coldest month between 18°C (64.4°F) and 0°C (32°F); hot summers; year-round precipitation (*f*), winter drought (*w*).	East coast location between 20° and 40° N and S latitudes; humid, onshore (monsoonal) air movement in summer, cyclonic storms in winter.	Southeastern United States; southeastern South America; coastal southeast South Africa and eastern Australia; eastern Asia from northern India through south China to southern Japan.	High humidity; summers like humid tropics; frost with polar air masses in winter; precipitation 65–250 cm (25–100 in), decreasing inland; monsoon influence (*w*) in Asia.	Mixed forests, some grasslands, pines in sandy areas; strongly leached red-yellow soils; high production with fertilization; rice, wheat, corn, cotton, tobacco, sugarcane, citrus.
Marine West Coast (*Cfb, Cfc*) Warmest month above 10°C (50°F); coldest month between 18°C (64.4°F) and 0°C (32°F); year-round precipitation; mild summers (*b*), cool summers (*c*).	West coast location under the year-round influence of the westerlies; warm ocean currents along some coasts.	Coastal Oregon, Washington, British Columbia, and southern Alaska; southern Chile; interior South Africa; southeast Australia and New Zealand; northwest Europe.	Mild winters, mild summers, low annual temperature range; heavy cloud cover, high humidity; frequent cyclonic storms, with prolonged rain, drizzle, or fog; 3–4 month frost period.	Naturally forested, green year round; strongly leached gray-brown soils; root crops, deciduous fruits, winter wheat, rye, pasture and grazing animals; coastal fisheries.

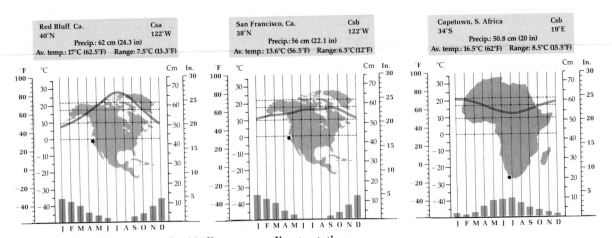

FIGURE 8.1   Climographs for Mediterranean climate stations.

clearly distinguishable from those of other climates. Even among agricultural practices, crops, recreational activities, and architectural styles of Mediterranean lands there are strong similarities.

WARM, DRY SUMMERS; MILD, MOIST WINTERS The major characteristics of the Mediterranean climate are a dry summer, a mild, moist winter, and abundant sunshine (90 percent of possible sunshine in summer and as much as 50 to 60 percent even during the rainy winter season). Summers are warm throughout the climate, but there are enough differences between the monthly temperatures in coastal and interior locations to recognize two distinct subtypes. The moderate summer subtype (*Csb*) has the lower summer temperatures associated with a strong maritime influence. The hot summer subtype (*Csa*) is located further inland and reflects an increased continental influence. Because the inland version has higher summer and daytime temperatures and slightly cooler winter and nighttime temperatures than its coastal counterpart, it has greater annual and diurnal temperature ranges as well. Compare, for example, the annual range for Red Bluff, California, an inland station, with that of San Francisco, a coastal station about 240 kilometers (150 mi) further south (Fig. 8.1).

Whichever the subtype, Mediterranean summers clearly show the influence of the subtropical highs. Weeks go by without a sign of rain, and evaporation rates are high. Effective precipitation is lower than actual precipitation, and the summer drought is as intense as that of the desert. Days are warm to hot, skies are blue and clear, and sunshine is abundant. The high percentage of potential insolation coupled with nearly vertical rays of the noon sun may drive daytime temperatures as high as 30°C to 38°C (86°F to 100°F) except where moderated by a strong ocean breeze or coastal fog.

Fog is common throughout the year in coastal locations and is especially noticeable during the summer. As moist maritime air moves onshore, it passes over the cold ocean currents that typically parallel west coasts in Mediterranean latitudes. The air is chilled, condensation takes place, and fog regularly creeps in during the late afternoon, is present through the night, and "burns off" during the morning hours. As in the desert, radiation is rapid at night and even in summer, temperatures are commonly only 10°C to 15°C (50°F to 60°F). People accustomed to the humid summer nights of the eastern United States do not recognize, without experience, the need to bring a sweater for an evening at Disneyland.

Winter is the rainy season in the Mediterranean climate. The average annual rainfall in these regions is usually between 35 and 75 centimeters (15 and 30 in), with 75 percent or more of the total rain falling during the winter months. The precipitation results primarily from the cyclonic storms and frontal systems common in the westerlies. Annual amounts increase with elevation and decrease with increased distance from the ocean. Only because the rain comes during the cooler months when evaporation rates are lower is there sufficient precipitation to make this a humid climate.

Despite the rain during the winter season there are often many days of fine, mild, weather. Potential insolation is still usually above 50 percent, and the average temperature of the coldest month rarely falls below 4°C to 10°C (40°F to 50°F). Frost is uncommon, and because of its rarity, many less hardy tropical varieties of fruits and vegetables are grown in these regions. Hence, when frost does occur, it can do great damage.

A very severe freeze lasting several days occurred in December, 1972, in the San Francisco Bay area. This killed great numbers of introduced (nonnative) plants, including the eucalyptus trees that had been planted in the Berkeley hills. The following summer this forest of huge dead trees posed an enormous fire threat to the community, in the tinderbox conditions of the summer drought. The threat was so great that government aid was sought to help in the removal of the trees. Interestingly enough, the native vegetation was totally unaffected by the freeze, indicating its adaptation to such conditions.

FIGURE 8.2 Typical chaparral vegetation in California. (R. Sager)

SPECIAL ADAPTATIONS The summer drought, not frost, is the great challenge to vegetation in Mediterranean regions. The natural vegetation reflects the wet-dry seasonal pattern of the climate. During the rainy season the land is covered with lush, green grasses that turn golden and then brown under the summer drought. Only with winter and the return of the rains does the landscape become green again. Much of the natural vegetation is **sclerophyllous** (hard-leaved) and drought-resistant. Like xerophytes, these plants have tough surfaces, shiny, thick leaves that resist moisture loss, and deep roots to help combat aridity.

One of the most familiar plant communities is made up of many low, scrubby bushes that grow together in a thick tangle. In the western United States this is called **chaparral** (Fig. 8.2). The most common of these low bushes in California is the manzanita, a tough shrub with crooked limbs that interlock to form an almost impenetrable barrier. Most chaparral is less than 25 years old because of the frequent fires that occur in this dry brush. People have often removed chaparral as a preventive measure against fires, yet the removal can have disastrous results, because the chaparral acts as a

check against erosion of soils during the rainy season. With the chaparral removed, soils wash or slide down hillsides during the heavy rains of winter, frequently taking homes with them.

Brush similar to the chaparral in California is called *mallee* in Australia, *mattoral* in Chile, and *maquis* in France. In fact, the "maquis" French Underground that fought against Nazi occupation in World War II literally meant *underbrush*.

Where trees appear in the Mediterranean climate they also respond to moisture conditions (Fig. 8.3). Because of their drought-resistant qualities, the needle-leaf pines are among the more common species. Groves of deciduous and evergreen oaks appear in depressions where moisture collects and on the shady north sides of hills where evaporation rates are lower. Where the summer drought is more distinct, the scrub and woodlands open up to parklands of grasses and scattered oak trees. Even the great redwood forests of northern California probably could not survive without the heavy fogs that regularly invade the coastal lands in summer (Fig. 8.4). These great trees demand a unique variation of the usual Mediterranean climate regime, a climatic variation that should

FIGURE 8.3 Mediterranean climates, as represented by this area along coastal California, often exhibit groves of mixed deciduous trees surrounded by grasslands. (R. Gabler)

probably not be considered Mediterranean at all, as it has more in common with the marine west coast climate further north.

The soils of the Mediterranean regions bear a close relationship to the varying vegetative cover, parent materials, and microclimates. In general the dry summers protect the soils from heavy leaching, so they are high in soluble nutrients. Much of the area around the Mediterranean Sea is composed of limestone. This parent material produces a distinctive, calcium-rich, red soil, generally known as *terra rosa*. In an intact condition these soils have excellent structure and are highly productive. However, destructive agricultural practices and overgrazing over thousands of years of human use have caused enormous erosion loss of the soil throughout this region. Today, bare white limestone is widely visible on hillsides in Spain, Italy, Greece, Crete, Syria, Lebanon, Israel, and Jordan, where woodlands once were rooted in mature soils. The situation is quite different in the Mediterranean climatic areas of California, where limestone is rare. Here the soils are derived from nonlimy marine deposits and have poor structure because of the unusual properties of sodium, which these soils contain in excess. The sodium is leached from such parent materials in more humid climates. The soils are dense clays, gluelike when wet and hard as concrete when dry. In the period of

Spanish domination the clay was formed into adobe bricks and used for building material. Such clay soils are accordingly referred to in California as *adobe*. These clay soils are also hazardous on slopes, for they absorb so much water during the wet season that they behave almost like a fluid, producing mud flows that frequently destroy roads and homes.

In all Mediterranean regions the most productive soils are found in the alluvial lowlands. Here farmers have made their special adaptations to climatic conditions. There is sufficient rainfall in the cool season to permit fall planting and spring harvesting of winter wheat and barley. These grasses originally grew wild in the eastern Mediterranean region. Grapevines, fig and olive trees, and the cork oak, which undoubtedly were also native to Mediterranean lands, are especially well adapted to the dry summers because of their deep roots and thick, well-insulated stems or bark. Where water for irrigation is available, an incredible diversity of crops may be seen (Fig. 8.5). These include, in addition to those already mentioned, oranges, lemons, limes, melons, dates, rice, cotton, deciduous fruits, various types of nuts, and countless vegetables. California, blessed with fertile valleys for growing fruits, vegetables, and flowers, as well as with snow meltwater for irrigation, is probably the most agriculturally productive of the Mediterranean regions (Fig. 8.5).

FIGURE 8.4 California redwoods (*Sequoia sempervirens*) may reach heights of 100 meters (330 ft) and live for thousands of years. (R. Sager)

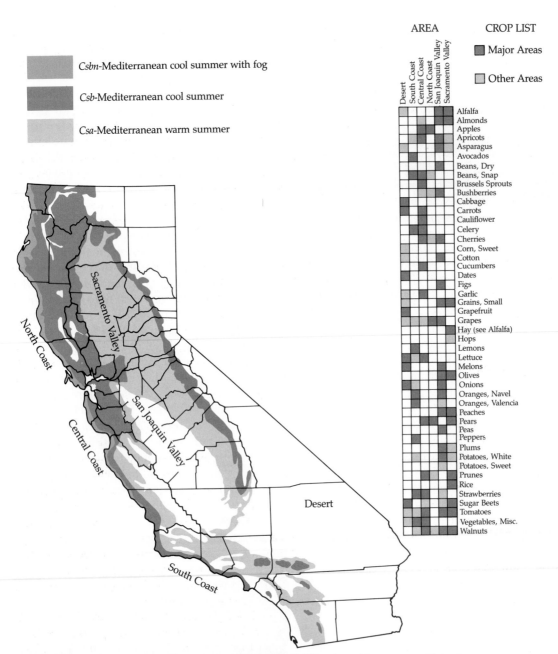

FIGURE 8.5 Areas of Mediterranean climate in California, with centers of production of selected commercial crops. (From Durrenburger, *Patterns on the Land*, 4th ed., 1972. Reprinted by permission of Mayfield Publishing Company, formerly National Press Books)

Even the houses of these regions show people's adaptation to the climate. Usually white or pastel in color, they gleam in the brilliant sunshine against clear blue skies. Many have shuttered windows to cut the glare of the sun and to keep the houses cool in summer. On the other hand, much less attention is paid to keeping a place warm during the cooler winter months. This is true even in the United States where many midwesterners and easterners are surprised at the lack of insulation in California homes and the small number and size of heating devices used.

The following quotation describes San Francisco and some of her Mediterranean flavor. Notice the references to Italy, to acropolis (the upper, fortified part of an ancient Greek city), to the green hills, and to the evening fog.

San Francisco put on a show for me. I saw her across the bay, from the great road that bypasses Sausalito and enters the Golden Gate Bridge. The afternoon sun painted her white and gold—rising on her hills like a noble city in a happy dream. A city on hills has it over flat-land places. New York makes its own hills with craning buildings, but this gold and white acropolis rising wave on wave against the blue of the Pacific sky was a stunning thing, a painted thing like a picture of a medieval Italian city which can never have existed. I stopped in a parking place to look at her and the necklace bridge over the entrance from the sea that led to her. Over the green higher hills to the south, the evening fog rolled like herds of sheep coming to cote in the golden city. I've never seen her more lovely.

Travels with Charley
John Steinbeck

HUMID SUBTROPICAL CLIMATE (*Cfa*)

The humid subtropical climate extends inland from continental east coasts between 15° to 20° and 40° N and S latitude (refer again to Table 8.1 and Plate 12). Thus it is located within approximately the same latitudes and in a similar transitional position as the Mediterranean climate but on the eastern instead of western continental margins. There is ample evidence of this climatic transition. Summers in the humid subtropics are similar to the humid tropical climates further equatorward. When the noon sun is nearly overhead, these regions are subject to the importation of moist tropical air masses. High temperatures, high relative humidity, and frequent convectional showers are all characteristics that they share with the tropical climates. In contrast, during the winter months, when the pressure and wind belts have shifted equatorward, the humid subtropical regions are more commonly under the influence of the cyclonic and frontal systems of the continental middle latitudes. Polar air masses can bring colder temperatures and occasional frost.

COMPARISON WITH THE MEDITERRANEAN CLIMATE

Like the inland version of the Mediterranean climate, the humid subtropical climate has mild winters and hot summers. But it has no dry season. Whereas the Mediterranean lands are under the strongest influence of the subtropical highs, the humid subtropical regions are located on the weak western sides of the subtropical high-pressure cells. Subsidence and stability are greatly reduced or are absent even during the summer months. The warm ocean currents that are commonly found along continental east coasts in these latitudes also moderate the winter temperatures and warm the lower atmosphere, producing steep lapse rates and instability. Furthermore, there is a modified monsoon effect, especially in Asia, but also in the southern United States. This effect increases summer precipitation as the moist, unstable tropical air is drawn in over the land.

As might be expected from the year-round rainfall, average annual precipitation for humid subtropical locations usually exceeds that for Mediterranean stations and may vary more widely as well. Humid subtropical regions receive anywhere from 60 to 250 centimeters (25 to 100 in) a year. Precipitation generally decreases inland toward continental interiors and away from the oceanic sources of moisture. Not surprisingly, these regions are noticeably drier the closer they are to steppe regions inland, toward their western margins.

Both the Mediterranean and humid subtropical climates receive winter moisture from cyclonic storms and frontal systems. As we have noted, the great contrast occurs in the summer when the humid subtropics receive much precipitation from convectional showers, supplemented in certain regions by a modified monsoon effect. In addition, because of the shift in the sun and wind belts during the summer months, the humid subtropical climates are subject to tropical storms, some of which develop into hurricanes (or typhoons), especially

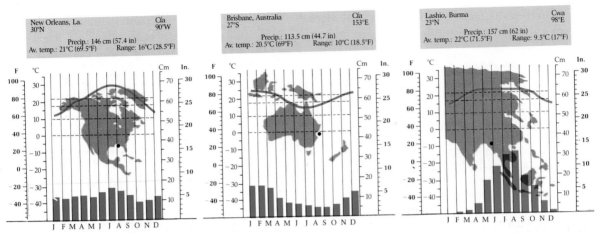

FIGURE 8.6 Climographs for humid subtropical stations.

in late summer. These three factors—the modified monsoon effect, convectional activity, and tropical storms—combine to produce a precipitation maximum in most of these regions in late summer. The climographs for New Orleans, Louisiana, and Brisbane, Australia, are illustrative of these effects (Fig. 8.6).

A subtype of the humid subtropical climate is found most often on the Asian continent, where the monsoon effect is most pronounced because of the magnitude of the seasonal pressure changes over this immense landmass. There, the low-sun period or winter season is noticeably drier than the high-sun period. High pressure over the continent blocks the importation of moist air, so that some months receive less than 3 centimeters (1.2 in) of precipitation. The Köppen symbol *w* is used to indicate a dry winter season. Thus there are subtropical climates with a dry winter that are symbolized by *Cwa*. The climograph for Lashio, Burma, illustrates the subtropical wet-dry *Cwa* climate.

Temperatures in the humid subtropics are much like those of the Mediterranean climates. Annual ranges are similar despite a greater variation among stations in the humid subtropical climate, primarily because the climate covers a far larger land area. Mediterranean climates record higher summer daytime temperatures, but summer months in both climates average around 25°C (77°F), increasing

to as much as 32°C (90°F) as maritime influence decreases inland. Winter months in both climates average around 7° to 14°C (45° to 57°F). Frost is a similar problem. The long growing season in the warmest humid subtropical regions enables farmers to grow such delicate crops as oranges, grapefruit, and lemons, but, as in the Mediterranean climates, they must be prepared with various means to protect their more sensitive crops from the danger of freezing. The growers of citrus crops in Florida have concentrated in the central lake district to take advantage of the moderating influence of nearby bodies of water.

The big difference between temperatures of humid subtropical and Mediterranean climates is how they feel. The summer temperatures in humid subtropical regions feel far warmer than they are because of the high humidity there. In fact, summers in this climate are oppressively hot, sultry, and uncomfortable. Nor is there the relief of lower night temperatures, as in the Mediterranean regions. The high humidity of the humid subtropical climate prevents much reradiation of heat at night. Consequently the air remains hot and sticky. Diurnal ranges, in winter as well as in summer, are far smaller in the humid subtropical than in the drier Mediterranean climates. But despite the relatively mild temperatures, humid subtropical winters seem cold and damp, again because of the high humidity.

Plate 11   Varying Landscapes Reflecting Different Climatic Controls

LATITUDE

Tropical *(A)* climate. Island of Jamaica.

Mesothermal, mild winter *(C)* climate. Summer in southern Spain.

Microthermal, severe winter *(D)* climate. Winter in Illinois.

Polar *(E)* climate. Glaciers in the Alaska Mountain Range.

ATMOSPHERIC PRESSURE

ALTITUDE

Arid *(B)* climate. Mojave Desert, southern California.

Highland *(H)* climate. Grand Tetons, near Jackson Lake, Wyoming.

(All photos by R. Gabler)

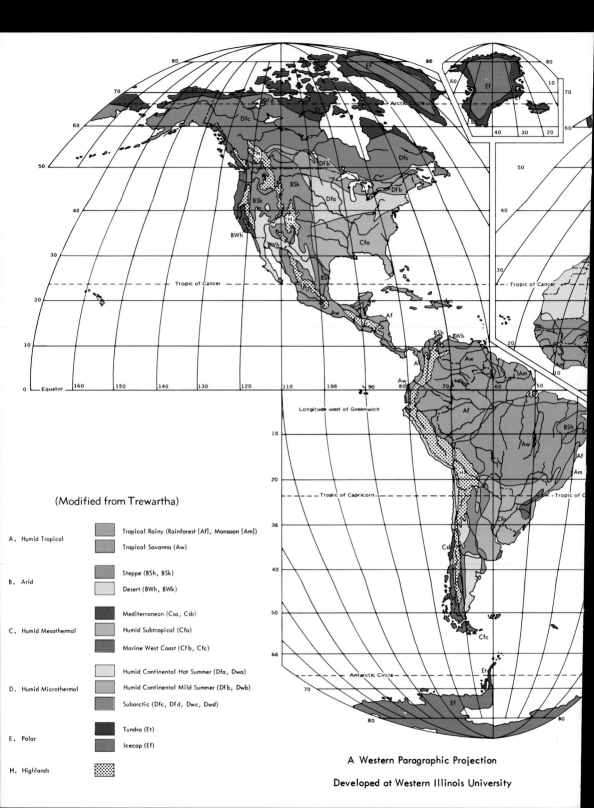

(Modified from Trewartha)

A. Humid Tropical

Tropical Rainy (Rainforest [Af], Monsoon [Am])

Tropical Savanna (Aw)

B. Arid

Steppe (BSh, BSk)

Desert (BWh, BWk)

C. Humid Mesothermal

Mediterranean (Csa, Csb)

Humid Subtropical (Cfa)

Marine West Coast (Cfb, Cfc)

D. Humid Microthermal

Humid Continental Hot Summer (Dfa, Dwa)

Humid Continental Mild Summer (Dfb, Dwb)

Subarctic (Dfc, Dfd, Dwc, Dwd)

E. Polar

Tundra (Et)

Icecap (Ef)

H. Highlands

A Western Paragraphic Projection

Developed at Western Illinois University

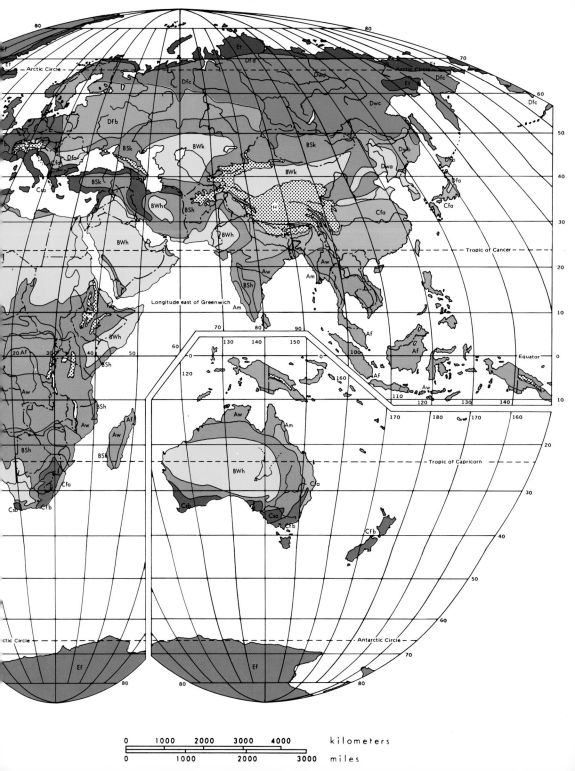

Arctic Circle

Et

Dfd

Dfc

Dwd

Dfc

Dwc

Dfc

60

Dfb

BSk

BWk

BSk

50

Dfc

Dfa

Dwb

Dfb

Dfa

Csa

BSk

BWh

BSh

BWk

Dwa

Cfa

Dfa

40

BWh

H

Cfa

30

BWh

Tropic of Cancer

BWh

Aw

20

Longitude east of Greenwich

BSh

Am

Aw

Am

10

70   80   90

20 Af

30   40   50

130   140   150

0   0

Af

Equator   0

60

BWh

120

160

110   100

Af

BSh

170

Af

Aw

Aw

Af

10

Aw

180   170   160

Af

Am

120   130   140

BSh

Aw

Af

Aw

20

BSh

BSk

Tropic of Capricorn

BWh

Cfa

30

Cfa

Csb   Cfb

Csb

Cfb

Csa

Cfb

40

Cfb

50

60

Antarctic Circle

70

ctic Circle

Ef

Ef

80

80

80

kilometers

| 0 | 1000 | 2000 | 3000 | 4000 |

miles

| 0 | 1000 | 2000 | 3000 |

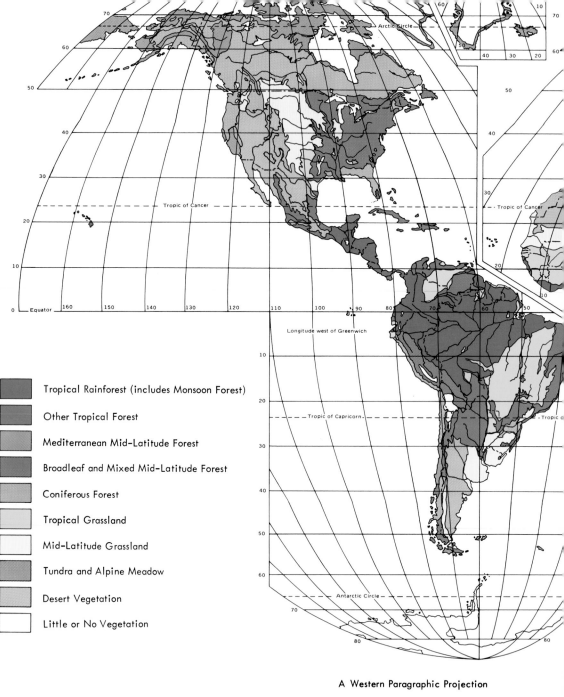

Tropical Rainforest (includes Monsoon Forest)

Other Tropical Forest

Mediterranean Mid-Latitude Forest

Broadleaf and Mixed Mid-Latitude Forest

Coniferous Forest

Tropical Grassland

Mid-Latitude Grassland

Tundra and Alpine Meadow

Desert Vegetation

Little or No Vegetation

A Western Paragraphic Projection

Developed at Western Illinois University

©1975 by Rinehart Press

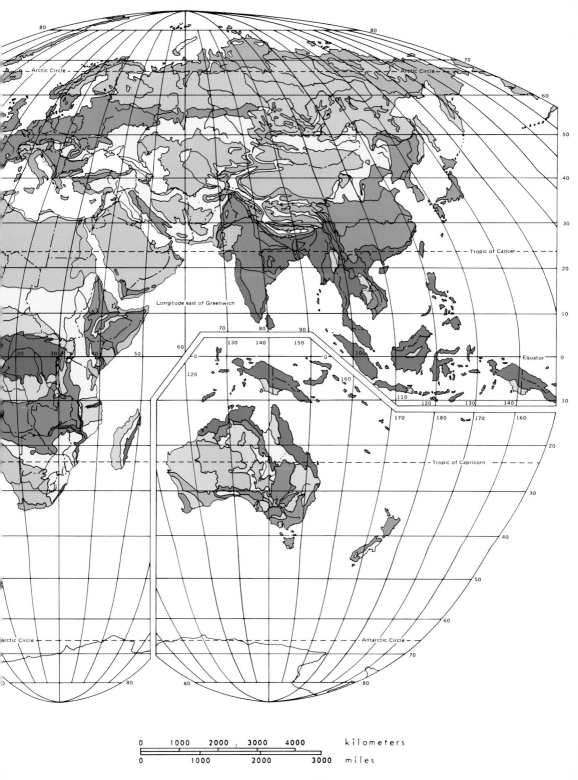

Arctic Circle — Arctic Circle —

80

70

60

50

40

30

Tropic of Cancer —

20

10

Longitude east of Greenwich

Equator 0

Tropic of Capricorn —

20

30

40

50

60

70

80

0	1000	2000	3000	4000		kilometers
0		1000		2000	3000	miles

Tropical rain forest (Af). Pineapple fields, island of Oahu, Hawaiian Islands. *(R. Gabler)*

n forest **(Af).** Amazon Basin, Peru.

Tropical savanna (Aw). Tanzania, Africa. *(R. Sager)*

anna **(Aw).** Central Mexico. *(R. Gabler)*

Desert (BW). North central Mexico. *(Courtesy Alphons*

Humid subtropical *(Cfa)*. Orange groves in central Florida. *(R. Gabler)*

Humid subtropical *(Cfa)*. Lakeshore near Hakone, Japan. *(R. Gabler)*

Marine west coast *(Cfb)*. Coast of British Columbia. *(R. Sager)*

Marine west coast *(Cfb)*. Northern coast of Scotland. *(R. Gabler)*

...oast near Barcelona, Spain

Humid continental, hot summer (Dfa). Summer in central Illinois. *(R. Gabler)*

Humid continental, hot summer (Dfa). Glaze covered trees during winter in central Illinois. *(R. Gabler)*

Humid continental, mild summer (Dfb). Fall in New England. *(R. Gabler)*

Subarctic (Dfc). Southern edge of Canadian shield near Georgian Bay. *(R. Gabler)*

Highland (H). High in the mountains of Wyoming.

Highland (H). Yosemite National Park, California.

A PRODUCTIVE CLIMATE   Vegetation generally thrives in humid subtropical regions with the abundant rainfall, high temperatures, and long growing season. In the wetter portions there are forests of broad-leaf deciduous trees, pine forests on sandy soils, and mixed forests. In the drier interiors near the steppe regions, forests give way to grasslands, which require less moisture. There is an abundant and varied fauna. A few of the common species are deer, bears, foxes, rabbits, squirrels, oppossums, raccoons, skunks, and birds of many sizes and species. Bird life in lake and marsh areas is incredibly rich (Fig. 8.7). Alligators inhabit the American swamps, such as the Dismal Swamp of North Carolina and the Okefenokee in Georgia.

Soils tend to be strongly leached, though not so thoroughly as their tropical counterparts. Their red and yellow colors are derived from the insoluble iron and aluminum compounds that remain in the upper layers. These soils tend to lack humus, because microorganisms act efficiently to break it down in the high temperatures. Only in limited areas are the soils more naturally fertile. In the drier grasslands, such as the pampas of Argentina and Uruguay, there is more humus and less leaching. This is South America's "bread basket." Such conditions occur in the Black Belt of Alabama and on the Black Prairies of Texas.

FIGURE 8.8   The subtropical regions of Japan are ideally suited for the production of rice. (Courtesy of Ronald W. Bradley)

Whatever the soils, the humid subtropical regions are of enormous agricultural value because of their favorable temperature and moisture characteristics. They have been used intensively for both subsistence crops, such as rice and wheat in Asia, and commercial crops, like cotton and tobacco in the United States. When we consider that this (with its monsoon phase) is the characteristic climate of south China, as well as of the most densely populated portions of both India and Japan, we realize that this climatic regime contains and feeds far more human beings than any other type (Fig. 8.8). The care with which agriculture has been practiced and the soil resource conserved over thousands of years of intensive use in eastern Asia is in sharp contrast to the agricultural exploitation of the past 200 years in the corresponding area of the United States. The traditional system of cotton and tobacco farming, in particular, devastated the land by exhausting the soil and triggering massive sheet and gully erosion. Over much of the old cotton and tobacco belt, extending from the Carolinas through Georgia and Alabama to the Mississippi delta, all of the topsoil or a significant part of it has been lost, and we see only the red clay subsoil and occasionally bare rock where crops formerly flourished (Fig. 8.9). In the remaining areas practices have had to change to conserve the soil that is left. Heavy applications of fertil-

FIGURE 8.7   Flamingos are well suited for the humid subtropical climate of central Florida. (R. Gabler)

FIGURE 8.9 Erosion of the topsoil in the humid subtropical climate of the southern Piedmont often exposes the clay subsoil. (R. Gabler)

izer, scientific crop rotation, and careful tilling of the land are now the rule.

Where forests still form the major natural vegetation, forest products, such as lumber, pulp wood, and turpentine, are important commercially. The long-leaf and slash pines of the southeastern United States have long been the world's leading source of naval stores, the resinous products of the pine tree (pitch, tar, resin, and turpentine). The absence of temperature and moisture limitations strongly favor forest growth. In Georgia, for example, trees may grow two to four times faster than in colder regions such as New England. This means that trees can be planted and harvested in much less time than in cooler forested regions, offering distinct commercial advantages.

The fact that living things thrive in the humid subtropical climate presents certain problems, however, for parasites and disease-carrying insects thrive along with other forms of life. The boll weevil, for example, has caused enormous damage to the cotton industry of the southern United States since its immigration from Mexico in 1888.

Agriculture and lumbering are not the only important industries in the humid subtropical climates. In the southeastern United States, livestock raising has been increasing greatly in importance. This trend began when breeds of beef and dairy cattle were developed that could tolerate the hot summers of these regions. Tropical breeds from Africa and India were crossed with European types to produce the desired characteristics. Because of the mild climate, pastures remain green throughout the year and cattle do not need to be housed indoors. In addition, the vegetation grows so rapidly that it requires far less land to support each animal than in cooler or more arid climates.

Despite the commercial advantages of this climate, people often find it an uncomfortable one in which to live. The development and spread of air conditioning helps to mitigate this problem. Where the ocean offers relief from the summer heat, as in Florida, the humid subtropical climate is an attractive recreation and retirement region (Fig. 8.10). The beauty of its more unusual features, such as its cypress swamps and forests draped with Spanish moss, has to be experienced to be fully appreciated.

MARINE WEST COAST CLIMATE (*Cfb, Cfc*)

Proximity to the sea and prevailing onshore winds make the marine west coast climate one

FIGURE 8.10 Daytona Beach, with its expansive area of white sand, is a major tourist destination. Note the cumulus clouds associated with the strong convection over land that produces the daytime sea breeze. (R. Gabler)

graph for Reykjavik, Iceland, represents the cool summer variety (*Cfc*).

OCEANIC INFLUENCES As they come from the oceans, the westerlies carry with them the moderating marine influence on temperature, as well as much moisture. In addition, warm ocean currents, such as the North Atlantic Drift, bathe some of the coastal lands in the latitudes of the marine west coast climate, further moderating climatic conditions and accentuating humidity. This latter influence is particularly noticeable in Europe, where the marine west coast climate extends along the coast of Norway to beyond the Arctic Circle (Plate 12).

In this climatic zone the marine influence is so strong that temperatures decrease little with poleward movement. Thus the influence of the oceans is even stronger than latitude in determining temperatures. This is obvious when we examine isotherms in the areas of marine west coast climates on a map of world temperatures (Figs. 3.19 and 3.20). Wherever the marine west coast climate prevails, the isotherms swing poleward, parallel to the coast, clearly demonstrating the dominant marine influence.

Another result of the ocean's moderating effect is that the annual temperature ranges in the marine west coast climates are relatively small, considering the latitude. For an illustra-

of the most temperate in the world. Thus it is sometimes known as the *temperate oceanic climate*. Found in those midlatitude regions (between 40° and 65°) that are continuously influenced by the westerlies, the marine west coast climate receives ample precipitation throughout the year. However, unlike the humid subtropical climate just discussed, it has mild (*b*) to cool (*c*) summers. The climographs for Bordeaux, France, and Stuttgart, Germany, in Figure 8.11 are representative of the mild summer marine west coast climate (*Cfb*), and the climo-

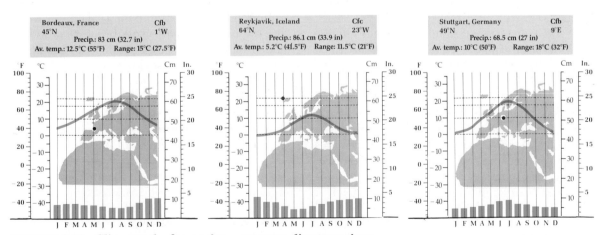

FIGURE 8.11 Climographs for marine west coast climate stations.

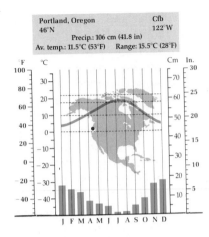

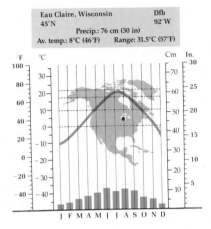

FIGURE 8.12 Effect of maritime influence on climates of two stations at the same latitude. Portland, Oregon (Cfb), exemplifies the maritime influences dominating marine west coast climates. Eau Claire, Wisconsin (Dfb), shows the effect of location in the continental interior.

tion of this, compare the monthly temperature graphs for Portland, Oregon, and Eau Claire, Wisconsin (Fig. 8.12). Though these two cities are at the same latitude, the annual range at Portland is 15.5°C, while at Eau Claire it is 31.5°C. The moderating effect of the oceans on the temperatures at Portland in both winter and summer is clearly in contrast to the effect of an inland continental position on the temperatures at Eau Claire.

Diurnal temperature ranges are also smaller than they are in other climatic regions at similar latitudes and in more arid climates. Heavy cloud cover and high humidity, both in summer and winter, diminish daytime heating, and also prevent much reradiation of heat at night. Consequently the difference between the daily maximum and minimum temperatures is small.

Of course these climographs and climate statistics are primarily averages, and this can be misleading. Marine west coast climatic regions experience the unpredictable weather conditions associated with the polar front. Occasional invasions of a tropical air mass in summer or a polar air mass in winter can move against the general westerly flow of air in these latitudes and produce surprising results. For example, under just such weather conditions, temperatures in Seattle, Washington, have reached a high of 38°C (100°F) and a low of −19°C (−3°F).

Despite the insulating effect of cloudy skies and high moisture content of the air, which slows heat loss at night, frost is a signifi-

cant factor in the marine west coast climate. It occurs more often, may last longer, and is more intense than in other mesothermal regions. The growing season is limited to 8 months or less, but even during the months when freezing temperatures may occur, only half the nights, or less, may experience them. The possibility of frost and the frequency of its occurrence increase inland far more rapidly than they do poleward, once more illustrating the importance of the marine influence.

As final evidence of oceanic influences, study the distribution of the marine west coast climate on Plate 12. Where mountain barriers prevent the movement of maritime air inland, this climate is restricted to a narrow coastal strip, as in the Pacific Northwest of North America and in Chile. Where the land is surrounded by water, as in New Zealand, or where the air masses move across broad plains, as in much of northwestern Europe, the climate extends well into the interior of the landmass.

CLOUDS AND PRECIPITATION The marine west coast has a justly deserved reputation as one of the cloudiest, foggiest, rainiest, and stormiest climates in the world. This is particularly true during the winter season. Rain or drizzle may last off and on for days, though the amount of rain received is small for the number of rainy days recorded. Even when rain is not falling, the weather is apt to be cloudy or foggy. Advection fog may be especially common and long-lasting in the winter months when air masses pass over warm ocean currents before striking the colder land. The cyclonic storms and frontal systems are also strongest in the winter when the subtropical highs have shifted equatorward. Conspicuous winter maximums in rainfall occur near the coasts and near boundaries with the Mediterranean climate. However, further inland a summer maximum may occur, as in Stuttgart, Germany (Fig. 8.11).

Though all parts of this climatic type receive ample precipitation, there is much greater station-to-station variation in precipitation averages than in temperature statistics. Precipitation tends to decrease very gradually as one moves inland, away from the oceanic source of moisture. It also decreases equatorward, especially during summer months, as the influence of the subtropical highs increases and the influence of the westerlies decreases. This can bring about periods of beautiful, clear weather, something rarely associated with this climate but not uncommon in our Pacific Northwest.

The most important factor in the amount of precipitation received is local topography. When a mountain barrier like the Cascades in the Pacific Northwest or the Andes in Chile parallels the coast, abundant precipitation, both cyclonic and orographic, falls on the windward side of the mountains. Valdivia, Chile, located windward of the Andes, receives an average of 267 centimeters (105 in) of precipitation a year. A similar location in Canada, Henderson Lake, British Columbia, averages 666 centimeters (262 in) of rain a year, the highest figure for the entire North American continent. During the colder Pleistocene Epoch these high precipitation amounts, falling largely as winter snow, produced large mountain glaciers. In many cases these came down to the sea, excavating deep troughs that now appear as steep-walled inlets or fjords. Fjord coasts are present in Norway, British Columbia, Chile, and New Zealand—all areas of marine west coast climates today (Fig. 8.13). In contrast, where there are lowlands and no major landforms of high elevation, precipitation is spread more evenly over a wide area, and the amount received at individual stations is more moderate (50 to 75 centimeters (20 to 30 in) annually). This is the situation in much of the Northern European Plain, extending from western France to eastern Poland.

Two aspects of precipitation are directly related to the moderate temperatures of this climatic regime. Snow falls infrequently, and

FIGURE 8.13 Scenic fjords, such as this one in Norway, were produced by glacial erosion during the Pleistocene Epoch. (S. Brazier)

FIGURE 8.14 Reliable precipitation makes a diversified type of agriculture possible in the marine west coast climatic areas, with emphasis upon grain, orchard crops, vegetables, and dairying. (R. Gabler)

when it does, it melts or turns to slush as soon as it hits the ground. Snow is especially rare in lowland regions of this climatic zone. Paris averages only 14 snow days a year; London, 13; and Seattle, 10. In addition, thunderstorms and convectional showers are uncommon, although they occur occasionally. Even in summer, surface heating is rarely sufficient to produce the towering cumulonimbus clouds.

CLIMATE AND CIVILIZATION   Some geographers, especially in Western Europe and the United States, have noted the association of the marine west coast climate with the development of modern, diversified agriculture and technology-based civilization. There is little doubt that this climate offered certain advantages. The narrow annual temperature ranges, mild winters, long growing seasons, and abundant precipitation all favor plant growth. Many crops, such as wheat, barley, and rye, can be grown further poleward than in more continental regions. Although the soils common to these regions are not naturally rich in soluble nutrients, humus readily accumulates and highly successful agriculture is possible with the application of natural or commercial fertilizers (Fig. 8.14). Root crops, such as potatoes, beets, and turnips; deciduous fruits, such as apples and pears; berries; and grapes join the grains

previously mentioned as important farm products. Grass in particular requires little sunshine, and pastures are always lush. The greenness of Ireland—the Emerald Isle—is evidence of these favorable conditions, as is the frequency of herds of fat beef and dairy cattle.

The magnificent forests that form the natural vegetation of the marine west coast regions have been a readily available resource. Some of the finest stands of commercial timber in the world are found along the Pacific Coast of North America, where pines, firs, and spruces abound, commonly exceeding 30 meters (100 ft) in height (Fig. 8.15). Europe and the British Isles were once heavily forested, but most of those forests have been cut down (even the famous Sherwood Forest) for building material and have been replaced by agricultural lands and urbanization.

A cause and effect relationship between climate and civilization or between the environment and human activity is considered invalid today. But at least one geographer seemed to have other views. Ellsworth Huntington described the climate of the British Isles as the best climate for human efficiency and stimulation. As he wrote in 1945,

The relative uniformity of the rainfall permits the farmers to know what they can count on. It lessens the danger of debt, want, and misery by rea-

FIGURE 8.15 Commercial timber is the outstanding natural resource of the marine west coast climatic region in North America. (R. Gabler)

son of unforeseen crop failures. . . . Here, too, other factors, including low temperature and unseasonable frosts in the north, are significant, but the good effect of storms is dominant. . . .

Such conditions stimulate the progress of civilization by encouraging people to try new methods. In places where drought and frost often ruin the crops, a feeling of despair or fatalism tends to prevail. Bad weather foils the farmer so often that when new methods are suggested he says, "What's the use? If God sends rain, the crop will be good. If he doesn't, we starve." People need the challenge of difficulty, but they become apathetic and fatalistic when the difficulties are too great. In thoroughly cyclonic regions, on the contrary, the farmer is almost certain to get at least a reasonably good crop each year. If he tries improved methods, the chances are large that his extra effort will not be wasted. This encourages a progressive attitude in contrast to the fatalistic, inert attitude so common in regions where unreliable rainfall causes sharp alternations between good crops and poor. Such psychological contrasts help to explain many historical events in which "national character," especially optimism and initiative versus fatalism and apathy, plays a part.

Mainsprings of Civilization
Ellsworth Huntington

Huntington's analysis may hold true for some, of course, but many others find the frequent gloom of clouds, fog, and drizzle downright depressing. And in winter, when the damp cold penetrates clothes, flesh, and bones, British weather can seem anything but comfortable.

HUMID MICROTHERMAL CLIMATIC REGIONS

Our definition of humid microthermal includes high enough temperatures during part of the year to have a recognizable summer and cold enough temperatures six months later to have a distinct winter. In between there are the two periods we call spring and fall when all life, and especially vegetation, makes preparation for the temperature extremes. Thus in this section we will be talking about climatic regions that clearly display four readily identifiable seasons.

However, seasonality is not the only explanation for why we often use the word *variable* when describing the humid microthermal climates. As Plate 12 indicates, these climates are generally located between 35°N and 75°N on the North American and Eurasian landmasses. Thus they share the fronts and storms, the variable weather, of the westerlies and the polar front with the marine west coast climate. But their position in the continental interiors and in high latitudes prevents them from experiencing the moderating influence of the oceans. In fact, the dominance of continentality in these climates is best demonstrated by the fact that they do not exist in the Southern Hemisphere where there are no large landmasses in the appropriate latitudes.

The recognition of three separate microthermal climates is based mainly on latitude and the resulting differences in the length and severity of the seasons (Table 8.2). Winters tend to be longer and colder toward the poleward margins because of latitude and toward interiors throughout the microthermal climates because of the continental influence. Summers inland are also inclined to be hotter, but they

become progressively shorter as the winter season lengthens poleward. Thus the three microthermal climates can be defined as humid continental, hot summer (*Da*); humid continental, mild summer (*Db*); and subarctic, with a cool summer (*Dc*) or, in extreme cases, a long, bitterly cold winter (*Dd*).

All microthermal climates have several features in common. By definition they all experience a surplus of precipitation over potential evaporation, and, with the exception of an area in Asia that experiences winter drought (*w*) because of the strong Siberian High, they have year-round rainfall (*f*). However, the greater frequency of maritime tropical air

TABLE 8.2 The *D* Climates

NAME AND DESCRIPTION	CONTROLLING FACTORS	GEOGRAPHIC DISTRIBUTION	DISTINGUISHING CHARACTERISTICS	RELATED FEATURES
Humid Continental Hot Summer (*Dfa, Dwa*) Warmest month above 10°C (50°F); coldest month below 0°C (32°F); hot summers; year-round precipitation (*f*), winter drought (*w*).	Location in the lower middle latitudes (35° to 45°); cyclonic storms along the polar front; prevailing westerlies; continentality; polar anticyclone in winter (*w*).	Eastern and midwestern U.S. from Atlantic coast to the 100th meridian; east central Europe; northern China, Manchuria, northern Korea, and Honshu.	Hot, often humid summers; occasional winter cold waves; rather large annual temperature ranges; weather variability; precipitation 50–115 cm (20–45 in) decreasing inland and poleward; 140–200 day growing season.	Broad-leaf deciduous and mixed forest; moderately leached graybrown soils in wetter areas; grasslands, prairie and highly fertile black soils in drier areas; "corn belt," soybeans, hay, oats, winter wheat.
Humid Continental Mild Summer (*Dfb, Dwb*) Warmest month above 10°C (50°F); coldest month below 0°C (32°F); mild summers; year-round precipitation (*f*), winter drought (*w*).	Location in the middle latitudes (45° to 55°); cyclonic storms along the polar front; prevailing westerlies; continentality; polar anticyclone in winter (*w*).	New England, the Great Lakes region, and south central Canada; southeastern Scandinavia; eastern Europe, west central USSR; eastern Manchuria and USSR; Hokkaido.	Moderate summers; long winters with frequent spells of clear, cold weather; large annual temperature ranges; variable weather with less total precipitation than further south; 90–130 day growing season.	Mixed or coniferous forest; moderately leached graybrown soils in wetter areas; grasslands, prairie and highly fertile black soils in drier areas; spring wheat, corn for fodder, root crops, hay, and dairying.
Subarctic (*Dfc, Dwc, Dfd, Dwd*) Warmest month above 10°C (50°F); coldest month below 0°C (32°F); cool summers (*c*), cold winters (*d*), year-round precipitation (*f*), winter drought (*w*).	Location in the higher middle latitudes (50° to 70°); westerlies in summer, strong polar anticyclone in winter (*w*); occasional cyclonic storms; extreme continentality.	Northern North America from Newfoundland to Alaska; northern Eurasia from Scandinavia through most of Siberia to the Bering Sea and the Sea of Okhotsk.	Brief, cool summers; long, bitterly cold winters; largest annual temperature ranges; lowest temperatures outside Antarctica; low (25–50 cm/10–20 in) precipitation; unreliable 50–80 day growing season; permafrost common.	Northern coniferous forest (taiga); strongly acidic soils; poor drainage and swampy conditions in warm season; experimental vegetables and root crops.

FIGURE 8.16   Map of the contiguous United States showing average annual number of days of snow cover. (After USDA)

masses in summer and continental polar air masses in winter, the monsoon effect, and strong summer convection combine to produce a precipitation maximum in summer. Although the length of time snow remains on the ground increases poleward and toward the continental interior (see Fig. 8.16), all three microthermal climates experience significant snow cover. This decreases the effectiveness of insolation and helps to explain their cold winter temperatures. Finally, the unpredictable and variable nature of the weather is especially apparent in the humid microthermal climates and is present throughout all microthermal regions.

With these generalizations in mind, compare the microthermal climates with the mesothermal climates we have just examined. Regions with microthermal climates have more severe winters, a lasting snow cover, shorter summers, shorter growing seasons, shorter frost-free seasons, a true four-season development, lower nighttime temperatures, greater average annual temperature ranges, lower rel-ative humidity, and much more variable weather than do the mesothermal climatic regions.

HUMID CONTINENTAL, HOT SUMMER CLIMATE (*Dfa, Dwa*)

Unlike the other two microthermal climates, the humid microthermal, hot summer climate is relatively limited in its distribution on the Eurasian landmass (Table 8.2). This is unfortunate for the people of Europe and Asia because it has by far the greatest agricultural potential and is the most productive of the microthermal climates. In the United States this climate is distributed over a wide area that begins with the eastern seaboard of New York, New Jersey, and southern New England and stretches continuously across the heartland of the eastern United States to encompass much of the American Midwest. It is one of the most densely populated, highly developed, and agriculturally productive regions in the world.

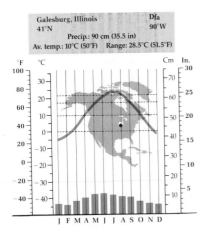

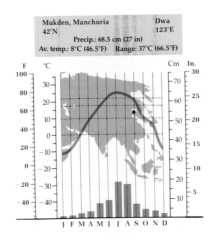

FIGURE 8.17 Climographs for humid continental, hot summer stations.

In terms of environmental conditions, the hot summer variety of microthermal climate has some obvious advantages over its poleward counterparts. Its higher summer temperatures and longer growing season permit farmers to produce a wider variety of crops. Those lands within the hot summer region that were covered by ice sheets are far enough equatorward so that there has been sufficient time for most negative effects of continental glaciation to be removed and primarily positive effects remain. Soils are inclined to be more fertile, especially under forest cover where the typical soil-forming processes are not as extreme and where deciduous trees are more common than the acid-associated pine. Of course, some advantages are matched by liabilities. The lower fuel bills of winter are often more than offset by the cost of air conditioning during the long, hot summers not found in other microthermal climates.

INTERNAL VARIATIONS From place to place within the humid continental, hot summer climate, there are significant differences in temperature characteristics. The length of the growing season is directly related to latitude. It varies from 200 days equatorward to as little as 150 days along poleward margins of the climate. In addition, the degree of continentality can have an effect upon both summer and winter temperatures and, as a result, upon temperature range. Ranges are consistently large, but they become progressively larger toward conti-

nental interiors. Especially near the coasts in this climatic region temperatures may be modified by a slight marine influence, so that summer temperatures are not so high nor winter temperatures so low as at inland locations at comparable latitudes. Large lakes may cause a similar effect. Even the size of the continent exerts an influence. Galesburg, Illinois, a typical station in the United States, has a significantly lower temperature range than Mukden, Manchuria, which is located in almost the same latitude but which experiences the greater seasonal contrasts of the Eurasian landmass (Fig. 8.17).

The amount and distribution of precipitation is also variable from station to station. As in all microthermal climates the total precipitation received decreases both poleward and inland (Fig. 8.18). A move in either direction is a move away from the source regions of warm maritime air masses that provide much of the moisture for cyclonic storms and convectional showers. This decrease can be seen in the average annual precipitation figures for the following cities, all at a latitude of about 40°N: New York (longitude 74°W), 115 centimeters; Indianapolis, Indiana (86°W), 100 centimeters; Hannibal, Missouri (92°W), 90 centimeters; and Grand Island, Nebraska (98°W), 60 centimeters. Most stations have a precipitation maximum in summer when the warm moist air masses dominate. As we have already seen, this is so intensified by the strong winter monsoon

FIGURE 8.18 Decreases of precipitation inland from coastal regions are clearly evident in this map of the eastern United States.

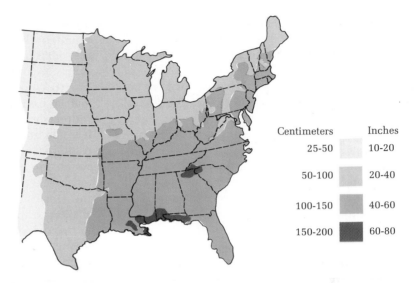

Centimeters		Inches
25-50		10-20
50-100		20-40
100-150		40-60
150-200		60-80

in Asia that regions there must be classified with a seasonal drought (*Dwa*).

As might be expected, vegetation and soils vary with the climatic elements, especially precipitation. In the wetter regions moderately leached, gray-brown soils predominate, and the natural vegetation is usually broad-leaf deciduous forest or mixed broad-leaf deciduous-coniferous forest. Where soils are sandy or especially acidic the trees are primarily coniferous. At one time in certain sections of the American Midwest tall prairie grasses grew where precipitation was sufficient to support forests, and in all the drier portions of the climate grasslands are the natural vegetation. The soils that developed under these grasslands are among the richest in the world.

SEASONAL CHANGES The four seasons are highly developed in the humid microthermal, hot summer climate. Each is distinct from the other three and has a character and life all its own. The winter is cold and often snowy; the spring is warmer with frequent showers that produce flowers, budding leaves, and green grasses; the hot, humid summer brings occasional violent thunderstorms; and the fall has spells of both clear and rainy weather, with mild days and frosty nights, in which the green leaves of summer turn to beautiful reds, oranges, yellows, and browns before falling to the ground.

Of course the most significant differences are between summer and winter. In all regions the summers are long, humid, and hot. The centers of the migrating low-pressure systems usually pass poleward of these regions, and they are dominated by tropical maritime air. So-called "hot spells" can go on day and night for a week or more with only temporary relief available from convectional thunderstorms or an occasional cold front. Asia, in particular, experiences the heavy summer precipitation associated with the monsoonal effect. Conditions are usually ideal for vigorous vegetative growth. In fact, they say the corn grows so fast in Iowa that you can hear it! And if this seems an exaggeration, one thing is certain. The summer heat and humidity are just the proper formula for insects: Mosquitoes, flies, gnats, and bugs of all kinds abound. "Putting up the screens" is a common early summer project.

Winters are not as severe as those further poleward, but average January temperatures are usually between −5°C and 0°C (23°F and 32°F) or below. And once again the averages only tell part of the story. There is invariably a prolonged invasion of cold, dry arctic air once or twice during the winter. This often occurs just after a storm has passed and the ground is covered with snow. The sky remains clear and blue for days at a time, the temperatures will stay near −18°C (0°F) and may occasionally dip to 30°C or 35°C (20°F or 30°F) *below zero* at night. The ground remains frozen for long periods, and there may be snow cover for several days, even weeks, at a time. However, these characteristics do not last continuously because the greatest frequency of cyclones occurs during winter and sudden weather contrasts are common. Cold air precedes warmer air, and thaw follows freeze. Vegetation remains dormant throughout the winter season but bursts into life again with the return of consistently warmer temperatures. Throughout its early growth it is in constant danger of late spring frost.

As should be apparent from this description, the atmospheric changes within seasons are just as significant as those between seasons. The humid continental, hot summer climate is the classic example of variable midlatitude weather. This is the domain of the polar front. Cyclonic storms are born as tropical air masses move northward and confront polar air masses migrating to the south. The daily weather in these regions is dominated by days of stormy frontal activity followed by the clear conditions of a following anticyclone. Above the land is a battlefield in which storms mark the struggles of air masses for dominance. The general circulation of the atmosphere in these latitudes continuously carries the cyclones and anticyclones toward the east. When the polar front is most directly over these regions, as it is in winter and spring, one storm and its associated fronts seem to follow directly behind another with such speed and regularity that the only safe weather prediction is that the weather will change.

HUMID CONTINENTAL, MILD SUMMER CLIMATE (*Dfb, Dwb*)

If you review the relative distributions of the humid continental, hot summer and mild summer climates on Plate 12, the close relationship between the two is unmistakable. Where one is found the other is found as well, and in each situation the mild summer climate invariably lies adjacent to and poleward from the hot summer climate.

In most instances the mild summer climate is essentially a more continental or severe winter version of its equatorward counterpart. It is characterized by distinct seasonality. There is significant climatic variation, particularly with respect to precipitation, from place to place within the climate. Variable weather is the rule, and storms along the polar front provide most of the precipitation within this climatic type. Of course there are differences between the neighboring climates. These are especially apparent when we examine certain aspects of temperature, growing season, vegetation, and human activity. A brief comparison of the humid continental, mild summer climate with the humid continental, hot summer climate should help to point out important similarities and differences.

MILD SUMMER-HOT SUMMER COMPARISON In the microthermal climates precipitation tends to decrease poleward; therefore the humid continental, mild summer climate tends to have less precipitation than the hot summer regions closer to the equator. Precipitation continues to decrease throughout this climatic type toward the poleward margins and from the coasts toward the arid interiors. As in its hot summer counterpart, the monsoon effect is strong enough in Asia to produce a dry winter season (Fig. 8.19).

Although it may seem a poor pun, as far as temperatures are concerned the major difference between the mild summer climate and its neighbor to the south is one of degree. Winters are more severe, longer, and colder. Summers, on the other hand, are not as long or hot. The combination of more severe winters and shorter summers makes for a growing season of between 90 and 130 days, which is 1 to 3 months shorter than in the hot summer climate. In addition, although overall precipitation totals—50 to 100 centimeters (20 to 40 in)—are generally lower, snowfall is greater,

FIGURE 8.19 Climographs for humid continental, mild summer stations.

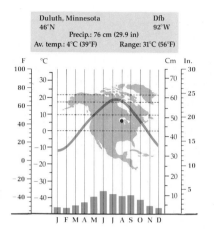

Duluth, Minnesota   Dfb
46°N   92°W
Precip.: 76 cm (29.9 in)
Av. temp.: 4°C (39°F)   Range: 31°C (56°F)

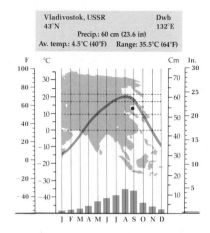

Vladivostok, USSR   Dwb
43°N   132°E
Precip.: 60 cm (23.6 in)
Av. temp.: 4.5°C (40°F)   Range: 35.5°C (64°F)

and snow cover is both thicker and longer lasting (Fig. 8.20).

The humid continental, mild summer regions exhibit seasonal changes just as clearly as the hot summer regions. Annual temperature ranges are generally larger. Vigorous polar and tropical air-mass interaction makes weather change a common occurrence. However, the more poleward position of the mild summer climate brings about a greater dominance of the colder air masses and explains why temperature variability is not as abrupt or as great as it is further south. Under normal conditions, tropical air is strongly modified by the time it reaches mild summer regions, and even in the high-sun season intrusions of warm humid air rarely last more than a few days at a time. By contrast, winter invasions of cold arctic air periodically bring several successive days or weeks of clear skies and frigid temperatures.

Climate, vegetation, and soils continue to show their interdependence. As in the humid continental, hot summer climate, the wetter regions of the mild summer climate are associated with a natural forest vegetation and moderately leached gray-brown soils. However, the deciduous trees so common in the hot summer climate, such as oaks, hickories, and maples, find it difficult to compete with the needle-leaf firs, pines, and spruces, especially toward the colder polar margins of these regions. In addition, the highly acidic soils tend to replace the gray-brown soils near the borders with the subarctic climate. In contrast, northward exten-

FIGURE 8.20 People, animals, and plants living in the humid continental, mild summer regions have learned to cope with, as well as enjoy, the beauty of abundant snow, which is present continuously for many months at a time. (U.S. Forest Service)

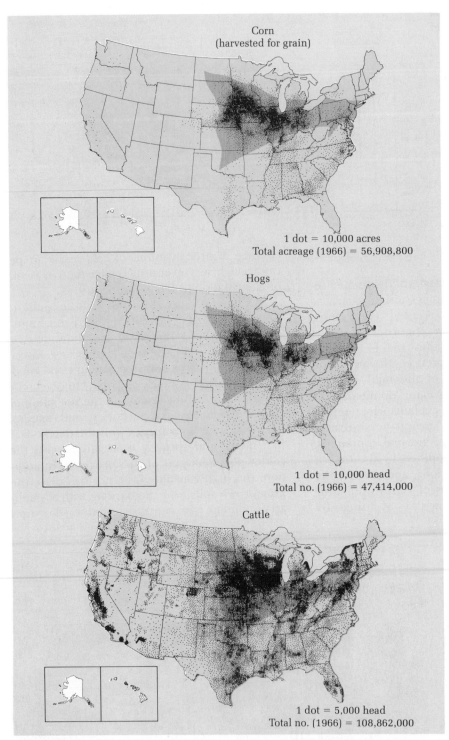

FIGURE 8.21 (Top) Corn harvested in the United States; (middle) hog production in the United States; (bottom) cattle production in the United States. All three maps are overlaid by boundaries of *Dfa* climate.

sions of the grasslands and of the rich soils of the hot summer regions are found in the drier portions of the mild summer climate.

HUMANS IN THE HUMID CONTINENTAL CLIMATES Perhaps the greatest contrast between the hot summer and mild summer, humid continental regions is exhibited in agriculture. Despite the unpredictability of the weather, the humid continental, hot summer agricultural regions are among the finest in the world. The favorable combination of long hot summers, ample rainfall, and highly fertile soils has made the American Midwest a leading producer of corn, beef cattle, and hogs (Fig. 8.21). Soybeans, which are native to similar climatic regions in northern China, are now second to corn throughout the Midwest as feed for animals and as a raw material for the food processing, plastics, and vegetable oil industries. Wheat, barley, and other grains are especially important in European and Asian regions, and winters are sufficiently mild so that fall-sown varieties may be raised in the United States. In the mild summer climate, on the other hand, a shorter growing season imposes certain limitations on agriculture and restricts the crops that can be grown. Farmers rely more on quick-ripening varieties, grazing animals, orchard products, and root crops. Spring wheat or other spring-sown grains must be raised, and corn does not have time to mature so, if produced, it is harvested green for fodder. Especially in

Europe, potatoes, beets, turnips, and cabbages are important. Dairy products—milk, cheese, butter, cream—are mainstays in the economies of Wisconsin, New York, and northern New England. The moderating effect of the Great Lakes or other water bodies permits the growth of deciduous fruits, such as apples, plums, and cherries.

The length of the growing season is the most obvious reason for the differences in agriculture between the two humid continental climates, but there is another climate-related reason as well. The great ice sheets of the Pleistocene Ice Age have had significant, though different, effects upon mild summer and hot summer regions. In the hot summer regions, the ice sheets thinned and receded, releasing the enormous load of soil and solid rock debris they had stripped off the lands nearer to their centers of accumulation. The material was laid down in a blanket hundreds of feet thick in the areas of maximum glacial advance. As the ice retreated northward, less and less debris was deposited, much being flushed away by meltwater streams. The more southerly hot summer region consequently has an undulating topography underlain by thick masses of glacial debris. The soils developed on this debris are young, only a little leached, and fertile. The more northerly mild summer region, on the other hand, mainly shows the effects of glacial erosion (Fig. 8.22). Rockbound lakes and marshy lowlands alternate with ice-scoured

FIGURE 8.22 While the effect of glacial action further south was to deposit material, here in Minnesota we see an area of glacial erosion, which reduces agricultural possibilities but greatly enhances recreational potential. (U.S. Forest Service)

rock hills. Soils are either thin and stony or waterlogged. Because of its lower agricultural potential large sections of this area remain in forest.

However, because of its wilderness character and abundance of lakes in basins produced by glacial erosion, recreational possibilities in a mild summer region far exceed those of a more subdued hot summer region. Minnesota calls itself the "Land of 10,000 Lakes," and in New England, lakes, rough mountains, and forest combine to produce some of the most spectacular scenery east of the Rocky Mountains.

SUBARCTIC CLIMATE
(*Dfc, Dwc, Dfd, Dwd*)

The subarctic climate is the furthest poleward and most extreme of the microthermal climates. By definition it has at least one month with an average temperature above 10°C (50°F), and its poleward limit roughly coincides with the 10°C isotherm for the warmest month of the year. As you may recall from our earlier discussion of the simplified Köppen system, forests cannot survive where at least one month does not have an average temperature over 10°C. Thus the poleward boundary of the subarctic climate is the poleward limit of forest growth as well.

As Plate 12 indicates, the subarctic climate, like the other microthermal climates, is found exclusively in the Northern Hemisphere. It covers vast areas of subpolar Eurasia and North America. Conditions are sufficiently varied over these great distances that the subarctic is divided into four climatic subtypes on the basis of winter temperatures and distribution of precipitation. The most severe winter subtypes (*Dfd, Dwd*) are found along the poleward margins or deep in the interior of the Asian landmass. As in the case of other microthermal and mesothermal climates, subarctic subtypes with a winter drought (*Dwc, Dwd*) are found in association with the Siberian high and its clear skies, bitter cold, and strong subsidence of air over interior Asia during winter. Other subarctic regions experience less severe winters (*c*) or year-round precipitations (*f*).

Further study of Plate 12 suggests some additional observations. Ocean currents tend to influence the distribution of the subarctic climate. Along the west coasts of the continents, especially in North America, the warm ocean currents modify temperatures sufficiently to permit the marine west coast climate to extend into latitudes normally occupied by the subarctic and to cause the subarctic to be found well beyond the Arctic Circle. Along east coasts where cold ocean currents help to reduce winter temperatures, the subarctic is situated further south. Also note that the development of the subarctic climate is not so extensive in North America as in Eurasia. This is because (1) the Eurasian continent is a larger landmass, which increases the effect of continentality, and (2) the large water surface of Hudson Bay in Canada provides a modifying marine influence inland, which tends to counter the effect of continentality there.

THE EFFECTS OF HIGH LATITUDE AND CONTINENTALITY   Subarctic regions experience short, cool summers and long, bitterly cold winters (Fig. 8.23). The rapid heating and cooling associated with continental interiors in the higher latitudes allow little time for the in-between seasons of spring and fall. At Eagle, Alaska, a *Dfc* station in the Klondike region of the Yukon River Valley, the temperature climbs 8°C to 10°C (15°F to 20°F) per month as summer approaches and drops just as rapidly prior to the next winter season. At Verkhoyansk in Siberia the change between the seasonal extremes is even more rapid, averaging 15°C to 20°C (30°F to 40°F) per month.

Because of the high latitudes of these regions, summer days are quite long and consequently nights are short. And the noon sun is as high in the sky during a subarctic summer as during a subtropical winter. The combination of moderately high angle of sun's rays and many hours of daylight produces some subarctic locations that receive as much insolation at the time of the summer solstice as the equator

FIGURE 8.23 Climographs for subarctic stations.

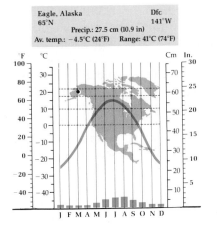

Eagle, Alaska   Dfc
65°N   141°W
        Precip.: 27.5 cm (10.9 in)
Av. temp.: −4.5°C (24°F)   Range: 41°C (74°F)

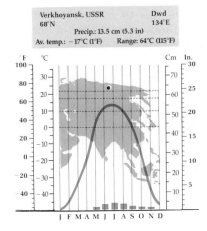

Verkhoyansk, USSR   Dwd
68°N   134°E
        Precip.: 13.5 cm (5.3 in)
Av. temp.: −17°C (1°F)   Range: 64°C (115°F)

does. As a result, temperatures during the one to three months of the subarctic summer usually average 10°C to 15°C (50°F to 60°F), though on some days they reach 20°C (68°F) and even approach 30°C (86°F). Thus, the brief summer in the subarctic climate is often pleasantly warm, even hot, on some days.

The winter season in the subarctic is bitter, intense, and lasts for as long as 8 months. Even Eagle, with its milder *Dfc* regime, has 8 months with average temperatures below freezing. In the Siberian subarctic the January temperatures regularly *average* −40°C to −50°C (−40°F to −60°F), and the coldest temperatures in the Northern Hemisphere—officially −68°C (−90°F) at both Verkhoyansk and Oymyakon; unofficially −78°C (−108°F) at Oymyakon—have been recorded there. In addition, the winter nights, with an average 18 to 20 hours of darkness that extend well into one's working hours, can be mentally depressing and can increase the impression of climatic severity.

As a direct result of the intense heating and cooling of the land, the subarctic has the largest annual temperature ranges of any climate. Average annual ranges for *Dfc* and *Dwc* stations vary from near 40°C (72°F) to over 45°C (80°F). The exceptions are near western coasts, where warm ocean currents and the marine influence may significantly modify winter temperatures. Average annual ranges for *Dd* stations are even greater than for their *Dc* counterparts. The climograph for Verkho-

yansk, which indicates a range of 64°C (115°F), is an extreme example.

Like subarctic temperature, subarctic precipitation is influenced by latitude and continentality. These climatic controls combine to limit annual precipitation amounts to less than 50 centimeters (20 in) for most regions and to 25 centimeters (10 in) or less in northern and interior locations. The low temperatures in the subarctic reduce the moisture-holding capacity of the air, even during the passage of an occasional cyclonic storm. Location toward the center of large landmasses or near lee coasts increases distance from oceanic sources of moisture. Finally, the higher latitudes occupied by the subarctic climate are dominated by the polar anticyclone, especially in the winter season. Much as the subtropical high-pressure systems restrict precipitation in the subtropical arid regions, subsidence and divergence of air in the polar anticyclone limit the opportunity for lifting and hence for precipitation in the subarctic regions. This high-pressure system also blocks the entry of moist air from warmer areas to the south.

Subarctic precipitation is cyclonic or frontal, and because the polar anticyclone is weaker during the warmer summer months most precipitation comes during that season. The meager winter precipitation falls as fine, dry snow. Though there is not as much snowfall as in less severe climates, the temperatures remain so cold for so long that the snow cover lasts for as long as 7 or 8 months. During this time there is almost no melting of snow, especially in the dark shadows of the forest.

A LIMITING ENVIRONMENT The climatic restrictions of subarctic regions place distinct limitations on plant and animal life and on human activity. The characteristic vegetation is coniferous forest, adapted to the severe temperatures, the physiological drought associated with frozen soil water, and the infertile soils. Except near the warmer southern limits of this climate, the soils are so acidic, poor in structure, and low in fertility that it is difficult to improve their productivity even through the additions of fertilizers. Seemingly endless tracts of spruce, fir, and pine thrive over enormous areas, untouched by humans (Fig. 8.24). In the

FIGURE 8.24   The majestic beauty of the subarctic forests is illustrated by this stand in northern Canada. (R. Gabler)

USSR the forest is called the *taiga,* which is the name sometimes given to the subarctic climate type itself.

In the warmer and wetter parts of the subarctic climate, deciduous trees such as willow, birch, and poplar are mixed with the coniferous trees. In the extremely severe climate of interior Siberia (*Dwd*), even the needle-leaf adaptation to cold is not sufficient protection against icy winter gales, and the only coniferous trees that can survive are the larches which shed their needles during the cold season. Poleward, the coniferous forest becomes progressively less dense, and the individual trees are smaller and thinner. Eventually, all that is left are patches of stunted trees surrounded by open tundra.

The brief summers and long cold winters severely limit the growth of vegetation in subarctic regions. Even the trees are shorter and more slender than comparable species in less severe climatic regimes. There is little hope for agriculture. The growing season averages 50 to 75 days and frost may occur even during June, July, or August. Thus in some years a subarctic location may have no truly frost-free season. Although scientists are working to develop plant species that can take advantage of the long hours of daylight in summer, only minimal success has been achieved in southern parts of the climate with certain vegetables, such as cabbage, and root crops such as potatoes.

A particularly vexing problem to man in subarctic regions is **permafrost,** a permanently frozen layer of subsoil and underlying rock that may extend to a depth of 300 meters (1000 ft) or more in the northernmost sections of the climate. Permafrost is present over much of the subarctic climate, but it varies greatly in thickness and is often discontinuous. Where it occurs the land is frozen completely from the surface down in winter. The warm temperatures of spring and summer, however, cause the top few feet to thaw out. Yet because the land beneath this thawed top layer remains frozen, water cannot percolate downward, and the thawed soil becomes sodden with moisture, especially in spring, when there is an abundant supply of water from the melting snow. Permafrost poses a problem to agriculture by preventing proper soil drainage. Seasonal freeze and thaw of the surface layer above the permafrost also pose special problems for construction engineers. The cycle causes repeated expansion and contraction, heaving the surface up and then letting it sag down. The effects of this cycle break up roads, force buried pipelines out of the ground, cause walls and bridge piers to collapse, and even swivel buildings off their foundations (Fig. 8.25).

With agriculture a questionable occupation at best, there is little economic incentive to draw humans to subarctic regions. Logging is unimportant because of small tree size. Even use of the vast forests for paper, pulp, and wood products is restricted by their interior location far from world markets. Miners exploit occasionally rich ore deposits, while isolated hunters, trappers, and fishermen, many of them Indians native to the subarctic regions, pursue the relatively limited wildlife. Fur-bear-

ing animals such as mink, fox, wolf, ermine, otter, and muskrat are of greatest value.

A final restriction to the human use of these regions is the relentless hordes of mosquitoes, flies, gnats, and other insects that come to life in the brief summer. Only the hardiest sportsman will brave such a challenge to seek the moose, elk, and deer of the open subarctic forest or the abundant fish in the lakes and streams.

POLAR CLIMATIC REGIONS

The polar climates are the last of Köppen's humid climatic subdivisions to be differentiated on the basis of temperature. These climatic regions are situated at the greatest distance from the equator, and they owe their existence primarily to the low annual amounts of insolation they receive. No polar station experiences a month with average temperatures as high as 10°C (50°F), and hence all

FIGURE 8.25 This photo illustrates the problem of construction in a permafrost area. The railroad fill caused the underlying permafrost to thaw, allowing the land to subside. (U.S. Dept. of the Interior, Geological Survey)

TABLE 8.3   The E Climates

NAME AND DESCRIPTION	CONTROLLING FACTORS	GEOGRAPHIC DISTRIBUTION	DISTINGUISHING CHARACTERISTICS	RELATED FEATURES
Tundra (ET) Warmest month between 0°C (32°F) and 10°C (50°F); precipitation exceeds potential evaporation.	Location in the high latitudes; subsidence and divergence of the polar anticyclone; proximity to coasts.	Arctic ocean borderlands of North America, Greenland and Eurasia; Antarctic Peninsula; some polar islands.	Summerless; at least 9 months average below freezing; low evaporation; precipitation usually below 25.5 cm (10 in); coastal fog; strong winds.	Tundra vegetation; tundra soils; permafrost; swamps or bog conditions during brief cool summers; life most common in nearby seas; Eskimos; mineral and oil resources; defense.
Ice-Cap or Frost (EF) Warmest month below 0°C (32°F); precipitation exceeds potential evaporation.	Location in the high latitudes and interiors of landmasses; year-round influence of the polar anticyclone; ice cover; elevation.	Antarctica; interior Greenland; permanently frozen portions of the Arctic Ocean and associated islands.	Summerless; all months average below freezing; world's coldest temperature; extremely meager precipitation in the form of snow, evaporation even less; gale-force winds.	Ice- and snow-covered surface; no vegetation; no exposed soils; only sea life or aquatic birds; scientific exploration.

are without a warm summer (Table 8.3). Trees cannot survive in such a regime, and in the regions where at least 1 month averages above 0°C (32°F) they are replaced by tundra vegetation. Elsewhere the surface is covered by great expanses of frozen ice. Thus there are two polar climatic types, tundra (ET) and frost or ice-cap (EF).

There are two important points to keep in mind in the discussion of polar climatic regions. First, these regions have a net annual radiation loss. That is, they give up more radiation or energy than they receive from the sun during a year; thus there is a radiation deficiency. The transfer of heat from lower to higher latitudes to make up for this deficiency is the driving force for much of the general atmospheric circulation. Without this compensating poleward transfer of heat from the lower latitudes, the polar regions would become too cold to permit any form of life, and the equatorial regions would heat to temperatures no organism could survive.

An equally important characteristic of polar climates is the unique pattern of day and night. At the poles, six months of relative darkness caused when the sun is positioned below the horizon alternate with six months of daylight during which the sun is above the horizon. Even when the sun is above the horizon, however, the sun's rays are at a very oblique angle, and little insolation is received for the number of hours of daylight. Moving outward from the poles, the length of continuous winter night and summer day periods decreases rapidly from 6 months to 24 hours at the Arctic and Antarctic Circles (66½° N and S). Here the 24-hour night or day occurs only at the winter and summer solstices, respectively.

TUNDRA CLIMATE (ET)

Compare the location of the tundra climate with that of the subarctic climate on Plate 12. You can see that although the tundra climate is situated closer to the poles, it is also along the periphery of landmasses, and, with the exception of the Antarctic Peninsula, it is everywhere adjacent to the Arctic Ocean. Even though temperature ranges in the tundra are large, they are not as large as in the subarctic because of the maritime influence. Winter temperatures, in particular, are not as severe in the tundra as they are inland (Fig. 8.26).

FIGURE 8.26 Climographs for tundra stations.

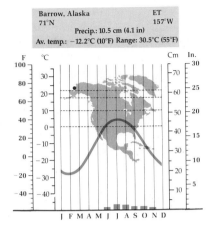

Barrow, Alaska   ET
71°N   157°W
Precip.: 10.5 cm (4.1 in)
Av. temp.: −12.2°C (10°F) Range: 30.5°C (55°F)

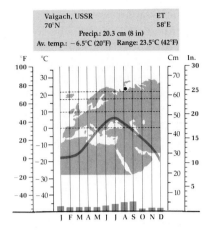

Vaigach, USSR   ET
70°N   58°E
Precip.: 20.3 cm (8 in)
Av. temp.: −6.5°C (20°F) Range: 23.5°C (42°F)

It almost seems inappropriate to call the unpleasantly chilly and damp conditions of the tundra's warmer season *summer*. Temperatures average around 4°C (40°F) to 10°C (50°F) for the warmest month and frosts regularly occur. The air does warm sufficiently to melt the thin snow cover and the ice on small bodies of water, but this only causes marshes, swamps, and bogs to form across the land because drainage is blocked by permafrost (Fig. 8.27). And out of this soggy landscape, known as **muskeg** in Canada and Alaska, swarm clouds of black flies, mosquitoes, and gnats. The one bright note in the landscape is provided by the enormous number of migratory birds that nest in the arctic regions at this time of year and feed on the insects. However, as soon as the shrinking days of autumn approach, these birds wisely depart for warmer climates.

Winters are cold and seem to last forever, especially in tundra locations where the sun is below the horizon for days at a time. The climograph for Barrow illustrates the low temperatures of this climate. Note that average

FIGURE 8.27 Permafrost regions, such as this area at the base of the Alaska Range, become almost impenetrable swampland during the brief Alaskan summer. Travel over land is only feasible in the winter season. (R. Gabler)

monthly temperatures are *below freezing* nine months of the year. The average annual temperature is −12°C (10°F).

The tundra regions exhibit several other significant climatic characteristics. Diurnal temperature ranges are small because insolation is uniformly high during the long summer days and uniformly low during the long winter nights. Precipitation is generally low, except in eastern Canada and Greenland, because of exceedingly low absolute humidity and the influence of the polar anticyclone. Icy winds sweep across the open land surface and are an added factor in eliminating the trees that might impede their progress. Coastal fog is characteristic in marine locations where cool polar maritime air drifts onshore and is chilled below the dew point by contact with the even colder land.

The low-growing tundra vegetation survives despite the forbidding environment. It consists of lichens, mosses, sedges, flowering herbaceous plants, small shrubs, and grasses. In particular, the plants have adjusted to the conditions associated with nearly universal permafrost, as the following passage indicates.

It seems strange that the plants . . . [willow-herb, wild rose, forget-me-not, hellebore, chives, valerian, thyme, vetches and others, which grow on the dunes of windblown sand found along the banks of some of the large rivers] should spring only from the dry sand of the dunes, but the apparent riddle is solved when we know that it is only the sand thus piled up, that becomes sufficiently warmed in the months of uninterrupted sunshine for these plants to flourish. Nowhere else throughout the tundra is this the case. Moor and bog, morass and swamp, even the lakes with water several yards in depth only form a thin summer covering over the eternal winter which reigns in the tundra, with destructive as well as with preserving power. Wherever one tries to penetrate to any depth, in the soil one comes—in most cases scarcely a yard from the surface—upon ice, or at least on frozen soil, and it is said that one must dig about a hundred yards before breaking through the ice-crust of the earth. It is this crust which prevents the higher plants from vigorous growth, and allows only such to live as are content with the dry layer of soil which thaws

in summer. It is only by digging that one can know the tundra for what it is: an immeasurable and unchangeable ice-vault which has endured, and will continue to endure, for hundreds of thousands of years.

From North Pole to Equator
Alfred Edmund Brehm

ICE-CAP CLIMATE (*EF*)

The ice-cap climate is the most severe and restrictive climate on earth. As Table 8.3 indicates, it covers large areas in both the Northern and Southern Hemispheres, a total of about 16 million square kilometers (6 million sq mi)—nearly twice the area of the United States. Because all average monthly temperatures are below freezing and because most surfaces are covered with glacial ice, no vegetation can survive in this climate. It is a virtually lifeless regime of perpetual frost.

Antarctica is the coldest place in the world, although Siberia sometimes has longer and more severe periods of cold in winter. Nevertheless, the world's coldest temperature, −88°C (−127°F), was recorded at Vostok, Antarctica. Consider the climographs for Little America, Antarctica, and Eismitte, Greenland, for a fuller picture of the cold ice-cap temperatures (Fig. 8.28).

The primary reason for the low temperatures of ice-cap climates is the minimal insolation received in these regions. Not only is little or no insolation received during half the year, but the sun's radiant energy that is received arrives at very oblique angles and consequently is spread over large areas. In addition, the perpetual snow and ice cover of this climate reflects nearly all of the incoming radiation. A further factor, both in Greenland and Antarctica, is elevation: The ice-caps covering both regions rise over 3000 meters (10,000 ft) above sea level (Fig. 8.29). Naturally this elevation contributes to the cold temperatures.

Precipitation is so meager in the ice-cap climate that regions within this regime are sometimes incorrectly referred to as *polar deserts*. However, because of the exceedingly low

FIGURE 8.28   Climographs for ice-cap stations.

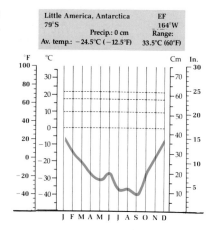

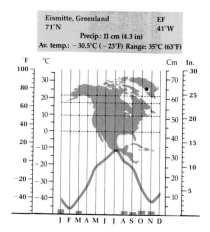

evaporation rates associated with the severely cold temperatures, precipitation still exceeds potential evaporation and the climate can be classified as humid. The annual precipitation surplus produces glaciers, which are a means of exporting snowfall similar to the way rivers export rainfall.

Scientists are carefully recording and studying atmospheric conditions in ice-cap regions, especially in Antarctica, because it is believed that this remote region may significantly influence worldwide weather patterns. But

their research has only recently begun and is exceptionally expensive to support. So far little is known for certain, and it is only assumed that the polar anticyclone severely limits precipitation there to the fine, dry snow associated with occasional cyclonic storms. Perhaps the most reliable statistics are those related to the polar winds. Mawson Base, Antarctica, for example, has approximately 340 days a year with gale-force winds 15 meters per second (33 mph) or over. Katabatic winds, which are caused by the downslope drainage of heavy

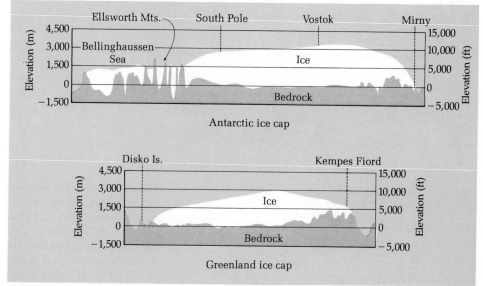

FIGURE 8.29   The Antarctic ice sheet resulting from the ice-cap climate of the south polar regions is as much as 4000 meters (13,200 ft) thick locally. Where it is thinnest it is floated by sea water to produce an ice shelf. The smaller Greenland ice sheet is about 3000 meters (10,000 ft) thick.

cold air accumulated over ice-caps, are common along the edges of the polar ice. The winds of ice-cap regions can result in "white-outs," periods of zero visibility due to blowing fine snow and ice crystals.

HUMAN ACTIVITY IN POLAR REGIONS

The climatic severity that limits animal life in polar regions to a few scattered species in the tundra is just as restrictive on human settlement. The celebrated Lapps of northern Europe migrate with their reindeer to the tundra from the adjacent forest during warmer months. They join the musk-ox, arctic hare, fox, wolves, and polar bear that manage to make a home there despite the prohibitive environment. Only the Eskimos of Alaska, northern Canada, and Greenland have in the past succeeded in developing a year-round life-style adapted to the tundra regime. Yet even this group relies less on the resources of the tundra than on the large variety of fish and sea mammals, such as cod, salmon, halibut, sea otter, seal, walrus, and whale, that occupy the adjacent seas.

As their communication with the rest of the world has increased and as they have become acquainted with alternative life-styles, however, the permanent population of Eskimos living in the tundra has greatly diminished, and life for those remaining has changed drastically. Some have gained a new economic security through employment at defense installations or at sites where they join skilled workers from outside the region to exploit mineral or energy resources (Fig. 8.30). However, the new population centers based on the construction and maintenance of radar and missile defense stations or, as in the case of Alaska's North Slope, on the production and transporting of oil, cannot be considered permanent. Workers depend on other regions for support and often inhabit this region only temporarily.

The ice-cap climate cannot serve as a home for man or other animals. Even the penguins, gulls, leopard seals, and polar bears are coastal inhabitants. It is without question the harshest, most restrictive, most nearly lifeless climatic zone on earth. Yet it is of strategic importance and of great scientific interest. It plays a role in controlling weather phenomena that affect middle latitudes; it offers evidence (in ice cores) of past climates and glacial geophysics; and its ice may cover important mineral deposits, especially coal. Its strategic value is so widely recognized that the world's nations have voluntarily given up claims to territorial rights

FIGURE 8.30 A truck convoy hauling heavy equipment across the Alaskan tundra in winter. This equipment in support of the Alaska Pipeline and North Slope oil must travel across the tundra when the surface is frozen. (Aleyska Pipeline Service Company)

FIGURE 8.31 McMurdo Station research base in Antarctica. (U.S. Navy)

in Antarctica, the largest ice-cap region, in exchange for cooperative scientific exploration on behalf of all humankind (Fig. 8.31).

HIGHLAND CLIMATIC REGIONS

As we saw in Chapter 3, temperature decreases with increasing altitude at the rate of about 6°C per 1000 meters (3.6°F per 1000 ft). Thus you might suspect that there are broad zones of climate in highland regions based on changes in temperature with elevation that roughly correspond to Köppen's climatic zones based on change of temperature with latitude. This is the case with one important exception. Seasons only exist in highlands if they also exist in the nearby lowland regions. For example, although zones of increasingly cooler temperature occur at progressively higher elevations in the tropical climatic regions (A), the seasonal changes of Köppen's C, D, or E climates are not present.

Elevation is only one of several controls of highland climates; exposure is another. Just as some continental coasts face the prevailing wind, so do some mountain slopes, while others are lee slopes or are sheltered behind higher topography. The nature of the wind, its temperature, and its moisture content depend on

whether the mountain is in a coastal location or deep in a continental interior, or at a high or low latitude within or beyond the reaches of cyclonic storms and monsoon circulation. In the middle and high latitudes mountain slopes and valley walls that face the equator receive the direct rays of the sun and are warm; poleward-facing slopes are shadowed and cool. West-facing slopes feel the hot afternoon sun, while east-facing slopes are sunlit only in the cool of the morning. This factor is known as **slope aspect** and affects where people live in the mountains and where particular crops will do best. The higher one rises in the mountains, the more important direct sunlight is as a source of warmth and energy for plant and animal life processes.

Complexity is the hallmark of highland climates. Every mountain range of significance is composed of a mosaic of climates far too intricate to differentiate on a world map, or even on a map of a single continent. Highland climates are therefore undifferentiated and are symbolized as H, signifying climate complexity. Highland climates are indicated on Plate 12 wherever there is marked local variation in climate as a consequence of elevation, exposure, and slope aspect. We can see that these regions are distributed widely over the earth but are particularly concentrated in Asia, central Europe, and western North and South America.

The areas of highland climate on the world map are cool, moist islands in the midst of the zonal climates that dominate the area around them (Fig. 8.32). Consequently, highland areas are also biotic islands, supporting a flora and fauna adapted to cooler and wetter conditions than those of the surrounding lowlands. This coolness is part of the highland charm, particularly where mountains rise cloaked with forests above arid plains, as do the Rocky Mountains and California's Sierra Nevada.

Highlands stimulate moisture condensation and precipitation by forcing moving air masses to rise over them. Where mountain slopes are rocky and forest-free, their surfaces grow warm during the day, causing upward convection, which often produces afternoon thundershowers. Mountains receive abundant precipitation and are the source area for multitudes of streams that join to form the great rivers of all of the continents.

There are few streams of significance whose headwaters do not lie in rugged highlands. Much of the stream flow on all conti-

nents is produced by the summer melting of mountain snowfields. Thus the mountains not only wring water from the atmosphere, but they also store much of it in a form that gradually releases it throughout summer droughts, when water is most needed for irrigation and for municipal and domestic uses.

PECULIARITIES OF MOUNTAIN CLIMATES

A general characteristic of mountain weather is its changeability from hour to hour as well as from place to place. Strong convection over mountains often causes clouds to form very quickly, leading to thunderstorms and longer rains that do not affect surrounding cloud-free lowlands. Where the cloud cover is diminished, diurnal temperature ranges over mountains are far greater than over lowlands. Since mountains penetrate upward beyond the densest part of the atmosphere, the *atmospheric effect* (see Chapter 3) is less developed in them than anywhere else on earth. Insolation, which is less impeded than in lower areas, strongly

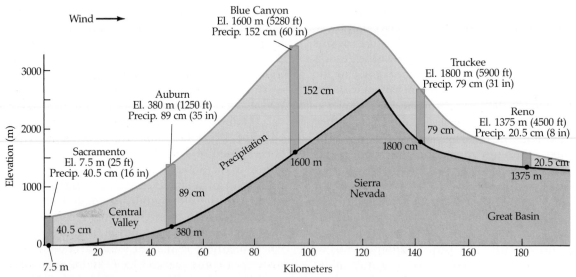

FIGURE 8.32 Variation in precipitation caused by uplift of air crossing the Sierra Nevada range of California from west to east. The maximum precipitation occurs on the windward slope, since air in the summit region is too cool to retain a large supply of moisture. Note the strong rain shadow to the lee that gives Reno a desert climate.

FIGURE 8.33 The last tree species found at the tree line are stunted, prostrate forms, which often produce an elfin forest. Where the trees are especially gnarled and misshapen by wind stress the vegetation is called krummholz (crooked wood). (R. Gabler)

warms surfaces in the daytime. Long-range re-radiation from the earth has less atmosphere to penetrate, so that much of it is lost outward, making the air temperature at night even cooler than the elevation alone would indicate. Since the atmospheric shield is thinnest at high elevations, plants, animals, and humans receive proportionately more of the sun's short-wave radiation at high altitudes. Violet and ultraviolet radiation are particularly noticeable; severe sunburn is one of the real hazards of a day in the high country.

In the middle and high latitudes mountains rise from mesothermal and microthermal climates into tundra and frost climates. The lower slopes of mountains are commonly forested with conifers, which become more stunted as one moves upward, until the last dwarfed tree is passed at the **tree line**—the line beyond which low winter temperatures and severe wind stress eliminate all forms of vegetation except those that grow low to the ground where they can be protected by a blanket of snow (Fig. 8.33). Where mountains are high enough the land surface will be permanently covered by snow or ice. The line above which summer melting is insufficient to remove all of

the preceding winter's snowfall is called the **snow line.**

In tropical mountain regions the vertical zonation of climate is even more pronounced, and both tree line and snow line occur at higher elevations than in middle latitudes. Any seasonal change is mainly restricted to rainfall, and temperatures are stable year-round, regardless of elevation. Each climatic zone has its own particular association of natural vegetation and has given rise to a distinctive crop combination where agriculture is practiced (Fig. 8.34). In Latin America four vertical climatic zones are recognized: *tierra caliente* (hot lands); *tierra templada* (temperate lands); *tierra fria* (cool lands); and *tierra helada* (frozen lands).

HIGHLAND CLIMATES AND HUMAN ACTIVITY

In midlatitude highlands, soils are poor, the growing season is short, and the winter snow cover is heavy in the conifer zone, which dominates the lower and middle mountain slopes. Therefore, little agriculture is practiced and permanent settlements in the mountains are

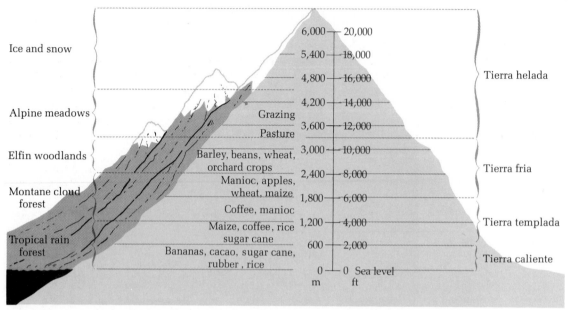

FIGURE 8.34 **Natural vegetation, vertical climatic zones, and agricultural products in tropical mountains. Note that this example extends from tropical life zones to the zone of permanent snow and ice. There is little seasonal temperature change in tropical mountains, which permits life forms sensitive to low temperatures to survive at relatively high elevations.**

few. However, as the winter snow melts off the high ground just below the bare rocky peaks, grass springs into life, and humans drive herds of cattle and flocks of sheep and goats up from the warmer valleys. The high pastures are lush throughout the summer, but in early fall they are vacated by the animals and their keepers who return to the valleys. This seasonal movement of herds and herders between alpine pastures and fixed habitations in the valleys was once very common in the European highlands (the Alps, Pyrenees, Carpathians, and the Scandinavian mountains) and is still practiced there on a reduced scale.

Otherwise, the midlatitude highlands serve mainly as sources of timber and of minerals formed by the same geologic forces that elevated the mountains, and as arenas for recreation—both in summer and in winter. Recreational use of the highlands is a relatively recent phenomenon, resulting both from new

interest in mountain areas and from new access to them by road, rail, and air routes that did not exist a century ago.

In contrast to poleward mountain regions, tropical highlands may actually experience more favorable climatic conditions and are often a greater attraction to human settlement than adjacent lowlands. In fact, large permanent populations are supported throughout the tropics where topography and soil favor agriculture in the vertical climatic zones. Highland climates are at such a premium in many areas that steep mountain slopes have been extensively terraced by man to produce level land for agricultural use. Spectacular agricultural terraces can be seen in Peru, Yemen, the Philippines, and many other tropical highlands. Where the climate is appropriate and population pressure is high, people have created a topography to suit their needs, although they have had to hack it out of mountainsides.

QUESTIONS FOR DISCUSSION AND REVIEW

1. What characteristics make the Mediterranean climate readily distinguishable from all others? What controls are responsible for producing these characteristics?

2. Summarize the special adaptations of vegetation and soils in Mediterranean regions. How have humans also adapted to these regions?

3. Compare the humid subtropical and Mediterranean climates. What are their most obvious similarities and differences?

4. What factors combine to cause a precipitation maximum in late summer in most of the humid subtropical regions?

5. What factors serve to make the humid subtropical climate one of the world's most productive? What are some of its handicaps?

6. How are temperature, precipitation, and geographic distribution of marine west coast regions linked to the controlling factors for this climate?

7. The marine west coast climate has long had its supporters and critics. Give reasons why you would or would not wish to live in such a climatic region.

8. Explain why the microthermal climates are limited to the Northern Hemisphere.

9. List several features that all humid microthermal climates have in common. How do these features differ from those displayed by the humid mesothermal climates?

10. Why is weather in the humid continental, hot summer climate so variable?

11. Describe the relationship between vegetation and climate in the humid continental, mild summer regions.

12. What are the major differences between the humid continental, hot summer, and mild summer climates? Contrast human use of regions occupied by the two climates.

13. How have past climatic changes helped to bring about differences in the configuration of the land between some hot summer and mild summer regions?

14. Using the climographs for Eagle, Alaska, and Verkhoyansk, USSR (Fig. 8.23), describe the temperature patterns of the subarctic regions.

15. What factors limit precipitation in the subarctic climates?

16. How does permafrost affect man's activities in both the subarctic and tundra regions?

17. Identify and compare the controlling factors of the tundra and ice-cap regions. How do these controlling factors affect the distribution of these climates?

18. What kinds of plant and animal life can survive in the polar climates? What special adaptations must this life make to the harsh conditions of these regions?

19. How do elevation, exposure, and slope aspect affect the microclimates of highland regions? What are the major climatic differences between highland regions and nearby lowlands?

20. Describe and compare the vertical zonation of highland climates in the tropics and in the middle latitudes. How are the climatic zones in tropical highlands related to agriculture?

21. How does human use of highland regions in the middle latitudes differ from that in the tropics? What special human adaptations have aided utilization of these regions?

9

ECOSYSTEMS

As was noted in Chapter 1, there is an increasing concern among all scientists as to the ability of the earth to indefinitely support future generations of humanity. What lasting effects will increased industrialization have on the water that animals in the next century will drink or on the atmosphere in which tomorrow's plants will grow? What will be the consequences for life forms if concrete and steel replace additional square kilometers of forest and field? What could happen to aquatic life if toxic wastes accumulate in the world's oceans? Can humans learn to work with, and not against, nature in order to sustain and improve life as it is known on planet earth today?

The answers to these and countless similar questions lie in that branch of science we earlier identified as *ecology*. Plants, animals, and the environments in which they live are interdependent, each affecting the other. Animal life would not exist without plants as basic food, and plants as we know them today could not survive without animals. Photosynthesis by plants requires carbon dioxide, and experts suggest that the earth's plants would remove all of the carbon dioxide from the atmosphere in

little more than a year if it were not for the respiration of animals and the burning of fuels. Together plants and animals must adapt to their environment. Humans alone have the intelligence and capacity to alter, either carelessly or deliberately, the plant-animal-environment relationship. It makes uncommonly good sense to examine the earth's *ecosystems* in order to better understand the impact that human beings have upon them.

ORGANIZATION WITHIN ECOSYSTEMS

It can be said that ecology is an old science. The great voyages of exploration that began in the fifteenth century carried colonists and adventurers to uncharted lands with exotic environments. The more scholarly observers within each group made careful note of the flora and fauna to be found in each new part of the world. It soon became apparent that certain plants and animals were found together and that they bore a direct relationship to the climate in which they lived. As information

East Africa: Vanishing Ecosystems (Donald Marshall collection, WIU International Programs)

about various world environments became more reliable and readily available, early biologists began to study plant communities and classify vegetation types. As the relationships of animals to these plant communities were recognized, naturalists in the early twentieth century began dividing the earth's life forms into *biotic associations*. Most recently the functional relationships of plants, animals, and their physical environment have been the primary focus of attention, and the concept of the ecosystem has become widely utilized.

Our definition of an ecosystem is both broad and flexible. The term can be used in reference to the earth system in its entirety or to any group of organisms occupying a given area and functioning together with their non-living environment (Fig. 9.1). An ecosystem

FIGURE 9.1 Relatively small ecosystems, such as the one associated with this pond in New Hampshire, were among the first to be studied because they have fairly distinct boundaries and less complicated relationships between the nonliving environment and the organisms that occupy it. (R. Gabler)

may be large or small, marine or terrestrial (on land), short-lived or long-lasting. It may even be artificial. When a farmer plants crops, spreads fertilizer, practices weed control, and sprays insecticides, a new ecosystem is created, but this does not alter the fact that plants and animals are living together in an interdependent relationship with the soil, rainfall, temperatures, duration of sunshine, and other factors that comprise the physical environment. The farmer will attempt to influence the nature of the ecosystem but will be the first to state that he has a limited effect on the final form it will take at harvest time (Fig. 9.2).

As previously noted, ecosystems are *open* systems. There is free movement of both energy and materials across their boundaries. They are usually so closely related to nearby ecosystems or so integrated with the larger ecosystems of which they are a part that they are not isolated in nature or readily delimited. Nevertheless the concept of the ecosystem is a valuable model for examining the structure and function of life on earth.

MAJOR COMPONENTS

Ecosystems may be many and varied, but the typical ecosystem has four principal parts or basic components. The first of these is the non-living or **abiotic** part of the system. This is the physical environment in which the plants and animals of the system live. In an aquatic ecosystem (a pond, for example), the abiotic component would include such organic or inorganic substances as calcium, mineral salts, oxygen, carbon dioxide, and water. Some of these would be dissolved in the water, but the majority would lie at the bottom as sediments—a natural reservoir of nutrients for both plants and animals. In a terrestrial ecosystem the abiotic component provides life-supporting elements and compounds in the soil, groundwater, and atmosphere.

The second and perhaps most important component of an ecosystem consists of the basic producers or **autotrophs** (*self-nourished*). Plants,

FIGURE 9.2 Whether because of early spring flood that delays planting, as in this photograph, or later crop damage from drought, blight, insects, or hail, the Midwest farmer has limited control over the artificial ecosystem for which he is responsible. (Illinois Department of Agriculture)

the most important autotrophs, are essential to all life on earth because they are capable of utilizing light energy from sunlight to convert water and carbon dioxide into organic molecules through the process known as *photosynthesis.* The sugars, fats, and proteins produced by plants through photosynthesis are the basic building blocks, the foundation, for the food supply that supports all other forms of life. It should be noted that some bacteria are also capable of photosynthesis and hence are classed as autotrophs along with plants.

A third component of most ecosystems consists of consumers or **heterotrophs** (*other nourished*). These are animals that are incapable of making their own food and that must survive by eating plants or other animals. Biologists classify heterotrophs in an ecosystem on the basis of their feeding habits. **Herbivores** eat only living plant material; **carnivores** eat other animals; **omnivores** feed on both plants and animals; and **detritivores** eat dead plant and animal material. Although they are not primary producers, animals make an essential contribution to the earth ecosystem of which they are a part. They utilize oxygen in their respiration and return as an end product to the atmosphere the carbon dioxide which is required for photosynthesis by plants.

We might assume that plants, animals, and a supporting environment are all that are required for a functioning ecosystem, but such is not the case. Plants rely on a perpetual supply of mineral nutrients for growth, and the process of weathering, which alters bedrock to usable soil minerals, is exceedingly slow. Without the fourth component of ecosystems, the **decomposers,** plant growth would soon come to a halt. The decomposers feed on dead plant and animal material, promote decay, and return mineral nutrients to the soil in a form that plants can utilize.

TROPHIC STRUCTURE

From the discussion of the autotrophs and heterotrophs it becomes apparent that there is a definite arrangement of the major components of an ecosystem. The components form a sequence in their levels of eating: Herbivores eat plants, carnivores may eat herbivores or other carnivores, and decomposers feed on dead plants and animals. The pattern of feeding in an ecosystem is called the **trophic structure,** and the sequence of levels in the feeding pattern is referred to as a **food chain.** The simplest food chain would include only plants and decomposers. However, the chain usually includes at least four steps, e.g., grass–field mouse–owl–fungi (plants–herbivore–carnivore–decomposer). More complex food chains may include six or more levels as carnivores feed on other carnivores, e.g., small fish eats floating plants–larger fish eats small fish–bear eats larger fish, etc.

Organisms within a food chain are often identified by their **trophic level,** or the number of steps they are removed from the autotrophs or green plants in a food chain (Table 9.1). Green plants occupy the first trophic level, herbivores the second, carnivores feeding on herbivores the third, and so forth until the decomposers or the last level is reached. Omnivores may belong to several trophic levels because they eat both the plants, which are the producers, and the animals, which are the consumers.

In reality, linear food chains do not operate in isolation; they overlap and interact to form a feeding mosaic within an ecosystem called a **food web** (Figs. 9.3 and 9.4). Both food chains and food webs merit careful study be-

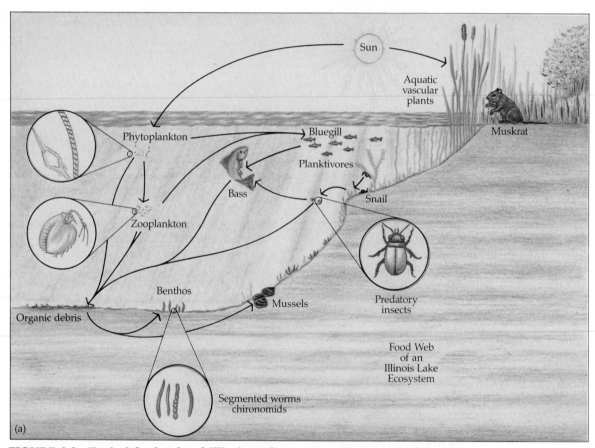

FIGURE 9.3 **Typical food webs of Illinois surface-water ecosystems: (a) lacustrine ecosystem and (b) riverine ecosystem. Phytoplankton, aquatic plants, and periphyton are the autotrophs or basic producers that occupy the first trophic level. Zooplankton, snails, and land animals like the muskrat are the herbivorous primary consumers, and at the top of each food web are carnivorous organisms such as the bass and the snapping turtle. Each organism in its turn is recycled by worms, microbes, and other detritivores after death.**

TABLE 9.1 Trophic Structure of Ecosystems Based on Eating Levels

ECOSYSTEM COMPONENT	TROPHIC LEVEL	EXAMPLES
Autotroph	First	Trees, shrubs, grass
Heterotroph	Second	Locust, rabbit, field mouse, deer, cow, bear
	Third	Praying mantis, owl, hawk, coyote, wolf, bear
	Fourth, etc.	Bobcat, wolf, hawk, bear
Decomposer	Last	Fungi, bacteria

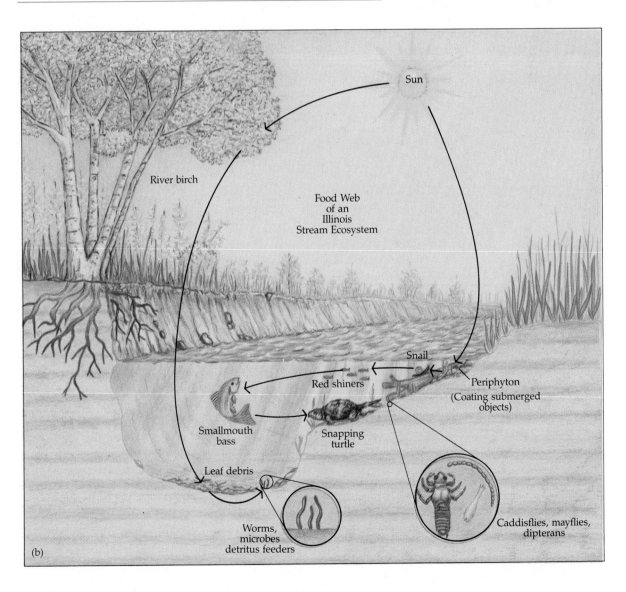

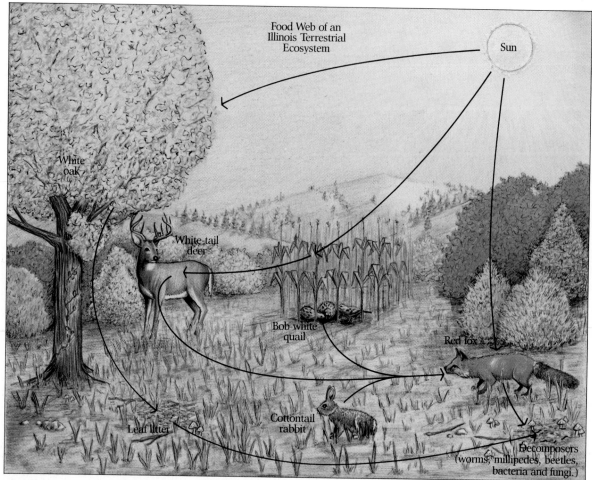

FIGURE 9.4 **Typical food web of an Illinois terrestrial ecosystem. The herbivorous deer, rabbit, and quail fall prey to the fox—or to man, the deadliest carnivore. At the end of the food web await the worms, millipedes, bacteria, and other decomposers.**

cause they can be used to trace the movement of food and energy from one level to another in the ecosystem.

ENERGY FLOW

When physical geographers study ecosystems they trace the movement or flow of energy through the system just as they do when they study energy flow in streams or glaciers. And just as they do with other systems, the laws of thermodynamics apply to ecosystems. For ex-

ample, as the first law of thermodynamics states, energy cannot be created or destroyed; it can only be changed from one form to another. Energy comes to the ecosystem in the form of sunlight, and some of this energy is used by plants in photosynthesis. This energy is stored in the system in the form of organic material in plants and animals. It flows through the system along food chains and webs from one trophic level to the next. It is finally released from the system when oxygen is combined with the chemical compounds of the organic material through the process of oxidation. Fire is one form of oxidation. Respiration, which involves the combination of oxygen with chemical compounds in living cells, and which can occur at any trophic level, is another.

The total amount of living material in an ecosystem is referred to as the standing crop or **biomass.** Because the energy of a system is stored in the biomass, a scientist measures the standing crop at each trophic level to trace the energy flow through the system. It is usually discovered that the standing crop decreases with each successive trophic level (Fig. 9.5). There are a number of explanations for this, each involving loss of energy. The first instance occurs *between* trophic levels. The second law of thermodynamics states that whenever energy is transformed from one state to another there will be a loss of energy through heat. Hence, when an organism at one trophic level feeds on an organism at another, not all of the food energy is utilized. Some is lost to the system. The second instance involves respiration, which occurs in plants and animals *within* each trophic level. As we have noted, respiration also involves loss of energy. Another instance occurs as heterotrophs expend significant amounts of energy to move about and feed, and at each successive trophic level the amount of energy required is greater. A deer may graze in a limited area, but the wolf that preys on the deer must hunt over a much larger territory. Whatever the reason for energy loss, it follows that as the flow of energy decreases with each successive trophic level the biomass also decreases. This same principle applies to agriculture. There was a great deal more biomass (and food energy) available in a field of corn than there is in the cattle that ate the corn.

PRODUCTIVITY

Productivity in an ecosystem is defined as the rate at which new organic material is created at a particular trophic level. **Primary productivity** refers to the formation of new organic matter through photosynthesis by autotrophs, whereas

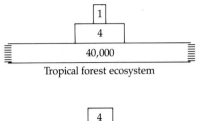

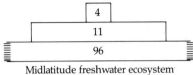

Biomass expressed as dry weight (gr/m²)

FIGURE 9.5 Trophic pyramids showing biomass of organisms at various trophic levels in two contrasting ecosystems. Dry weight is used to measure biomass because the proportion of water to total mass differs from one organism to another.

secondary productivity refers to the rate of formation of new organic material at the heterotroph level.

For the purpose of ecological studies primary productivity is usually described as **gross** or **net;** gross productivity is the total amount of organic material formed through photosynthesis, while net productivity is that which accumulates after some is released as heat during the plant's own respiration. It is the net primary productivity that is available as food for heterotrophs.

PRIMARY PRODUCTIVITY Once we accept the fact that plants are at the first level of almost all food chains, it becomes extremely important to determine the potential for primary productivity in various ecosystems. Just how efficient are plants at producing new organic matter through photosynthesis? The answer to this question depends on a number of variables. Photosynthesis requires sunlight and we have already seen that the receipt of solar energy, which is dependent upon the duration of sunlight and the angle of the sun's rays, differs widely with latitude. Even the process of photosynthesis is affected by factors such as soil moisture, temperature, the availability of mineral nutrients, and the age and species of the

TABLE 9.2   Net Primary Productivity of Selected Ecosystems*

TYPE OF ECOSYSTEM	NET PRIMARY PRODUCTIVITY, $g/(m^2\ year)$	
	NORMAL RANGE	MEAN
Tropical rain forest	1000–3500	2200
Tropical seasonal forest	1000–2500	1600
Midlatitude evergreen forest	600–2500	1300
Midlatitude deciduous forest	600–2500	1200
Boreal forest (taiga)	400–2000	800
Woodland and shrubland	250–1200	700
Savanna	200–2000	900
Midlatitude grassland	200–1500	600
Tundra and alpine	10–400	140
Desert and semidesert scrub	10–250	90
Extreme desert, rock, sand, and ice	0–10	3
Cultivated land	100–3500	650
Swamp and marsh	800–3500	2000
Lake and stream	100–1500	250
Open ocean	2–400	125
Upwelling zones	400–1000	500
Continental shelf	200–600	360
Algal beds and reefs	500–4000	2500
Estuaries	200–3500	1500

*From R. H. Whittaker: *Communities and Ecosystems.* 2nd ed. New York, Macmillan, 1975.

individual plants. Furthermore, the net primary productivity, which is most significant to the higher levels of the food web, is dependent upon the relative rates of photosynthesis and respiration.

There have been a number of studies of productivity in ecosystems, but they have usually been concerned with measuring the standing crop or the net biomass at the autotroph level (Table 9.2). Wherever figures have been compiled noting the efficiency of photosynthesis, the efficiency has been surprisingly low. Most studies indicate that less than 5 percent of the available sunlight is used to produce new biomass in ecosystems. For the earth as a whole the figure is probably less than 1 percent. Despite this fact, the net primary productivity of the entire earth ecosystem is enormous. It is estimated to be in the range of 170 billion metric tonnes (a metric tonne is about 10 percent greater than a U.S. ton) of organic matter annually. Even though oceans cover approximately 70 percent of the earth's surface, slightly over two thirds of net annual productivity is from terrestrial ecosystems and less than a third from oceanic ecosystems. Perhaps even more surprising is the fact that humans use less than 1 percent of the earth's primary productivity as plant food.

Table 9.2 illustrates the wide range of net primary productivity displayed by various ecosystems. The latitudinal control of insolation and the subsequent affect on photosynthesis can easily be recognized when comparing figures for terrestrial ecosystems. There is a noticeable decrease in productivity from tropical ecosystems to those in middle and higher latitudes. Even the tropical savannas, which are dominated by grasses, produce more biomass in a year than the boreal forests, which are found in the colder climates.

The reasons for differences among aquatic or water-controlled ecosystems are not quite as apparent. Swamps and marshes are especially well supplied with plant nutrients and therefore have a relatively large biomass at the first trophic level. On the other hand, depth of water has the greatest impact on ocean ecosystems. Most nutrients in the open ocean sink to the bottom beyond the depth where sunlight will penetrate and photosynthesis is possible. Hence the most productive ocean ecosystems are found in the shallow water of estuaries, continental shelves, or coral reefs or in areas where ocean movements carry nutrients nearer to the surface.

Some artificial ecosystems associated with agriculture can be fairly productive when compared with the natural ecosystems that they have replaced. This is especially true in the warmer latitudes where farmers may raise two or more crops in a year or in arid lands where irrigation supplies the water essential to the growth of an unusually large standing crop. However, Table 9.2 indicates that mean productivity for cultivated land does not approach that of forested land and is just about the same as that of middle-latitude grasslands. Where most quantitative studies have been undertaken, it has been shown that agricultural ecosystems are significantly less productive than natural systems in the same environment. Moreover, mechanized agriculture frequently requires more energy input than is gained in net primary productivity.

SECONDARY PRODUCTIVITY As we have seen, secondary productivity results from the conversion of plant materials to animal substances. We have also noted that the ecological efficiency, or the rate of energy transfer from one trophic level to the next, is low (Fig. 9.6). The efficiency of transfer from autotrophs to heterotrophs varies widely from one ecosystem to another. The amount of net primary productivity actually eaten by herbivores may range from as high as 15 percent in some grassland areas to as low as 1 or 2 percent in certain forested regions. In ocean ecosystems the figure may be much higher, but there is a greater loss during the digestion process. Once the food is eaten, energy loss through respira-

d. Net carnivore productivity: 0.4 g (0.15% of a; 6.7% of c)
c. Net herbivore productivity: 6 g (2.2% of a; 20.7% of b)
b. Every year herbivores eat: 29 g (10.7% of a)
a. Net primary productivity (plants): 270 g (dry weight)

FIGURE 9.6 Productivity at the autotroph, herbivore, and carnivore trophic levels as measured in a Tennessee field. The figures represent productivity for one square meter of field in one year. Note the extremely small proportion of primary productivity, which reaches the carnivore level of the food chain. (From Turk, J., Turk, A., and Arms: *Environmental Science.* **3rd ed. Philadelphia, Saunders College Publishing, 1984.)**

tion or as a result of body movement reduces secondary productivity to a small fraction of the biomass available as net primary productivity. Most authorities consider 10 percent to be a reasonable estimate of ecological efficiency for both herbivores and carnivores.

There are some interesting implications in the figures we have just cited. If herbivores consume only a low percentage of available net primary productivity, and if both herbivores and carnivores have ecological efficiencies of only 10 percent, the relationship of standing crop at the first trophic level to standing crop among carnivores at the third trophic level is several thousand to one. It obviously requires a huge biomass at the autotroph level to support one animal that eats only meat. As human populations grow at increasing rates and agricultural production lags behind, it is indeed fortunate that human beings are omnivores. It may well become necessary for humans to minimize the energy losses between trophic levels and adopt a more vegetarian diet (Fig. 9.7).

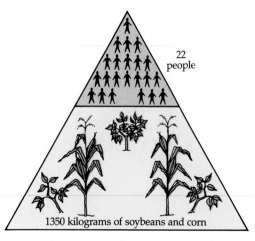

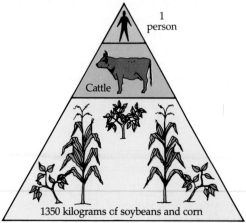

FIGURE 9.7 The triangles illustrate the advantages of a vegetarian diet as we contemplate another century of rapid population growth. It is fortunate that humans are omnivores and can choose to eat grain products. The same 1350 kilograms of grain that will support only one person, if converted to meat, will support 22 people if cattle or other animals are omitted from the food chain. (From Turk, J., Turk, A., and Arms: *Environmental Science*. 3rd ed. Philadelphia, Saunders College Publishing, 1984.)

ECOLOGICAL NICHE

There are a surprising number of species in all but the smallest ecosystems or those severely restricted by adverse environmental conditions. Yet each organism performs a specific role in the system and lives in a certain location, described as its **habitat.** The combination of role and habitat for a particular species is referred to as its **ecological niche.** A number of factors influence the ecological niche of an organism. Some species are **generalists** and can survive on a wide variety of food. The bear, as an omnivore, is pleased with berries for breakfast, honey for lunch, and fish for dinner. On the other hand, the silkworm and the koala bear are **specialists.** The former requires mulberry leaves and the latter eats only the leaves of the eucalyptus tree. Specialists do well when their particular food supply is abundant, but they cannot adapt to changing environmental conditions, and the generalists are in the majority in ecosystems. Although it may show distinct preferences in eating habits, the broader ecological niche of a generalist permits it to survive on alternative food supplies.

The physical characteristics of a species' habitat are important in identifying its ecological niche. This is especially true of plants and animals with little mobility. As an example, certain organisms living in a coastal swamp may require special conditions of soil moisture, water salinity, or level of water table. Each change of habitat within the swamp of a few centimeters above or below the water table or a fraction of a percent in salinity will reveal a new organism filling a quite different ecological niche. The habitats of carnivores or of herbivores that migrate to graze on seasonal grasses can cover vast areas, but even in these instances the food-gathering territory of each species is essential to identifying its ecological niche.

Some generalists among species occupy an ecological niche in one ecosystem that is quite different from that which they occupy in another. As food supply varies with habitat, so varies the ecological niche. Humans are the extreme example of the generalist for in some parts of the earth they are carnivores, in some parts herbivores, and in some parts omnivores. It is also true that different species may occupy the same ecological niche in habitats that are similar but located in separate ecosystems. As an example, horses grazed the steppes of Asia when bison roamed the prairies of North America. Yet, when horses were introduced to North America they did well in the prairie grasslands.

SUCCESSION AND CLIMAX COMMUNITIES

Up to this point we have been discussing ecosystems in general terms. But in the remainder of this chapter we will note that it is the species that occupy the ecosystem that give the ecosystem its character. At least for terrestrial ecosystems it is the autotrophs, the plant species at the first trophic level, that most easily distinguish one ecosystem from another. All other species in an ecosystem depend upon the autotrophs for food, and the association of all living organisms determines the energy flow and the trophic structure of the ecosystem. It should also be noted that the species that occupy the first trophic level are greatly influenced by climate; thus again we see the interconnections between the major earth subsystems.

If the plants that comprise the biomass of the first trophic level have been allowed to develop naturally, without obvious interference from or modification by man, the resulting association of plants is called **natural vegetation.** These plant associations or **plant communities** live in harmony because each species within the community has different requirements in relation to major environmental factors such as light, moisture, and mineral nutrients. If two species within a community were to compete, one would eventually eliminate the other. The species forming a community at any specific place and time will be an aggregation of those that together can adapt to the prevailing environmental conditions.

SUCCESSION

Natural vegetation of a particular location develops in a sequence of stages involving different plant communities. This developmental process, known as **succession,** usually begins with a relatively simple plant community. The *pioneer* community begins to alter the environment—the microclimate, the topsoil, the subsoil structure, and the ability of the soil to retain and to store moisture. As a result the species structure of the ecosystem does not remain constant. In time, the alterations of the environment become sufficient to allow a new plant community (a community that could not have survived under the original conditions) to appear and eventually to dominate the original community. The process continues with each succeeding community rendering further changes to the environment.

A common plant succession in the southeastern United States is depicted in Figure 9.8. After agriculture has ceased, weeds and grasses are the first vegetative types to adapt to the somewhat adverse conditions associated with bare fields. These low-growing plants will stabilize the topsoil, add organic matter, and in general pave the way for the development of hardwood brush such as sassafras, persimmon, and sweet gum. During the brush stage the soil will become richer in nutrients and organic matter, and its ability to retain water will increase. These conditions encourage the development of pine forests, the next stage in this vegetative evolutionary process. Pine forests thrive in the newly created environment and will eventually dwarf and dominate the weeds, grasses, and brush that preceded.

Ironically, the dominance of the pine forest leads to its demise. Pine trees require much sunlight if their seeds are to germinate. When competing with low-lying brush, grasses, and weedy annuals there is no problem in getting enough sunlight, but once the pines dominate the landscape, their seeds will not germinate in the shade and litter that their dense foliage creates. Thus, pines eventually will give way to hardwoods, such as oak and hickory, whose

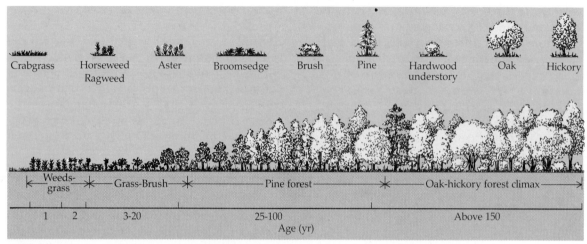

Crabgrass | Horseweed Ragweed | Aster | Broomsedge | Brush | Pine | Hardwood understory | Oak | Hickory

|← Weeds-grass →|← Grass-Brush →|← Pine forest →|← Oak-hickory forest climax →|

1   2   3-20   25-100   Above 150

Age (yr)

FIGURE 9.8 A common plant succession in the southeastern United States. Each succeeding vegetation type alters the environment in such a way that species having more stringent environmental requirements can develop. (After Odum)

seeds can germinate under those conditions. In this example, then, the end result is an oak-hickory forest. If the changes continue uninterrupted, it is estimated that the complete succession will take approximately 150 years (this time could be more or less in other ecosystems).

THE CLIMAX COMMUNITY

The theory of plant succession has been widely accepted since it was introduced by an American ecologist early in this century. However, some of the original ideas have undergone considerable modification. As initially proposed, succession was considered to be an orderly process which included various *predictable* steps or phases and ended with a dominant vegetative cover that would remain in balance with the environment until disturbed by human activity—or until there were major changes in the environment itself over geologic time. The final step in the process of succession has been referred to as **climax vegetation** or the **climax community.** It was generally agreed that such a community was self-perpetuating and had reached a state of equilibrium or stability with the environment. In our illustration of plant succession in the southeastern United States,

the oak-hickory forest would be considered the climax community.

Succession is still recognized as a useful model and a valuable tool in the study of ecosystems. Why then have some of the original ideas been challenged and what is the most recent thought on the subject? For one thing, early proponents of succession emphasized the predictable nature of the theory. One plant community would follow another in regular order as the species structure of the ecosystem evolved. But it has been demonstrated on many occasions that changes in ecosystems do not occur in such a rigid fashion. More often than not, the movement of species into an area to form a new community occurs in a random fashion and may be largely a function of chance events.

Even the idea of a single climax community has been challenged. Many scientists today no longer believe that there is only one type of climax vegetation possible for each major climatic region of the world. Some suggest that one of several different climax communities might develop within a given area; one might be influenced directly by climate, while another might be more affected by drainage conditions, the availability of certain nutrients in the soil, or topography. Most recently, ecologists have come to view plant communities, and the ecosystems upon which they are based, as the expression of all the various environmental factors functioning together. Each particular habitat is unique and constantly changing, and resultant plant communities must constantly adjust to these changes. The dominant environmental influence may be climate, but climates are altering in a cyclical fashion. By the time the species structure of a plant community has adjusted to the new climatic conditions, the climate may have changed again. Because of the dynamic nature of each plant community's habitat, some scientists argue that there can be no climax vegetation, that no one community can exist in equilibrium with the environment for an indefinite period of time.

ENVIRONMENTAL CONTROLS

It is apparent from the preceding paragraphs that the plants and animals occupying a particular ecosystem at a given time are those species that are most successful in adjusting to the unique environmental conditions that comprise their habitat. Each species of living organism has a range within which it can adapt to a given environmental factor. For example, some plants can survive under a wide range of temperature conditions, while others have narrow temperature requirements. Biogeographers and other ecologists refer to this characteristic as an organism's **tolerance** for a particular environmental condition. The ranges of tolerance for a species will determine where on earth that species may be found, and species with wide ranges of tolerance will be the most widely distributed.

We will discover later in this chapter that climate has the greatest influence over natural vegetation if we study plant communities on a worldwide basis. The major types of terrestrial ecosystems, or **biomes,** are each associated with specific ranges of temperature and critical precipitation characteristics, such as annual amounts and seasonal distribution. However, at the local scale there are other environmental factors that are equally as important as climate. In certain areas a plant's range of tolerance for seemingly minor environmental factors such as the acidity of the soil, the drainage of the land, or the salinity of the water may be the critical factor in determining whether that plant is a part of the ecosystem. The discussion that follows serves to illustrate how the major environmental controls influence the organization and structure of ecosystems.

CLIMATIC FACTORS

Of all the various climatic factors that influence the ecosystem, conditions of sunlight are often the most critical. Sunlight is the vital source of energy for photosynthesis in plants and it can also act as a control on the migration of animals. The competition for light makes forest trees grow taller and limits growth on the forest floor to plants, such as ferns, which can tolerate shade conditions. The *quality* of light is important, especially in mountain areas where plant growth may be severely retarded by ultraviolet radiation. This radiation does significant damage in the thin air of higher elevations but is effectively screened out by the denser atmosphere of lower elevations. Light *intensity* affects the rate of photosynthesis and hence the rate of primary productivity in an ecosystem. The more intense light of the low latitudes produces a higher energy input and greater biomass in the tropical forest than the less intense light of the higher latitudes produces in Arctic regions. The *duration* of daylight, in association with the changing seasons, has a profound effect upon the flowering of plants, the activity patterns of insects, and the mating habits of animals.

A second important climatic control of ecosystems is water, or more specifically, the availability of water. Most organisms require large amounts of water to survive. Plants require water for germination, growth, and reproduction; and most plant nutrients are dissolved in soil water before they are absorbed by plants. Some plants are adapted to living completely under water (seaweed); some, like mangrove and bald cypress, form forests rising from coastal marshes and inland swamps; others thrive in the constantly wet rain forest. Certain tropical plants drop their leaves and become dormant during dry seasons, while others store water received during periods of rain in order to survive seasons of drought. Plants of the desert, such as the cacti, are especially adapted to obtain and store water when it is available, while minimizing their water loss from transpiration. Animals too are severely restricted when water is in short supply. Warm-blooded animals usually require more water than cold-blooded animals because the former use water to regulate body heat. In arid regions animals must make special adaptations to environmental conditions. Many become inactive during the hottest and driest seasons and most leave their burrows or the shade of plants and rocks only at night. Others like the camel can travel for great distances and live for extended periods of time without a water supply.

Organisms are affected less by temperature variations than by sunlight and water availability. Most plants can tolerate a wide range of temperatures, although each species has optimum conditions for germination, growth, and reproduction. These functions can be impeded by excessively high and excessively low temperatures. Temperatures may also have indirect effects on vegetation. For example, high temperatures will lower the relative humidity, thus increasing transpiration. If a plant's root system cannot extract enough moisture from the soil to meet this increase in transpiration, the plant will wilt and eventually die. Both warm-blooded and cold-blooded animals make adjustments to temperature conditions. Some warm-blooded animals develop a layer of fat or fur and shiver to protect themselves against the cold. In hot periods they may sweat or lick their fur in an attempt to stay cool. In true hibernation the body temperature of a warm-blooded animal roughly changes in response to outside temperatures and behaves in a fashion similar to that of cold-blooded animals. Even

FIGURE 9.9 Krummholz vegetation at the upper reaches of the subalpine zone on Pennsylvania Mountain in the Mosquito Range of the Colorado Rockies. The healthy green vegetation has been covered by snow much of the year and has been protected from the bitterly cold temperatures associated with gale-force winter winds. Note the flag trees, which give a clear indication of wind direction. (R. Gabler)

cold-blooded animals such as the desert rattlesnake move in and out of shade in response to temperature change.

Although most significant in areas such as deserts, polar regions, coastal zones, and highlands, wind can also serve as a climatic control. Wind may cause direct injury to vegetation or may have an indirect affect by increasing the rate of evapotranspiration. To prevent loss of water in areas of severe wind stress, plants will twist and grow close to the ground in order to minimize the degree of their exposure (Fig. 9.9). In severe winter climates they are better off buried by snow than exposed to icy gales. In some coastal regions the shoreline may be devoid of trees or other tall plants, and where trees do grow they are often misshapen or swept bare of leaves or branches on their windward sides (Fig. 9.10).

SOIL AND TOPOGRAPHY

In terrestrial ecosystems, the soils in which plants grow supply all of the moisture and minerals that are transformed into plant tissues.

FIGURE 9.10 Windswept trees along the California coast near Carmel. Strong winds from off the Pacific Ocean have bowed trees inland and polished limbs on their exposed windward sides. (R. Gabler)

Soil variations are among the most conspicuous influences on plant distribution and often produce sharp boundaries in vegetation type. This is partly a consequence of the varying chemical requirements of different plant species and partly a reflection of other factors, such as soil texture. In a particular area clay soil may retain too much moisture for certain plants, while sandy soil retains too little. The clay soil will also tend to be richer in mineral nutrients but will warm more slowly in the spring than will the sand. All of these factors may be reflected in the plant cover. It is well known that pines thrive in sandy soils, grasses in clays, cranberries in acid soils, and wheat and chili peppers in alkaline soils. The subject of soils and their influence on vegetation will be explored more thoroughly in Chapter 10.

In an earlier discussion of highland climatic regions, we learned that topography influences ecosystems indirectly by providing an almost unlimited number of microclimates within a relatively small area. Plant communities vary significantly from place to place in hilly and mountainous areas in response to the differing nature of these climatic conditions. Some plants thrive on the sunny south-facing slopes of highland areas in the northern hemisphere, while others survive on the colder north-facing slopes that are most often in

shade. Luxuriant forests tower above the well-watered windward sides of mountain ranges such as the Sierra Nevada and Cascade, while semiarid grasslands cover the leeward sides. Each major increase in elevation produces a different mixture of plant species that can tolerate the lower ranges of temperature found at the higher altitudes. Each change in the degree of slope even in lowland areas can affect drainage and the retention of soil moisture—two nonclimatic environmental controls that are nonetheless related to topography.

BIOTIC FACTORS

Although they might tend to be overlooked as environmental controls, other plants and animals may be the critical factors in determining whether a given organism is a part of an ecosystem. Some interactions between organisms may be beneficial to both species involved, while others may have an adverse effect on one or both. Because most ecosystems are suitable to a wide variety of plants and animals, there is always competition between species and among members of a given species to determine which organisms will survive. The greatest competition occurs between species that occupy the same ecological niche, especially during the earliest stages of life cycles when organisms are most vulnerable. Among plants the greatest competition is for light. Those trees that become dominant in the forest are those that grow the tallest and partially shade the plants growing beneath them. Other competition occurs underground where the root development of the strongest plants crowds out the weaker and where there is always a struggle for soil water and plant nutrients.

Interactions between plants and animals and competition both within and between animal species also have significant effects upon the nature of an ecosystem. Animals are often helpful to plants during pollination or the dispersal of seeds, and plants are the basic food supply for many animals. The simple act of grazing may help to determine the species that make up a plant community. During dry peri-

VIEWPOINT: CAN HUMANS ALTER THE ENVIRONMENT?

Of deep concern to the scientific world are the drought and famine that plague portions of Africa, especially south of the Sahara in a region known as the Sahel. These devastating events are linked to *desertification*—the encroachment of arid conditions into populated areas of the semiarid lands bordering deserts. In 1985, the southward encroachment of the Sahara was estimated at 7 meters (7.5 yd) per year. What is even more alarming is that desertification (and its associated consequences) in the Sahel is not new, having occurred periodically over the past two decades. Although desertification is a complex phenomenon, one common element associated with the present episode, and an earlier episode of desertification in the 1960s, is the role of human activity.

During the 1960s modern technology and improved health practices had combined to cause a shift from nomadism to a more sedentary way of life and a doubling of the number of animals and people in the Sahel. This apparently upset the delicate ecological balance that once existed in this semiarid region. The increased population cut down what few trees there were for firewood, consumed enough water to lower the water table, and farmed on marginally acceptable land. These activities, combined with

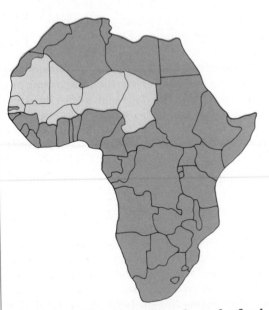

Location of the Sahel. Although much of sub-Saharan Africa has periodically experienced drought conditions, the Sahel has received the most news coverage.

overgrazing by the animals, contributed to a baring and drying of the soil, which made it very susceptible to wind erosion. Many climatologists believe these human use patterns only enhanced the desertification process that was initiated by natural causes. However, others theorize that desertification was not only enhanced but initiated and perpetuated by the human-induced change in the ecological system.

The rationale of this latter group of scientists begins with the assumption that overgrazing decreased the plant cover, which in turn increased the earth's albedo (solar energy reflected from the earth's surface). With more energy lost by reflection, the temperature of the Sahel decreased. As a consequence of this cooling, the atmosphere subsided, which suppressed convection and the associated rainfall. The reduced rainfall caused plants to die, and this combined with continued overgrazing and animal trampling to further increase the albedo. This cycle—overgrazing, increased albedo, less rainfall, plant death, increased albedo, and overgrazing—continued until there was not enough grass to support animal herds, which died off rapidly. The population of the Sahel, so dependent on the herds, suffered accordingly.

By 1972 desertification resulted in the encroachment of the Sahara on thousands of square kilometers of once livable land, the loss of 20 to 50 percent of the animal herds, and, most tragic of all, the death of nearly 5 percent of the inhabitants of the Sahel. With the reduced herds, the grass began to grow back, so that by 1974 the process started to reverse. But this reversal was quite brief and desertification, linked to human activity, escalated during the late 1970s, becoming once again a major problem in the region today. Whether human activity causes or only enhances the desertification in the Sahel is academic. The fact remains that human activity does, in one way or another, unintentionally contribute to the process that results in the encroachment of the Sahara into the region, along with its devastating results.

Other unforeseen climatic or more local microclimatic changes may occur when human beings alter their natural environments significantly. The deforestation of large areas, such as the Amazon, may reduce humidity. The drainage of marshlands, as in the Netherlands and California's San Joaquin Valley, may contribute to greater diurnal temperature changes and an increase in radiation fog. Extensive irrigation projects and man-made lakes may similarly affect air temperature and humidity patterns.

The most drastic changes to the atmosphere are related to urbanization and industrialization. The air quality within large urban areas is usually poor, with the concentration of atmospheric contaminants five to ten times greater than in the surrounding rural areas. The temperatures of large cities are also higher than those of rural areas, especially at night when the minimum temperature can be 5°C to 8°C (9°F to 15°F) higher in the city. These so-called "urban heat islands" result from the multiple heat-absorbing surfaces of the buildings, as well as from the internal heating of the structures. Thus, convection currents over the city make precipitation more likely, and condensation will readily occur around the pollutant particles in the air over the city.

The irregular profile of the city tends to increase frictional drag, which reduces the overall wind speed. However, there is a tunneling effect when strong winds blow along streets with tall buildings on each side. More static air, combined with pollutants, may result in smog conditions, with all the allied health effects: respiratory diseases, especially emphysema, lung cancer, and bronchitis. Smog also kills vegetation and can eat away at buildings and monuments. The Parthenon in Athens, Greece, has suffered more deterioration from atmospheric pollution in the last 25 years than it had in the previous 2500 years!

The climatic changes so far discussed tend to be local or regional rather than global. However, there is growing concern that human activity, especially that associated with the large industrialized centers of the Northern Hemisphere, may induce more widespread changes in climate. Climatologists are divided in their assessment of this global climatic change. One group, which subscribes to the *greenhouse effect,* points to the fact that the carbon dioxide content of the atmosphere has increased significantly in the twentieth century because of the burning of carbon-rich fossil fuels. They point out that, while the incoming short-wave solar radiation can penetrate this increased carbon dioxide layer, the layer blocks the escape of long-wave radiation from the earth's surface. The mechanism is similar to that of a greenhouse, with the same results: The earth's lower atmosphere (the interior of the greenhouse) is heated, which may lead to melting glaciers and rising sea levels.

Another group of climatologists, however, points to the pollutant particles that are spewed out by our industrial plants and automobiles. These, they argue, block the incoming solar radiation and could lead to a global cooling trend. Whether this or the greenhouse effect will be dominant in the long run remains to be seen.

Sulfur dioxide, an industrial air pollutant, combines with oxygen and water droplets to produce mild sulfuric acid, which results in "acid rain." Rain as acid as vinegar (pH = 3) has been observed. In addition to harming vegetation, acid rain has resulted in the wholesale elimination of fish in lakes in the Adirondack Mountains of New York and in Scandinavia.

So far, we have looked at inadvertent or unintentional changes in climate that may be caused by human activities. Human modification of the climate may, on the other hand, be quite intentional and planned. Perhaps the best known of these is cloud seeding, often undertaken to increase the amount of local precipitation. Silver iodide is dropped into supercooled clouds—clouds below 0°C (32°F)—to induce a phase change from the liquid to the solid state. This change of state facilitates the coalescence of cloud droplets into raindrops, thus increasing the amount of rain from a specific cloud. Studies indicate that cloud seeding may increase rainfall by approximately 10 percent. The principles of cloud seeding can be used to dissipate fog over airports, with a high rate of success.

Other experiments in weather modification are directed at hail and lightning suppression and hurricane control. These also involve cloud seeding with the intent to induce condensation, releasing latent heat and thus dissipating some of the fury of these phenomena and their resulting damage. Already, however, legal questions have arisen; for instance, if clouds are successfully seeded in New Mexico, can farmers in Texas sue for crop losses, or if a hurricane is diverted from the Florida coast and strikes Mississippi instead, can Mississippi sue the cloud seeders?

The problem is that the atmosphere is a single entity, and to meddle with it in one place (reversing ocean currents, "black-topping" the Arctic, or reversing major rivers in order to modify air masses above them—all of which have been suggested) is inevitably to produce repercussions elsewhere. Should we risk major modifications without knowing more fully how our atmospheric engine really works?

ods, herbivores may be forced to graze an area more closely than usual and the taller plants are quickly grazed out. It is the plants that grow close to the ground, or that are nonpalatable, or that have the strongest root development that survive. Hence, grazing is a part of the natural selection process, but serious overgrazing rarely results because wild animal populations increase or decrease with the available food supply. To be more precise, the number of animals of a given species will fluctuate between the maximum number that can be supported when its food supply is greatest and the minimum number required for reproduction of the species. For most animals predators are also an important control of numbers. Fortunately, when the predator's favorite species is scarce, it will seek an alternative species for its food supply, and it would be rare for an animal to be excluded from an ecosystem through predation.

HUMAN IMPACT ON ECOSYSTEMS

Throughout human history we have modified the natural development of ecosystems. Except in regions too remote to be altered significantly by civilization, humans have eliminated much of the earth's natural vegetation. Farming, intentional and unintentional fire, irrigation, grazing of domesticated animals, afforestation and deforestation, road building, urban development, intentional flooding, raising and lowering of water tables, open-pit mining, and the filling in of coastal marshlands are just a few ways in which humans have modified the plant communities around them.

We have, in fact, so changed the vegetation in some parts of the world that we can characterize classes of cultivated vegetation cared for by humans. For example, the flowers, shrubs, and grasses that we have domesticated to decorate our living areas are one form of vegetation; the grains, vegetables, fruits, and so forth that we raise for our own food are another; and the vegetation we cultivate to feed the animals we eat is a third. Because of their concern with the functioning of the natural en-

vironment, physical geographers are usually more interested in natural plant communities than in the study of cultivated or cultural (i.e., related to humans) vegetation. For this reason our focus in the remaining pages of this chapter will be on the major terrestrial ecosystems of the earth as we assume they would appear without human modification.

CLASSIFICATION OF TERRESTRIAL ECOSYSTEMS

The geographic classification of natural vegetation involves as much difficulty as the classification of any other complex phenomena influenced by a variety of factors. However, plant communities are among the most highly visible of natural phenomena, and, therefore, they can be categorized on the basis of form and structure or gross physical characteristics. Of course, the nature of the natural vegetation changes from place to place in a gradational manner, just as temperature and rainfall do, and while distinctly different types are apparent, there may be broad transition zones between them. Nevertheless, over the world there are distinctive recurring plant communities indicating a consistent botanical response to systematic controls, which are essentially climatic. It is the dominant vegetation of these plant communities that we recognize when we classify the earth's major terrestrial ecosystems (biomes).

All of the earth's biomes can be categorized into one of four easily recognized types: forest, grassland, desert, and tundra. Because vegetation adapted to cold climates may occur at high altitudes in any latitude as well as in high-latitude regions in general, the last of the major types is often referred to as arctic *and* alpine tundra by biogeographers. For some, this simple four-part classification might be adequate, but biogeographers would find it relatively useless, since it is a gross oversimplification. The forests of the equatorial lowlands are an entirely different world from those of Siberia or of New England, and the original grasslands of Kansas bore little resemblance to those of the Sudan or Kenya. Hence, the four major types of ecosystems can be subdivided into distinctive biomes, each of which is an association of plants and animals of many different species. Pure stands of particular trees, shrubs, or even grasses are extremely rare and are limited to small areas having peculiar soil or drainage conditions.

The earth's major biomes are mapped on Plate 13 on the basis of the dominant associations of natural vegetation that give each its distinctive character and appearance. The direct influence of climate and the indirect influence of soils on the distribution of these biomes is immediately apparent. Temperature (or latitudinal effect upon temperature and insolation)

and the availability of moisture are the key factors in determining the location of major biomes on the world scale (Fig. 9.11).

FOREST BIOMES

Forests are easily recognized as associations of large, woody, perennial tree species, generally several times the height of a man, and with a more or less closed canopy of leaves overhead. They vary enormously in density and physical appearance, some being evergreen and either needle- or broad-leaf, others dropping their leaves to reduce moisture losses during dry seasons or periods when soil water is frozen. Forests are found only under condi-

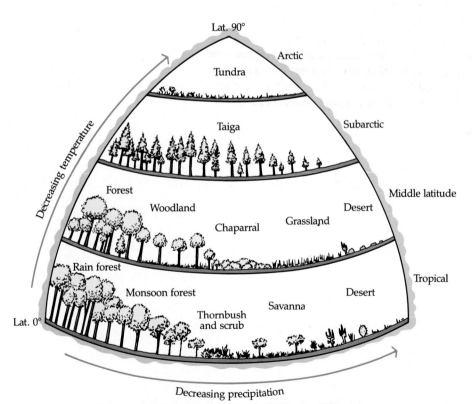

Influence of latitude and moisture on distribution of biomes

FIGURE 9.11 Schematic diagram showing distribution of earth's major biomes as they are related to temperature (latitude) and the availability of moisture. Within the tropics and middle latitudes there are distinctly different biomes as total biomass decreases with decreasing precipitation.

tions where the annual moisture balance is positive—where moisture availability considerably exceeds potential evapotranspiration in the growing season. Thus, they occur in the tropics, where either the ITC or monsoonal circulation brings plentiful rainfall, and in the midlatitudes, where precipitation is associated either with cyclonic depressions along the polar front or results from summer convectional rainfall.

Tropical and midlatitude forests have evolved different characteristics in response to the nature of the physical limitations in each area. In general, tropical forests have developed in the less restrictive of the contrasting forest environments. Temperatures are always high, though not extreme, in the humid tropics, encouraging rapid and luxuriant growth. Midlatitude forests, on the other hand, must be adapted to combat either seasonal cold (ranging from occasional frosts to subzero temperatures) or seasonal drought (that may occur at the worst possible time for vegetative processes).

TROPICAL FORESTS

The forests of the tropics are far from uniform in appearance and composition. They grade poleward from the equatorial rain forests, which support the earth's greatest biomass, to the last scattering of low trees that overlook seemingly endless expanses of tall grass or desert shrubs on the tropical margins. We have subdivided the tropical forests into three distinct biomes: the tropical rain forest, the monsoon rain forest, and other tropical forest types, primarily thornbush and scrub. Of course, there are gradations between the different types as well as distinctive variations that are found in individual localities only.

TROPICAL RAIN FOREST   In the equatorial lowlands dominated by Köppen's tropical rain-forest climate, the only physical limitation for vegetation growth is competition between adjacent species. The competition is for light. Temperatures are high enough to promote constant growth and water is always sufficient.

Thus we find forests consisting of an amazing number of evergreen tree species of rather similar appearance, since special adaptations are not required (Fig. 9.12). A cross section of the forest often reveals concentrations of leaf canopies at several different levels. The trees composing the distinctive individual tiers have similar light requirements—lower than those of the higher tiers, but higher than those of the lower tiers. Little or no sunlight reaches the forest floor, which may support ferns, but is often rather sparsely vegetated. The forest is literally bound together by vines, **lianas,** that climb the trunks of the forest trees and intertwine in the canopy in their own search for light. Aerial plants may cover the limbs of the forest giants, deriving nutrients from the water and the plant debris that falls from higher levels.

The forest trees commonly depend upon grotesquely flared or buttressed bases for support, since their root systems are shallow. This is a consequence of the richness of the surface soil and the poverty of its lower levels. The rain-forest vegetation and soil are intimately associated. The forest litter is quickly decomposed, its nutrients being released and almost immediately reabsorbed by the forest root systems, which consequently remain near the surface. Tropical soils that maintain the amazing biomass of the rain forest are fertile only as long as the forest remains undisturbed. Clearing the forest interrupts the crucial cycling of nutrients between the vegetation and the soil; the copious amounts of water percolating through the soil leach away its soluble constituents, leaving behind only inert iron and aluminum oxides that cannot support forest growth. The present rate of clearing threatens to obliterate the magnificent worldwide tropical rain forests within the foreseeable future. The largest remaining areas of unmodified rain forests are in the upper Amazon Basin, where they cover hundreds of thousands of square kilometers.

Within the areas of rain forest and extending outward beyond its limits along streams are patches or strips of jungle. Jungle consists of an almost impenetrable tangle of vegetation, contrasting strongly with the rela-

FIGURE 9.12 A rain forest near El Salto Falls in eastern Mexico. The dense nature of the forest canopy effectively conceals the vast number of different evergreen tree species and relatively open forest floor. (R. Gabler)

tively open nature (at ground level) of the true rain forest. It is often composed of secondary growth that quickly invades the rain forest where a clearing has allowed light to penetrate to the forest floor. Jungle commonly extends into the drier areas beyond the forest margins along the courses of streams. There it forms a gallery of vegetative growth closing over the watercourse and hence has been called **galeria forest**.

MONSOON RAIN FOREST In areas of monsoonal circulation there is an alternation between dry winters, when the dominant flow of air is from the land to the sea, and wet summers, when the atmospheric circulation re-

verses, bringing moist air onshore along tropical coasts. The summer rainfall may be very high, hundreds of centimeters where air is forced upward by topographic barriers. In any case, it is sufficient to produce a forest that, once established, remains despite the dry winter season. Monsoon forests themselves may have discernible tiers of vegetation related to the varying light demands of different species, and they are included with the tropical rain forest on Plate 13. However, the number of species is less than in the true rain forest, and the overall height and density of vegetation is also somewhat less. Some of the species are evergreen, but many are deciduous.

OTHER TROPICAL FORESTS, THORNBUSH, AND SCRUB Where seasonal drought has precluded the development of true rain forest, or where soil characteristics prevent the growth of such vegetation, variant types of tropical forests have developed. These tend to be found on the subtropical margins of the rain forests and on old plateau surfaces where soils are especially poor in nutrients. The vegetation included in this category varies enormously but is generally low-growing in comparison to rain forest, without any semblance of a tiered structure, and is denser at ground level. Commonly it is thorny, indicating defensive adaptation against browsing animals, and it shows resistance to drought in that it is generally deciduous, dropping its leaves to conserve moisture during the dry winter season. Ordinarily grass is present beneath the trees and shrubs. As we move away from the equatorial zone we find the trees more widely spaced and the grassy areas becoming dominant. This outward movement brings us into the tropical savannas, which will be discussed later.

MIDLATITUDE FORESTS

The forest biomes of the midlatitudes are different from those of the tropics because the dominant trees have evolved mechanisms to withstand periods of water deprivation due to both low temperatures and annual variations in

FIGURE 9.13 The distinctive sclerophyllous evergreen vegetation type encountered wherever hot, dry summers alternate with rainy winters. Species are deep-rooted to probe far below the surface for moisture and small-leaved to resist moisture loss. Oaks commonly occupy relatively damp sites, such as the gullies and windward slopes seen here. (S. Brazier)

precipitation. Evergreen and deciduous plants are present, equipped to cope with seasonal extremes not encountered in tropical latitudes.

MEDITERRANEAN SCLEROPHYLLOUS WOODLAND

Surrounding the Mediterranean Sea and on the west coasts of the continents between approximately 30° and 40° N and S latitude, we have seen that a distinctive climate exists, Köppen's mesothermal hot and dry summer type (*Csa*). Here annual temperature variations are significant, but freezing temperatures are rare. Thus it is possible for the vegetation to be evergreen. However, there is little or no rainfall during the warmest months, when maximum growth normally occurs. Thus plants must be drought-resisting during their period of highest activity. This requirement has resulted in the evolution of a distinctive vegetation that is relatively low-growing, with small hard-surfaced leaves and roots that probe deeply for water. The leaves must be capable of photosynthesis with minimum transpiration of moisture. The general look of the vegetation is a thick scrub, called *chaparral* in the western United States and *maquis* in the Mediterranean region (Fig. 9.13). Wherever moisture is concentrated in depressions or on the cooler north-facing hill slopes, deciduous and evergreen oaks occur in groves. Drought-resisting needle-leaf trees, especially pines, are also part of the overall vegetation association. Thus the vegetation is a mosaic related to site characteristics and microclimate. Nevertheless, the similarity of the natural vegetative cover in such widely separated areas as Spain, Turkey, and California is astonishing. (Note the location of Mediterranean midlatitude forest, Plate 13.)

BROAD-LEAF DECIDUOUS FOREST

The humid regions of the middle latitudes experience a seasonal rhythm dominated by warm air in the summer and invasions of cold polar air in the winter. To avoid frost damage during the colder winters and to survive periods of total moisture deprivation when the ground is frozen, trees whose leaves have large transpiring surfaces drop these leaves and become dormant, coming to life and producing new leaves only when the danger period is past (Fig. 9.14). A large variety of trees have evolved this mechanism: Certain oaks, hickory, chestnut, beech, and maples are common examples. The seasonal rhythms produce some beautiful scenes, particularly during the periods of transition between dormancy and activity: The sprouting of

(a)

(b)

FIGURE 9.14 The appearance of hardwood forests in midlatitude regions with cold winters changes dramatically with the seasons. The green leaves of summer, (a), change to reds, golds and browns in fall, (b), and drop to the forest floor in winter, (c). Leaf dropping in areas of cold winters is a means of minimizing transpiration and moisture loss when the soil water is frozen. (R. Gabler)

(c)

new leaves in the spring and the brilliant coloration of the fall as chemical substances draw back into the plant for winter storage.

The trees of the deciduous forest may be almost as tall as those of the tropical rain forests and, like them, produce a closed canopy of leaves overhead or, in the cold season, an interlaced network of bare branches. However, lacking a multistoried structure and having lower density as a whole, the midlatitude deciduous forests allow much more light to reach ground level. Forests of this type are the natural vegetation in much of western Europe, eastern Asia, and the eastern United States. To the north and south they merge with mixed forests composed of broad-leaf deciduous trees and conifers. (Broad-leaf deciduous forests and mixed forests are included together on Plate 13 as Mixed Midlatitude Forest.) Both the broad-

leaf deciduous and mixed forests have been largely logged off or cleared for agricultural land, and the original vegetation of these regions is rarely seen.

BROAD-LEAF EVERGREEN FOREST   Beyond the tropics, broad-leaf trees remain evergreen (active throughout the year) on a large scale only in certain Southern Hemisphere areas. Here the maritime influence is too strong to permit either dangerous seasonal drought or severely low winter temperatures. Southeastern Australia and portions of New Zealand, South Africa, and southern Chile are the principal areas of this type. In the Northern Hemisphere broad-leaf evergreen forest may once have been significant in eastern Asia but has long since been cleared for cultivation. Limited areas occur in the United States in

Florida and along the Gulf Coast as the belt of evergreen oak and magnolia.

MIXED FOREST   Poleward and equatorward the broad-leaf deciduous forests in North America, Europe, and Asia gradually merge into mixed forests, including needle-leaf coniferous trees, normally pines. In general, where conditions permit the growth of broad-leaf deciduous trees, coniferous trees cannot compete successfully with them. Thus, in mixed forests the conifers, which are actually more adaptable to soil and moisture deficiencies, are found in the less hospitable sites: in sandy areas, on acid soils, or where the soil itself is thin. The northern mixed forests reflect the transition to colder climates with increasing latitude; eventually conifers become dominant in this direction. The southern mixed forests are more problematical in origin. In the United States they are transitional to pine forests situated on sandy soils of the coastal plain. In Europe and Asia they coincide with highlands probably entirely dominated by conifers during a stage in the plant succession that began with the change of climatic environments at the end of the ice ages, some 10,000 years ago.

CONIFEROUS FOREST   The coniferous forests occupy the frontiers of tree growth. They survive where other species cannot endure the climatic severity and impoverished soils. The hard, narrow needles of coniferous species transpire much less moisture than do broad leaves, so that needle-leaf species can tolerate conditions of physiologic drought (unavailability of moisture because of excessive soil permeability, a dry season, or frozen soil water) without defoliation. Pines, in particular, also demand little from the soil in the form of soluble plant nutrients, particularly basic elements such as calcium, magnesium, sodium, and potassium. Thus they grow in sandy places and where the soil is acid in character. As a whole, conifers are particularly well adapted to regions having long, severe winters combined with summers warm enough for vigorous plant growth. Because all but a few exceptions retain their leaves (needles) throughout the year, they are ready to begin photosynthesis as soon as temperatures permit, without having to produce a new set of leaves to do the work.

Thus we find a great band of coniferous forests (the **boreal forests** or **taiga**) dominated by spruce and fir species, with pines on sandy soils, sweeping the full breadth of North America and Eurasia northward of the 50th parallel of latitude, approximately occupying the region of Köppen's subarctic climate. Conifers differ from other trees in that their seeds are not enclosed in a case or fruit but are carried naked in cones. All are needle-leaved and drought-resisting, but a few are not evergreen. Thus a large portion of eastern Siberia is dominated by larch, which produces a deciduous, coniferous forest. In this area January mean temperatures may be −35° to −51°C (−30° to −60°F). Here we encounter the most severe winter climate in which trees can maintain themselves, and even needle-leaf foliage must be shed for the vegetation to survive. Hardy broad-leaf deciduous birch trees share this extreme climate with the indomitable larches.

Extensive coniferous forests are not confined to high-latitude areas of short summers. Higher elevations in midlatitude mountains of the Northern Hemisphere have forests of pine, hemlock, and fir (Fig. 9.15). The forests along the sandy coastal plain of the eastern United States are there in part because of the sandy soils present, but they may also reflect a stage in plant succession that in time will lead to domination by broad-leaf types. Similarly, a temporary stage in the postglacial vegetation succession of the Great Lakes area included magnificent forests of white pine and hemlock that were completely logged off during the late nineteenth century.

A more permanent coniferous forest occupies the west coast of North America extending from southern Alaska to central California. It is made up of sequoias, Douglas fir, cedar, hemlock, and farther north, Sitka spruce. Many of the California sequoias are thousands of years old and more than 100 meters (330 ft) high. The southern regions experience summer drought, and farther north sandy, acidic, or coarse-textured soils dominate.

FIGURE 9.15 Evergreen coniferous forest is the characteristic vegetation of higher-altitude regions in the middle latitudes. The needle-leaf trees are an adaptation to the physiologic drought of the winter season, which is longer and more severe at higher elevations than it is near sea level. (R. Gabler)

GRASSLAND BIOMES

Grasses, like conifers, appear in a variety of settings and are part of many diverse plant communities. They are, in fact, an initial form in most plant successions. However, there are enormous, continuous expanses of grasslands on our earth. In general, it is thought that grasses are dominant only where trees and shrubs cannot maintain themselves either because of excessive or deficient moisture in the soil. On the global scale grassland biomes are located in continental interiors where most if not all of the precipitation falls in the summer. Two great geographical realms of grasslands are generally recognized: the tropical and the midlatitude grasslands. However, it is difficult to define grasslands of either type using any specific climatic parameter, and geographers suspect that human interference with the natural vegetation has caused expansion of grasslands into forests in both the tropical and middle latitudes.

TROPICAL SAVANNA GRASSLANDS

The tropical grassland biome differs from grassland biomes of the midlatitudes in that it ordinarily includes a scattering of trees; this is implied in the term **savanna,** as it is presently used (Fig. 9.16). In fact, the demarcation between tropical scrub forest and savanna is seldom a clear one. The savanna grasses tend to be tall and coarse, with bare ground visible between the individual tufts. The related tree species generally are low-growing and wide-crowned forms, having both drought- and fire-resisting qualities, indicating that fires frequently sweep the savannas during the winter drought season. Indeed, the role of fire in maintaining savannas (suppressing tree invasions) and the role of man in creating or expanding the savannas at the expense of the forest has been a topic of continuing discussion among geographers. Savannas occur under a variety of temperature and rainfall conditions but generally fall within the limits of Köppen's *Aw* type. Commonly they occur on red-colored soils leached of all but iron and aluminum oxides, which become bricklike or slaggy when dried. They likewise coincide with areas in which there is a dramatic fluctuation in the level of the water table (the zone below which all soil and rock pore space is saturated by water). The up-and-down movement of the water table may in itself inhibit forest development, and it is no doubt a factor in the peculiar chemical nature of savanna soils.

FIGURE 9.16 The Savanna biome of eastern Africa. This classic landscape of grasses and scattered trees supports dwindling herds of grazing animals as land use patterns change. (Donald Marshall collection, WIU International Programs)

MIDLATITUDE GRASSLANDS

The zone of transition between the midlatitude deserts and forests is occupied by the midlatitude grasslands. On their dry margins they pass gradually into deserts in Eurasia and are cut off westward by mountains in North America. However, on their humid side they terminate rather abruptly against the forest margin, again raising questions as to whether their limits are natural or have been created by human activities, particularly the intentional use of fire to drive game animals. The midlatitude grasslands of North America, like the African savannas, formerly supported enormous herds of grazing animals—in this case antelope and bison, which were the principal means of support of the American Plains Indians.

Like the savannas, the midlatitude grasslands were diverse in appearance. They too consisted of varying associations of plant species that were never uniform in composition. The grasses were as much as 3 meters (10 ft) tall in the more humid sections, as in Iowa, Indiana, and Illinois, but only 15 centimeters (6 in) high on the dry margins from New Mexico to Montana. Thus the midlatitude grassland biomes are usually divided into tall-grass prairie and short-grass steppe, often with a zone of mixture recognized between. Unlike growth in the savannas, the germination and growth of midlatitude grasses are attuned to the melting of winter snows, followed by summer rainfall. Although the grasses may be annuals that complete their life cycle in one growing season or perennials that grow from year to year, they are dormant in the winter season. Also, unlike the savannas, the soils beneath these grasslands are extremely rich in organic matter and soluble nutrients. As a consequence, most of the midlatitude grasslands have been completely transformed by agricultural activity. Their wild grasses have been replaced by domesticated varieties—wheat, corn, and barley—thus they have become the "breadbaskets" of the world.

PRAIRIE GRASSLANDS   The tall-grass prairies were an impressive sight; in some better-watered areas they made up endless seas of grass moving in the breeze, reaching higher than a horse's back. Flowering plants were conspicuous, adding to the effect. Unfortunately, this scene, which inspired much vivid description by those first encountering it, is no longer visible anywhere. The tall-grass prairies, which once reached continuously from Alberta to

Texas, have been destroyed. Their tough sod, formed by the dense grass root network, defeated the first wooden plows, which had served well enough in breaking up the forest soils. But the steel plow, invented in the 1830s, subdued the sod and was aided by the introduction of subsurface tile for draining the nearly flat uplands and by the simultaneous appearance of well-digging machinery and barbed wire. These four innovations transformed the prairie from grazing land to cropland.

In North America the tall-grass prairie pushed as far eastward as Lake Michigan. Why trees could not invade the prairie in this relatively humid area remains an unanswered question. Farther west shallow-rooted grass cover is fully understandable because the lower soil levels, to which tree roots must penetrate for adequate support and sustenance, are bone dry. In such areas trees can survive only along streams or where depressions collect water.

Beyond North America, tall-grass prairies were found on a large scale in a discontinuous belt commencing in Hungary and extending through the Soviet Ukraine and central Asia to Manchuria, and in the pampas of Uruguay and Argentina. Today all of these areas have been given over to agriculture. The factor that seems to account best for the tall-grass prairie—precipitation that is both moderate and variable in amount from year to year—is the principal hazard in the use of these regions as farmland. However, this hazard becomes much more important in the areas of steppe grasslands.

STEPPE GRASSLANDS West of the 100th meridian in the United States and extending across Eurasia from the Black Sea to Manchuria, roughly coinciding with the areas of Köppen's *BS* climates, are vast, nearly level grasslands composed of a mixture of tall and short grass, with short-grass steppes becoming dominant in the direction of lower annual precipitation totals (Fig. 9.17). On the Great Plains between the Rocky Mountains and the prairies, the steppe zone more or less coincides with the zone in which moisture rarely penetrates more than 60 centimeters (2 ft) into the soil, so that the subsoil is permanently dry. Moving toward the drier areas, the grassland vegetation association dwindles in diversity and, more conspicuously, in height to less than 30 centimeters (1 ft). This is a consequence of reduction in numbers of tall-growing species and higher abundance of shallow-rooted and lower-growing types. In their natural conditions the

FIGURE 9.17 Steppe vegetation of the Nebraska Sand Hills. Although the tall-grass prairies have been completely transformed by man, vast areas of steppe vegetation remain because of their low and unpredictable precipitation, which makes them hazardous for agriculture. Once they supported immense herds of bison and antelope; now they are grazing land for beef cattle. (R. Gabler)

steppes of North America and central Asia supported higher densities of grazing animals—bison and antelope in the former, wild horses in the latter—than did the prairies. Indeed, it is suspected that the specific plant association of the steppe regions may have been a consequence of overgrazing under natural conditions. The short-grass steppes cannot be cultivated without irrigation, so they remain the domain of wide-ranging grazing animals, but today's animals are domesticated cattle, not the thundering self-sufficient herds of wild species that formerly made these plains one of the earth's marvels.

DESERT

Eventually drought may become too severe even for the hardy grasses. Where evapotranspiration demands greatly exceed available moisture throughout the year, as in Köppen's desert climates, either special forms of plant life have evolved or the surface is bare. Plants that actively combat drought are present, equipped to probe deeply or widely for moisture, to reduce moisture losses to the minimum, or to store moisture when it is available. Other plants evade drought by merely lying dormant, perhaps for years, until enough moisture is available to assure successful growth and reproduction. The desert biome may be recognized in two ways: by the presence of plants that are either drought-resisting or drought-evading (Fig. 9.18). Plants that have evolved mechanisms to combat drought are known as *xerophytes*. They are perennial shrubs whose root systems below ground are much more extensive than their visible parts or that have evolved tiny leaves with a waxy covering to combat transpiration. They may have leaves that are needlelike or trunks and limbs that photosynthesize like leaves or that have expandable tissues or accordion-like stems to store water when it is plentiful (the succulent cacti). They may be plants that can tolerate excessively saline water or shrubs that shed their leaves until sufficient moisture is available for new leaf growth. The nonxerophytic vegetation consists mainly of short-lived annuals that germinate and hurry through their complete life cycle of growth, leaf production, flowering, and seed dispersal in a matter of weeks when triggered by moisture availability. Like other species, these ephemeral plants also require days of a certain length, so they appear only in particular months; therefore, the month-to-month and year-to-year variation in form and appearance of vegetation is enormous. The similar life forms and habits of the different plant species found in the deserts of widely separated continents is a remarkable display of repeated evolution to ensure survival in similar climatic settings.

FIGURE 9.18 The saguaro cactus is one of the most spectacular of all desert plants, having evolved mechanisms to store water and resist evaporation losses. (R. Sager)

ARCTIC AND ALPINE TUNDRA

Proceeding upward in elevation and poleward in latitude, we finally come to the region in which the growing season is too brief to permit tree growth. We enter a vast realm dominated by subfreezing temperatures and thin snow cover much of the year, so that the ground is frozen to depths of hundreds of meters. Only the top 36–60 centimeters (15–25 in) thaw during the short summer interval. Still, vegetation survives here and in fact forms a nearly complete cover over the surface. Such vegetation must be equipped to tolerate frozen subsoils (permafrost), icy winds, a low sun angle, summer frosts, and soil that is waterlogged during the short growing season. The result is tundra—a mixture of grasses, flowering herbs, sedges, mosses, lichens, and occasional low-growing shrubs (Fig. 9.19). Most of the plants are perennials that produce buds close to or beneath the soil surface, protected from the wind. Many of the plants show xerophytic adaptation, such as small, hard leaves, in response to extreme physiologic drought resulting from wind stress. The effect of wind is evident by the fact that the less-exposed valleys within the tundra region may be occupied by coniferous woodlands.

In a band of varying width reaching across northern Alaska, Canada, Scandinavia, and the Soviet Union several types of tundra are recognized: bush tundra, consisting of dwarf willow, birch, and alder, which grow along the edge of the coniferous forest; grass tundra, which is hummocky and water-soaked during the summer; and desert tundra, in which expanses of bare rock may be completely

FIGURE 9.19 Close examination of the tundra biome reveals a mixture of grasses, sedges, and rushes. All vegetation, including the occasional shrub or stunted tree, grows close to the ground to avoid the extreme drought and low temperatures associated with wind stress. (R. Gabler)

FIGURE 9.20 The alpine tundra vegetation of the mountainous western United States is similar to the arctic tundra that covers large regions in the higher latitudes. During the winter it will be completely covered by snow, but in the summer it is often a colorful landscape featuring a variety of flowering plants. (R. Gabler)

covered by colorful lichens. In the ice-free areas of Antarctica only desert tundra occurs. Alpine conditions are not exactly like those in the arctic latitudes for the deeper snow cover of the high mountains prevents the development of permafrost, and the summer sun results in considerably more evaporation (Fig. 9.20). Microclimate becomes an important control of vegetation because of the varying exposures to sun and wind. Nevertheless, the short growing season and severe wind stress produce an overall plant community similar to that in the arctic regions.

QUESTIONS FOR DISCUSSION AND REVIEW

1. Give reasons why the study of ecosystems is important in the world today.

2. Give several examples of natural ecosystems within easy driving distance of your own residence. What artificial ecosystems are in your community or nearby?

3. What are the four basic components of an ecosystem? Which do you consider the most important and why?

4. Define *trophic structure, trophic level, food chain, food web*. How can an organism belong to more than one trophic level? Give an example.

5. How can the laws of thermodynamics help us to understand the functioning of ecosystems? How does the *biomass* at one trophic level usually compare in weight with that at the next level?

6. What is the difference between *gross* and *net productivity*? Which is most important to the ecosystem?

7. What factors are most critical in affecting the net primary productivity of a terrestrial ecosystem; of an aquatic ecosystem?

8. What is an organism's *habitat*? its *ecological niche*? Why are generalists in the majority in ecosystems?

9. Define *natural vegetation; climax community*. In what ways has the original theory of succession been modified?

10. What are the important environmental controls of ecosystems? Which are most important on a worldwide basis?

11. What are the four major types of earth biomes?

Why is this classification inadequate for biogeographers?

12. What important climatic characteristics are related to each of the major biome types?

13. Describe a true tropical rain forest. How does such a forest differ from jungle? How are the two related?

14. What are the distinctive features of chaparral vegetation? What climatic conditions are associated with chaparral?

15. What conditions of climate or soil might be anticipated for each of the following in the middle latitudes: broad-leaf deciduous forest; broad-leaf evergreen forest; needle-leaf coniferous forest?

16. How is the larch tree adapted to the climate of eastern Siberia? What trees in your area display similar adaptations?

17. What factors contribute to the development and maintenance of savannas?

18. How have xerophytes adapted to desert climatic conditions?

10

SOILS AND SOIL DEVELOPMENT

I n an urban society few people outside of science and agriculture give much thought to their nation's heritage of varied, fertile soils. This is a serious oversight in an age of environmental awareness and concern. Soil is a dynamic natural body capable of supporting a vegetative cover. Composed primarily of weathered minerals and varying amounts of water, oxygen, and organic materials, soil covers most land surface with a fragile mantle that, along with water and air, ranks as an indispensable resource.

Soil contains chemical solutions, gases, organic refuse, and an active fauna. The complex physical, chemical, and biological processes that take place among the various components of soil are an integral part of its dynamic character. As an active body, soil responds to changes in climate (especially to temperature and moisture), to land surface configuration, to vegetative cover and composition, and to animal activity. The result is that soils evolve and mature toward a state of equilibrium with the environment that is subject to change in response to any alterations in that environment. Thus soil serves as an outstanding example of the integration of the earth's subsystems.

The formation of soils depends on a large number of factors. But the dominant influence of climate on soils is unmistakable when viewed on a worldwide scale. The climate-soil relationship, as well as the association of soils with climate-controlled vegetation, were both obvious as climatic regions were considered in Chapters 7 and 8.

PRINCIPAL SOIL COMPONENTS

W hat actually makes up soil? What does the scoop of a bulldozer contain when it shovels up a load of soil? What does it take to support the earth's varied vegetation?

INORGANIC MATERIALS

Soil is made up of both insoluble mineral material, that is, minerals that will not dissolve in water to form a solution, and of soluble minerals or chemicals in solution. The most common minerals found in soils are combinations of the most common elements of the earth's crust: silicon, aluminum, oxygen, and iron. In

Tropical soil under sugar cane, Oahu, Hawaii. (R. Gabler)

fact, most of the known chemical elements are found in soils in some form. Some of these occur in chemical combinations; others are found in the air and water that are also part of soil. We have already learned that a large number of these elements are necessary to sustain the flora and fauna of the earth's ecosystems. Important among these, in addition to the four elements listed above, are carbon, hydrogen, nitrogen, sodium, potassium, zinc, copper, and iodine.

The elements and chemical compounds that form a part of soil material come from many sources. Some are derived from the weathering of underlying rocks or from accumulations of loose sediments. Others enter in solution in water or as a part of the chemical structure of water itself. Still others are part of the air found in soils or are derived from organic activity, some of which helps to disintegrate rocks, releases gases, or creates new chemical compounds.

Because plants need many chemical elements for growth, a knowledge of a soil's mineral and chemical content is necessary if we are

to determine its productive potential. Frequently we can rectify a deficiency in a specific element through fertilization and thereby increase the productivity of a given soil.

SOIL WATER

Plants need air and water to function, live, and grow. They depend primarily upon the soil in which they are rooted for their supply of these two necessities (Fig. 10.1). Soil water is not pure oxygen combined with hydrogen but is actually a solution bearing traces of many soluble nutrients. Not only does soil water supply the moisture necessary for the chemical reactions that sustain life, but it also provides the nutrients in such a form that they can be extracted and used.

The original source of soil moisture is, of course, precipitation. When precipitation falls on the land surface, it is either absorbed or it runs off downslope to a stream that eventually channels it into a larger body of water. The water absorbed by the soil washes over and through various soil materials, dissolving some

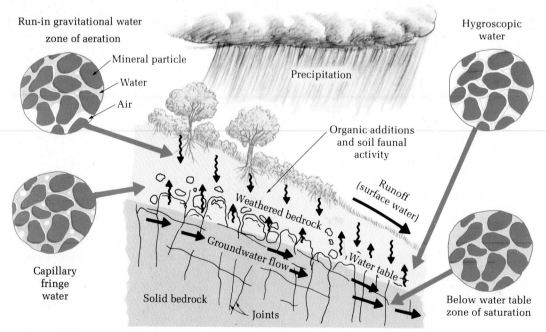

FIGURE 10.1   The relation of soil to environmental factors.

Plate 15 Soil Profiles of Humid and Polar Regions

Tropical latosol (oxisol), central Puerto Rico.

Intrazonal grumusol (vertisol), Lajas Valley, Puerto Rico.

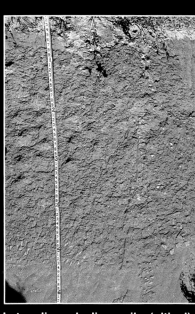

Latosolic-podzolic soil (ultisol), North Carolina Piedmont.

Gray-brown podzolic soil (alfisol), southern Michigan.

Podzol (spodosol), northern New York.

Tundra soil (inceptisol), northern Alaska.

Plate 16 World Map of Major Soil Classes

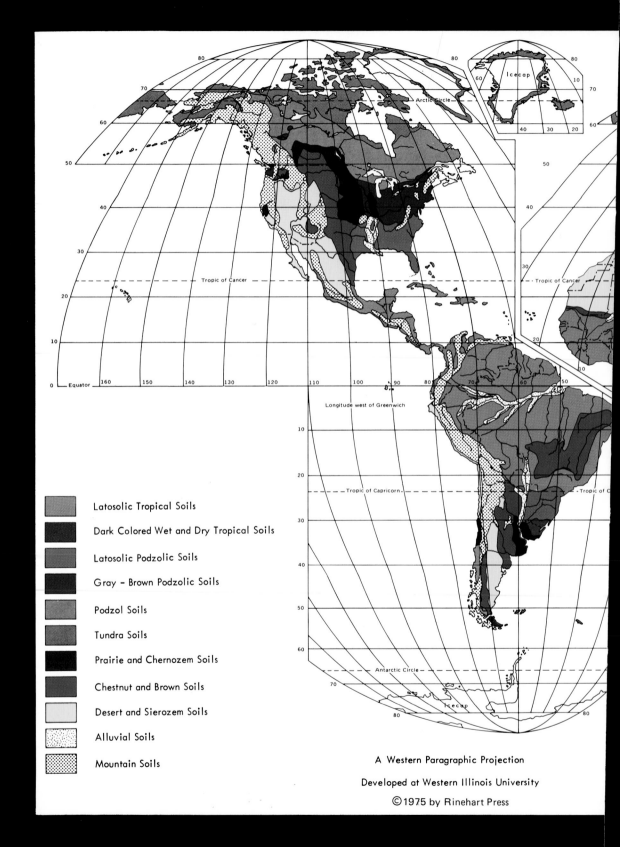

Latosolic Tropical Soils

Dark Colored Wet and Dry Tropical Soils

Latosolic Podzolic Soils

Gray – Brown Podzolic Soils

Podzol Soils

Tundra Soils

Prairie and Chernozem Soils

Chestnut and Brown Soils

Desert and Sierozem Soils

Alluvial Soils

Mountain Soils

A Western Paragraphic Projection

Developed at Western Illinois University

©1975 by Rinehart Press

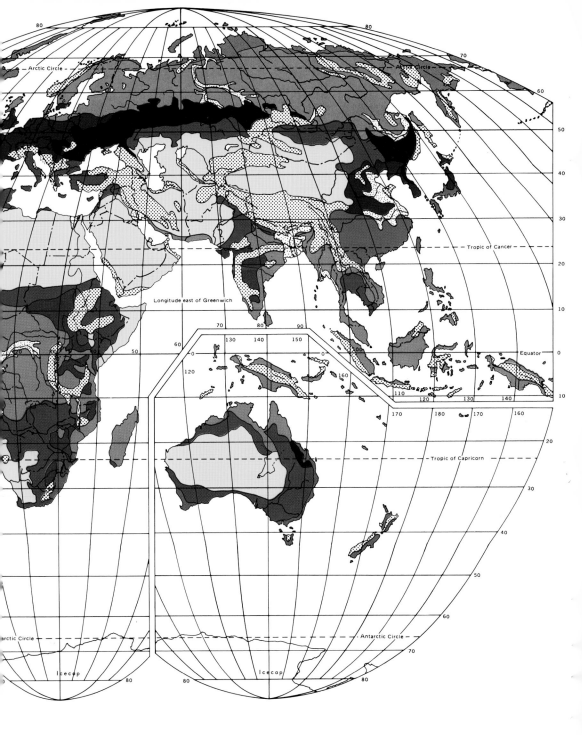

Longitude east of Greenwich

Tropic of Cancer

Equator

Tropic of Capricorn

Antarctic Circle

Icecap

```
        0    1000   2000   3000   4000        k i l o m e t e r s
        |----|------|------|------|
        0          1000         2000        3000    m i l e s
```

Prairie soil (mollisol), central Iowa.

Chernozem soil (mollisol), southeastern South Dakota.

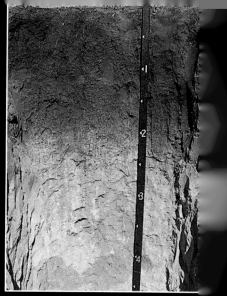

Reddish chestnut soil (mollisol), eastern New Mexico.

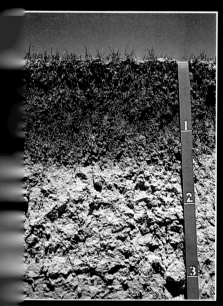

Brown soil (mollisol), eastern Colorado.

Red desert soil (aridisol), southern New Mexico.

Intrazonal solonchak (aridisol), central Nevada.

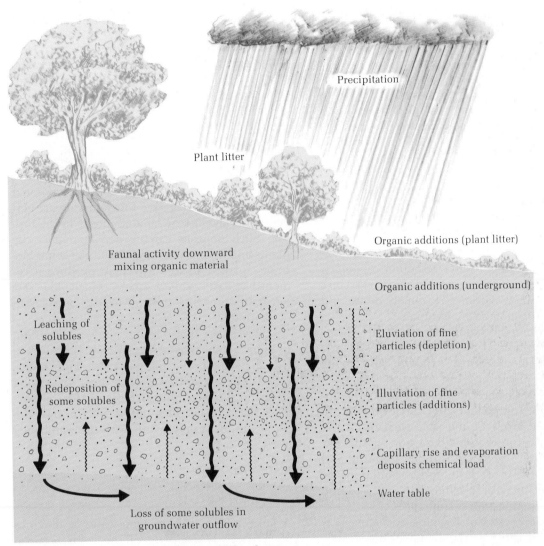

Precipitation

Plant litter

Organic additions (plant litter)

Faunal activity downward
mixing organic material

Organic additions (underground)

Leaching of
solubles

Eluviation of fine
particles (depletion)

Redeposition of
some solubles

Illuviation of fine
particles (additions)

Capillary rise and evaporation
deposits chemical load

Water table

Loss of some solubles in
groundwater outflow

FIGURE 10.2   Processes affecting soil development.

of these materials and carrying them through the soil.

Water is found in soil in different circumstances. The water that percolates downward through the soil, pulled by gravitation force, is called **gravitational water.** Gravitational water moves downward through the spaces between the individual soil particles and the clumps of soil toward the water table and the so-called **zone of saturation.** The zone of saturation, as its name implies, begins at the level below which all the spaces between the soil particles and clumps are filled with water. As a consequence the water cannot percolate any further.

As we might expect, the amount of gravitational water in a soil is related to several conditions, including the amount of rainfall that has occurred, the length of time since the last rainfall, the ease with which the water moves through the soil, and the amount of space available for water storage.

Gravitational water functions in the soil in several ways (Fig. 10.2). First, it has important effects upon the color, structure, and texture of a soil. In addition, as gravitational water moves down through a soil, the water takes with it the finer particles from the topsoil. This removal of soil components from the topsoil is called **eluviation.** Eventually as the gravitational water percolates downward, it begins to deposit some of the clay and silt particles it has removed from the topsoil. Such deposit in the subsoil of soil components is called **illuviation.** Gravitational water therefore serves as an agent of transportation and mixing as it moves soil particles from one level to another. The result of eluviation is that the texture of the topsoil becomes coarser as the finer particles are removed. Consequently, the topsoil's ability to retain water is reduced, while illuviation enhances the subsoil's water retention capability. The process may be carried to such extremes that the subsoil becomes extremely dense and compact, forming a clay **hardpan.**

Gravitational water also affects the chemical composition of a soil and as a result also affects its color, texture, structure, and ability to provide plant nutrients. As gravitational water moves downward, it dissolves the soluble inorganic soil components and carries them into the deeper levels of the soil, perhaps to the zone of saturation. This depletion of the nutrients in the upper soil is called **leaching** and when extensive, as it is under conditions of heavy precipitation, leaching can rob a topsoil of all but the most inert substances.

The processes of leaching and eluviation are a major cause of the characteristic stratification found in soils. The upper portions are composed of coarse material and are somewhat impoverished in soluble nutrients. Both fine material and some of the substances dissolved from the upper soil come to rest in the lower portion, which becomes dense, and sometimes is strongly colored by accumulated iron compounds.

Some soil water is held to the surface of the individual soil particles and soil clumps by surface tension (the same property that causes small droplets of water to form rounded beads instead of spreading out in a thin film). This soil water, called **capillary water,** serves as the storage supply of water for plants. Capillary water can move in all directions through soil, its migration being determined by its tendency to move from areas with more water to areas with less. Thus, during the periods between rainfalls when there is no gravitational water flowing through the soil, capillary water can move upward or horizontally to supply plant roots with the necessary moisture and dissolved nutrients.

When capillary water moves upward in a soil toward the surface, it carries with it minerals from the subsoil. If this water is evaporated in the upper soil layers, the minerals are left behind as alkaline or saline deposits in the topsoil. Such deposits can be detrimental to the plants and animals existing in the soil. Lime deposits formed in this way may produce a cement-like layer, called *caliche*, which, like clay hardpan, prevents the downward percolation of rainwater.

Soil water is also found in a very thin film, invisible to the naked eye, which is bound to the surface of all soil particles. This water, called **hygroscopic water,** is held by strong electrical forces to the surfaces of the soil particles. Because it does not move through the soil it cannot supply plants with the water they need.

SOIL AIR

A large part of soil, in some cases as much as 50 percent, is made up of the voids between individual soil particles and between clumps or aggregates of soil particles. These spaces, when not filled with water, are filled with air. Soil air is much like the air of the atmosphere above the earth's surface, though it is likely to have less oxygen, more carbon dioxide, and a fairly high relative humidity because of the presence of capillary and hygroscopic water.

For most of the microorganisms and plants that live in the soil, soil air supplies the oxygen and carbon dioxide necessary for life processes. Thus the problem with water-saturated soils is not so much the excess water itself but the fact that with all the pores filled with water there is no supply of air. It is because of the lack of air, then, that many plants and animals find it difficult to survive in water-saturated soils.

ORGANIC MATTER

In addition to various chemical compounds, both soluble and insoluble, and to air, other gases, and water, soil also contains organic matter. The decayed remains of plant and animal material, partially transformed by bacterial action, are called **humus.** Humus is important to soils in several ways, but it is most important for its role as a catalyst in the chemical reactions by which plants extract nutrients from the soil and for its role in restoring minerals to the soil. Humus also improves soil structure, making it more workable and increasing its capacity to retain water. Humus serves, too, as a source of food for the enormous variety of microscopic organisms that live in soil.

Living organisms in soil range from countless microscopic bacteria to good-sized earthworms, rodents, and other burrowers. Many of these animals are useful in the development and enrichment of soils. They are important in the creation of humus from inert plant litter and in mixing organic material deeper into the soil. In addition, we cannot ignore the chemical and mechanical roles of the plants and their root systems, which are an integral part of the soil system.

Each of the soil constituents is important in determining the characteristics of a particular soil variety. Soils vary from place to place both locally and over the earth as do the proportions of their constituents. For example, soils in midlatitude grasslands normally have a very high proportion of organic debris or humus; those in deserts are baked dry, have very little water, and are rich in soluble constituents such as lime and salt; tropical soils have a noticeably high iron and aluminum oxide content. Knowledge of a soil's water, mineral, and organic components and their proportions can help determine its potential productivity and what the best use for that particular soil might be.

CHARACTERISTICS OF SOIL

Soils have many physical properties that are useful in describing and differentiating among them. The most important include color, texture, structure, acidity or alkalinity, and capacity to hold and transmit water and air.

COLOR

Color, if not the most important physical attribute of soil, is at least the most visible. Many laymen who know next to nothing about the constituents of soil or its formation processes are aware of the variations in color that exist from place to place. The well-known red clay of Georgia is not far from Alabama's belt of black soil. Soil colors vary from black to brown to reds, yellows, grays, and near whites, and each of these colors offers a clue to the physical and chemical characteristics of the particular soil.

For example, humus or decomposed organic matter is black or brown, and soils that are high in humus content are generally black or dark brown. As the humus content of soil decreases, because of either low organic activity or leaching, the color gradually fades to light brown or gray. A large proportion of humus in a soil usually indicates that the soil is highly fertile, for humus acts as a catalyst in the complex chemical reactions that allow plants to obtain nutrients from the soil. For this reason dark brown or black soils are spoken of as *rich* and are considered superior soils. It should be noted, however, that this is not always the case, for there are some black or dark brown soils with little or no humus content that get their dark color from other factors.

The red and yellow colors of soils are usually due to the presence of iron compounds. In moist climates a light gray or white soil indicates that iron has been removed and aluminum oxides are present, while in dry climates the same color usually indicates a high proportion of salts.

Soil colors are useful in providing clues to the physical and chemical characteristics of

VIEWPOINT: CAN LAND MANAGEMENT SAVE OUR SOILS?

New threat to a desert environment. (U.S. Department of Interior, Bureau of Land Management)

People have obviously affected soils to a great extent through their agricultural practices: stirring and churning up the earth, altering soil horizons, adding fertilizers, irrigating, providing means for drainage, and, most important of all, changing the vegetative cover with which the soil had been organically linked for thousands of years. Such agricultural practices have resulted in both the enrichment and depletion of soils and in some cases in their total destruction and removal. Primitive people started the process by setting fires to drive game animals. In clearing forests to create grazing land, pastoralists furthered it. Now former forests have been turned into grass or shrublands; grasslands have been transformed into deserts. Each such change alters the soil-forming factors.

But agriculture is not the only way in which we affect soils. Surface mining, housing tracts, highway construction, and the building of dams and irrigation systems are further examples of ways in which people alter and sometimes totally destroy this valuable resource. In fact, beyond the shrinking limits of the rain forests in the tropics and in mountain regions of the midlatitudes, there may be no soil in natural equilibrium with its present human-influenced environment.

SOIL EROSION

One of the most serious results of human intervention in the soil system is soil erosion. It is true that there has always been some soil erosion, especially on steeper slopes. In many cases, however, this natural erosion of the soil surface has been offset by the soil renewal carried out below by the soil-forming processes. Ever since people first entered the fields of agriculture, forestry, and mining, however, soil erosion has been accelerated many times over.

The problem of soil erosion is a serious one. For example, when a farmer fails to maintain his soil's fertility through such means as crop rotation and the use of fertilizers, the soil becomes exhausted. If something is not done to reverse the process (for example, let the field lie fallow for a few years), the soil will be so destroyed that it will be incapable of supporting any vegetative cover at all. Without vegetation to protect the soil and roots to bind it, erosion by wind or water will occur, and the land will become useless to the farmer. Such ruin is tragic when it happens to one farmer, but the results can be disastrous when they occur to large numbers of farmers. Whole economies and even entire civilizations have been seriously hurt if not destroyed by the failure to maintain soil fertility.

LAND MANAGEMENT

It is possible to prevent or control erosion in many ways, all of which, however, require that some attention be paid to the man-soil relationship. For example, we can stop overgrazing and overcropping of land, save the natural vegetation cover where possible, plant cover crops where none exist, practice contour plowing, terracing, and gully control, use fertilizers, and practice strip cropping and crop rotation.

It is also possible that, with the careful land-use planning now being carried on in the United States and in some other nations, we can control much of the erosion and preserve our vital soil resource. The Soil Conservation Service (SCS) of the USDA has developed a land classification system by which lands have been categorized according to their potential for productivity (see Table). The uses recommended—wildlife preserve, recreation, urban-industrial development, cropland, pasture, forestry, etc.—are dependent upon their rating with the SCS. Land zoning can provide similar soil protection and conservation. In Orange County, California, as well as around the outskirts of London, England, land has been zoned to prevent urban sprawl from taking over some of the area's best croplands. Recently steps have been taken to counter a new danger to soils: man's invasion of delicate natural environments with a variety of off-road vehicles, such as snowmobiles and "dirt bikes." These conveyances, often used irresponsibly, scour trails through vegetation, rut exposed soil, and accelerate slope erosion. In California there are now areas set aside to either admit, forbid, or restrict the use of off-road vehicles.

Land-Capability Classification

LAND CLASS	LAND-CAPABILITY AND USE PRECAUTIONS	PRIMARY USES	SECONDARY USES
	GROUP 1—LANDS SUITABLE FOR CULTIVATION		
I	Excellent land, flat, well drained. Suited to agriculture with no special precautions other than good farming practice.	Agriculture	Recreation Wildlife Pasture
II	Good land with minor limitations such as slight slope, sandy soils, or poor drainage. Suited to agriculture with precautions such as contour farming, strip cropping, drainage, etc.	Agriculture Pasture	Recreation Wildlife
III	Moderately good land with important limitations caused by soil, slope, or drainage. Requires long rotation with soil-building crops, contouring or terracing, strip cropping or drainage, etc.	Agriculture Pasture Watershed	Recreation Wildlife Urban-industrial
IV	Fair land with severe limitations caused by soil, slope, or drainage. Suited only to occasional or limited cultivation.	Pasture Tree crops Agriculture Urban-industrial	Recreation Wildlife Watershed
	GROUP II—LANDS NOT SUITABLE FOR CULTIVATION		
V	Land suited to forestry or grazing without special precautions other than normal good management.	Forestry Range Watershed	Recreation Wildlife
VI	Suited to forestry or grazing with minor limitations caused by danger from erosion, shallow soils, etc. Requires careful management.	Forestry Range Watershed Urban-industrial	Recreation Wildlife
VII	Suited to grazing or forestry with major limitations caused by slope, low rainfall, soil, etc. Use must be limited, and extreme care taken.	Watershed Recreation Wildlife Forestry Range Urban-industrial	
VIII	Unsuited to grazing or forestry because of absence of soil, steep slopes, extreme dryness or wetness.	Recreation Wildlife Watershed Urban-industrial	

Source: Modified from land-classification system of U.S. Soil Conservation Service. Department of Agriculture.

soils, as well as in making the job of soil differentiation easier, but of course, color alone does not answer all of the important questions about a soil's qualities or potential for use.

TEXTURE

Soil texture varies according to the size of the particles that make up the soil. In **clay** soils the particles have diameters which are less than 0.002 mm (soil scientists universally use the metric system). The particles of **silt** soils are defined as being between 0.002 mm and 0.05 mm. **Sandy** soils have particles with diameters between 0.05 mm and 2.0 mm. Individual soil particles with diameters larger than 2.0 mm are regarded as inert gravel or rock fragments and technically are not soil particles. Since no soil is made up of particles of uniform size, the proportion of particles in various size ranges determines the texture of the soil.

For example, a soil composed of 50 percent silt-sized particles, 45 percent clay, and 5 percent sand would be considered a silty clay. A triangular diagram (Fig. 10.3) has been developed to show different classes of soil texture and the percentage ranges of each **soil grade** (as sand, silt, and clay are called) within each class. Point A within the silty clay class represents the example just given. A second soil sample (B) that is 20 percent silt, 30 percent clay, and 50 percent sand would be referred to as a sandy clay loam. The **loam** soils, which occupy the central areas of the triangular diagram, are those in which no one of the three grades of soil particles dominates over the other two. Interestingly, it is the loam soils that are best suited to the support of plant life.

Soil texture is important in that it helps determine the capacity of a soil to retain moisture and air, both of which are necessary for plant growth. Soils with a greater proportion of larger particles are well aerated and allow wa-

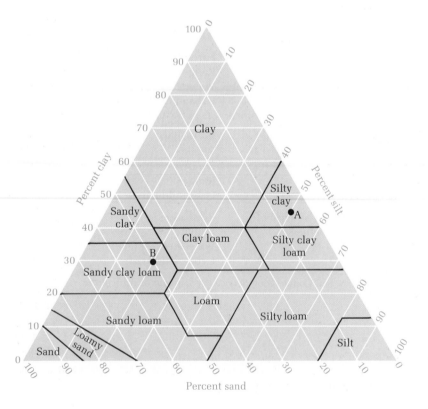

FIGURE 10.3 The texture of any soil can be represented by a dot on this diagram. Texture is determined by sieving the soil to determine the percentage of particles falling into the size ranges for clay, silt, and sand. (USDA)

ter to pass through (or infiltrate) the soil more quickly, sometimes so quickly, in fact, that plants are unable to make use of the water. Clay soils present the opposite problem: They transmit water very slowly, become water-logged, and are deficient in air. An important part of cultivation is the aeration of soil. This is accomplished by plowing, disking, harrowing, or shoveling the soil in order to open up its structure and allow it to breathe.

STRUCTURE

In most soils the individual mineral particles aggregate into larger distinctive masses or clumps, known as **peds,** which give the soil a particular structure. The structure of a soil is an important factor in its workability. Structure also influences a soil's **permeability** (the ability to allow water to pass through) and its **porosity** (the amount of space between soil particles and between clumps). Permeability and porosity in turn affect soil drainage as well as the transmission of water, nutrients, and air, and significantly affect the availability of these elements to living things in the soil. As a further complication, soils of similar textures may have different structures. Consequently one may be more productive than another.

Soil structure may be influenced by such outside factors as the moisture regime and the nature of the nutrient cycle by which plants and the soil constantly interchange chemicals, keeping certain ones in the system while others are leached away. We have all seen the structural change in certain soils from when they are wet to when they have been baked dry in the sun. Soil structure can also be influenced by such human actions as plowing, cultivation, irrigation, and even fertilization. The addition of certain fertilizers, as well as the existence of lime or decayed organic debris in a soil, affects structure through chemical means that encourage first the clumping of individual soil particles and then the maintenance of those groups or clumps. Excess sodium and magnesium work the other way, causing clay soils to be

structureless "glue" when wet, or concrete-like when dry. On the other hand, the development of a definite structure is hindered by the absence of some of the smaller soil particles. It is for this reason that sandy beaches and deserts, made up as they are of the larger soil particles, have no apparent soil structure. This also explains why some soil has more structure below the layers closest to the surface, for the smaller particles of the topsoil have been removed to lower layers by water traveling down through the soil.

Scientists have classified soil structure according to the various forms assumed. These range from columns, prisms, and angular blocks, to nutlike spheroids, laminated plates, crumbs, and granules (Fig. 10.4). For our purposes it is most important to know merely that both massive and fine structures are less useful than aggregates of intermediate size and stability, which permit ideal drainage and aeration.

ACIDITY AND ALKALINITY

What makes a soil fertile or infertile are the many complex chemical processes and exchanges that take place in soils and plant systems. The general nature of the soil chemistry is usually expressed as the degree to which the soil departs from chemical neutrality toward either acidity or alkalinity (baseness).

Soil acidity or alkalinity is important since it helps determine the availability of nutrients to plants and ultimately controls plant growth. Plants receive virtually all of their necessary nutrients in solution. That is, a plant is unable to absorb nutrients unless they are dissolved in liquid. However, when the soil moisture lacks some degree of acidity, the soil water has little ability to dissolve minerals and release their nutrients. As a result, even though the nutrients are in the soil, plants may not have access to them. To correct this alkalinity, which is more common in arid soils than in any others, and to make the soil more productive, a farmer can flush the soil by irrigating under conditions of good drainage.

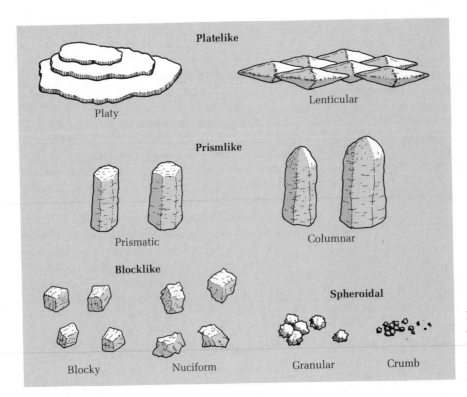

FIGURE 10.4 Classification of soil structure on the basis of soil peds. (Courtesy of Fred Caspall)

As might be expected, a strongly acidic soil is also detrimental to plant growth. In a soil that is too acidic, the soil moisture dissolves nutrients, which become leached away before they can be obtained by plant roots. This makes the soil still more acid and accentuates the problem. It is, moreover, a situation occurring in soils and climates that are often otherwise suitable to plant growth. Luckily it can be corrected by the addition of lime to the soil.

The acidity or alkalinity of a soil is measured on a scale of 0 to 14, called the **pH scale.** This actually is a measure of the hydrogen ions present in the soil moisture. Low pH values indicate acid soil moisture while high values indicate alkaline conditions. A pH of 7 indicates neutrality—the soil is neither acidic nor alkaline. Soil scientists have shown that most complex plants will grow only in soils whose pH is between pH 4 and pH 10. The optimum pH for plant productivity varies with the plant itself. Vegetation has evolved around the world in a variety of climates and is in equilibrium with the soil environment in which it is native. Thus barley can tolerate alkaline soils, but camellias and rhododendrons prefer more acid conditions. Those who bring a diversity of plants into their gardens must be prepared to provide soils of varying pH values for their specimens. This is achieved through **fertilization.**

In addition to affecting plant growth through the availability of nutrients, the acidity or alkalinity of a soil also affects the microorganisms functioning in the soil. Like plants, the microorganisms are highly sensitive to a soil's pH, and each has its optimum situation.

SOIL PROFILE DEVELOPMENT

Soils begin to develop when either rocks or deposits of loose material are colonized by simple plant and animal life. Once the organic processes of life and death begin to take place among mineral particles or disintegrated rock, differences begin to develop from the surface down through the soil parent material. This vertical differentiation comes about originally from such simple factors as the gradual accumulation of organic matter at the surface and the removal of fine particles and dissolved matter from the top layers by downward percolating water, followed by the deposition of these materials at a lower level. As climate, vegetation, animal life, and steepness of slope affect soil formation over time, this vertical differentiation becomes more and more apparent. Often, especially in middle latitudes, mature soils exhibit a vertical zonation into distinct layers or **horizons** that are distinguished by their different physical and chemical properties. Even when soils have formed horizons, the boundaries between different zones may not be distinct.

In mature soils in the middle latitudes there are three generally recognizable horizons. Nearest the surface is the *A* horizon. Typically this horizon has the greatest concentration of organic matter and humus. This layer is sometimes referred to as the **zone of depletion,** since soil water removes particles from it in suspension (eluviation) and solution (leaching). Below this is a **zone of accumulation,** the *B* horizon, where much of the material removed from the *A* horizon is deposited. Except in soils that have a high organic content and in which there is a lot of mixing, the *B* horizon generally has little humus. The *C* horizon is the weathered **parent material** from which the soil is developed, fragments of the bedrock directly beneath, or transported and deposited material. The *C* horizon does not reflect the movements of matter and the organic activity in the higher zones. The lowest layer, sometimes called the *D* horizon, is unchanged bedrock or unmodified material transported to the site by water, wind, or glacial action. Particular horizons in some soils may not be as well developed as others, and certain horizons may be missing altogether. In other cases definite vertical differentiation can be noted within a particular horizon.

The vertical cross section of a soil from its surface down to the parent material from which it is formed is called the **soil profile** (Fig. 10.5). The differences among the infinite vari-

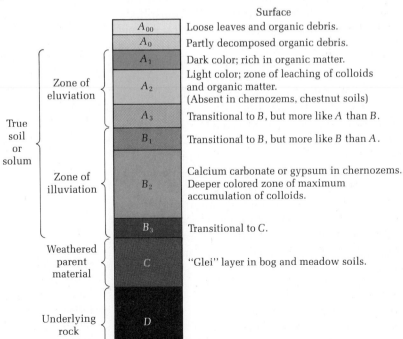

FIGURE 10.5 Soils are categorized by the degree of development and physical characteristics of their horizons. (After USDA, 1938)

ety of soils that exist are apparent in an examination of their profiles. For this reason soil scientists have grouped and classified soils in large part on the basis of differences exhibited in the soil profiles and their horizon development (Plates 15 and 17).

FACTORS AFFECTING SOIL FORMATION

Many agents and influences are involved in the formation of soil. Because of the variations that exist among soil components as well as the effects wrought by the changing character of the various agents and controls of soil formation, no two soils are identical in all their characteristics.

Perhaps it would be useful at this point to draw a parallel between soil and the atmosphere, as both are dynamic systems acted upon by many interrelated agents. The atmosphere is made up of various gases, water vapor, dust, etc.; likewise, soil is composed of organic and inorganic materials, water, and air. We note changes in the state of the atmosphere from time to time and place to place by noting changes in the elements of weather. Variations in soil appear as lateral changes in soil characteristics or in the vertical development of the soil profile. Weather patterns are controlled by the climatic factors in much the same way that soil characteristics are affected by the agents and controls of soil formation that we are now going to examine. The factors controlling the formation and distribution of different types of soils are parent material, organic processes, climate, land surface configuration, and time. Of these, the parent material is distinctive as the raw material. The other factors are the machinery that produces soil from this raw material.

PARENT MATERIAL

All soil is derived from the weathered fragments of rock material. If this weathered material accumulates *in place* through rock decay,

that is, through the physical and chemical breakdown of the bedrock directly beneath the soil, we refer to the weathered fragments as **residual parent material.** If the rock fragments from which a soil is formed have been carried to the site by streams, waves, winds, or glaciers to form a new deposit, this mass of sediment, which will eventually develop a surface soil, is called **transported parent material.** It is the development and action of organic matter through the life cycles of living things that are primarily responsible for differentiating soil from its fragmentary rock source or parent material, which will always be present beneath it.

Parent material is one of the agents or factors of soil formation that helps to determine a soil's characteristics. It differs, however, from the other factors because it is the original substance with which the whole process of soil formation begins. The parent material is the raw material on which the processing factors of climate, organic activity, surface configuration, and time are imposed to manufacture a soil.

Parent material varies in the degree to which it influences the characteristics of the soil derived from it. Some parent materials, such as sandstone, which contains mainly the extremely stable mineral quartz, are far less subject to weathering and change than others. Soils that develop from these parent materials demonstrate a high level of similarity with their parent source. On the other hand, some chemically reactive bedrock materials are easily weathered, and the soils that develop from them, like those on transported material, are apt to show a greater correlation with soils of similar climate than with those of similar parentage but different climate. Thus it is common to find that one or more of the other soil-forming agents or controls may have a far greater influence on the soil than does the parent material. In fact, on a worldwide basis, climate and the associated plant communities cause stronger and larger variations in soil characteristics than do parent materials. The differences among soils based on variations in parent material are visible, however, on a local level and are of more than casual interest to the soil scientist and agriculturist.

The influence of a soil's parentage is affected by time. As a soil evolves and matures, the influence of parent material on its characteristics declines. Thus a younger or less mature soil, provided the other agents such as climate are similar, will show more similarity to its parent material than will a more mature one.

Both residual and transported parent material affects the soil that develops from it in very specific ways. First, the chemicals and nutrients available to the plants and animals living in a soil are derived from the soil's parent material (Fig. 10.6). Thus, a parent material deficient in calcium will produce a soil deficient in calcium. Its natural fauna and plant cover will be of a type that requires little calcium. Only the artificial addition of lime (calcium carbonate) can correct such a deficiency. Likewise, a parent material rich in aluminum, such as granite or basalt, will produce a soil also rich in aluminum. In fact the main source of metallic aluminum is the bauxite ore found in tropical soils where almost everything else has been leached away.

The size of the particles that result from weathering of the parent material is a prime factor in the determination of soil texture and structure. A rock material such as sandstone, which contains little clay and is weathered into relatively coarse fragments, will make up a soil of coarse texture. The parent material, therefore, exerts an important influence on the availability of air and water to a soil's living population.

ORGANIC ACTIVITY

Plants and animals effect soil formation in many ways. The life processes of the dominant plants are as important to the soil as its microorganisms, the microscopically small plants and animals that abound in most soils.

At the most general level, the completeness of the vegetative cover affects erosion rates. A forest of any type, because it provides a protective canopy over the soil and produces a mulch of litter on the surface, keeps rain from beating on the soil surface and increases the proportion of rainfall entering the soil rather than running off on its surface. Vegetation can also affect the evaporation rate. A scanty vegetative cover will allow greater evaporation of soil moisture than will thick protective vegetation. This evaporation in turn affects

FIGURE 10.6 Despite the strong leaching under a wet tropical climate, Philippine soils remain high in nutrients because their parent material is recent volcanic ash and young alluvium. (Courtesy of Reece A. Jones)

the movement of capillary water toward the surface. Furthermore, the nature of the plant community determines the nutrient cycles that are a part of soil formation. Certain elements are absorbed by plants and then returned to the soil after the plants die and are decomposed. These exchanges vary among types of plants, as some use more of certain chemicals than others. Soluble nutrients that are not used by plant cover are soon lost in the leaching process, which impoverishes the soil (Fig. 10.7). The roots of larger plants affect the soil structure as well by making it more porous and by absorbing water as well as various plant nutrients from the soil.

All parts of vegetation (their leaves, bark, branches, flowers, and root networks) contribute to the organic content of a soil when the plants die. Logically the organic content of soil varies with the nature of the associated plant life. A grass-covered prairie is able to supply a far greater abundance of organic matter than the incomplete surface cover found in a desert region. There is some question, however, as to whether forests or grasslands (with their thick root network) furnish the soil with greater organic content. There is no question that many of the grassland regions of the world, like the American prairies, provide some of the richest and most fertile soils for cultivation, in part because of the high amount of organic debris that is naturally present.

The process of decay, aided by bacterial action, transforms organic matter into the jelly-like mass called humus. As we noted earlier, humus is important to soils in many ways. For one, it serves as the primary food supply for the microorganisms in the soil. Humus also affects soil structure by enhancing the water retention capabilities and workability of the soil. As it is further acted upon by microorganisms, humus returns to the soil organic and inor-

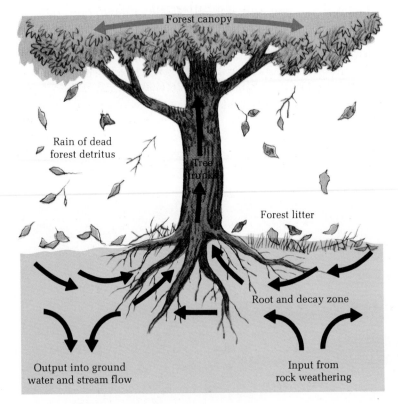

Forest canopy

Rain of dead forest detritus

Tree trunk

Forest litter

Root and decay zone

Output into ground water and stream flow

Input from rock weathering

FIGURE 10.7 The nutrient cycle in a forest. In a wet climate, if trees do not take up soluble nutrients in the soil, they are flushed away in groundwater and lost permanently. The more demanding the forest vegetation, the richer is the resulting soil. Pines are notoriously undemanding of soluble bases, and consequently in pine forests bases are lost by leaching and are not replaced by the vegetation, resulting in soil deterioration.

ganic materials necessary for further plant life. Consequently in most soils there is a direct correlation between humus content and fertility.

Bacteria, which are microscopic life forms, are perhaps the most important of the microorganisms that live in the soil and contribute to its formation. Bacteria feed on the organic matter and humus of the soil and by this process break down the debris of living things into their organic and inorganic components as well as allowing the formation of new organic compounds, all of which are then available for the promotion of further plant life.

It is difficult to estimate the number of bacteria, fungi, and other microscopic plants and animals that live in the soil, though some have suggested as many as one billion per gram of soil. Whatever the number, it is enormous. It is no surprise, then, to learn that the activities and remains of these microorganisms, minute though they are individually, add considerably to the organic content of a soil.

In addition to the microscopic animals already referred to, earthworms, nematodes, ants, termites, wood lice, centipedes, burrowing rodents, snails, and slugs stir up the soil, mixing mineral components from the lower levels with organic components from the upper portion. Earthworms are especially important in soil formation for they take soil in, pass it through their digestive tracts, and excrete it in casts. The process not only helps to mix up the soil, but it also changes the texture, structure, and chemical quality of the soil. Charles Dar-

win, in the late 1800s, estimated that earthworm casts produced each year would equal as much as 10 to 15 tons per acre. As for the number of earthworms themselves, sampling suggested that the weight of the earthworms beneath a pasture in New Zealand equaled the weight of the sheep grazing above them.

CLIMATE

In Chapter 7 we have already seen the marked similarities between the distribution of major soil classes as shown on Plate 16 and the distribution of climatic types as indicated on Plate 12. The correlation between climatic and soil types on these two maps demonstrates in a highly visible way that on a large scale climate is a major factor in soil variation. There are, of course, many variations among soils that are apparent on a smaller than global scale. The differences that are apparent on a smaller scale show the influence of such other factors as parent material, land surface configuration, vegetation type, and time.

Temperature clearly affects the activity of soil microorganisms. This activity in turn affects the rate of decay and decomposition of organic matter. In the hot equatorial regions the great activity of soil microorganisms precludes any thick accumulation of organic debris or humus. Figure 10.8 shows that the amount of partially decomposed organic matter and humus in a soil increases as we move into the middle latitudes from the equatorial zone. In

FIGURE 10.8 Relationship of temperature to production and destruction of organic matter in the soil. (After M. W. Senstius)

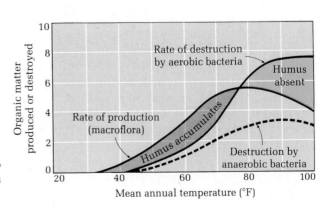

the mesothermal *C* and microthermal *D* climates the activity of soil microorganisms slows enough to allow the accumulation of rich layers of decaying organic matter and humus. However, moving further poleward into colder regions, the combination of retarded microorganism activity combined with limited plant growth results in the accumulation of only thin layers of undecomposed or partially decomposed organic matter.

Temperature also affects the rate of chemical reactions that take place in soil, many of which make available nutrients necessary for plant growth. Chemical activity tends to increase and decrease directly with temperature. As a result, the parent material of soils in the hot equatorial regions is transformed to a far greater degree by chemical means than is the parent material in colder zones. Because chemical activity in soils ceases when soil moisture is frozen, the weathering of parent material in most polar climates is almost totally mechanical.

Temperature affects soil indirectly in its influence on the nature of vegetation that develops in a particular region. We know that particular vegetation associations are adapted to certain temperature regimes. The soil in which this vegetation grows often reflects the character of the plant cover as a result of the nutrient cycles that keep both the vegetation and the soil in chemical equilibrium.

Moisture conditions affect the development and character of soils more clearly than any other factor. Without precipitation and consequently soil water and the chemicals dissolved therein, plant life is impossible. The absence of plant life greatly diminishes the organic content and thereby the fertility of a soil.

We have already discussed the effects of both gravitational and capillary water on soil structure, texture, color, and development. As precipitation is the original source of all soil water (disregarding the minor contribution of dew), the amount of precipitation received by a soil affects the rate and degree of leaching, eluviation, and illuviation that occur, and thereby the rate of soil formation and horizon development.

When considering the effect of precipitation on the quantity and movements of soil water, we should note that the evaporation rate is a factor as well. Thus salt and gypsum deposits from the upward migration of capillary water are more extensive in hot, dry regions, like the southwest United States, where the evaporation rate is high, than in colder dry regions (Fig. 10.9).

Just as temperature affects soil development indirectly through its effects on vegetation, so too does the moisture regime of a region. Where the warm season is dry the soils are alkaline. When the summers are warm and extremely wet, with drier winters following, intense chemical weathering and fluctuating water tables produce a peculiar type of soil crust, known as *laterite*. This crust is common in the tropics, where it is quarried for building material.

LAND SURFACE CONFIGURATION

The slope of the land as well as its aspect (the direction in which it faces) affects soil development both directly and indirectly. Steep slopes are generally better drained than more gentle ones. They are also subject to more rapid runoff of surface water. As a consequence there is less infiltration of water on steeper slopes, which retards the soil-forming processes. This retardation inhibits the development of mature layered soils, sometimes to the extent that no soil at all will develop over the parent material. In addition, rapid runoff on steep slopes often erodes their surfaces as fast or faster than soil can develop on them. On more gentle slopes, because there is less runoff and more infiltration, more water is available for soil-forming purposes and for the support of vegetation, and erosion is not as extensive as on steeper slopes. In fact, the rate of erosion on gently rolling hills is often just enough to offset the ongoing production of soil from the underlying parent material. It is often gently rolling land that allows mature soils of the most ideal characteristics to develop.

FIGURE 10.9 Harmful concentrations of chemical substances are frequently encountered in arid regions. Shown here is a surface salt crust resulting from upward capillary rise of water from a water table that is saline and close to the surface. Evaporation at the surface deposits the salt. (U.S. Geological Survey)

Valley floors and flatlands often are very poorly drained. When this is the case and the water level in the soil is near the surface, gravitational water is unable to percolate downward, and capillary water may in fact move salt and alkaline substances to the surface in harmful concentrations. This condition is a constant danger in the irrigation of flatlands because irrigation tends to raise the water table. Artificial drainage ditches must be provided to lower high water tables in such instances.

The direction a slope faces affects its microclimate. Thus a north-facing slope in the Northern Hemisphere has a microclimate that is colder and wetter than a south-facing exposure, which receives the sun's rays more directly and as a consequence is warmer and drier. Local variations in soil depth, texture, and profile development result directly from such differences between microclimates.

Land surface configuration also affects the development of soils indirectly through its effect on vegetation. Where a steep slope prevents the development of a mature soil to support abundant vegetation, the poor quality of the plant cover diminishes the amount of organic debris available for the soil.

TIME

We have spoken of the evolution of a soil toward a state of dynamic equilibrium with its environment. A soil is mature when it has reached such a state of equilibrium. Young soils are still in the process of alteration to achieve equilibrium. The mark of a mature soil is a well-developed and stable horizon structure. A young or immature soil, on the other hand, generally has poorly developed horizons or none at all. Very old soils may have horizons that are so well developed as to present problems. Such soils frequently contain dense pans or crusts in their *B* horizons. These may consist of eluviated clay or chemically concentrated calcium carbonate (lime), silica, or oxides of iron and aluminum. Soils on ancient flat surfaces in the tropics, where soil formation is rapid, frequently present such problems. To use old soils agriculturally, their crusts may have to be broken up with dynamite, or less dramatically, by deep plowing.

Another effect of time is that the more mature soil is decreasingly influenced by its parent material and increasingly reflects its climatic environment and vegetative cover. In fact, as our comparison of Plates 12 and 16 demonstrated, the influence of climate is usually the most apparent of all controls on soil development at a global scale, provided the soil has had sufficient time to reach maturity.

The importance of time in soil formation is especially clear in the case of soils developed upon transported parent materials. Depositional surfaces are in most cases quite young and have not been exposed to the effects of the atmosphere as long as have those parts of the earth's surface that have been worn down by erosion over periods of tens of millions of years. Recent deposits of transported materials have not yet been leached of their soluble nutrients, nor has their soil developed undesirable characteristics. Deposition is occurring today in a variety of settings: on the floodplains of large rivers, where the material accumulating is known as *alluvium;* downwind from dry areas, where dust settles out of the atmosphere to form deep blankets of *loess;* and in volcanic regions that are occasionally showered by ash and veneered by lava. Ten thousand years ago glaciers withdrew from vast areas, leaving behind a veneer of *till* or *outwash,* some of it plastered over the land surface by the advancing ice and other portions let down or washed out in great sheets as the ice melted. In terms of agricultural productivity the world's best soils are found on young alluvium, loess, volcanic ash and lava, and certain types of glacial drift. Somewhat older deposits of similar materials

are less productive. They have already been leached of vital nutrients and have developed unfavorable structures, similar to those of the soils on older erosional surfaces that constitute most of the earth's land surface.

There is no fixed amount of time that it takes for a soil to become mature. This is because of the number and variability of the factors that affect soil formation. Generally, though, it takes hundreds to thousands of years for a soil to reach maturity.

SOIL-FORMING REGIMES

By now it should be clear that there are infinite possible combinations of the factors of soil formation that function together to produce soils of all descriptions. Nevertheless, an examination of the world's soils reveals that they can be separated into a limited number of general types. The characteristics that differentiate these major types can be attributed to their different soil-forming regimes, each resulting from a combination of different processes. The differences between these soil-forming regimes are primarily the result of climatic differences and indirectly the result of differences in plant cover.

At the broadest scale of generalization, there are three primary soil-forming regimes that relate to climatic differences. These are laterization, podzolization, and calcification (Fig. 10.10). On a local scale other processes become important, but they are of lesser significance in the worldwide distribution of soils, and for that reason we will briefly mention only two of them. Soils may be formed by any one of these processes or by a combination of two or more, primarily depending on the climatic regime but also on surface configuration, vegetation, and parent material.

LATERIZATION

Laterization is a soil-forming process that occurs in humid tropical and subtropical climates as a result of the high temperatures and abun-

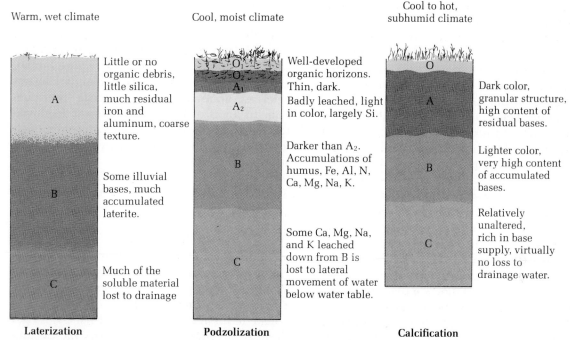

FIGURE 10.10 Profile development in the three major soil-forming regimes. (Courtesy of J. W. Vining)

dant precipitation. These effects produce rapid breakdown of rocks and decomposition of nearly all minerals. Soils produced by laterization are known as **latosols.** Despite the dense vegetation that is typical of these climates, there is little humus incorporated into the soil because of the rapid decomposition of plant litter and enormous numbers of microorganisms in the soil. Because of the abundance of moisture, eluviation and leaching of all but iron and aluminum oxides play dominant roles in the formation of latosols. Fine soil particles as well as most minerals and bases are removed from the top soil horizon (the *A* horizon) except for iron and aluminum compounds, which are insoluble in the soil solution primarily because of the absence of organic acids from humification. As a result, the topsoil is reddish in color, coarse in texture, and tends to be porous. In contrast to the *A* horizon, the *B* horizon has a heavy concentration of illuviated materials. Where the tropical forest vegetation remains, the soluble nutrients released in the weathering process

are quickly absorbed by the vegetation, which eventually returns them to the soil, where they are reabsorbed by plants. This rapid nutrient cycle prevents the total leaching away of the bases, so the soil is only moderately acidic. Removal of the vegetation permits total leaching of bases, resulting in the formation of crusts of iron and aluminum compounds, as well as accelerated erosion of the *A* horizon.

Laterization can take place year round because of the lack of distinct seasonal variation in temperature or precipitation. Because of this continuous activity and because of the strong weathering of the parent material, latosols may develop to depths of as much as 8 meters (25 feet).

PODZOLIZATION

Podzolization occurs in its purest form in the high middle latitudes where the climate (*Dfb, Dfc*) is moist with short, cool summers and long, severe winters. The typical coniferous forest of this climate is an integral part of the process of podzolization.

Where temperatures are low much of the year the activity of microorganisms is reduced enough that humus is allowed to accumulate; however, because of the small number of animals living in the soil there is little mixing of this humus below the soil surface. Leaching and eluviation by a strongly acidic soil solution remove the soluble bases and aluminum and iron compounds from the A horizon. The remaining silica gives a distinctive ash-gray color to the A horizon (*podzol* is derived from a Russian word meaning *ashy*). Because most coniferous trees require a minimum amount of bases, they return little basic material to the soil. The needles they drop are acidic chemically. This contributes to the acidic quality of the soil. Indeed, it is difficult to say whether the soil is acidic because of the vegetative cover, or whether the vegetative cover is adapted to the acidic soil.

Podzolization can take place beyond the typical cold, moist climate when the parent material is highly acidic, as on the sands common along the East Coast of the United States. The pine forests that can grow in such acidic conditions return acids to the soil, promoting the process of podzolization.

CALCIFICATION

The third distinctive soil-forming process is called **calcification.** In contrast both to laterization and podzolization, which require humid, moist climates, calcification demands climates where evapotranspiration exceeds precipitation. In areas of low precipitation little eluviation takes place, nor is leaching strong enough to remove the bases released by weathering or accumulated on the surface as wind-transported material. As a result, materials that would be leached to the water table in a more humid climate become concentrated in the soil. The most important of these is calcium carbonate ($CaCO_3$). In addition, the upward movement of capillary water is strong during dry periods, which occur frequently in semiarid climates. The capillary water evaporates in the B horizon, leaving its burden of calcium carbonate in the form of hard nuggets. Over a period of tens of thousands of years the calcium carbonate may form the concrete-like layer known as caliche.

Calcification becomes important in the climatic regimes where moisture penetration is shallow. The subsoil is too dry to support tree growth, and shallow-rooted grass is the primary form of vegetation (Fig. 10.11). Calcifica-

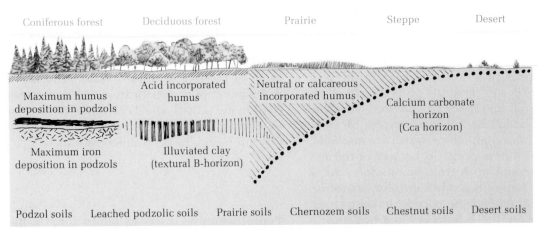

FIGURE 10.11 Relationship between rainfall and depth of calcium carbonate leaching, showing relation to other soil-forming regimes. The vertical scale is greatly exaggerated. (Bridges, *World Soils*, p. 18)

tion is enhanced by the fact that grass uses calcium, drawing it up from the lower soil layers and then returning it to the soil when the grass dies. The grasses and their dense root networks provide large amounts of organic matter in the soil, which typically is mixed deep into the soil by the numerous animals found there. Thus midlatitude grassland soils are rich both in bases and in humus and are the world's most productive agricultural soils.

It is interesting that a soil transect across the United States from Illinois to Colorado shows a gradual westward *decrease* in soil humus content and the thickness of the *A* horizon and an *increase* in the prominence of calcium carbonate at the base of the *B* horizon. The most productive soils are found about midway in this transect, where both humus and base status are relatively high. The deserts of the Far West have no humus, and the rise of capillary water leaves not only calcium carbonate but even more soluble sodium chloride (salt) at the surface.

REGIMES OF LOCAL IMPORTANCE

Two additional soil-forming processes are important enough to merit consideration. Both are characteristic of areas with locally poor drainage, although they occur under strikingly different climatic conditions. The first, *salinization,* occurs in stream valleys, interior basins, or other low-lying areas in desert regions. Surface water accumulates in these locations after spring snowmelt in nearby highlands or after the occasional desert downpour. Rapid evaporation of this water leaves behind the high concentration of soluble salts in the soil that characterizes the soil-forming regime (see Fig. 10.9 for an extreme example). The second, **gleization,** occurs in poorly drained areas under cold and wet environmental conditions. It is common in low-lying areas of the Tundra climate and in swampy regions of midlatitude climates with long, cold winters. The end product of gleization is a soil with a heavy surface accumulation of humus and with a blue-gray layer of thick, gummy, water-saturated clay di-

rectly beneath. Unreduced iron in the early stages of decomposition imparts the blue-gray color to the soil.

SOIL CLASSIFICATION

Soils, like other complex phenomena that vary from place to place and type to type, are difficult to classify. The literature on soil classification is enormous, and several different classification schemes are currently in use. Because they are interested in causes and effects rather than mere description, geographers have tended to favor soil classification schemes that stress the relationships of soils to the soil-forming factors. While parent material and vegetation may cause the major soil differences in a county, for instance, on a worldwide scale the major differentiating factor is gross climate. We have seen that climates are arranged in latitudinal zones as a consequence of temperature gradients, with departures from the general pattern in areas of low precipitation and where temperature gradients are disturbed by such factors as altitude and warm or cold ocean currents. The worldwide pattern of soil types is likewise zonal, with local variations from the overall pattern. This fact was first made clear by Russian soil scientists led by V. V. Dokuchaiev, whose final classification of soils in 1900 stressed the relationships of soil types to major vegetation regions—the most obvious indicators of zonal climatic differences. Of the 13 soil types recognized by Dokuchaiev, 7 were strictly related to climate-vegetation zones, with the remainder being either transitional types or developed under abnormal circumstances. The classification officially adopted by the United States Department of Agriculture (USDA) in 1938 combined Dokuchaiev's concept of great soil groups with the American system of soil types. The 1938 system employed a genetic classification, and the general importance of climate in the development of soils produced close correlation between climates and soils on a world scale.

While the earlier USDA classification is useful on a global scale, it is seldom used by soil scientists in the United States today. The Soil Conservation Service of the USDA in 1960 presented an entirely new descriptive system based solely on intrinsic properties of the soil itself. The **Comprehensive Soil Classification System** (CSCS) has also been referred to as the **7th Approximation System** because it was the seventh in a series of attempts to achieve an approximation to an ideal classification. The CSCS is an empirical system and observed physical characteristics of each soil are fundamental to its classification. There is only limited correspondence between CSCS soil maps and maps of climate because various interactions between soil processes can produce virtually identical soils in the CSCS system under totally different climatic conditions.

The Comprehensive Soil Classification System uses original descriptive terms derived from Latin roots, dispensing with the well-established but linguistically inconsistent terminology and variable criteria of the Russian-American system. Employment of the CSCS permits modern agriculturists to identify and classify significant differences among soils found in areas as limited in size as a farm or a field. This detailed knowledge is essential to the implementation of agricultural technology involving fertilization, irrigation, soil drainage, weed control, and the application of insecticides.

In an introductory physical geography course it would seem especially important to emphasize environmental phenomena as they are distributed on a global scale. Hence, because the earlier USDA classification of soils is so closely related to climate on a worldwide basis, our map of world soils (Plate 16) employs the 1938 USDA system. Our comparison of the world maps of climate, vegetation, and soils in Chapter 7 demonstrated the value of utilizing the soil classes of the traditional system when we examined the varied world climatic regions that serve as human environments. As far as possible our table of zonal soils, which draws on the earlier system, includes a notation of the CSCS equivalents. And both classification systems are briefly introduced in the text that follows because the CSCS is universally used by soil scientists but the 1938 system is more geographic. Please keep in mind as you read that the two systems are based on totally different criteria, and the detailed nature of the CSCS permits only limited correlation of CSCS soil orders with the distribution of soils as mapped on Plate 16.

MAJOR SOIL ORDERS OF THE RUSSIAN-AMERICAN SYSTEM

One of the basic differences among soils is based on their degree of horizon development. Three major orders of soils were thus recognized in the 1938 USDA classification system: zonal, intrazonal, and azonal. **Zonal soils** have well-developed profiles whose characteristics show the strong influences of climate and vegetation. These are the soils that are listed in Table 10.1 and are included in Plate 16. They are a consequence of the major soil-forming regimes just described. As a result, zonal soils are distributed in zones strongly associated with climatic and vegetation regions.

Intrazonal soils are also fairly well developed and exhibit definite characteristics. In contrast to zonal soils, however, intrazonal soils show strong influence by parent material, slope, drainage, or other local factors. Consequently, intrazonal soils do not have orderly distributions but occur in response to local conditions.

Azonal soils are those that for reasons of immaturity show little or no horizon development. Because of this it is not possible to classify azonal soils in the same way we can classify zonal and intrazonal soils on the basis of their horizon characteristics.

ZONAL SOILS The major soil groups are zonal and are therefore associated with climate and vegetation. Because of this we were able to mention soils and their distribution when we studied climatic regions in Chapters 7 and 8. Table 10.1, which describes individual soil classes, should allow you to differentiate among zonal soils and to understand to some extent the factors that produce likenesses and differences in the major soil types. In addition, frequent reference to Plates 15, 16, and 17 should help you to become more familiar with the soil classes of the earlier USDA system.

Table 10.1 shows that the zonal soils fall into one or the other of two general categories: pedalfers and pedocals. **Pedalfers** are soils that tend to be relatively enriched in aluminum and iron (*ped,* ground; *al,* aluminum; *fe,* iron) because of the loss of soluble bases. They are soils of humid, usually forested, regions in which rainfall exceeds evaporation, so that leaching is active. **Pedocals** (*cal,* calcium) are soils of semiarid regions in which potential evapotranspiration exceeds precipitation, so that calcium carbonate is concentrated in the soil. This implies that they also are rich in the other basic nutrients. Thus pedalfers include those soils formed by laterization and podzolization, and pedocals those soils formed by calcification.

AZONAL SOILS The most important and useful of the azonal soils are the alluvial soils. **Alluvium** is the fragmented weathered material eroded from one location, transported by moving water (streams, rivers), and deposited in a new location. Alluvial soils, then, are the soils that develop in this transported material. It is difficult to generalize about alluvial soils, especially immature ones, for their parentage can be so varied. Nevertheless, a few valid generalizations can be made.

Alluvial soils found in areas subject to flooding are usually unable to develop a mature profile. Even if they are agriculturally productive, potential flooding and improper drainage may be inhibiting factors. Also, as alluvial soils are usually the result of recent deposits, there has not been sufficient time for the action of all the soil-forming processes. Where flooding and improper drainage are not immediately inhibiting factors, however, alluvial soils, despite their incomplete or immature profiles, can be very productive.

Soils of mountain regions are both extremely varied and commonly azonal. In Chap-

ter 8 we saw that highland or mountain *H* climates are more dependent on altitude and exposure for their character than they are on latitude. Mountain climates are actually made up of many microclimates whose vertical zonation, because of altitude, resembles the global zonation of climates resulting from latitude. It follows that mountain soils will likewise exhibit a zonation with altitude.

One final note about mountain soils should be mentioned. On steep slopes the soil is removed by erosion as fast as it forms. This in itself is a limiting factor to soil profile development. The result is that any soils that are able to form on these steep slopes are extremely shallow and poorly developed.

INTRAZONAL SOILS Intrazonal soils that have formed in response to local conditions of the environment have minor importance on a global scale and need not be considered by the physical geographer in his examination of world patterns of distribution. As a matter of interest, however, the intrazonal soils are hydromorphic (waterlogged soils), calcimorphic or rendzina (distinctive soils derived from limestone), and halomorphic (soils that are saturated with salts or alkalis).

THE USDA COMPREHENSIVE SOIL CLASSIFICATION SYSTEM

Developed over a period of more than a decade by soil scientists of the USDA, the Comprehensive Soil Classification System was introduced in 1960 and has been refined and modified since that time. The system identifies six levels of classes in descending categories beginning with 10 *soil orders,* 47 *suborders,* 185 *great groups,* and ever-increasing numbers of *subgroups, families,* and *series.* The detail with which soil can be classified utilizing the CSCS is apparent when one realizes that there are some 10,000 different soil categories recognized under this system in the United States alone.

Classification within the CSCS is based directly on the characteristics and composition of a particular soil rather than on its origin or its

TABLE 10.1 Classification of Zonal Soils in the 1938 USDA System (CSCS Equivalents in Parentheses)

SOIL TYPE	SOIL-FORM-ING PROCESS	ASSOCIATED CLIMATE AND PRE-CIPITATION (INCHES)	ASSOCIATED VEGETATION
PEDALFERS (IRON-ALUMINUM-ACCUMULATING)			
Tropical latosols (Oxisols)	Strong laterization	*Af, Am, Aw* > 60	Tropical rain forest, savanna, thorn forest
Latosolic-Podzolic (Ultisols)	Laterization and podzolization	*Am, Cfa* 30–60	Tall-tree savanna, tropical monsoon forest, subtropical pine forest
Gray-brown podzolic (Alfisols)	Moderate podzolization	*Dfb, Dwb* 30–50	Mixed hardwood and coniferous forest
Podzols (Spodosols)	Strong podzolization	*Dfb, Dfc* 20–50	Coniferous forest
Tundra (Entisols, inceptisols, histosols)	Gleization	*ET* 10–20	Tundra grasses, shrubs
TRANSITIONAL			
Prairie soils (Mollisols)	Transitional: podzolization to calcification	*Cfa, Dfa, Dfb* 25–40	Tall-grass prairie
PEDOCALS (LIME-ACCUMULATING SOILS)			
Chernozems (Mollisols)	Calcification	*BS* transition to *Cfa, Dfa, Dfb* 15–20	Mixed tall- and short-grass prairie
Chestnut and brown (Mollisols)	Calcification	*BS* 8–20	Short grass (steppe)
Desert and sierozem (Aridosols)	Calcification and salinization	*BW* < 10	Desert shrubs

stage of development, as is the case with the 1938 USDA system. An entirely new vocabulary of names, derived from root words of classical languages like Latin, Arabic, and Greek, has been developed to label the many different soils in the system. The names, like the system, are both precise and consistent in the sense that

they are chosen to describe the particular characteristics that distinguish one soil from another and always cause that soil to be classified in the proper category.

When examining a soil for classification under the CSCS particular attention is paid to certain horizons or layers that serve to charac-

ORGANIC CONTENT	BASE STATUS	PRODUCTIVITY	DISTINCTIVE FEATURES
Very low	Low	Low	Strong leaching of bases and silica; humus destroyed by intense activity of microorganisms; usually red in color; may develop laterite crust
Low	Low	Requires fertilization	Less leached than tropical latosols; red to yellow in color; low in bases and humus
Moderate	Moderate	Moderately good with fertilization	Only moderately leached; organic A horizon developed, gray-brown; yellow to red-brown B horizon
Acid surface litter	Low	Low	Strongly leached of bases; high acid A horizon, gray at base; B horizon colored yellow to red by relocated iron; very strong horizonation
Variable	Low	Low	Waterlogged and acid—may include peat; little profile development; weak activity of soil organisms; subsoil permanently frozen
High	High	High	Neither leached nor calcified—neutral reaction; rich in humus and bases; A horizon dark brown, exceptionally deep; excellent structure
High	High	High	Black A horizon (thinner than prairie soil); calcium carbonate nodules at base of B horizon; neutral to slightly alkaline pH; subject to occasional drought
Moderate	High	High with irrigation	Becomes lighter in color as precipitation declines, because of decreasing organic content; lime accumulation may produce hardpan (caliche); moisture supply unreliable to deficient; fertile when irrigated, otherwise grazed
Very low	High	High with irrigation and fertilization	Poor profile development; low organic content; calcareous hardpan common; moisture deficient

terize the soil. Some of these horizons are situated below the surface **(subsurface horizons),** and some, called **epipedons,** are surface layers that usually exhibit the darker shading associated with the presence of organic material (humus). Examples of some of the more common horizons, which serve to illustrate how names were chosen to represent actual soil properties, may be found in Table 10.2.

Although there are few overall similarities between the CSCS and the USDA 1938 classification system, the relationships that do exist in the highest classes will be emphasized in the following discussion of the ten orders in the

TABLE 10.2   Selected Common Horizons in USDA Comprehensive Soil Classification System*

Oxic horizon (from *oxygen*)
A subsurface layer that contains hydrated oxides of iron and aluminum. Found in tropical and subtropical climates at low elevations.

Argillic horizon (from Latin: *argilla*, clay)
Usually formed beneath the *A* horizon by illuviation, this layer contains a high percentage of accumulated silicate clays.

Ochric epipedon (from Greek: *ochros*, pale)
A surface horizon that is light in color and very low in organic matter or very thin.

Albic horizon (from Latin, *albus*, white)
Commonly the *A2* horizon overlying a spodic horizon, this layer is usually sandy and very light in color due to the removal of clay and iron oxides.

Spodic horizon (from Greek: *spodos*, wood ash)
Usually underlying the *A2* horizon, the spodic horizon is dark in color because of the illuviation of humus, aluminum oxides, and often, iron oxides.

Mollic epipedon (from Latin: *mollis*, soft)
Very high in content of basic substances (calcium, magnesium, potassium), the mollic epipedon is a relatively thick and dark-colored surface layer.

Calcic horizon (from *calcium*)
A subsurface horizon rich in accumulated calcium carbonate or magnesium carbonate.

Salic horizon (from *salt*)
A layer of soil material at least 6 inches thick and containing at least 2 percent salt. Common in desert basins.

Gypsic horizon (from *gypsum*)
A subsurface soil horizon rich in accumulated calcium sulfate (gypsum).

*This table includes only some of the more common horizons, listed in the order identified in the text. Those called *epipedons* occur near the surface, while those called *horizons* are subsurface layers.

newer system. Three of the CSCS orders—**Entisols, Inceptisols,** and **Histosols**—would more likely be considered azonal or intrazonal soils rather than zonal soils in the 1938 system (although all three may be found in Tundra regions, as noted in Table 10.1).

Entisols are soils that lack horizons, usually because of their recent development. They are often associated with the constant erosion of sloping land in mountain regions or the frequent deposition of alluvium in river floodplains. However, they can also occur in areas of heavy sand accumulation where soil horizons are not easily developed. Entisols are quite similar to the azonal soils of the 1938 classification, and, like their counterparts, they can be found in many climatic regions and represent a full range of agricultural productivity.

Inceptisols are young soils with weak horizon development. Although these soils are found in humid areas from the tropics to polar regions, the processes of *A* horizon depletion (eluviation) and *B* horizon deposition (illuviation) are just beginning. Like Entisols, Inceptisols may form on recent alluvium. They are also common in tundra climates of the Arctic or highland regions, in areas of recent continental glaciation, and in tropical regions with recent accumulations of volcanic ash. Their usefulness for agriculture depends largely on their origin and associated climatic characteristics.

Histosols usually develop in poorly drained areas such as swamps, meadows, or bogs. They are largely composed of slowly decomposing plant material. The waterlogged conditions of the soil deprive bacteria of the oxygen necessary to prevent accumulation of organic matter. Histosols in the upper middle latitudes are commonly acidic and of low fertility, although these soils in peat bogs and properly drained lake beds may be used to produce special crops such as cranberries and garden vegetables. In the 1938 USDA classification Histosols would be considered intrazonal soils.

Six of the seven remaining CSCS soil orders can be broadly related to the zonal soils of the 1938 USDA system. As noted in Table 10.1, **Oxisols** are identical to tropical latosols. They are almost entirely leached of soluble bases and are characterized by a horizon of iron and aluminum oxides (i.e., *oxic horizon*). They develop in the humid tropics, although they also extend into savanna and tropical thorn forest regions as well. Oxisols represent the end product of the laterization process and retain their natural fertility only as long as the soils and forest cover maintain their delicate equilibrium—the trees returning to the forest floor the leaf mold to provide the few plant nutrients in the soil required to sustain growth.

Ultisols correspond roughly to the red and yellow latosolic-podzolic soils of the earlier system. Like the Oxisols, they are common in the wet tropics but extend into subtropical regions as well. For example, they are the most common soil order of the humid subtropical southeastern United States. Predominant colors are red and yellow due to the accumulation of iron and aluminum oxides in the *A* horizon. These soils are also characterized by a subsurface clay-enriched horizon called an *argillic horizon* (Table 10.2). Ultisols are acidic, although less so than Oxisols. The few bases present are

found near the surface and have been derived from leaf mold deposited by the trees that form the typical natural vegetation. Agriculture may be quite successful on these soils, but there is a need for continuous application of fertilizers coupled with the persistent threat of soil erosion.

Alfisols predominate in large areas of the humid continental midlatitudes but may be found in subtropical and tropical regions as well. There is usually an accumulation of clay minerals (*argillic horizon*) in the *B* horizon, but an even more important classification characteristic is the presence of a light-colored *ochric epipedon* (Table 10.2). Alfisols most closely resemble the gray-brown podzolic soils of the earlier USDA classification. They are only moderately leached of soluble bases and, especially in the humid continental climates, can be quite productive agriculturally. They form the soils of the eastern "Corn Belt" of the United States but are also found in some of the better farming areas of Europe, which have marine west coast or Mediterranean climate.

Spodosols closely correspond to the zonal podzol soils of the 1938 classification system. They are readily identified by their strong horizon development. A white or light gray *A* horizon (i.e., *albic horizon*) is covered with a thin black layer of partially decomposed humus and overlies a *B* horizon enriched by relocated organic material and iron and aluminum compounds (i.e., *spodic horizon*). The Spodosols are strongly leached and highly acidic. They are found in regions of humid continental mild summer (*Dfb*) and subarctic (*Dfc*) climates in association with the coniferous forests that serve as the natural vegetation. They have limited agricultural potential not only because of their lack of fertility but also because of the severe restrictions of the climates with which they are associated.

Mollisols include among their suborders many of the most agriculturally suitable soils on earth. They correspond to the transitional and most of the productive pedocal zonal soils of the traditional USDA classification system (Table 10.1). Mollisols have the advantages of a generous supply of bases, especially calcium, abundant plant root humus associated with grasslands as natural vegetation, and a soft granular structure conducive to cultivation. The characteristic horizon of a Mollisol is a *mollic epipedon,* a thick, dark-colored surface layer, rich in organic matter. Mollisols are associated with the subhumid and semiarid regions of the subtropics and middle latitudes. Although they originally developed under the grazing lands for countless herds of antelope, bison, and horses, they now support most of the grain production from domesticated grasses (wheat, rye, oats, corn, barley, and sorghum). In regions of adequate precipitation (the *prairie soils* of the 1938 system) the combination of soils and climate are unexcelled for agriculture. In areas of lesser precipitation, periodic drought is a constant threat and the temptation of fertile soils has been the downfall of many a farmer (see the Viewpoint in Chapter 7).

The last soils of the CSCS soil orders that can be directly correlated with the older USDA classification system are the **Aridisols.** These are the soils of desert regions, and they correspond to the reddish desert and sierozem soils of the earlier system. They may be the end product of either calcification or salinization. Although ordinary horizon development is weak because of lack of water movement in the soil, there is often subsurface accumulation of calcium carbonate (*calcic horizon*), salt (*salic horizon*), or calcium sulfate (*gypsic horizon*). Due to a lack of surface vegetation, there is little or no plant root humus in the soil. Aridisols are highly alkaline, but with appropriate irrigation and drainage to remove excess salts, these soils can produce bountiful harvests of whatever crops the limitations of the associated climate will permit.

The soils of the only CSCS soil order yet to be discussed, **Vertisols,** have little or no horizon development and have no equivalent in the traditional USDA classification system. They are found in tropical and subtropical climatic regions with drought periods of sufficient length to permit the soils to dry and shrink. This causes deep vertical cracks to appear. Vertisols also have a high content of clay minerals that readily absorb soil water during wet periods. As the soil moistens, adjacent blocks swell and push against one another, closing the cracks. The heaving soil may cause displacement of sidewalks or other surface objects. Vertisols are dark in color, are high in bases, and contain considerable organic material derived from the grasslands or savanna vegetation with which they are normally associated. Although they harden when dry and become gummy and difficult to cultivate when swollen with moisture, vertisols can be agriculturally significant. The use of modern mechanized farm machinery has permitted the black Vertisol belt of Texas to become one of the world's leading cotton-producing regions.

Regardless of their composition, origin, or state of development, the earth's soils remain one of our most important and vulnerable resources. Even the word *fertility,* so often associated with soils, has a meaning that takes into consideration the usefulness of these soils to man. Soils are fertile in reference to their effectiveness in producing specific vegetation. Some soils may be fertile for corn and others for potatoes. There are soils that retain their fertility only as long as they remain in delicate equilibrium with their vegetative cover. But in every instance the significance of the soil's fertility is only of consequence to those human beings who would make use of the soil resource.

It is clearly the responsibility of all of us who enjoy the agricultural end products of farm, ranch, and orchard, or who simply appreciate the beauty of forest and field to recognize and help protect our valuable soils. Although space does not permit a thorough examination of the problems of soil erosion, soil depletion, and land mismanagement, we should be conscious of their existence in the world today (Fig. 10.12). At the same time we should be aware that for each of these problems there are reasonable solutions (Fig. 10.13). Maintaining soil fertility and usefulness is a serious challenge to humanity and one of the essentials in our continuing struggle to live in harmony with the natural environment.

FIGURE 10.12 Gully erosion is one of the more spectacular examples of poor agricultural practices. It produces permanent alteration of the landscape and guarantees that the original productivity of the land cannot be regained. (USDA)

FIGURE 10.13 The contour farming techniques utilized on this farm are an excellent example of conservation methods designed to preserve the soil resource.

QUESTIONS FOR DISCUSSION AND REVIEW

1. Why is soil an outstanding example of the integration of the earth's subsystems?

2. Describe the different circumstances in which water is found in soil.

3. Define eluviation and illuviation. What is the effect of each if carried to an extreme?

4. Under what conditions does leaching take place? What is the effect of leaching on the soil and consequently on the vegetation it supports?

5. How can capillary water contribute to the formation of caliche? What is the effect of caliche on drainage?

6. How might soil air differ from air in the atmosphere? What is the effect on life when air is excluded from water-saturated soils?

7. How is humus formed? What relation does humus have to soil fertility?

8. What conclusions can you draw from the color of the soil in your area? How might color relate to fertility?

9. How is texture used to classify soils? Use Figure 10.3 to describe the texture of a typical clay loam soil.

10. Describe the ways scientists have classified soil structure.

11. What is the pH scale? What pH range indicates soil suitable for most complex plants?

12. What is meant by a soil profile? What are the general characteristics of each horizon in a soil profile? How are soil profiles important to scientists?

13. How does transported parent material differ from residual parent material? List those factors which help to determine how much effect the parent material will have on the soil.

14. What are the most important effects of parent material on soil?

15. List a number of ways in which humus is important to soils.

16. What effect do plants of the legume family have on soil fertility?

17. How does the presence of earthworms alter soil?

18. Describe the various ways in which temperature and precipitation are related to soil formation.

19. The Bonneville Salt Flats in Utah are well known as a natural soil formation that provides a perfect surface for auto racing. How do you suppose these salt flats were formed?

20. Describe the three major soil-forming regimes.

21. What are the chief differences between the three major USDA 1938 soil orders? Why are physical geographers most concerned with zonal soils?

22. Match soil orders of the USDA 1960 comprehensive system with zonal soils of the 1938 system. Which has no equivalent?

23. Which soil orders of the 1960 system have the most agricultural potential? Why?

24. Where would you rank soils among a nation's environmental resources? Give your opinion of the overall value of soils in the United States and the extent to which these soils are preserved and protected.

11

LANDFORMS AND EARTH STRUCTURE

LANDFORM CLASSIFICATION

The study of landforms, called **geomorphology,** is an important subdivision of physical geography. Landforms are the surface expression of the lithosphere and owe their development to processes originating from both within and outside the earth's surface. We have already distinguished among the large subsystems that make up the total earth system: the atmosphere, the hydrosphere, the biosphere, and the lithosphere. Up to this point we have primarily examined the atmosphere and, to a lesser extent, the biosphere. In the seven chapters that remain, we will concentrate our attention on the lithosphere and, particularly in Chapter 17, on the hydrosphere as well. It will become increasingly obvious as we proceed that all the subsystems are interrelated—we cannot focus on one without noting the influences upon it of all the others. For example, as we identify typical landforms and discover the forces that have produced them and how they are presently being changed, we will quickly recognize the roles of the atmosphere, hydrosphere, and biosphere in helping to create them.

There is enormous variety in the earth's landforms: hills, mountains, valleys, plains, plateaus, cliffs, and canyons, each found in a multitude of sizes, shapes, and combinations. However, although no two landforms are identical, there are similarities among them in size, shape, height, depth, slope, or surface material. These similarities provide a basis for developing a system of landform classification. Classifying landforms is difficult because of this variety of forms, which exists on all scales from sand ripples to mountain chains. Another problem of classification is the fact that landforms, unlike climate, vegetation, and soils, are not arranged in systematic and predictable patterns. Instead, they are largely a consequence of geologic processes and movements that have varied in both intensity and location with the passage of time. Thus, we can predict the distribution of landforms only if we have a knowledge of the full geologic history of the earth and a knowledge of each locality on the earth.

The classification used here has been adapted from one devised by Edwin H. Hammond, a noted American physical geographer and geomorphologist. Our classification is intended to produce a map that differentiates the

The Andes, Peru. (S. Brazier)

319

gross contrasts in surface configuration over the earth as a whole. Its categories are very broad and descriptive. This classification is only a first step in the study of landforms, for it is not concerned with origins or processes.

The system we will use as an introduction to landform classification is based on (1) the general nature of **slopes** (inclination of land surfaces from the horizontal) and (2) **local relief** (difference in elevation between the highest and lowest points in a specified area). Slope can range from dead flat (an angle of 0°) in the case of level plains to vertical (an angle of 90°) in the form of cliffs. Local relief is an especially good criterion for landform classification because it can be quantified, and quantities can be compared. Furthermore, local relief suggests the dimensions of the landforms in question. However, local relief does not completely describe land surface. Where the local relief is 150 meters (500 ft), several possible landscapes exist: many hills; an area of gently rolling plains with one large distinctive hill; a relatively flat plateau or tableland interrupted here and there by deep canyons; or a sharp escarpment or cliff dropping down to a plain below. To clarify the description of an area when its local relief is known, the geographer must state the nature of its overall slope characteristics.

The limits of a general landform classification system are as follows. First, there is the problem of relative size or dimension. For example, foothills of the Alps with a local relief of 300 meters (1000 ft) would look like mountains if we found them on the Kansas plains or even on the rolling country of Ohio and Illinois. Thus, the broader region must be considered to see how a particular landform fits within it.

Second, the criteria used in classification are not absolute, and for the purposes of coherence within the system, the parameters must be flexible. As with any classification system it is difficult to establish boundaries between major types. A system of worldwide landform classification is less accurate, less comprehensive, and must be more flexible than a climate classification because of difficulties in quantifying the size, shape, and arrangement of highly diverse features.

With the two criteria we are using—amount and kind of slope and local relief—we can distinguish four major surface types. **Mountains** consist mostly of steep slopes and have a great deal of local relief. **Hills** have steep slopes but less local relief than mountains. **Plains** are characterized mainly by gentle slope and small local relief. Finally, there are **plains with localized high relief.** This last type of terrain includes two contrasting landscapes. In one, the gently sloping land is at low elevation and the relief features rise above the plains; we call these **plains with hills and mountains.** In the other, the plains may lie at high elevations with the relief cut below them in the form of valleys, or they may overlook lower adjacent plains areas. This type of terrain is called a **tableland** or **plateau.** The area around the Grand Canyon known as the Colorado Plateau is a classic example of a tableland. The difference in the profiles of tablelands and plains with hills and mountains is illustrated in Figure 11.1.

PLAINS

Plains are lands with a maximum of gentle slopes and a minimum of local relief (Fig. 11.2). (The world distribution of plains and other major land surface types is indicated on Plate 19.) There is considerable variation, however, within this definition, so that some plains, like California's Central Valley, are nearly flat; others, such as those of Iowa, are rolling or undulating; and still others, such as the Georgia Piedmont, are quite rough and stream-dissected, though without the steep slopes characteristic of hills. Most plains, however, are relatively level and have a flat appearance and some slight relief. Plains are usually found at low elevations, though along the east front of the Rocky Mountains the "high plains" rise well above 1500 meters (5000 ft).

Physical geographers distinguish between the different types of plains on the basis of their origin, location, or the distinctive nature

White Mountains, monadnocks on the New England Upland.

Old till plains of west central Illinois.

Mississippi River floodplains.

Badland remnants on the South Dakota plains.

Youthful Yellowstone River falls and gorge cut in a lava plateau.

Plains with scattered volcanic hills and mountains in western Arizona.

Grand Canyon in the Colorado tableland.

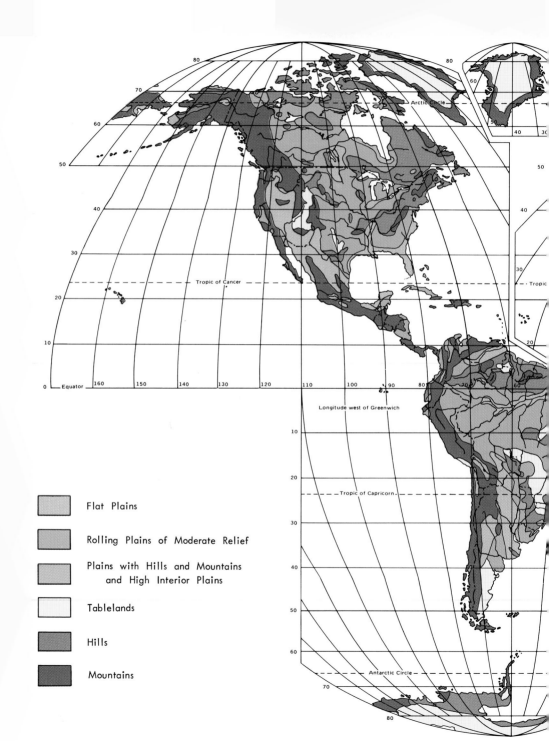

Flat Plains

Rolling Plains of Moderate Relief

Plains with Hills and Mountains
    and High Interior Plains

Tablelands

Hills

Mountains

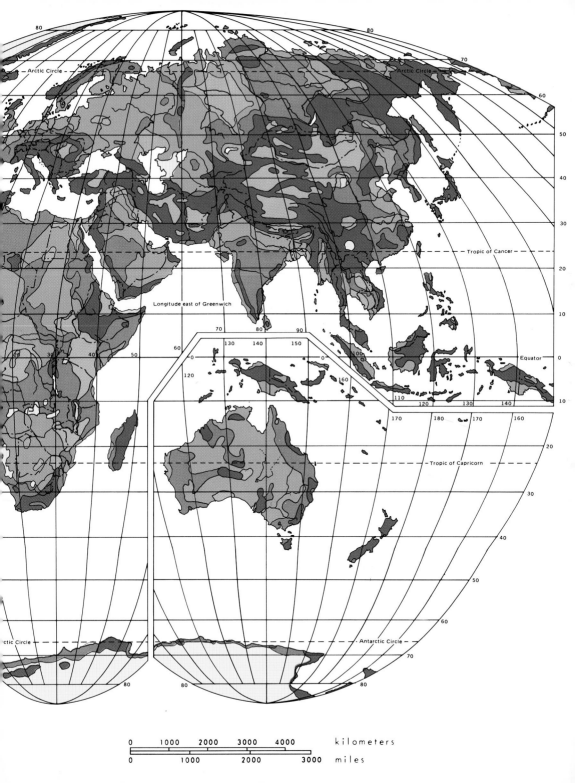

Arctic Circle

Arctic Circle

Tropic of Cancer

Longitude east of Greenwich

Equator

Tropic of Capricorn

Antarctic Circle

Arctic Circle

Antarctic Circle

kilometers

miles

0 1000 2000 3000 4000

0 1000 2000 3000

ake in the glaciated ntains.

Mount Rainier, premier peak of the volcanic Cascades.

Glaciated peaks of the fault bl Grand Tetons.

Upper valley of Yosemite in the fault block, Sierra Nevada.

ata surrounding the cture of the Black

Catskill region of the Appalachian hill lands.

Rolling hills of the "Driftless" r gion of northwest Illinois.

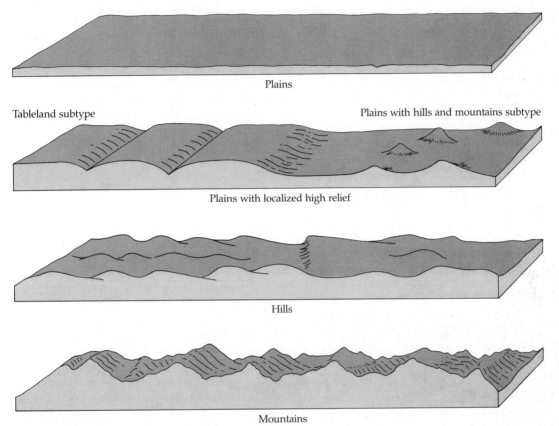

Plains

Tableland subtype

Plains with hills and mountains subtype

Plains with localized high relief

Hills

Mountains

FIGURE 11.1   Profiles of different landscape types.

FIGURE 11.2   The High Plains near Denver, Colorado. (S. Brazier)

of their surface features (Plate 18). Some plains are formed primarily by erosion over long periods of time by streams or glaciers. Others may be characterized by material deposited by water, glaciers, or the wind. Climate can have a major effect on surface appearance, for instance when plains in humid areas are contrasted with those in arid regions. Plains differentiated according to their location include **coastal plains,** which are recently emerged from the sea and found near continental shorelines, and **interior plains,** which are far from the oceans in the continental interiors. Within the latter group may be found **lacustrine plains,** which are extremely flat former lake beds.

MOUNTAINS

Mountains are the highest lands on earth, although they account for the smallest amount of land surface (Fig. 11.3). Generally massive and rugged, rising abruptly in steep slopes from the land below, mountains can be awe-inspiring regions of incredible beauty. According to our criteria, mountains are the opposite of plains. They have a maximum of steep slopes, high lo-

cal relief, and small summit areas at elevations thousands of meters higher than adjacent nonmountainous areas. It is impossible to set a more specific elevation or local relief requirement for mountains, for an area is considered mountainous in relation to its surroundings.

Mountains, like plains, can be distinguished by their origin (Plate 20). The tilted fault blocks of California's Sierra Nevadas or Wyoming's Grand Tetons differ strikingly in appearance from the folded Alps or the Appalachian Ridge and Valley of the eastern United States. These mountains, in turn, differ from the volcanic Cascades of Washington and Oregon and from the domed ranges of the Colorado Rockies.

Mountains are also affected by time and the persistent forces of gradation. Mountain ranges uplifted far in the past are generally more rounded and less severe than the young rugged mountains formed during more recent periods of earth history. Although mountain building is a result of complex processes that have operated constantly since the earth was formed, geologists have identified peak periods of such activity. As the geologic timetable (Table 11.1) indicates, these *orogenies* occurred in

FIGURE 11.3   Mountain topography, Wilbur Creek, Glacier National Park. (R. Gabler)

TABLE 11.1 Outline of Earth History

ERA	PERIOD	EPOCH	DISTINCTIVE FEATURES	MILLIONS OF YEARS AGO	PERCENT OF EARTH'S HISTORY	MOUNTAIN-BUILDING EPOCHS (OROGENIES)
CENOZOIC	Quaternary	Recent	Modern Man	.01	.04	Alpine and Cascadian
CENOZOIC	Quaternary	Pleistocene	Early man; glaciation	2	.04	Alpine and Cascadian
CENOZOIC	Tertiary	Pliocene	Large carnivores	13	1.3	Alpine and Cascadian
CENOZOIC	Tertiary	Miocene	Abundant grazing mammals	25	1.3	Alpine and Cascadian
CENOZOIC	Tertiary	Oligocene	Large running mammals	36	1.3	Alpine and Cascadian
CENOZOIC	Tertiary	Eocene	Modern types of mammals	58	1.3	Laramide (Rocky Mtns.)
CENOZOIC	Tertiary	Palocene	First placental mammals	63	1.3	Laramide (Rocky Mtns.)
MESOZOIC	Cretaceous		First flowering plants; climax of dinosaurs, followed by extinction	135	3.6	Laramide (Rocky Mtns.)
MESOZOIC	Jurassic		First birds, first true mammals; many dinosaurs	180	3.6	Laramide (Rocky Mtns.)
MESOZOIC	Triassic		First dinosaurs; abundant cycads and conifers	230	3.6	Appalachian
PALEOZOIC	Permian		Extinction of many kinds of marine animals, including trilobites; continental glaciation in Southern Hemisphere	280	8.0	Hercynian (Europe)
PALEOZOIC	Carboniferous	Pennsylvanian	Great coal swamps, conifers; first reptiles	345	8.0	
PALEOZOIC	Carboniferous	Mississippian	Sharks and amphibians; large-scale trees and seed ferns	345	8.0	Acadian
PALEOZOIC	Devonian		First amphibians; fishes very abundant	405	8.0	
PALEOZOIC	Silurian		First terrestrial plants	425	8.0	Caledonian Taconic (Taconian)
PALEOZOIC	Ordovician		First fishes; marine invertebrates	500	8.0	Caledonian Taconic (Taconian)
PALEOZOIC	Cambrian		First abundant record of marine life; trilobites and brachiopods dominant	600	8.0	
	PROTEROZOIC	Precambrian	Limited evidence of abundant algae		87.0	Archaean
	ARCHEOZOIC	Precambrian				

the Eurasian landmass (Archean, Caledonian, Hercynian, and Alpine) at slightly different times than they occurred in North America (Taconic, Acadian, Appalachian, Laramide, and Cascadian).

As you study the geologic timetable, you will note that each great mountain-building revolution seems to have coincided with major evolutionary developments in the animal kingdom. You will also discover later in Chapter 12 that the periods of mountain-building activity are closely related to the major events of plate tectonics (continental drift). *Tectonic activity* describes those processes that derive their energy from within the earth's crust and result in major vertical or horizontal movements of crustal rocks. The relationship between biological change and tectonic activity seems logical when one considers the significant climatic and associated environmental alterations that must accompany major changes in crustal location, elevation, and configuration.

The immensity of geologic time over which these events occurred is difficult to grasp. If we took a 24-hour day to represent the whole of earth history, the first 21 hours would be consumed by Precambrian time, an era about which we know very little. The Quaternary Period would take less than 35 seconds, and human beings as we know them would be on the scene for less than a single second.

HILLS

It is difficult to develop a precise definition of hills. Low mountains in one place often resemble in elevation, relief, and slope the high hills of another place. As a general rule hills are less rugged than mountains. They have more level land, larger summits, and shorter and less steep slopes (Fig. 11.4).

Like mountains, hills vary in origin (Plate 20). Many hills have been created by downward stream cutting in an uplifted surface. Some hills have developed as a direct result of tectonic forces but have never reached the size or magnitude of mountain ranges. Others are remnants of former mountains, but the long-continued activity of the various weathering and erosional processes has gradually reduced the mountains to more subdued forms. Still other hills are foothills or spurs on the margins of mountain ranges.

FIGURE 11.4 The Pennine Hills of England. (S. Brazier)

PLAINS WITH SOME LOCAL HIGH RELIEF

Some terrain types resemble plains in that a majority of their area is composed of gentle slopes, yet they cannot be classified as plains because they include significant areas of local high relief (Plate 18). Two different situations are apparent: plains with isolated hills and mountains and tablelands or plateaus cut by steep canyons or ending in escarpments that drop to land at a lower elevation (Fig. 11.5).

PLAINS WITH HILLS AND MOUNTAINS There are three relatively common ways by which plains with hills and mountains have been formed. In the first, erosional processes have reduced all but a few remnants of a formerly hilly or mountainous region to plains of low relief. The remnant hills or mountains have survived because they are more resistant to erosion. Second, plains with hills or mountains can develop where relatively flat surfaces have been roughened by local volcanic activity. The development of the volcanic cone, Paricutín, from a Mexican cornfield in a matter of several months, is a striking example. Third, plains with mountains of a different type may occur where the surface has been ruptured by geologic forces along great fractures (faults), causing some crustal blocks to rise and others to sink. This usually produces a succession of sediment-filled basins divided by more or less parallel mountain ranges. Examples can be found in the Great Basin of Nevada and in portions of adjacent states as well as in Tibet and Iran.

TABLELANDS Most tablelands are the result of general uplifting of the land, which triggers stream erosion and eventual dissection of the uplifted area. Some of these tablelands may be bordered by steep escarpments formed along the faults where the uplift has occurred. Others, like the Tibetan Plateau or the Bolivian Altiplano, are really high-elevation plains that are completely ringed by mountains. Such tablelands may have little internal dissection by

streams. These illustrate the difficulty of landform classification, because, except for their relatively gentle slopes throughout, they might be classified as plains with hills and mountains. A third type of tableland, seen in India, Iceland, portions of Africa, and the Columbia Plateau of the northwestern United States, is composed of lava. In these regions immense floods of molten lava drowned the preexisting relief and accumulated in layers hundreds of meters in depth. The elevation of these volcanic tablelands results from the buildup of the lava mass, flow on top of flow. Subsequently, streams such as the Snake and Columbia Rivers in Idaho, Oregon, and Washington have cut through these volcanic piles, sometimes exposing the old surfaces beneath them.

Other factors may help create the distinctive tableland or plateau scenery. Aridity, for example, discourages the dissection and erosion of the high-level tablelands by many rivers and streams, which would otherwise hasten the modification of these regions into hills or low-level plains. Often plateaus are cut through by only one or two major streams. The Colorado Plateau, dissected by the Colorado River and several tributaries, is an example. Sometimes a plateau is capped by a layer of rock that is especially resistant to erosion. This layer creates a tableland, with a very flat upper surface and an abrupt rim. On the Colorado Plateau the rock cap is frequently a permeable sandstone that resists erosion because it absorbs precipitation, thereby preventing surface runoff. A lava cap produces much the same result. Caprock plateaus are commonly fringed with **mesas** and **buttes**—flat-topped remnants of the tableland that become detached as the edges of the tableland are eaten away by stream erosion.

LANDFORMS AS PART OF THE EARTH SYSTEM

We have already seen that the distribution and shapes of landmasses influence climate significantly. Likewise, major landforms like mountain ranges affect

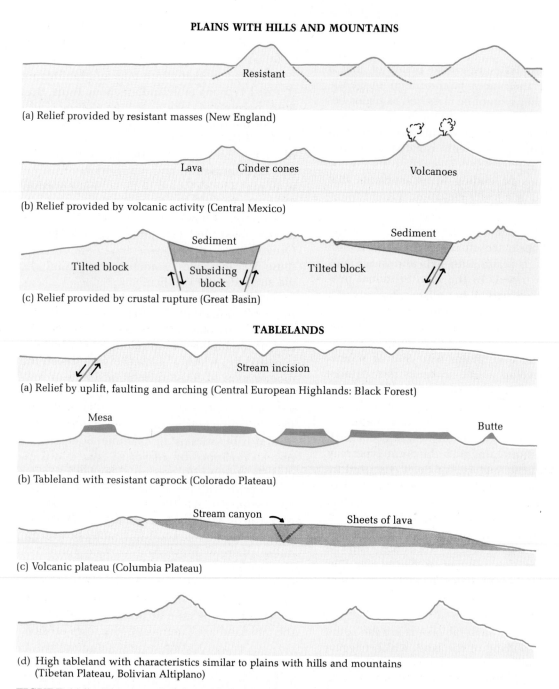

PLAINS WITH HILLS AND MOUNTAINS

Resistant

(a) Relief provided by resistant masses (New England)

Lava   Cinder cones   Volcanoes

(b) Relief provided by volcanic activity (Central Mexico)

Sediment   Sediment

Tilted block   Subsiding   Tilted block
                block

(c) Relief provided by crustal rupture (Great Basin)

TABLELANDS

Stream incision

(a) Relief by uplift, faulting and arching (Central European Highlands: Black Forest)

Mesa   Butte

(b) Tableland with resistant caprock (Colorado Plateau)

Stream canyon   Sheets of lava

(c) Volcanic plateau (Columbia Plateau)

(d) High tableland with characteristics similar to plains with hills and mountains
    (Tibetan Plateau, Bolivian Altiplano)

FIGURE 11.5 Diagram of plains with local high relief.

climate by forcing moisture-carrying winds to rise, producing heavy rainfall and creating dry conditions beyond. This in turn affects vegetation, soils, and animal life. Another example of the interrelationship between landforms and climate is the effect of slope and terrain on soil development. It is difficult for soils to mature on steep mountainsides. Very flat terrain also affects soil development by causing poor drainage.

There are also many ways in which landforms and people are interrelated. We will look first at some of the influences landform configuration has on human life and then at some of the ways people in turn affect the land on which they live.

LANDFORMS AND PEOPLE

It would be unlikely today to find a physical geographer who would suggest that landforms can determine the course of human events, but even the average student recognizes that landforms are one of the major aspects of the earth system with which human beings must live on a daily basis. The character and quality of the land and its resources (slope, altitude, roughness, soil, minerals, and water) are significant influences on the character and quality of human life. The distribution of population over the earth shows the importance of land surface configuration to people. All over the world people have tended to congregate along coastlines that allow easy access to the sea, especially where there are good harbors, and on plains where climate and soils have allowed agricultural development. They have avoided mountainous areas, remote hill lands, and isolated tablelands because of the restriction these landforms impose on agriculture, commerce, and communication.

It is not easy to generalize about the positive or negative aspects of one landform or another. Some coastal areas with good natural harbors encourage the development of a fishing industry, as in New England, Alaska, Norway, and Japan. Direct access to the sea also furthers trade, communication, and international interests. Yet in coastal, alluvial, and delta regions, the advantages of fertile soil and easy water transportation may be somewhat offset by the hazard of frequent floods and severe coastal storms such as those experienced by the citizens of Bangladesh at the mouth of the Ganges-Brahmaputra river system.

Broad, well-watered plains stretching to the interiors of continents often encourage agriculture and efficient transportation networks. Such is the case in China, northern Europe, and the American Midwest, but not in Amazonia, where climate is a formidable handicap. Broad, flat plains areas also offer little protection from foreign invasion during periods of world conflict, as the suffering citizens of Belgium, the Netherlands, and Poland have rediscovered each time armies have marched across the North European Plain.

Landforms other than plains all possess significant restrictions on human inhabitation. The poor soils, severe climates, steep slopes, and high elevations of mountain regions severely limit agriculture. Mountains also form barriers to population expansion, transportation, trade, communication, and the spread of ideas and cultures. Rugged hill country communities are often isolated from each other, self-sufficient, and antagonistic toward outsiders. The hill lands of Appalachia and the Ozarks are examples in the United States. The steep cliffs, canyons, ravines, and escarpments of tablelands always present severe problems in transportation, communication, trade, and immigration. In many plains areas with hills or mountains, the highlands are extensive enough to disrupt communications or are high enough to produce a rain shadow to their lee.

However, it may be just as significant that people can live and prosper in all landform regions by overcoming the handicaps or by taking advantage of what others might consider severe limitations. We have already noted in earlier chapters the climatic advantages of highlands in tropical areas. Terracing has turned steep slopes into productive fields in both tropical and middle latitudes where population pressure has forced people away from adjacent plains. The Hopi and Zuni Indians of the Colorado Plateau built their pueblo villages where they were easy to defend and where springs emerging at the foot of a cliff watered their fields. The Swiss have utilized the rugged nature of their Alpine terrain to help maintain their neutrality through several wars. It is undoubtedly safe to say that landforms exert a significant influence over world patterns of settlement and human activity but that the resourcefulness and persistence of human beings provide many exceptions to this influence.

PEOPLE AFFECT LANDFORMS

Although the changes people make in the natural landscape seem negligible compared to all the changes made by natural forces, on a local level human alterations of natural landforms can be highly significant. In fact, human alteration of the land is sometimes so complete and the changes so familiar that the human role can go unnoticed. For example, some reservoirs are so artfully created that only the existence of a dam at one end hints at their human origin. Coastal lands that are filled or drained and then used for crops or housing are often not recognizable as landforms created by humans.

People shape the earth in many ways, with both positive and negative results. Agriculture is the oldest of human activities that have significantly altered landforms. Entire forests have been cleared for farming, and usually erosion has been greatly accelerated. In some parts of the world massive gullies now exist where once there were gently rounded slopes covered by tall trees. However, in other areas farming societies have terraced hill and mountain slopes to create additional agricultural space while actually slowing the erosion process. In lowlands and coastal areas humans have literally created dry land for farming purposes by diking and draining lakes, marshes, and tidelands. Such is the case in the polders of the Netherlands and on Florida's Gulf Coast.

Humans have had a major impact on the land through construction and urbanization. Watch a crew build a stretch of highway: They dig out the land, fill it in, build it up, flatten it out, change stream beds, and finish it all off with an icing of asphalt. Urbanization brings to mind acres of land covered with concrete, tar, asphalt, steel, buildings, sewage grates, and manhole covers. Except for small, preserved parkland areas, the natural land of the urban center is completely invisible. Often even the shape of the land—the mounds and depressions—has been eliminated by the bulldozer. And as humans affect the natural landscape, they affect the other earth systems as well: Soils, vegetation, and animal life are altered or totally eliminated.

In some regions of the world where mineral and fuel resources are concentrated the entire physical landscape may be changed. Mountains are reduced to piles of mining wastes; plains become hill lands of pit and soilbank; gaping holes appear in the earth where gently rolling land previously existed. Although deep shaft mining leaves its mountains of waste material, open-pit or strip mining takes the greatest toll in the natural landscape. Noisy and ugly, surface mining gouges up the earth, creates vast craters, greatly affects natural drainage, wipes out the local vegetation, increases erosion, either removes or buries productive soils, and destroys animal habitats.

Perhaps the most detrimental effect of humans on landforms is associated with waste disposal. The deposition of solid wastes on the ground, in pits, in canyons (as in the Los Angeles area), or as landfill in coastal areas (such as the shores of San Francisco Bay) drastically alters the physical environment. Mounds of human garbage replace depressions, cover level land, or eliminate coastal marshes. If the waste happens to be toxic or radioactive, there may be negative effects on the physical landscape that will last for centuries, and water supplies may be seriously damaged by the leakage of such toxic materials beneath the ground.

LANDFORM ANALYSIS

Though humans have been and continue to be an effective force in altering the shape of the land, the work of humans is minute in contrast to the natural forces that have been active over the immense span of geologic time. Thus in order to understand and explain such different landforms as Death Valley, Mount St. Helens (shown in Fig. 12.9), the Grand Canyon, Cape Cod, the Mississippi Delta, and the glacial plains of Illinois, one must understand something of the nature and arrangement of the materials composing them, or their **structure.** It is also necessary to understand the processes by which these materials have developed into distinctive landforms (known as **geomorphic processes)**. Finally, because the geomorphic processes are evolutionary, we must know what **stage** of development a particular landform has achieved. Knowing the stage helps us understand what the landform was like in the past and what it will probably be like in the future. William Morris Davis (1850–1934), an American geographer, was the first to emphasize the importance of landform stage, near the beginning of the twentieth century. Others have suggested that the stage of landform evolution is in large part a consequence of the rate and nature of vertical movement of the earth's crust from place to place.

At one time scientists believed that the earth's shapes were created in great cataclysms, that the Grand Canyon, for example, split open one violent day and has remained that way ever since. This theory is called **catastrophism.** For almost two centuries, however, geographers, geologists, and other earth scientists have accepted the theory of **uniformitarianism**—the idea that forces are operating today in the same fashion as they have for millions of years.

In the formation of landforms there is an earthwide continuous struggle between the forces that elevate, disrupt, and create inequalities in the earth's surface and the processes that wear down, fill in, and tend to level the surface. The processes that roughen the earth's surface gain their energy from the earth's interior and are called **tectonic processes** (from the Greek: *tekton*, carpenter, builder). The leveling processes are known as **gradational processes.** Gradation includes the fragmentation and chemical breakdown or **weathering** of earth materials, which makes them removable; **mass wasting,** the movement downslope of the weathered materials due to the pull of gravity; the actual removal or **erosion** of the materials; their **transportation** by agents such as wind, water and ice; and their **deposition** at lower elevations. The opposing forces, one roughening and the other smoothing the earth's surface, are constantly in conflict.

Though we will consider first the tectonic processes and then the processes of gradation, the shape of the land is actually the result of continuous interaction between the internal tectonic forces and the external gradational processes. The Grand Canyon, for example, is the result of the lifting of a plateau by tectonic processes, combined with the gradational action of water on the land, which has carved out the canyon itself. As a ranger at Grand Canyon National Park once suggested, "Not only is the knife (the river) cutting down through the slab of butter (the rocks), but the slab of butter is also being pushed up against the knife."

When we see the Grand Canyon today we must remember that this is only an early stage in its evolution. In time the water will remove more and more of the uplifted land, the shape of the canyon will change, and perhaps additional uplifting will occur as well. In some regions, there is a seemingly endless cycle of uplift and wearing down, of elevation and leveling. Elsewhere the tectonic forces have long remained dormant, the leveling process has dominated, and gradation is almost complete.

THE DEEP STRUCTURE OF THE EARTH

Scientists still know relatively little about the interior of the earth. Increased knowledge of its structure, composition, and the processes going on within will help answer questions about crustal motion, potential earthquakes, the formation of mineral deposits, and the origins of the continents and of the earth itself.

The earth has a radius of about 6400 kilometers (4000 mi), but scientists have been able to penetrate, and examine directly, only its thin outer skin. Through direct means such as mining and drilling, we have gained a very limited knowledge of the earth's interior. The lure of gold has taken prospectors to a depth of 3.2 kilometers (2 mi) in South African mines, and drilling for oil has penetrated to more than three times that distance. These explorations have been helpful in providing knowledge about the earth's uppermost layers, but they are really only scratches in the surface of the earth.

Most of what we have learned about the interior structure and composition of the earth has been deduced through indirect means. The most important tool that scientists have used to gain such indirect knowledge is the behavior of vibratory earthquake waves and other shock waves (usually generated by man-made explosions) as they pass through the earth. Such vibratory (seismic) waves can be recorded by an instrument called a **seismograph.**

There are three different types of seismic waves, which travel at various speeds in materials of different densities and states. These are usually labeled the 'P' (primary) waves, which travel fastest and arrive first at the seismograph recording a quake; the 'S' (secondary) waves, which travel more slowly; and the 'L' (longitudinal) waves, which travel along the surface and cause the damage associated with severe quakes. Repeated patterns of the 'P' and 'S' waves on the seismograph suggest that these waves are refracted, or bent, as they meet marked changes of density in the materials within the earth's interior. Such information, supplemented by studies of the earth's magnetism and gravitational pull, suggests a series of layers, or zones, in the earth's internal structure. These zones, from the innermost to the surface, are known as the core, the mantle, and the crust (Fig. 11.6).

THE EARTH'S CORE

The central core of the earth, with a radius of about 3360 kilometers (2100 mi) is believed to be composed primarily of iron and nickel. The core material is under enormous pressure, several million times the atmospheric pressure at sea level. Because the outer core, some 2400 kilometers (1500 mi) thick, screens out the seismic 'S' waves that will not travel through liquids, the outer core is assumed to be molten in spite of the very high pressure that it must be under. This seismic behavior could only occur if outer core temperatures are also extremely high, perhaps as much as 2200°C (4000°F).

However, the innermost core of the earth appears to be solid. Scientists explain the solid state of the inner core in this way: The melting point of a material depends not only on temperature but on pressure as well. The pressure on this innermost part of the earth is so great that the inner core remains solid; that is, its melting point has been raised to a temperature above even the high temperatures found there. The outer core, on the other hand, though its temperatures are lower, also is under less pressure and can exist in a molten state.

Scientists have deduced that the density of the earth's core is about 12.5 grams per cubic centimeter. This density would balance the much lower density of the earth's mantle (3.3 to 5.5) and crust (2.8) and provide an explanation for the earth's overall density of 5.5. This high density of the earth's core is one reason that scientists believe that iron and nickel are its primary components.

THE EARTH'S MANTLE

The mantle is about 2885 kilometers (1800 mi) thick and constitutes about 80 percent of the earth's total volume. Earthquake waves that pass through the mantle indicate that this part of the earth's interior is for the most part a rigid, dense solid, in contrast to the molten outer core that lies beneath it. Scientists are unsure as to the exact appearance, consistency, density, or temperature of the mantle. They do agree that the most common mineral of the mantle is probably olivine, an iron magnesium silicate. Small crystals of this mineral are common in lavas erupted in oceanic areas, such as Hawaii.

Despite its overall solid character, the mantle contains layers, or zones, of differing strength and rigidity. The uppermost layer of the mantle combines with the crust to form the **lithosphere,** the rigid outer portion of the earth that earth scientists agree is divided into individual units or plates. The term *lithosphere* has traditionally been used to describe the entire solid earth (see p. 3). In recent years, however, earth scientists have used the term in this more precise way to describe the material

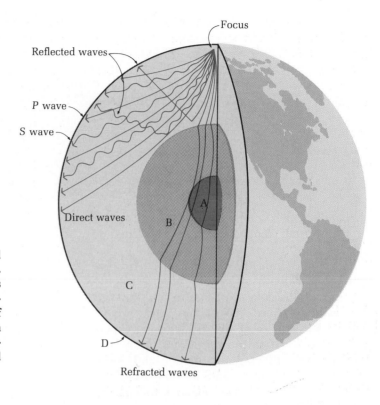

FIGURE 11.6 The earth's internal structure as revealed by seismic waves. (a) The presence of discontinuities is revealed by the refraction of 'P' (primary) seismic waves and the inability of 'S' (secondary) waves to pass through earth's liquid outer core. (b) Cross section through the earth's structural zones.

A. Inner core
B. Outer core
C. Mantle
D. Crust

(a)

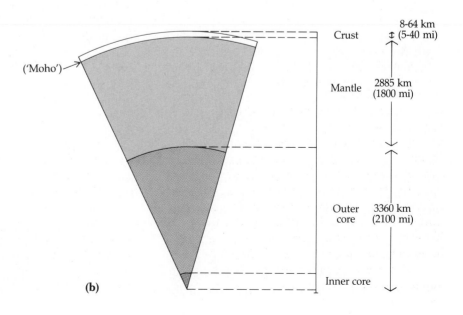

(b)

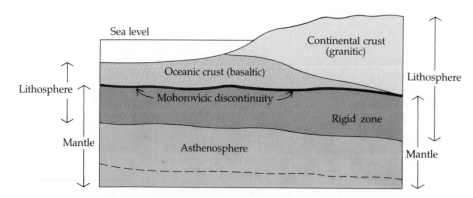

FIGURE 11.7 Two distinctive types of crust, oceanic and continental, lie over the Mohorovicic discontinuity (Moho). These, together with the upper part of the mantle above the asthenosphere, form the lithosphere.

of the crustal plates and upper rigid mantle (Fig. 11.7).

Located immediately beneath the lithosphere, at a depth of 64–240 kilometers (40–150 mi), is a thick layer of plastic mantle called the **asthenosphere** (Greek: *asthenos*, weak). The plastic quality of the asthenosphere permits its material to move both vertically and horizontally, dragging the lighter and more rigid lithospheric plates with it. Many earth scientists now believe that the major source of energy for tectonic forces comes from movement within the asthenosphere produced by thermal convection currents originating deep within the mantle.

THE MOHOROVICIC DISCONTINUITY

The upper level of the mantle at its interface with the crust appears to be marked by a significant change of density, indicated by an abrupt increase in the velocity of earthquake waves at this junction. Scientists have labeled this zone as the **Mohorovicic** discontinuity, or **Moho,** for short, after the Yugoslavian geophysicist who first detected it in 1909. The Moho does not lie at a constant depth around the earth. In fact, it tends to be the mirror image of surface topography, being deepest under highland areas and rising to within 8 kilometers (5 mi) of the ocean floor (Fig. 11.7). In the early 1960s, the United States sponsored Project Mohole, which was an attempt to drill through the earth's crust to the Moho and beyond, to obtain a sample of the material of the

mantle. Only experimental drilling in the ocean floor, where the crust is thinnest, had been done when, in 1966, federal funds were cut off and the project had to be abandoned. More recently, the Soviet Union has drilled about 13 kilometers (8 mi) deep in the Kola peninsula near Finland in attempts to penetrate the earth's crust and recover samples of the mantle.

THE EARTH'S CRUST

The only portion of the earth that earth scientists have direct knowledge of is the crust, the outer 1 percent of its radius. Not only are we in daily contact with the surface of this layer, but we have also been able to penetrate and sample it to depths of several kilometers in different parts of the world. The earth's crust is the outer portion of the lithosphere and is of primary importance in understanding the earth's landforms. The earth's deep interior (the core and mantle) are of concern in physical geography only in so far as they are responsible for, and can help to explain, changes in the lithosphere. It is the crust on which we live and that forms the ocean floors and continents. It is the rocks and debris of the crust from which soils are formed and which we penetrate in search of mineral wealth.

The earth's crust is less dense than either the core or the mantle. It is much thinner, as well, a relatively thin veneer that has sometimes been compared to an eggshell. It varies in thickness from 8 kilometers (5 mi) in the ocean basins to as much as 64 kilometers (40 mi) under some mountain systems, the average thickness on the continents being about 32–40 kilometers (20–25 mi).

The earth's crust is thought to be relatively rigid and brittle in comparison with the mantle. It responds to stresses by fracturing, wrinkling, and being elevated or depressed into domes and basins. Two layers can be distinguished in the earth's crust. The lower layer is **oceanic crust** composed of basaltic rock, which constitutes the ocean floors all around the earth. It is more basic chemically than the upper layer exposed on the continents, is darker, and has a higher density (3.0). This oceanic crustal layer of basic rock used to be called **sima** because its most common minerals are compounds of silica (Si) and Magnesium (Ma). Basalt itself also appears in great lava outflows on all of the continents.

The upper layer, which is **continental crust** and used to be called **sial** (Si for silica, Al for aluminum), is the material of the continents. It is less dense (2.8), more acidic chemically, and lighter in color than the oceanic crust. Although every known rock type of every geologic age may be found on the continental masses, the density of the continental crust averages out to the density of granite, a common rock on all continents. Thus the continental crust is regarded as granitic, in contrast to the basaltic oceanic crust (Fig. 11.7).

THE COMPOSITION OF THE EARTH'S CRUST

The tectonic and gradational processes that work to create the variety of landforms over the earth's surface produce vastly different results from place to place. The earth's crust is composed of a variety of rocks and minerals that respond in different ways and at different rates to the earth-shaping processes. Therefore the physical geographer should have an awareness of the different types of rocks and their primary characteristics, especially their responses to the tectonic and gradational processes.

ROCKS AND MINERALS

A rock is an aggregate (a whole made up of parts) of mineral particles. Each mineral in a rock remains separate and retains its own distinctive properties, which determine the properties of the rock itself. Though there are usually several different minerals in a rock, as in the case of both granite and basalt, a few rocks, such as limestone or quartzite, may be composed totally of particles of a single mineral. Rocks are the materials of the lithosphere. Rocks are lifted, pushed down, and deformed by the tectonic processes; they are weathered and eroded by the gradational forces, to be deposited as sediment elsewhere.

A solid rock layer is called **bedrock.** Above the bedrock is usually a layer of decomposed rock called **regolith.** Above this regolith may be a soil. A trench cut into the earth's surface does not always reveal these three layers. On a mountain slope, for example, running water may remove weathered material as fast as it forms, so that the bedrock is left exposed. Such exposed bedrock is called an **outcrop.**

MINERAL CLASSIFICATION

The most common elements of the earth's crust (and so of the minerals and accordingly the rocks that make up the crust) are oxygen and silicon, followed by aluminum and iron, and the bases: calcium, sodium, potassium, and magnesium. As you can see in Table 11.2, a mere eight chemical elements, out of the more than 100 known, account for almost 99 percent of the weight of the earth's crust. The most common minerals are combinations of these eight elements.

Minerals are naturally occurring inorganic substances. They are well-defined combinations of atomic elements, and each can be characterized by its unique chemical formula. Each mineral has other distinctive qualities as well: a particular color, luster, hardness, tendency to fracture, and specific gravity. Minerals are usually crystalline in nature, although this may be evident only when they are viewed

TABLE 11.2   Most Common Elements in the Earth's Crust

ELEMENT	PERCENTAGE OF THE EARTH'S CRUST BY WEIGHT
Oxygen (O)	46.60
Silicon (Si)	27.72
Aluminum (Al)	8.13
Iron (Fe)	5.00
Calcium (Ca)	3.63
Sodium (Na)	2.83
Potassium (K)	2.70
Magnesium (Mg)	2.09
Total	98.70

Source: J. Green, "Geochemical Table of the Elements for 1953," *Bulletin of the Geological Society of America* 64 (1953).

through a microscope. Thus, in addition to other uniform characteristics, mineral crystals have consistent geometric forms that express their atomic structure (Fig. 11.8). The atomic elements composing each mineral are held together by electrical bonds; to be stable, each mineral must have a balance between positive and negative charges. Accordingly, minerals are lattice-like structures held together either by covalent atomic bonding or by the electrostatic attraction between oppositely charged ions of the elements included (for example: Na+ and Cl− combine to produce common salt). Bonding affects the breakdown of minerals, and thereby of rocks. Minerals whose internal bonds are weakest are most easily altered. Ions may leave or be traded within their struc-

ture, producing physical changes. These characteristics of bonding are the chemical basis of weathering.

There are many discrete families of minerals, because of the ease of certain elements to combine with a variety of others. The most active of these elements are silicon, oxygen, and carbon. Consequently, the most common mineral groupings are the silicates, oxides, and carbonates. The **silicates** form by far the largest and most important group, constituting 92 percent of the earth's crust. They are created by the cooling of molten magma (a melt containing all the elements from which minerals form), which causes the crystallization of certain minerals at successively lower temperatures. All of these are compounds of oxygen and silica and of one or more metals and/or bases. Olivine is one of the first silicate minerals to crystallize (at high temperatures) and quartz is one of the last (at relatively low temperatures). This relationship resembles their relative stability in a rock. In the weathering of granite, which may include a variety of silicates, olivine is one of the first minerals to decompose and quartz one of the last.

The **oxides** do not form masses of rock. Oxides formed by crystallization occur in veins, sometimes large veins (one example is magnetite, an iron ore). The more common oxides are actually the product of weathering: Oxygen that combines with other substances, such as iron or aluminum, is introduced by water entering the structure of an iron- or aluminum-

(a)

(b)

FIGURE 11.8 The geometric arrangement of the atoms composing a mineral determines its crystal form. (a) Feldspar crystal. (b) Quartz crystals.

bearing mineral. Oxides, such as goethite (iron rust) and some hematite, are a product of mineral alteration.

The **carbonates** are of both organic and inorganic origin. Carbon's ability to form complex compounds makes this element an important building block in nature. In combination with oxygen and calcium, carbon produces calcite, the mineral composing limestone, one of the more common rock types. Calcite added to magnesium produces the mineral dolomite, which itself forms rock masses. The carbon of limestone and dolomite is frequently derived from organic material in the form of microscopic marine organisms.

The only mineral groups in which oxygen is not an important constituent are halides and sulfides. In the halides, which may form rock masses, chlorine plays the role usually played by oxygen, combining with the base, sodium, to form halite (common salt). In the sulfides, which occur only as veins, sulfur acts like oxygen, combining with iron, for example, to form pyrite ("fool's gold"). Many of the earth's deep metallic mineral deposits are sulfide ores, whereas those near the surface have been changed to oxides by proximity to atmospheric oxygen.

The only remaining group of minerals are those consisting of single elements uncombined with any others. These are rare and valuable and include gold, silver, copper, sulfur, and carbon (in the form of graphite and diamonds). Most metals, however, occur as oxides, carbonates, or sulfides—ores that must be refined at considerable cost.

THE CLASSIFICATION OF ROCKS

Although the number of minerals making up most of the rocks of the lithosphere are limited, they are combined in so many different ways that the variety of rock types is enormous. Nevertheless all rocks can be categorized as one of three major types, based on their origin. These rock types are igneous, sedimentary, and metamorphic.

IGNEOUS ROCKS

Igneous rocks are formed when molten rock-forming material cools and solidifies. Below the earth's surface this melt is called **magma.** The igneous material with which we are most familiar is **lava,** the molten material spewed forth by volcanoes at temperatures of as much as 1090°C (2000°F). Lava is the surface form of magma.

Molten material that emerges at the earth's surface and solidifies is called **extrusive** igneous rock, or volcanic rock (after Vulcan, the Roman god of fire). If the rising magma does not break through to the surface but solidifies *within* the rocks below it, such rock is known as **intrusive** igneous rock. When intrusive igneous masses are large and deep, they are sometimes referred to as **plutonic** rocks (after Pluto, the Roman god of the underworld).

Different igneous rocks have little in common except their method of formation: the crystallization and solidification from molten material. Igneous rocks vary in chemical composition, tendency to fracture, texture, crystalline structure, and the presence or absence of layering. Nevertheless they may be grouped or classified in terms of their crystal size or texture as well as their chemical composition.

Extrusive rocks, and some intrusive rocks that have pushed close to the surface, undergo rapid cooling and solidification. This forms fine-grained igneous rock, such as basalt or rhyolite (Fig. 11.9). Only small crystals are produced under conditions of rapid cooling, because there is little time for crystal growth prior to solidification. In some instances cooling is so rapid that the resulting rock has a glassy texture, as in obsidian (volcanic glass).

Plutonic rocks cool far more slowly because the surrounding masses of rock retard the loss of heat from the molten magma. Slow cooling allows more time for larger crystal formation prior to solidification. Rocks formed in this manner are coarse-grained, with crystals often as long as 3 centimeters or more. Granite and gabbro illustrate this coarse texture (Fig. 11.10).

The chemical composition of igneous rocks varies from acidic (rich in light minerals, especially silica) to basic (low in silica, rich in heavy minerals, such as compounds of iron and magnesium—the ferromagnesium minerals). Granite, an acidic, coarse-grained, plutonic rock, has the same chemical and mineral composition as rhyolite, a fine-grained extrusive rock. Likewise, gabbro is the coarse-grained, plutonic version of the fine-grained basalt, both having been formed from the same chemically basic magma.

Many igneous rocks are jointed. **Joints** are cracks within a rock structure, which develop as an igneous rock shrinks in volume during its formation. Basalt is famous for its hexagonal columnar jointing. Devil's Postpile, California, and Devil's Tower, Wyoming, are formed of huge hexagonally jointed basalt columns (Fig. 11.11).

SEDIMENTARY ROCKS As their name implies, **sedimentary rocks** are derived from accumulated sedimentary material that is transformed into rock (lithified) by compaction and/or cementation. After having been deposited in layers, the sediment is compacted by the pressure of the material above it, expelling water and reducing pore space. Cementation also occurs when silicon dioxide, calcium carbonate, or iron oxide accumulates in the remaining pores between the particles of sediment. Together, the processes of compaction and cementation transform the sediment into a solid, coherent layer of rock. The sedimentary materials (cobbles, pebbles, sand, silt, or clay) are debris particles eroded from any previously existing rock, transported, and deposited on land, a lake bottom, or the ocean floor. Rocks formed from such rock debris are called **clastic** rocks (Latin: *clastus*, broken). By far the most common clastic rocks are those formed on the ocean floors from slowly sinking and settling continental and marine sediments.

Common clastic rocks are conglomerate, sandstone, siltstone, and shale (Fig. 11.12). Conglomerate is a solid mass of cemented boulders, cobbles, pebbles, or gravel, often with sand filling in the spaces between large particles. It is a hard rock, relatively resistant to

weathering. Sandstone is formed by the cementation of fine grains of quartz. It is usually granular, porous, hard, and also resistant to weathering. However, its qualities are largely determined by the cementing material. When sandstone is cemented by substances other than silica (such as calcite or iron oxide), it is more easily weathered. Siltstone is similar to sandstone, though much finer-textured, being formed of the much smaller particles of silt. Shale is produced from the compaction of clay. The resulting stone, finely bedded, is smooth-textured and nonporous. It is also brittle, easily cracked, broken, or flaked. These clastic sedimentaries may be further classified as marine or terrestrial sedimentaries, depending on their origin. Thus marine sandstones have been formed originally in coastal zones, while terrestrial sandstones have originated in desert conditions on land.

Sedimentary rocks may also be formed from the remains of organisms, both plants and animals. Such rocks are **organic** sedimentaries. Coal, for example, was formed through the accumulation and compaction of decayed vegetation in acid, swampy environments where water-saturated ground prevents oxidation. The initial transformation of such material produces peat, which, when subjected to burial and further compaction, is lithified to produce coal. Unlike other types of rock deposits, the world's greatest coal deposits originated during a particular geologic time, some 300 million years ago, known as the Pennsylvanian Period (called the Carboniferous Period in Europe—see Table 11.1).

Other organic sedimentary rocks have been formed from organisms growing in lakes and seas. As the skeletal remains of shellfish, corals, and microscopic floating organisms drifted to the bottom of such water bodies, they became cemented and compacted together to form *shell and coral limestones*. Frequently, however, these calcium carbonate ($CaCO_3$) materials were dissolved into the water and later became precipitated or separated out from the water. In this way *secondary limestones* were formed, including chalk as seen in the White

FIGURE 11.9 Fine-grained igneous rocks. (S. Brazier)

(a) basalt

(b) rhyolite

(c) obsidian

FIGURE 11.10 Coarse-grained igneous rocks. (S. Brazier)

(a) granite

(b) gabbro

FIGURE 11.11 Hexagonal columnar jointing in igneous rocks. (R. Gabler)

(a) Devil's Postpile National Monument, California

(b) Devil's Tower National Monument, Wyoming

Cliffs of Dover on the English Channel (Fig. 11.13a,b).

Limestone, therefore, may vary from a complex of visible shells or skeletal material of various sizes and types to a smooth-textured, crystalline mass. Where magnesium is an important constituent along with calcium carbonate, the rock is called *dolomite* (Fig. 11.13c).

Other mineral salts that have been precipitated out from evaporating sea water or lake basins have formed a variety of sedimentary deposits often useful to man. These include flint or chert, gypsum (or alabaster in its purest form), halite (common salt), and borates, which are important in hundreds of products from fertilizer, fiber-glass, and pharmaceuticals to detergents and photographic chemicals.

Because the nature of the original sedimentary materials has varied over time, sedimentary rocks display distinctive layering or **stratification.** For example, a period of time during which coarse-textured sands have been deposited may be followed by a period during which fine clays have been laid down. At other times organic deposits may predominate. Each different type of sediment becomes a different rock type, resulting in differentiated **strata** or beds. The **bedding planes,** or boundaries between the differing rock types, indicate changes in the nature of the deposits but no

FIGURE 11.12 Clastic sedimentary rocks. (S. Brazier)

(a) conglomerate

(b) sandstone

(c) shale

FIGURE 11.13   Organic sedimentary rocks. (S. Brazier)

(a) marine limestone

(b) The White Cliffs of Dover are made of chalk

(c) dolomite.

real break in the sequence of deposition (Fig. 11.14a). Where a marked mismatch occurs between beds, the surface of contact between the rocks is called an **unconformity** (Fig. 11.14b). This indicates an interruption in the sequence of deposition during which erosion has removed earlier deposits before deposition has been renewed.

Within some sedimentary rocks, especially sandstones and shales, even finer "micro-bedding" may occur (Fig. 11.14c). Thus we may find **lamination planes,** parallel to the major bedding planes, in laminated sandstones or shales (Latin: *lamina,* thin leaf). Lamination planes are evidence of minute variations in deposition, for example, with each tide on a shore or during flood periods of a river's flow.

Another form of micro-bedding is called **cross-bedding,** characterized by a pattern of zig-zags at an angle with the main bedding, often reflecting shifts of direction by winds over desert dunes (Fig. 11.14d).

Many sedimentary rocks are *jointed,* since they cracked during the process of drying out after their initial formation in water bodies. (Fig. 11.14e,f). Limestone, especially, may be massively jointed, and the impressive slabs of rock at Arches National Monument, Utah, owe their form to vertical joints in great beds of sandstone.

Structures such as bedding planes, lamination planes, and joints are important in the formation of different physical landscapes, since these structures are weak points in the

FIGURE 11.14 Structures in sedimentary rocks. (S. Brazier)

(a) Bedding plane divides sandstones (above) from shale (below) at bottom of the Grand Canyon.

(b) an unconformity at the bottom of the Grand Canyon

(c) laminated shales

(d) cross-bedding in sandstones

(e) joints in the eroded limestones of northern England

(f) joints in sandstone of Arches National Monument, Utah

rock, where weathering and erosion can attack. Joints can allow water to penetrate deeply into some rock masses, causing them to be removed at a faster rate than more solid rock masses nearby.

METAMORPHIC ROCKS The word **metamorphic** means "changed," and that is just what metamorphic rocks are. Enormous heat and pressure deep in the earth's crust, often associated with tectonic activity, can totally reconstitute rock, changing it into a new product. Usually the resulting rock is harder and more compact, has a crystalline structure, and is more resistant to weathering than before.

Metamorphism occurs most commonly where crustal materials are forced down to lower levels by tectonic processes or where molten magma is rising through the crust, giving off heat and also solutions and gases that can modify the rock already present. The major effect of metamorphism is to either wholly or partially fuse or melt the rock being affected, so that it can be deformed, or *flow* slightly but without becoming molten magma. This process causes mineral crystals to recrystallize with an orientation that reflects the direction of flow. Such metamorphism produces rocks whose minerals are segregated in wavy bands, the effect being known as **foliation** (Fig. 11.15a,b,c). Where the banding is very fine, the individual minerals have a flattened, "platy" structure; the rocks tend to flake along these bands. Such rocks are called **schists.**

FIGURE 11.15 Metamorphic rocks. (S. Brazier)

(a) schist

(b) gneiss

(c) slate

(d) marble

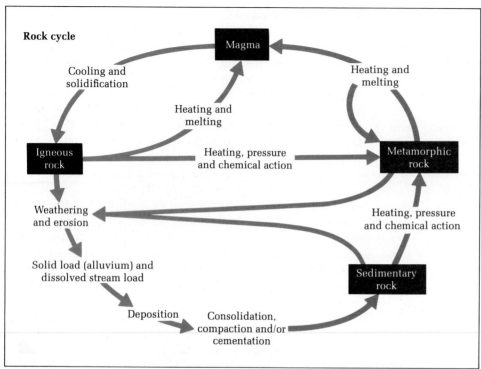

Rock cycle

Magma

Cooling and
solidification

Heating and
melting

Igneous
rock

Heating and
melting

Heating, pressure
and chemical action

Metamorphic
rock

Heating and
melting

Weathering
and erosion

Heating, pressure
and chemical action

Solid load (alluvium) and
dissolved stream load

Sedimentary
rock

Deposition

Consolidation,
compaction and/or
cementation

**FIGURE 11.16   The Rock Cycle, "No trace of a beginning, no prospect of an end,"
stated by James Hutton (1726–1797).**

Where the bands are broad, the rock is extremely sound and is known as **gneiss** (pronounced "nice"). Coarse-grained rocks such as granite generally recrystallize as gneiss, whereas fine-grained rocks may produce schists. Some shale produces a more massive metamorphic rock known as **slate,** which exhibits a tendency to break apart or **cleave** along flat surfaces.

Rocks originally composed of one dominant mineral are not foliated by metamorphism. Limestone is merely reconstituted into much denser **marble;** the impurities in the limestone produce beautiful variegated colors (Fig. 11.15d). Silica-rich clastic rocks such as sandstone are fused in solid sheets of quartz, known as **quartzite.** Quartzite is brittle but almost inert chemically. Thus it is virtually immune to chemical weathering and commonly forms cliffs and rugged mountain peaks, such as those in the Canadian Rockies.

THE ROCK CYCLE   Like landforms, rocks do not remain in their original form indefinitely but instead are always in the process of transformation. There is, in fact, a cycle of rock formation that has no obvious beginning and no predictable end (Fig. 11.16). When magma is cooled, igneous rocks are formed. Igneous rocks can return to a molten condition (magma) through the addition of heat, or they can be changed into metamorphic rock through the application of heat, pressure, and/or chemical action, or their weathered particles may form the basis of sedimentary rocks. Sedimentary rocks can be formed from the weathered particles of either igneous or metamorphic rocks. Finally, metamorphic rocks can be created out of either igneous or sedimentary rock. In addition, metamorphic rocks can be heated sufficiently to become magma.

QUESTIONS FOR DISCUSSION AND REVIEW

1. Define four major land surface types using the criteria of slope and local relief.

2. Describe some of the ways in which different kinds of plains are classified. How are plains formed?

3. Study Table 11.1. In what ways do orogenies relate to major segments of earth history and important events in the evolution of life?

4. How might mountains be involved in the origin of hills?

5. What is the difference between a plateau and a mesa? Where are two major plateau regions in the United States?

6. How has land configuration affected population distribution in your area? (Compare Plate 19 with Plate 2.)

7. How do human effects on landforms compare with the effects of natural forces? What forms does human intervention on the land surface take?

8. How do open-pit and strip mining affect the natural environment? What are some ways in which their effects can be countered?

9. What are some of the currently proposed solutions to the problem of solid waste disposal? Describe some of the advantages and disadvantages of each.

10. What are the differences between tectonic and gradational processes?

11. Identify the major zones of the earth from the center to the surface. How do these zones differ from one another? What is the special significance of the asthenosphere?

12. What is the Mohorovicic discontinuity?

13. Define the difference between continental crust, oceanic crust, lithosphere, and asthenosphere.

14. List the most common elements in the earth's crust. What is a mineral? What is a rock?

15. Describe the three major classifications of rock and the means by which they are formed, and give an example of each.

12

TECTONIC PROCESSES

PLATE TECTONICS: A UNIFYING THEORY

Scientists in all disciplines are constantly searching for broad explanations that shed light on the detailed facts, recurring patterns, and interrelated processes that they observe and analyze. The theories of evolution and natural selection proposed by Charles Darwin and his scientific colleagues in the midnineteenth century comprised just such an explanation. Those theories ultimately revised the thinking of scientists in many fields and still guide the research of biologists and other life scientists today.

But what are the underlying explanations for the earth sciences? We have already mentioned the general acceptance of uniformitarianism—the idea that the earth as we know it today is only the current product of orderly processes that have been operating in a similar fashion for millions of years. Are there similar hypotheses that explain how and why these processes work? Is there one broad theory that can help to explain such diverse subjects as the growth of continents, the movement of bedrock along the San Andreas fault, the location

of great mountain ranges, the pattern of temperatures in ocean basin rock, and the violent eruption of Mount St. Helens? The answer is the theory of **plate tectonics.** This theory has created a major impact on the earth sciences today just as the theory of evolution affected the life sciences over a century ago.

CONTINENTAL DRIFT

Most of us at one time or another have noted on a globe or map of the world that South America and Africa look as if they fit together. In fact, if a globe were made into a spherical jigsaw puzzle, several of the widely separated large landmasses could be made to fit without large gaps or overlaps (Fig. 12.1). Is there a scientific explanation?

In the early twentieth century Alfred Wegener, a German meteorologist, hypothesized that all the continents had actually once been connected in one, or possibly two, large landmasses. Then, for some reason, these supercontinents broke apart, and their fragments (the present continents) moved to their existing positions. Evidence for Wegener's belief existed in the fact that early plant fossils found on the

---

The Mount St. Helens eruption of May 18, 1980. (J. Stewart Lowther, University of Puget Sound)

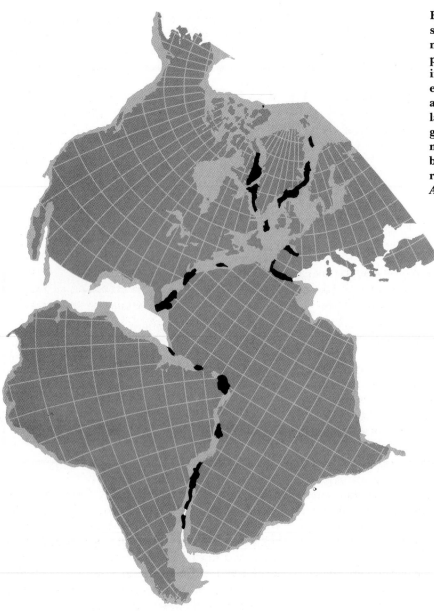

FIGURE 12.1 This figure shows the basis for Wegener's continental drift hypothesis. Note the manner in which the continental edges facing one another across the width of the Atlantic Ocean can be fitted together. (From "The Confirmation of Continental Drift" by Patric M. Hurley, Copyright © 1968 by *Scientific American*, Inc.)

different continents were related in ways that could not be attributed to chance. The continents must once have been joined so as to allow the spread of those early land plants. Furthermore, evidence of climatic changes (for instance, glacial deposits in the Sahara Desert and tropical fossils in Antarctica) could be explained best by the movement of the large landmasses from one climatic zone to another.

The reaction of the scientific community to Wegener's proposal was one ranging from skepticism to outright ridicule. A major objection to his hypothesis, which Wegener admitted, was that he could provide no explanation for the breakup, nor could he provide an explanation for the energy that would have been needed to propel huge landmasses through the rigid crust and across vast oceans.

Wegener did propose that perhaps the crust was not so rigid as had been believed. If, instead, it were plastic under certain conditions, and if the lighter material of the continents "floated" on the heavier rocks in which they are embedded, then the movement of the continents away from each other might be possible. Nevertheless, Wegener was still unable to find a believable propelling mechanism for continental drift.

SUPPORTING EVIDENCE

It was over a half a century before Wegener's seemingly fanciful proposal concerning the movement of continents received serious consideration by a majority of earth scientists. And in that time World War II had become history along with the rapid advances in science and technology that often accompany an all-out war effort. Research in oceanography and geophysics in particular had been aided by the development of sonar and radioactive dating, by the improvement of magnetometers, and by the coming of the computer age. Earth scientists were flooded with new evidence that portions of the lithosphere (including the continents) had indeed been on the move.

As one example, scientists were originally unable to explain the unusual orientation of magnetic fields in many igneous rocks that had cooled from the molten state millions of years before. Minute iron minerals within magma orient themselves like tiny compass needles pointing to the magnetic pole. This orientation is locked into the rock as it solidifies and is known as *paleomagnetism*. Rocks of different ages, or on different continents, show magnetic orientations at a variety of angles to the earth's magnetic field as it is today. At first it was assumed that the earth's magnetic poles had wandered, but this could not explain varying paleomagnetism in rocks of the same age.

A far better explanation emerged when the poles were accepted close to their present locations and the continents were then moved by computer simulation backward in time so that the magnetic orientation of the rocks was made to coincide with the earth's field during past periods of earth history. The ancient magnetism indicated an almost perfect fit of the continental jigsaw puzzle some 200 million years ago.

Supporting evidence for crustal movement came from a variety of additional sources. New fossil discoveries indicated that certain reptiles and land plants, earlier known to have been found in Australia, India, South Africa, and South America, were to be found in Antarctica as well. The plants or animals found in each instance were so similar and specialized that they could not have developed without land bridges between their now separate locations. Continental fit was discovered to be even better a few hundred meters below sea level along the continental shelves. Mountain ranges on the different continents were carefully matched by radioactive-dating techniques and were shown to be continuous when the continents were joined. Even climatology continued its contribution. Evidence of ancient glaciation in Brazil and South Africa or of tropical forest climate (coal measures) in Alaska and Antarctica could only be explained by crustal movement.

CONVECTION AS THE MECHANISM

The real key to a modern hypothesis was to be found on the ocean floor. First, mapping of the ocean basins revealed a system of midocean ridges that had configurations remarkably similar to the outlines of the continents. Second, it was discovered that parallel bands of rocks with similar magnetic properties extended one after the other in identical fashion on both sides of the ridges in the Atlantic and Pacific oceans. Third, scientists made the surprising discovery that although some continental rocks were nearly 4 billion years old, rocks on the ocean floor are all geologically young—they have been in existence less than 200 million years. Fourth, the oldest ocean floor rocks are nearest the continents and the youngest ones are nearest the midocean ridges. Finally, temperatures of rocks on the ocean floor vary significantly;

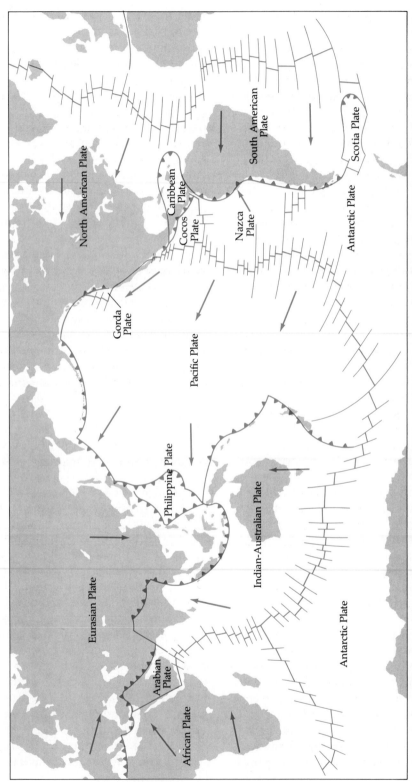

FIGURE 12.2   The major tectonic plates and their general direction of movement. Most tectonic activity occurs along the plate boundaries where plates separate, collide, or slide past one another.

they are highest near the ridges and become progressively lower farther away.

Only one logical explanation emerged to fit all of the new evidence. It became apparent that hot new oceanic crust is being formed at the midocean ridges and older, cooler oceanic crust is disappearing near the margins of the ocean basins. Earth scientists labeled this phenomenon **sea floor spreading.** Most now believe that the emergence of this new oceanic rock is associated with the movement of great sections or plates of the lithosphere away from the midocean ridges. The plates move at an average rate of a few centimeters per year, above the plastic asthenosphere of the mantle. Ocean floors are not permanent, and the young age of ocean basin rock is the result of crustal renewal at the ridges and the movement of this crustal material as part of lithospheric plates toward crustal destruction at the margins of the continents.

Plate tectonics, the modern version of continental drift, also includes a plausible explanation of the mechanism involved in sea floor spreading and the movement of the lithosphere. The mechanism is **convection,** the transfer of heat to the earth's surface from deep within the mantle. New molten material rises toward the surface and is expelled at the midocean ridges as part of huge subcrustal convection cells. The cellular motion is continued as crustal material moves away from the ridges and is completed as older oceanic crust is consumed in the great trenches that often mark the boundaries where continent and ocean basin plates meet.

TECTONIC PLATE MOVEMENT

The theory of plate tectonics suggests that the lithosphere (the crust and the rigid upper mantle) consists of as many as 20 rigid plates (Fig. 12.2). All plates move as distinct units—in some places traveling away from each other (diverging), in other places sliding past each other (moving laterally), and elsewhere coming together (converging). Seven are considered major plates and are of continental or oceanic pro-

portions. Five are of minor size, although they have apparently maintained their own identity and direction of movement for some time. The remaining seven or eight are quite small and are all in the active zones marking the boundaries between major plates. Although the largest plate, the Pacific plate, is primarily oceanic, portions of continents are deeply imbedded in the surfaces of all other major plates and are carried along by plate movement.

Once set in motion the shifting of the larger tectonic plates relative to one another provides the explanation for many landform features on the earth's surface. This is of particular interest to the physical geographer because through tectonic plate theory it is now possible to understand the paleogeography of the planet and the worldwide distributional patterns and close relationship among such diverse phenomena as earthquakes, volcanic activity, zones of crustal movement, and major landforms. Let us briefly examine the three ways in which tectonic plates relate to one another as a result of movement along their boundaries.

PLATE DIVERGENCE Tectonic plates diverge or pull apart along the midocean ridges as a direct result of sea floor spreading (Fig. 12.3). Tension-producing forces cause the oceanic crust to thin out and weaken. Shallow earthquakes are often associated with the crustal stretching, and magma wells up from the asthenosphere, forming new crustal ridges and ocean floor as the plates move away from each other. The formation of new crust in these areas gives the label *constructive plate margins* to these zones. In this way the Atlantic Ocean floor was formed as the South American and African continents were driven apart. "Oceanic" volcanoes like those of Iceland, the Azores, and Tristan da Cunha mark such boundaries.

Though most divergence zones are midocean ridges, they may also be associated with continents. The best example is the well-known rift valley system of East Africa, stretching from the Red Sea to Lake Malawi. The entire

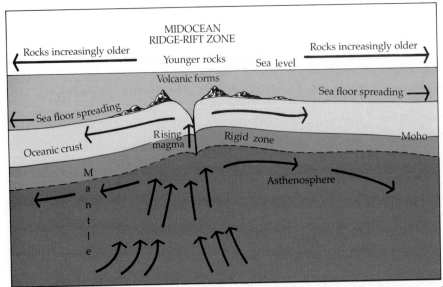

FIGURE 12.3   Diagram of plate divergence. There is formation of new crustal material that moves to the surface and pushes older rock progressively outwards from the separating plate boundary. This is a constructive plate boundary.

system is characterized by a series of crustal blocks, which have moved downward in respect to the crust on either side, with lakes frequently occupying the depressions. Measurable widening of the Red Sea suggests that it may indeed be a "proto-ocean" between Africa and the Arabian Peninsula just as was the young Atlantic between Africa and South America less than 200 million years ago.

PLATE CONVERGENCE   When tectonic plates converge there is a maximum of crustal activity. Despite the exceedingly slow rate of plate movement, which averages 2–5 centimeters (1–2 in) per year, the incredible energy involved causes the brittle crust to crumple as one plate overrides the other. The denser plate is forced deep below the surface in a process called **subduction.** This usually occurs where oceanic crust meets continental crust, and the denser oceanic crust descends below the lighter continental crust (Fig. 12.4). Such is the situation along the Pacific coast of South America where the oceanic Nazca plate subducts beneath the South American plate, or in Japan

where the Pacific plate dips under the Eurasian plate. Crustal rocks are being lost in these regions, known as *destructive plate margins*.

Deep oceanic trenches form where the crust is dragged downwards toward the mantle, as in the Peru-Chile trench and the Japanese trench. Frequently hundreds of meters of sediments, eroded from the adjacent landmasses, are carried into these trenches, later to form sedimentary rock. When the rock becomes squeezed and contorted between the colliding plates, it is heavily folded and metamorphosed. Many great mountain ranges, such as the Andes, have been formed by such processes at convergent plate margins.

A subducting plate is heated as it plunges downwards into the mantle. Its rocks are melted and the resultant hot magmas begin to migrate upwards along fissures and zones of weakness in the overriding plate. Where the magma reaches the surface, it forms a series of volcanic peaks in these same mountain areas, like the Cascades in Washington and Oregon or the Andean volcanoes. Later still, as the supply of magma increases, the volcanoes may

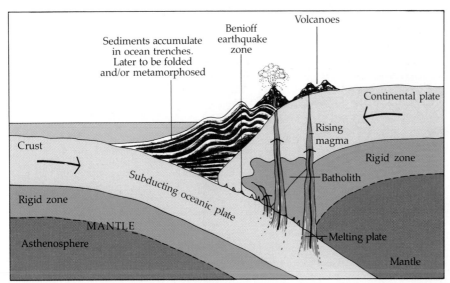

FIGURE 12.4 Diagram of subduction and plate convergence. Where the denser oceanic plate is overridden by the less dense continental plate, the denser plate is forced downward in a process called subduction. This is a destructive plate boundary, since crustal rocks are recycled into the earth's interior.

form major island chains separating the continents from the ocean trenches. The Aleutians, Japan, and the Marianas are all examples of island arcs near oceanic trenches that border the Pacific plate.

As the subducting plate grinds downward, enormous friction is produced, which explains the development of major earthquakes in these regions. These so-called "Benioff zones" are named after the seismologist Hugo Benioff, who first plotted the existence of earthquakes occurring at a steep angle along the descending edge of the subducting plate (Fig. 12.4).

Where two continental plates collide, massive folding and crustal block movement occur, rather than volcanic activity. This causes crustal thickening and again major mountain ranges may form. The Himalayas, the Tibetan Plateau, and other high Eurasian ranges were formed in this way, where the continental Indian plate slammed into Eurasia some 40 million years ago and the Alps were formed when the African plate collided with the Eurasian plate (Fig. 12.5).

Thus the zones of plate convergence mark the locations of the more spectacular landforms on our planet: huge mountain ranges, volcanoes, and ocean trenches. At last their locational patterns can be understood within the framework of tectonic plate theory.

LATERAL MOVEMENT A third type of plate contact occurs when plates neither pull apart nor converge but instead slide laterally past each other as they move in opposite directions. Such a plate boundary exists along the famous San Andreas fault in California (Fig. 12.6). In fact, the Baja California peninsula, and southern California, an area west of the fault, are actually a part of the Pacific plate. The land to the east of the fault, on the other hand, is part of the North American plate. In the area of the fault, the Pacific plate is moving laterally northwestward in relation to the North American plate at a rate of about 8 centimeters (3 in) a year. If this movement continues, in a few million years Los Angeles will eventually move to the position of San Francisco on its

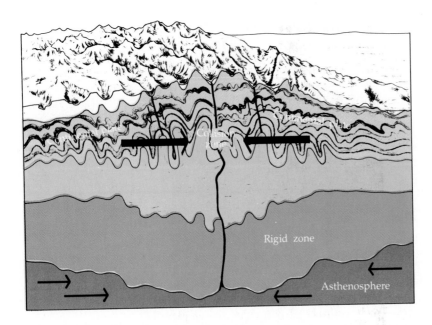

FIGURE 12.5 Diagram of plate collision. Where two continental plates collide, massive mountain building occurs as the crust is thickened. The Alps and Himalayas were formed when Africa and India slammed into Eurasia.

way to final disappearance in the Aleutian Trench. Margins along which the plates are sliding sideways are known as *passive plate margins,* since crustal material is neither being produced nor destroyed, but simply being displaced.

Another type of lateral plate movement on a somewhat smaller scale is found in areas of plate divergence. As the plates pull apart they usually do so along a series of minor cracks or fissures that tend to form at right angles to the major zone of plate contact. These cross-hatch plate boundaries along which lateral movement takes place are called **transform faults.** Such transform faults are common along midocean ridges, but examples can also be seen elsewhere, as in the border between the Pacific and Gorda plates (Fig. 12.6). They are probably caused by variable speeds of plate motion, causing stress and eventually movement of small portions of the plate. The most rapid plate spreading is on the East Pacific rise where plate motion is over 17 centimeters (5 in) per year.

GROWTH OF CONTINENTS

The origin of continents themselves remains a problem. It is clear that the individual conti-

nents tend to have a core area of very old igneous and metamorphic rocks. These cores are usually areas of relatively low relief, due to their lack of tectonic disturbances over an immense period of time. These ancient crystalline rock areas are called **continental shields.** The Canadian, Scandinavian, and Siberian Shields are outstanding examples. Around the peripheries of the exposed shields, flat-lying, younger sedimentary rocks indicate the continued presence of a stable and rigid mass beneath, as in the American Midwest, western Siberia, and much of Africa. Continents appear to grow outward by mountain building around the sediment-covered margins of the ancient shield areas. This process is clearly related to the plate tectonic concept, for the descent of lithosphere in an ocean trench around a continent generates new magma that produces plutons and volcanism on the continental edge. Simultaneously, the continental shelves tend to be buckled by the subduction process, and oceanic rocks are peeled off and piled against the continental margin. Except where two continents have collided to become one, tectonism is generally restricted to continental edges, where plutonic injections, surface volcanism, and the squeezing up of marine formations all add new material to the continental mass.

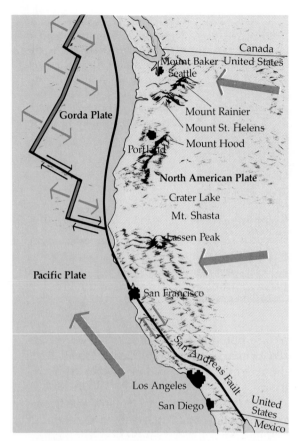

FIGURE 12.6 Diagram of lateral plate movement in western North America. The area to the west of the San Andreas Fault system is creeping northward in relation to the rest of western North America. The movement occasionally takes the form of sudden slippage along the plate boundary, producing major earthquakes.

THE MOLTEN TECTONIC PROCESSES

Volcanism refers to the rise of magma and its cooling above the surface. It also includes the extrusive rocks and landforms created from this surface activity. **Plutonism** refers to igneous action at depth below the earth's surface. It also includes the intrusive rocks and rock masses formed from the magma. Just as we differentiate between extrusive and intrusive igneous rocks, we also differentiate between volcanism and plutonism.

VOLCANOES

There is no spectacle in nature as awesome as an explosive volcanic eruption. The most violent eruptions are so cataclysmic that eyewitnesses may not survive to describe them.

Volcanic eruptions vary greatly in character. Consequently the volcanic landforms that result are extremely diverse. The variation in eruptive style and in resulting landforms is mainly the result of chemical differences in the magma feeding the eruption. Some magmas have cooled before eruption, so that heavy minerals have crystallized, while light minerals, those high in silica, remain dissolved. In such cases, a great deal of gas may already have separated from the magma, building up pressure in the magma chamber below the surface. The explosive release of these gases produces a violent eruption. If, on the other hand, the magma is very hot, crystal formation will not yet have begun, and the gases will still be dissolved in the magma itself. If such material leaks or is forced to the surface, the eruption is not explosive, although enormous amounts of highly fluid lava may be produced. Fluid layers become **pahoehoe** or ropy lavas, while the more viscous, pasty lavas are called **aa** or blocky lavas—both terms of Hawaiian origin (Fig. 12.7).

In addition to lava flows and gaseous materials, most volcanic eruptions hurl molten magma and solids of various sizes into the air. These pyroclastic materials (Greek: *pyros,* fire; *clastus,* broken) vary from huge volcanic "bombs" to cinders, ash, and fine volcanic dust. In highly explosive eruptions the volcanic dust may be hurled into the atmosphere to be carried by the jet stream around the world. It was volcanic dust from Krakatoa, west of Java, after an 1883 eruption, that caused temperature declines and spectacular sunsets as far away as London, England.

There are four basic types of volcanoes: composite cones, shield volcanoes, plug domes, and cinder cones. Many volcanoes consist of a mix of lava flows and pyroclastic materials, and therefore are known as **composite cones.** Most of the well-known volcanoes of the world are of

FIGURE 12.7 Two major types of lava on the island of Hawaii. On the right is the highly fluid lava, which forms pahoehoe or ropy lava. On the left is the more viscous pasty lava which forms aa or blocky lava. (R. Sager)

this type—Fujiyama in Japan, Kilimanjaro in East Africa, Cotopaxi in Ecuador, Vesuvius in Italy, Mount Rainier in Washington, and Mount Shasta in California (Fig. 12.8). Composite cones are composed of great volumes of both lava and ash. The ash provides steepness; the lava cements the structure together. Composite cones, also known as *stratovolcanoes* (consisting of stratified pyroclastic material and

lava), are potentially dangerous. They are composed primarily of andesite lava derived from acidic magma, which produces explosive eruptions.

On May 18, 1980, residents of the Pacific Northwest, and soon after people throughout the United States, were shocked by the realization that every age in history is a volcanic age. Mount St. Helens, a stratovolcano in southern

(a)

(b)

FIGURE 12.8 (a) Cotopaxi in Ecuador is a classic composite cone formed along a subduction boundary in the Andes. (S. Brazier) (b) Mt. Shasta, California, is a composite cone in the Cascade Range. Shastina is a secondary crater to the right of the main cone. (R. Sager)

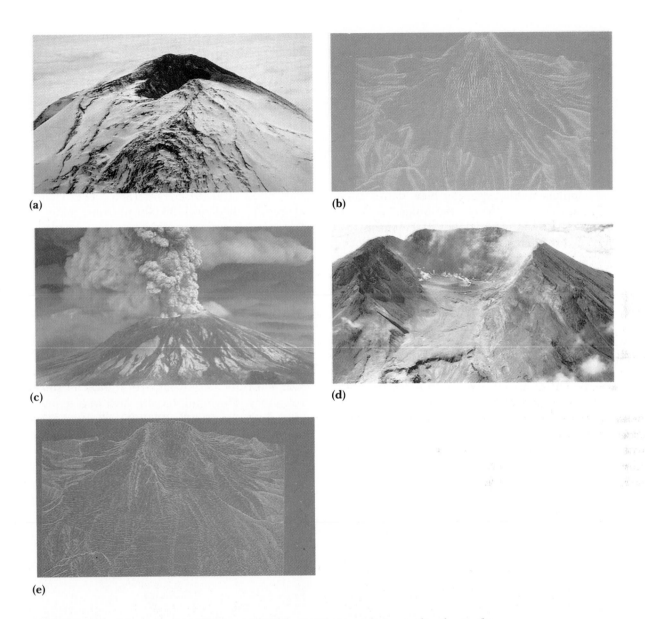

(a)

(b)

(c)

(d)

(e)

FIGURE 12.9   The eruption of Mount St. Helens. The summit 1 month prior to the
main eruption (a) showing where small eruptions had blown a small crater through
the ice- and snow-covered summit. (J. Stewart Lowther, University of Puget Sound) A
computer-generated digital terrain model on May 15, 1980 (b), just 3 days prior to
the main eruption. (C. D. Miller, U.S. Geological Survey) Mount St. Helens in violent
eruption (c), on May 18, 1980. (J. Stewart Lother, University of Puget Sound) Summit of the mountain after the eruption (d). Note that the north summit was completely
obliterated after the lateral blast. A mudflow remnant emerges from the crater while
gases still rise along the crater walls. (GeoPhoto Publishing Co., Alan L. Mayo, University of Colorado, Colorado Springs) A digital terrain model several days after the
main eruption (e) indicates the large section of the summit that was destroyed. (C. D.
Miller, U.S. Geological Survey)

Washington, which had been smoldering and venting steam and ash for several weeks, exploded with incredible force. A menacing bulge had been growing on the side of the mountain, and earth scientists warned of a possible major eruption, but no one could forecast the magnitude of the blast. Within minutes nearly 400 meters (1300 ft) of the mountain's north summit had disappeared into the sky and down the mountain side (Fig. 12.9).

Much of the explosion blew debris laterally outward from where the bulge had been. An incredible storm cloud of intensely hot steam, noxious gas, and volcanic ash, traveling at a speed of more than 320 kilometers (200 mi) per hour, obliterated forests, lakes, streams, and camping sites for a distance of nearly 32 kilometers (20 mi). Monstrous mudflows of ash and melted ice choked streams and valleys and added to the devastation by engulfing everything in their paths. Over 500 square kilometers of magnificent forest and recreation land were destroyed. Hundred of homes were buried or badly damaged; choking ash several centimeters thick covered nearby cities; untold numbers of wildlife were killed; and more than 60 human beings lost their lives in the eruption. It was a minor event in earth history, but a sharp reminder of the awesome power of natural forces.

Some of earth's worst natural disasters have occurred in the shadows of composite cones. Mount Vesuvius, Italy, killed over 20,000 people in the cities of Pompeii and Herculaneum in 79 A.D. Mount Etna, on the Italian island of Sicily, destroyed 14 cities, killing over 20,000 people in 1669. Today, Mount Etna remains Europe's most dangerous volcano. The greatest volcanic eruption in recent history was the explosion of Krakatoa in the Dutch East Indies (now Indonesia) in 1883. The violent eruption killed over 36,000 persons, many by tsunamis (seismic sea waves) as they swept the coasts of Java and Sumatra. In 1985 the composite cone Nevado del Ruiz, in the center of Colombia's coffee-growing region, erupted and melted its snowcap, sending torrents of mud and debris down its slopes, burying cities and villages with a death toll in excess of 23,000.

Another type of volcano is the **shield volcano**, composed of lavas and relatively little ash or cinders (Fig. 12.10). The gentle cones of Hawaii illustrate best what are known as shield volcanoes. They emit basaltic lava with temperatures of over 1090°C (2000°F). There is some minor escape of gases and steam, which hurls the lava into the air a few hundred meters, and some buildup of cinders (lava clots that congeal in the air), but the major feature is the outpouring of very fluid basaltic lava in flows only

FIGURE 12.10   The gentle cones on the island of Hawaii best illustrate shield volcanoes. Numerous craters in (a) emit hot fluid basaltic lava. (S. Brazier) As the lava flows downslope in (b) it cools from its molten orange color to the black color of pahoehoe basalt. (U.S. Geological Survey)

FIGURE 12.11 Plug dome volcanoes extrude stiff acidic lava forming steep volcanoes composed of solid lava. Mt. Lassen, a plug dome in northern California, is the southernmost volcano of the Cascade Range and was last active between 1914 and 1921. (GeoPhoto Publishing Co., Alan L. Mayo, University of Colorado, Colorado Springs)

a meter or so deep. The accumulation of flow on top of flow builds broad structures with very gentle slopes.

By contrast, **plug dome** volcanoes extrude extremely stiff acidic lava that fills the initial pyroclastic cone without flowing beyond it (Fig. 12.11). They are characteristically steep-sided volcanoes with broad summits composed of solid lava. Spikes of hard lava often project at the summit. Their vents repeatedly jam with congealed lava and then are cleared in cataclysmic explosive eruptions. In 1903 Mount Pelee, a plug dome on the French West Indies island of Martinique, destroyed in a single blast all but one resident of a town of 30,000. Mt. Lassen, in California, is a large plug dome that has been active in this century. Other plug domes have erupted in Japan, Guatemala, and the Aleutian Islands.

The smallest type of volcano, known as a **cinder cone,** produces little lava and consists largely of pyroclastics (Fig. 12.12). Cinder cones are illustrated by Craters of the Moon, Idaho, and Sunset Craters, Arizona. The remarkable cinder cone called Paricutín in 1943 grew from a crack in a Mexican cornfield to a height of 92 meters (300 ft) in five days, and to over 360 meters (1200 ft) in a year.

Occasionally an explosive composite volcano may expel so much material in a violent eruption that its summit collapses, producing a vast crater, or **caldera.** This crater subsequently fills with water. The best known is Crater Lake in Oregon, a circular body of water 10 kilometers (6 mi) across and almost 610 meters (2000 ft) deep, surrounded by near-vertical cliffs as much as 610 meters (2000 ft) high. The pres-

FIGURE 12.12 The smallest type of volcano, known as a cinder cone, consists largely of pyroclastics. Sunset Crater, Arizona, is typical of this type of volcano. (S. Brazier)

FIGURE 12.13 Crater Lake, Oregon, is probably the best known caldera in the United States. Calderas are formed when a violent eruption and summit collapse form a vast crater. Wizard Island, in Crater Lake, is a secondary crater formed after the main eruption and collapse. (R. Sager)

ent rim crests at about 2440 meters (8000 ft). A new cone has built up from the floor of Crater Lake to peak above the surface as Wizard Island (Fig. 12.13). Yellowstone National Park is the site of three ancient calderas. Krakatoa in Indonesia and Santorini (Thera) in the Greek Islands are also major examples of calderas. Other calderas can be found in the Philippines, the Azores, Japan, and Italy, most of them occupied by deep lakes.

DISTRIBUTION OF VOLCANISM

Most volcanoes are concentrated in several well-defined zones along plate boundaries, while a few volcanic regions are distributed more sporadically (Fig. 12.14). The midocean ridge system where crustal plates are diverging is entirely volcanic, as are the oceanic islands on the midocean ridge, such as the Azores and Iceland. Volcanoes also occur where continen-

tal plates are spreading; examples along the East African Rift Valleys include Mt. Kilimanjaro and Mt. Kenya.

A chain of volcanoes almost completely circles the Pacific Ocean and is known as the Pacific Ring of Fire. These explosive volcanoes erupt andesitic lava. When the oceanic crust descends below the continental crust and into the oceanic trench systems, it melts and is reconstituted into magma. The magma moves up to the surface under the continental borders to produce plutons and volcanism above them. In fact, wherever there is descent of crustal material into a trench system, volcanoes occur in the vicinity. This process produces the volcanoes of the Andes, Japan, the Aleutians, the Kuril Islands, and the Kamchatka Peninsula of the Soviet Union.

Another volcanic belt marks the line of collision between the northward-moving Southern Hemisphere crustal plates and the Eurasian plate. On this line are located the volca-

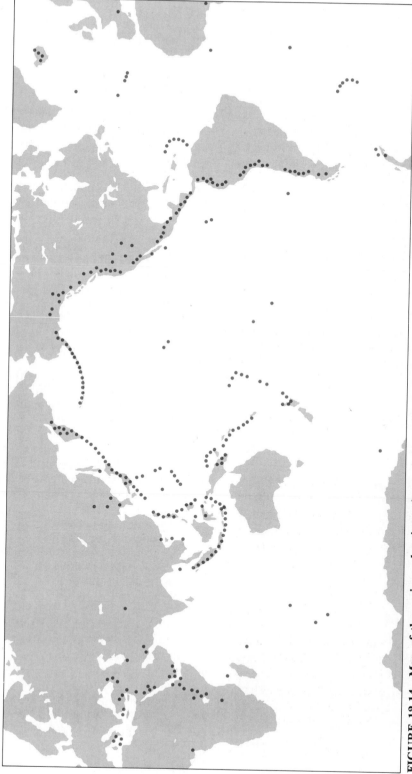

FIGURE 12.14 Map of the major volcanic regions of the world. Note the strong correlation with the plate boundaries in Figure 12.2. The intraplate "hot spots" are the only volcanic areas not along plate boundaries.

VIEWPOINT: HAVE WE FOUND ATLANTIS?

The myth of the long-lost continent of Atlantis has intrigued scholars and charlatans alike for centuries, ever since Plato first described this "island of Atlantis [where] there was a great and wonderful empire which had rule over the whole island and several others." Plato described Atlantis in much detail, stating often that the story was true, not myth, and that it was based on information from Egyptian priests. What is most fascinating about Atlantis is that this great empire did not die out slowly, nor was it overtaken by invaders. As Plato tells it, ". . . afterwards there occurred violent earthquakes and floods; and in a single day and night of misfortune all . . . men in a body sank into the earth, and the island of Atlantis disappeared, and was sunk beneath the sea."

The search for Atlantis has been conducted in many parts of the world including the Amazon rain forest, the Canary Islands, the Azores, and beneath the sea near the Mid-Atlantic Ridge. None of these searches, however, has resulted in anything as positive as what has been found in Plato's backyard, the Mediterranean Sea.

Thirty-five hundred years ago the Minoan kingdom, whose center was at Knossos on the island of Crete, was a powerful nation of traders, sailors, merchants, farmers, and craftsmen. Minoans traded wheat, obsidian, copper, pottery, and other goods with their own colonies and with other peoples around the Mediterranean. Slave labor helped to build terraced vineyards, multistoried buildings, and great palaces on the Minoan islands. The people developed a sophisticated written language, and their sensitivity of artistic expression is unrivaled among ancient civilizations, as evidenced by the superb frescoes unearthed by archeologists on the island of Santorini.

We have not always known about the Minoans. In fact, evidence of this ancient society was not uncovered until the early part of this century. But as scientists learned more and more about this highly organized and sophisticated culture, it became evident that at the very height of its brilliance nearly the entire civilization had abruptly disappeared. What kind of disaster could have been so powerful, so all-encompassing as to destroy a civilization as strong and widespread as the Minoan?

Largely because of the suddenness of the destruction and the comparative strength of the Minoan civilization, scholars have ruled out such human factors as revolution and invasion. Instead they have looked for a natural disaster, a cataclysmic event that would have wiped out the cities and resources to such an extent as to make recovery impossible.

It has long been known that the island of Santorini, 112 kilometers (70 mi) from Crete, is actually the remnant of a much larger, round volcanic island that rose nearly 1500 meters (5000 ft) above the sea. But recently geologists have been able to date the volcanic ash from Santorini and have discovered that it exploded in a massive, violent eruption at about the same time that archeologists believe the Minoan civilization abruptly ended, sometime around 1450 B.C.

To those who doubt that the eruption of one volcano could destroy a civilization as strong as the Minoan, geologists point to the eruption of Krakatoa in the East Indies in 1883. When the 445-meter (1460-ft) Krakatoa exploded, it hurled rocks a distance of 96 kilometers (50 mi). The roar of the eruption shook buildings nearly 800 kilometers (500 mi) away and could be heard 4800 kilometers (3000 mi) away. Pumice up to 4 meters (18 ft) high floated on the sea over 160 kilometers (100 mi) away. The explosion poured huge quantities of dust 48 kilometers (30 mi) up into the atmosphere. This dust circled the earth many times and caused abnormally red sunsets, a reduction in incoming solar radiation, and changes in the weather for months and possibly years.

Minoan fresco, Knossos Palace, Crete. (S. Brazier)

When Krakatoa's explosion had spent itself, the unsupported sides of the volcano collapsed into the center, creating a huge crater 185–275 meters (600–900 ft) deep. The surrounding sea rushed in to fill this new caldera and then rushed out again in giant tsunamis, which traveled at speeds of 80 kilometers per hour (50 mph) and smashed against nearby coasts with successive waves over 30 meters (100 ft) high. These waves destroyed nearly 300 towns and drowned more than 36,000 people. When the eruption was over and the seas had calmed, nearly 26 square kilometers (10 sq mi) of land with an average elevation of 215 meters (700 ft) had been replaced by the caldera.

In contrast to Krakatoa, the volcanic cone of Santorini was nearly 1500 meters (5000 ft) high, and its eruption 3500 years ago destroyed 130 square kilometers (50 sq mi) of land. The Santorini explosion spread burning ash 30–60 meters (100–200 ft) thick (versus less than 30 centimeters on Krakatoa) over the remainder of the island. Wind carried the ashes from Santorini over an area of 208,000 square kilometers (80,000 sq mi). For these reasons and others, geologists estimate that the explosion of Santorini may have had four times the force of Krakatoa's 1883 eruption. They further estimate that the tsunamis caused by the eruption were as much as 1.5 kilometers high at their vortexes and reached velocities of 320 to 480 kilometers per hour (200 to 300 mph). It is likely that these waves were still as much as 45 meters (150 ft) high when they slammed into the coast of Crete and the center of the Minoan civilization 112 kilometers (70 mi) away. There is evidence that they even drowned the port city of Ugarit in Syria over 960 kilometers (600 mi) away and deposited pumice at Jaffa 900 kilometers (560 mi) away and over 5 meters above sea level.

The devastation caused by such an eruption is difficult to comprehend. Obviously an event of this magnitude must have killed enormous numbers of people. In addition, the thick layers of ash that were spread over Santorini and neighboring islands like Crete must have made those lands unfit for agriculture and probably uninhabitable for years after.

When Santorini erupted in 1450 B.C., it was not the first time that a volcano had spewed forth its innards. And today in the middle of the deep caldera of Santorini three new peaks, the active tops of the volcano, stand as islands in the bay. Any day, one of these may erupt.

Actually the entire Mediterranean is an area of geologic instability. Volcanoes like Etna and Vesuvius dot the area and erupt periodically. The region is also subject to earthquakes like the devastating one that struck southern Italy in the winter of 1980. This intense tectonic activity occurs along a fracture zone where the Eurasian continental plate and the African plate are colliding with each other. The tensions and pressure that build up as a result create a great deal of volcanic and seismic activity.

Today it is widely theorized that the powerful eruption of the volcanic island of Santorini, combined as it was with tsunamis and earthquakes, was the precipitating factor in the decline of the Minoan civilization. What is not known—and what may never be because we have only Plato's hearsay description of Atlantis to go on—is whether the Minoan civilization was actually the legendary Atlantis. Yet it is this possibility that adds a special flavor to the search for the ancient Minoan kingdom.

Steep caldera edge of Santorini (S. Brazier).

noes of the Mediterranean region, Turkey, Iran, and Indonesia.

The movement of the earth's major crustal plates controls the distribution of most volcanism, as it does the tectonic processes of folding and faulting, which we will discuss next. However, some volcanic areas do not occur near plate boundaries but appear randomly within plates. The Hawaiian Islands, the Galapagos Islands and Yellowstone National Park are examples of these intraplate "hot spots." Scientists are still searching for the cause of the hot spots.

Not all volcanism involves erupting volcanoes. Some continental areas are covered with enormous accumulations of lava, called **lava plateaus** or **flood basalts**, that consist of hundreds of overlapping flows. In past geologic periods, basaltic magmas welled up quietly through many separate **fissures** and flowed across the landscape, often engulfing it to thousands of meters deep. The Columbia Plateau in Washington, Oregon, and Idaho, covering 520,000 square kilometers (200,000 sq mi), is a major example of such a lava plateau as is most of India's Deccan Plateau.

PLUTONS

The variety in shapes, sizes, and forms of solidified magma that result from plutonic activity is enormous, but when first formed most have no effect on the shape of the earth's surface. Plutons refer to those forms of intrusive igneous bodies that are especially deep-seated so that it is only after thousands and sometimes millions of years of erosion that some of these forms become exposed at the surface and thus become a part of the landscape. What usually happens is that the plutonic form, being composed of granite or some similar material, is eventually exposed by erosion and either stands higher or is reduced lower than the materials around it, depending on their relative resistance (Fig. 12.15).

An individual pluton exposed at the surface by erosion is known as a **stock.** A stock is usually limited in area to a few tens of square

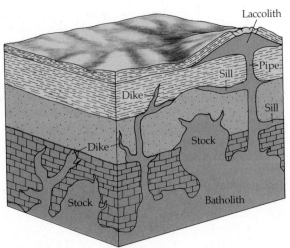

FIGURE 12.15 Igneous rocks crystallize in a variety of forms both above (volcanic) and below (plutonic) the earth's surface. The major igneous features are shown here.

kilometers. The largest of the forms created by plutonic volcanism is the **batholith.** Batholiths are enormous, complex masses of solidified magma, usually granite. A batholith is composed of many individual plutons that push aside some of the rocks of the crust while melting and digesting others. Batholiths vary in size; some are as much as several hundred kilometers across and thousands of meters thick. Batholiths form the core of most major mountains, primarily because uplift of the ranges has caused older covering rocks to be eroded away. These older rocks are preserved where the degree of uplift has been less. The Sierra Nevada, Idaho, Rocky Mountain, Coast, and Baja California batholiths cover areas of hundreds of thousands of square kilometers of western North America.

When magma forces its way into cracks and between layers, disturbing but not digesting the surrounding rock, and coming near to the surface, it is referred to as an **igneous intrusion.** A **laccolith** is formed when molten magma forces its way horizontally between layers of rock. The resulting mound on the earth's surface is often compared to a blister, the magma beneath the surface layers being comparable to the water beneath the skin of a blis-

FIGURE 12.16 A dike is formed when plutonic activity forces magma to cut across the general trend of the surrounding rocks. (S. Brazier)

ter. In a cross section of the crust, a laccolith resembles a mushroom, because the mound is usually connected with a source of magma by a pipe or stem. Like batholiths, laccoliths often form the core of mountains or hills after millions of years of erosion have worn away the surface covering of sedimentary rocks. The Henry, LaSal, and Abajo Mountains in Utah are formed from laccoliths.

Smaller but no less interesting landforms created by plutonic activity may be exposed at the surface through erosion. When magma forces its way toward the surface, it is some-

times able to spread out along a vertical fracture in the crust across the general trend of the surrounding rocks. If it solidifies, a wall-like sheet of magma will be formed, known as a **dike** (Fig. 12.16). When exposed by erosion, dikes appear as flat walls of igneous rock rising above the countryside. At Ship Rock, in New Mexico, dikes many kilometers in length rise sharply to over 90 meters (300 ft) above the plateau. Sometimes magma intrudes between and parallel with the layers of rock and solidifies in a horizontal sheet, called a **sill** (Fig. 12.17). The Palisades, along New York's Hud-

FIGURE 12.17 Sills form where magma intrudes between parallel layers of rock. Often they form resistant caprock tops on exposed plutonic areas. (S. Brazier)

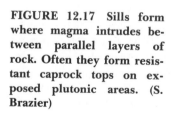

FIGURE 12.18 Ship Rock, New Mexico, is a volcanic neck. Volcanic necks are formed when the resistant subsurface pipe of the volcano is exposed by erosion. (S. Brazier)

son River, are an example. Ship Rock itself is a **volcanic neck:** the exposed (formerly subsurface) pipe that fed a volcano situated above it about 30 million years ago (Fig. 12.18).

THE DIASTROPHIC TECTONIC PROCESSES

Diastrophism refers to the bending, folding, warping, and fracturing of the solid earth's crust that are largely a product of the conflicting motions of the earth's crustal plates. Most of our information about diastrophism comes from direct observation of the rocks of the crust—their structure and arrangement. Sedimentary rocks are especially useful to study, since we know that when they are first formed the layers in these rocks are nearly horizontal and the youngest layers are always on top of older layers. If the layers of sedimentary rocks appear tilted, bent, or displaced, then we can assume some kind of diastrophism has taken place.

Geologists describe the tilting of the beds as the **dip.** Dip can be measured as an angle from the horizontal. The **strike** of such beds is their direction, compass-wise, at right angles to

the dip. Thus we might say that certain layers of outcropping rocks have a dip of 35° and are striking NE/SW (Fig. 12.19).

The earth's crust seems to have been subject to tectonic pressures and tensions through all of its history, though during some periods the stresses seem to have been greater than at others. The crust has responded to these stresses by wrinkling, warping, and sometimes fracturing, by being pushed up, or by sinking down. Most of these changes have occurred slowly over millions of years; some have been rapid and cataclysmic. The variety of responses, often appearing in combination with one another, have yielded a variety of complex structural configurations. In addition, the gradational processes act on these structures as soon as they begin to appear, working to wear them down and fill them in, to level the landscape.

Tectonic movements may involve uplift or depression of large sections of the crust or may result from either compression, which tends to shorten and thicken the crust, or tension, which tends to stretch and thin the surface. Generally speaking, compression usually results in wrinkling of the crust, which may eventually shear or slide over itself horizontally if the

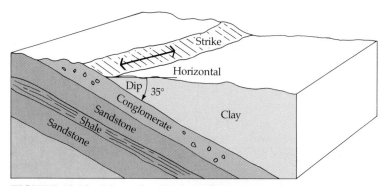

FIGURE 12.19   Dip and strike in sedimentary rocks. Dip is the angle that rock layers tilt, measured from the horizontal. Strike is the compass direction of the line marking the intersection of the rock layers with the earth's surface. The strike is at right angles to the dip. In this diagram, the dip is 35° SE, and the strike is NE/SW.

forces are great enough. Tension causes the crust to crack or fracture and then to collapse because of the loss of support.

WARPING OF THE EARTH'S CRUST

The broad and gentle deformation of the earth's crust over large areas is called **warping.** The Florida and Yucatan Peninsulas were raised to their present levels from below the ocean by crustal warping. Evidence for this warping in Florida is found in the number of marine fossils in the sedimentary rocks there. The extensive Colorado Plateau has been uplifted thousands of feet in similar fashion. The North Sea has been created by the slow downward warping of the crust and the consequent inundation of low plains by the sea. The crust beneath Hudson Bay is rising slowly, so that the bay eventually will disappear. A final example occurs around Lake Superior, where one shore is rising while the other is slowly sinking. The last two examples are a consequence of the disappearance of continental ice sheets, whose sheer weight depressed these areas hundreds of meters. With the disappearance of the ice, the areas are rising to their former levels.

This type of movement is related to **isostasy**—the concept that the earth's crust is floating in hydrostatic equilibrium in the denser material of the mantle. The effect of isostasy is that addition of weight (in the form of sediment, glacial ice, or a large body of water) causes the crust to sink slightly, whereas removal of weight (by erosion, deglaciation, or disappearance of large bodies of water) allows the crust to rise or to float a bit higher. Much broad crustal warping occurring today is clearly related to isostasy, but the kind of warping that raised the Colorado Plateau is not and remains unexplained.

Crustal warping is an important process of landform change, especially since it affects broad parts of the earth's surface. Although a form of diastrophism, it is unrelated to tension or compression of the earth's crust but proceeds from vertical movements probably related to lateral transfer of material in the asthenosphere.

FOLDING

Wrinkling of the earth's crust, known as **folding,** usually occurs in response to slow lateral compression. If you spread a cloth on a table and push it from one side, the folds of various kinds that result are much like the folds in the earth's crust that result from certain kinds of pressure exerted on it. Placing both hands

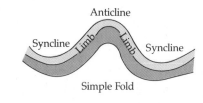

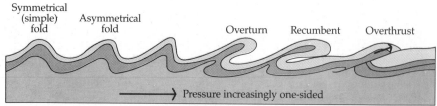

Increasingly distorted folds

FIGURE 12.20 The major types of folds.

some distance apart on the cloth and moving one hand toward you and the other away produces an even larger number of folds. Thus, if underlying rocks move laterally past one another, compressional folding occurs.

Folds produced naturally in rock may be very small, covering an area of a few centimeters, or they may be enormous, the vertical distance between the crests and troughs being measured in kilometers. Folds can be tight or broad, symmetrical or asymmetrical (Fig. 12.20). Much of the Appalachian Mountain system is an example of simple folding on a large

scale. In contrast, the Alps were formed by highly complex folding, in which folds are overturned, sheared off, and piled on top of one another. In any case, almost all mountain systems exhibit some degree of folding, though the folding is usually found in connection with some faulting (fracturing of the crust).

In a series of simple folds, the upfolds are called **anticlines** and the troughs are called **synclines** (Fig. 12.21). The flanks of the folds between the anticlinal crests and synclinal troughs are the fold limbs. Folds may be symmetrical where the compressional forces are

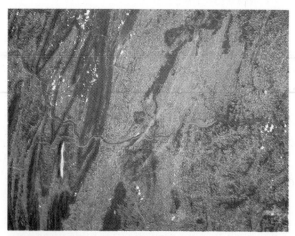

FIGURE 12.21 (a) An infrared satellite view of the large-scale folding of the Appalachian Mountains in Pennsylvania, forming typical ridge and valley topography. (NASA) (b) An anticline and syncline in the sedimentary rocks of southern England. (S. Brazier)

FIGURE 12.22 (a) Severe compressional forces during the formation of the Andes caused this overturned fold in Ecuador. (S. Brazier) (b) In complex mountains, such as the Canadian Rockies, the folds have been broken and sheared, forming spectacular overthrusts. (R. Gabler)

somewhat even from both sides. As these forces become stronger from one side, the folds become more asymmetrical, until the folds become overturned (Fig. 12.11b). Some may be so lopsided that the fold lies horizontally: these are known as **recumbent folds.** Ultimately the bending rocks will not yield further and the upper part of the fold shears or breaks and slides over the lower. Such a development is called an **overthrust:** Major overthrusts occur along the Rocky Mountain front and Southern Appalachians. Recumbent folds and overthrusts are important in the formation of complex mountains such as the Andes, northern Rockies, Alps, and Himalayas (Fig. 12.22).

FAULTING

Some rocks are too rigid to bend into folds and rupture instead. Both tension and compression cause rock masses to fracture and move differentially with respect to one another. When slippage or displacement of the crust occurs along a fracture plane, the fracture is called a **fault** (Fig. 12.23). The instantaneous movement along a fault during an earthquake varies from fractions of a centimeter to several meters. The maximum horizontal displacement along the

San Andreas fault in California during the 1906 San Francisco earthquake was over 6 meters (21 ft); a similar vertical displacement occurred during the Owens Valley earthquake in California in 1872. The cumulative displacement along a major fault over thousands to millions of years may be tens of kilometers vertically and hundreds of kilometers horizontally, though the vast majority of faults show offsets of much less magnitude.

Vertical displacement along a fracture occurs when one part of the crust moves up or drops down in relation to another. Movement is parallel to the *dip* of the fracture plane extending into the earth. Thus this type of movement is known as **dip slip.** Faults that have vertical displacement resulting from tension are called **normal faults;** those resulting from compression are called **reverse faults.** When compression pushes one crustal slab up and over another, the displacement is called a **thrust fault.**

The steep escarpment of the higher block where vertical displacement occurs is called a **fault scarp.** Such fault scarps account for some of the most spectacular mountain walls, especially in the western United States. The east face of the 645 kilometer-long Sierra Nevada

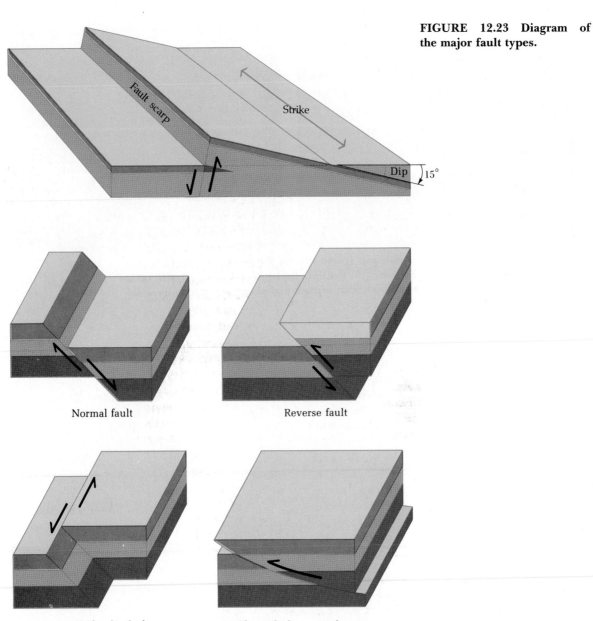

FIGURE 12.23 Diagram of the major fault types.

Normal fault

Reverse fault

Strike-slip fault

Thrust fault or overthrust

of California is a classic example of a fault scarp rising steeply 3350 meters (11,000 ft) above the desert. In contrast, the west side of the Sierras (the "back slope") descends very gently over a distance of 100 kilometers (60 mi) through rolling foothills. The range is a great slab tilted to the west, presenting its steep edge to the east. The equally dramatic Grand Tetons of Wyoming rise in the same way out of the plains to the east (Fig. 12.24). The Colorado Plateau steps down to the Great Basin by a series of similar fault scarps, this time facing westward in southern Utah and northern Arizona.

Let us consider for a moment the effects on other physical systems of such uplifting as occurred in the Sierra. Stream erosion is accelerated by the increase in slope. Precipitation on

FIGURE 12.24 The high fault scarp of the Grand Tetons in Wyoming. (R. Gabler)

the windward side is increased because of orographic lifting. The lee side of the raised land is drier than before, because it now lies in the rain shadow of the mountains. Increased precipitation and the lower temperatures at higher elevations combine to change the climate of the raised land significantly, which then changes vegetation, soils, and animal life as well. Soils are also affected by an increase in surface run-off and erosion. The process of uplift extended over several million years, so these changes have been very gradual. The Sierra is continuing to rise at a rapid rate, geologically speaking—perhaps a centimeter each year. In addition, it should be remembered that the gradational agents (water, ice, wind, and gravity, alone or in combination) begin working to level the land as soon as lifting begins. The Sierra, for instance, does not appear now as it once did, for it has been carved and etched by tens of thousands of years of glaciation, stream erosion, and gravitational mass movement of weathered material.

The displacement along some faults is horizontal rather than vertical. All such faults extend vertically into the earth's crust, but the slippage is parallel to the surface trace, or *strike,* of the fault. This horizontal movement creates a **strike-slip fault.** It is visible in the horizontal displacement of roads, railroad tracks, fences, and stream beds. The San Andreas Fault, which runs through much of California, is a strike-slip fault (Fig. 12.25).

Often a series of parallel faults is encountered. Where the block between two faults is elevated above the land to either side because it has been pushed up or because the land to either side has dropped down, the raised portion is called a **horst.** The central European Highlands north of the Alps contain many horsts, as does the Great Basin of Nevada. The great Ruwenzori Range of East Africa is a horst, as is the Sinai Peninsula between the fault troughs in the gulfs of Suez and Aqaba.

The opposite of a horst is a **graben**—a depression between two facing fault scarps that may occur when the block between two faults drops down or when the land to either side of the faults is uplifted. Several interconnected grabens form major lowlands called **rift valleys.** Classic examples of grabens are Death Valley in California, the middle Rhine Valley, the rift valley of East Africa in which are located Lakes Tanganyika and Malawi, among others, and the large Jordan rift valley, where the Dead Sea lies at a level some 390 meters

FIGURE 12.25 The San Andreas Fault in central California. This type of horizontal displacement is known as strike-slip faulting and can be seen by the offset stream valleys crossing the fault line. (U.S. Geological Survey)

(1280 ft) below that of the Mediterranean Sea, which is only 64 kilometers (40 mi) away.

Immense regions may be affected by crustal shattering along faults, producing a mosaic of horsts, grabens, and tilted blocks. The outstanding example is the Great Basin extending eastward from California to Utah and southward from Oregon to Arizona, including all of Nevada (Fig. 12.26).

Faulting, like folding, is an important part of landform development, especially mountain building. In fact, the two—faulting and folding—often occur in the same areas, sometimes in the vicinity of volcanism, the other important mountain builder. These processes are ongoing ones, as attested to by associated continuing earthquake activity.

EARTHQUAKES

The most prominent evidence of present-day diastrophism, **earthquakes** are vibrations of the earth that occur when the accumulating strain of slow crustal deformation is suddenly released by displacement along a fault. An earthquake does not cause displacement. Rather, the shock waves of an earthquake signify the release of energy caused by the movement of crustal blocks past one another. It is the L-waves (longitudinal), which pass along the crustal surface, that cause the damage and loss of life that we associate with major earthquakes. The point at which an earthquake originates is the **focus,** which may occur anywhere from the surface to a depth of 640 meters (400 mi). The earthquake **epicenter** is the point on the earth's surface that lies directly above the focus: This is where the strongest shock is normally felt.

Most earthquakes are so slight that we cannot feel them, and they produce no visible damage. Usually they occur deep within the earth; no displacement is visible at the surface. Others are strong enough to rattle a few dishes, while a few are so strong as to topple buildings, break power lines and water pipes, and trigger rock and landslides. Aftershocks may follow the main quake as the crustal movement settles. Geophysicists are currently investigating the possibility that foreshocks may alert us to major earthquakes, though evidence is presently inconclusive.

The **Richter scale** of earthquake magnitudes, based on the energy released, is a measure of ground motion as recorded on seismo-

THE DIASTROPHIC TECTONIC PROCESSES

FIGURE 12.26   Death Valley, California, is a classic example of a graben, a depression between two facing fault scarps. Death Valley is the lowest point of elevation in North America. (S. Brazier)

graphs. Every increase of one number in magnitude (e.g., from 6 to 7) means the ground motion is 10 times greater. (The actual energy released may be 30 times greater.) The extremely destructive 1906 San Francisco earthquake was rated at 8.3, which was also the approximate magnitude of the devastating earthquakes in the Owens Valley (California) in 1872 and Alaska in 1964. The 1971 San Fernando Valley earthquake in California caused large buildings to collapse, although the magnitude was only 6.6. In 1984 an earthquake at Coalinga, California, measured 6.5 on the Richter scale and did similar damage. In contrast, the Peruvian earthquake of 1970 (magnitude 7.8) collapsed whole villages and obliterated others under enormous avalanches of rock and ice falling from Andean peaks. The human death toll was 50,000 lives. Mexico's disastrous earthquake in 1985 (magnitude 8.1) was par-

ticularly destructive in the crowded high-rise sections of Mexico City where over 9000 lives were lost in hundreds of collapsed buildings.

While most earthquakes are related to major known faults and are in mountainous regions, the one most widely felt in North America was not. It occurred as one of a series during the period from 1811 to 1812, was centered near New Madrid, Missouri, and was felt from Canada to the Gulf of Mexico and from the Rocky Mountains to the Atlantic Ocean. Fortunately, the area was not densely settled at that stage in history. Though they are not common, recent small earthquakes have occurred in New England, New York, and the Mississippi Valley. Probably no region on earth is what could be called "earthquake proof." The map in Figure 12.27 shows world regions where earthquakes are most common. Note the especially strong correlation of earthquakes to plate boundaries.

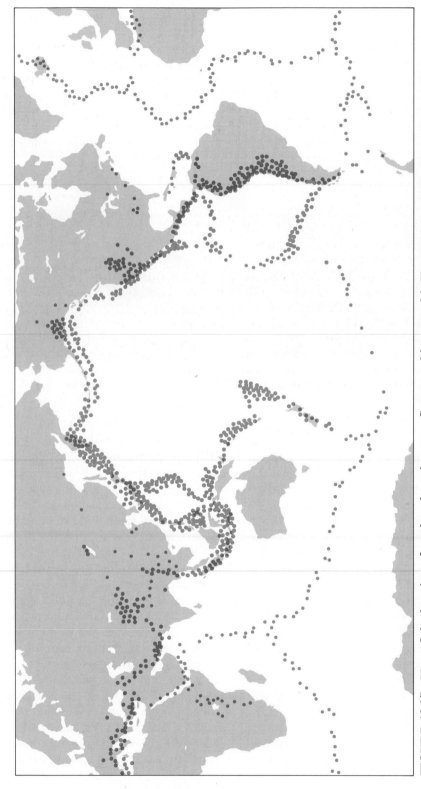

FIGURE 12.27 Map of the location of earthquake epicenters. Compare this map with Figure 12.2, which shows the crustal plates and demonstrates the strong relationship of earthquakes to plate boundaries. Note also the close relationship to the location of volcanic regions in Figure 12.14, which are also mainly concentrated on the margins of tectonic plates.

DIASTROPHIC LANDFORMS

The original structural forms that result from diastrophism vary from microscopic joints to mountain ranges. However, none of these remain in original form because of the leveling action of the agents of gradation. It is important to note the difference between the major landform types discussed in the previous chapter and the structural configurations mentioned here. An anticline may form a surface ridge but frequently does not because of erosional activity. For example, Nashville, Tennessee, occupies a topographic valley, yet it is sited in the remains of a structural dome. Only where the anticline is composed of extremely resistant rock does it long persist as a ridge. The important distinction is that *hill* or *ridge* refers to surface configuration, whereas the term *anticline* signifies a geologic structure.

QUESTIONS FOR DISCUSSION AND REVIEW

1. What evidence did Wegener rely on in the formulation of his theory of continental drift? What evidence did he lack? What evidence has since been found to support his theory?

2. Describe the theory of plate tectonics. How is it related to convection; to sea floor spreading? In what three ways do tectonic plates move in relation to one another?

3. What type of plate boundary is found near the Andes, along the San Andreas Fault, in Iceland, and in Hawaii?

4. What types of volcanoes are especially dangerous? What were the major causes of the eruption of Mount St. Helens?

5. What factors affect the distribution of volcanoes?

6. What is a batholith? How do plutonism and volcanism differ from one another?

7. What are the major types of diastrophism that affect the earth's crust? What causes them?

8. Draw a diagram of folding, showing anticlines, synclines, and thrusts.

9. What causes an earthquake? What is the relationship between the focus and the epicenter of an earthquake?

10. What type of tectonic activity is most responsible for the Sierra Nevada, the Ridge and Valley section of the Appalachians, and the Alps?

11. Describe the world distribution of volcanic activity and of earthquake epicenters. Relate this to plate boundaries.

13

GRADATION, WEATHERING, AND MASS MOVEMENT

GRADATION AND TECTONISM

The gradational processes work to smooth out the major differences in elevation resulting from tectonic processes. Gradation fills in depressions in the land and wears down areas of higher elevation. In some places gradation also involves building up the land surface by creating depositional landforms such as alluvial fans, deltas, natural levees (Chapter 14), and glacial moraines (Chapter 16). The gradational forces are sometimes called *exogenetic*, a term that refers to their origin and action at the earth's surface. In this sense they can be contrasted to the tectonic or *endogenetic* forces that originate below the earth's surface.

Although the ultimate result of gradation is a leveling and smoothing of the surface, the various stages between the original uplift by the tectonic forces and the final result of gradation may include landforms even more irregular than those produced by tectonism alone. For example, the surface of a plateau shortly after uplift may be relatively smooth, although its elevation has been raised thousands of feet. Gra-

dation, working through its primary agent, running water, gradually begins to cut up the land; valleys develop, separated by mountains or hills. Eventually, as the moving water wears away more and more of the raised portions of the land, the irregularities in the surface originally created by water erosion are reduced. Thus many variably shaped landforms are actually the result of the gradational reductions in differences in elevation.

We have been speaking here of a cycle of landform development in which the land is raised by tectonic processes and then worn down or leveled by gradation. To complete the cycle the land surface must be reduced to a level and form beyond which no further modification by the gradational processes is possible. This is, however, a hypothetical model of what actually happens. In reality, gradation is an ongoing process, and its agents are constantly at work in all places where there are differences in elevation. Furthermore, the forces of diastrophism and volcanism do not wait for gradation to be completed before they act again, but may occur at any time, creating new forms to be acted on by the gradational processes. Thus

Bryce Canyon, Utah. (S. Brazier)

we see many of the Earth's subsystems at work: tectonic action producing new landforms, such as fault scarps, folded mountains, and volcanoes, on which gradational forces work to create reduced slopes. These slopes themselves will again be altered by future tectonic movements, which renew the cycle once again.

THE GRADATIONAL PROCESSES

Gradation includes the picking up and removal (erosion) of loose material, its transportation to another location, and its deposition there. **Degradation,** or the wearing down of the land, is accomplished by erosion and transportation acting together to remove material, thereby lowering the elevation of the land. **Aggradation,** on the other hand, is the filling in of depressions or the raising of land elevation, and is accomplished by deposition. The combined result of degradation and aggradation—of erosion, transportation, and deposition—is the gradational reduction of irregularities in elevation (Fig. 13.1).

The ultimate effect of gradation is to reduce the land surface to **base level**—a surface so flat that the erosional forces can no longer affect it. The ultimate base level is the level of the sea, below which the land surface cannot be lowered by the normal gradational forces. If there were no further tectonic activity, the surfaces of the continents would eventually be reduced to sea level by gradation.

The removal of material from one place and its deposition in another place is accomplished by gradational agents, the most important of which are water, ice, and wind. Chapter 14 includes a discussion of how water below the land surface and in streams acts as a gradational agent. Chapter 15 focuses on both water and wind as gradational agents in arid regions. Chapter 16 includes an examination of ice as an agent of gradation, and Chapter 17 includes a discussion of gradation by waves in coastal zones. There are also less important gradational agents. For example, the actions of humans, animals, and plants often work to level the land.

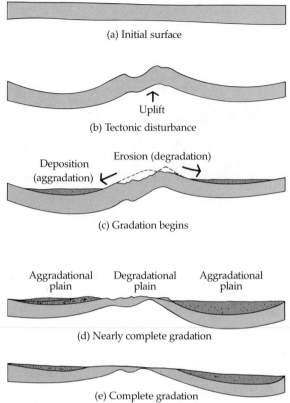

FIGURE 13.1 **The process of gradation.**

Gravitation and solar radiation provide the energy for gradational agents. Gradation may also be accomplished by the force of gravity acting alone. Gravity may be the primary cause of the movement of rock and weathered debris downslope to lower elevations. Such removal and deposition is called **mass wasting** or **mass movement.** In the final part of this chapter we will consider some of the varieties of mass movement.

The individual gradational agents usually do not work alone on a particular landform. More often they combine, sometimes sequentially, to level the land. For example, the awe-inspiring landscape of Wyoming's Grand Tetons is to a large extent the result of ice action, but stream erosion and mass wasting have also been involved (Fig. 13.2). However, in order to simplify the explanation of the gradational

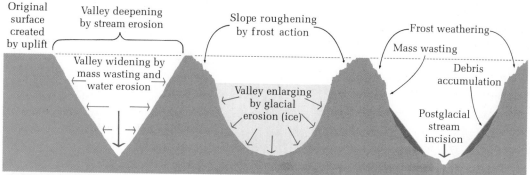

FIGURE 13.2   The scenery of most high mountain regions has been formed by at least three separate phases of gradational development: stream incision and valley development during uplift; glacial enlargement of former stream valleys, with intense frost weathering above the level of the ice; and postglacial weathering and mass wasting with some stream incision.

forces and processes and their resultant landforms, each agent and its major landforms will be discussed separately.

WEATHERING

Weathering includes various processes by which rocks at or near the Earth's surface are disintegrated or decomposed in preparation for their removal and transportation. Weathering eventually breaks down even the earth's most massive rocks into components small enough to be moved downslope by one or more of the gradational agents.

Some earth scientists include weathering in the concept of gradation. However, weathering of the earth's rocks can occur without gradation. Weathering does not by itself level the land; it only fragments rocks so they can be moved, thus allowing gradation to take place. The definition of weathering is the breakdown of surface rock material *in place,* involving little or no movement of the rock material by the agent of weathering itself. However, it is important to note that practically no gradation will result without initial preparation by weathering.

Earth scientists usually divide weathering into two basic types. **Physical** or **mechanical weathering** disintegrates rocks without altering their chemical composition. **Chemical weathering** decays rock by a variety of chemical reactions. Both types of weathering normally take place at or near the surface of the earth, though evidence of weathering has been found as far as 185 meters (600 ft) below the surface. Variations in the depths to which weathering will occur, as well as variations in the type and rate of weathering, depend primarily on (1) the structure and composition of the rocks, (2) climate, especially temperature and moisture conditions, (3) the configuration of the land surface, and (4) the vegetative cover.

PHYSICAL WEATHERING

The fragmentation or mechanical disintegration of rocks by physical weathering is important to gradation in two ways. First, the smaller rock pieces are more easily removed and transported by one of the gradational agents. In this way mechanical disintegration aids gradation. Second, the fragmentation of a large rock into smaller ones encourages chemical weathering, by increasing the rock's surface area and thus by exposing more of the rock to possible chemical decomposition.

The joints and fractures that develop in igneous, sedimentary, or metamorphic rocks are areas of weakness that encourage physical weathering by making fragmentation easier.

FIGURE 13.3 Jointed sandstones, Joshua Tree National Monument in California's Mojave Desert. The closer the spacing of joints in any type of rock, the more rapid the weathering, both physical and chemical. (S. Brazier)

Chemical weathering also proceeds faster along joint planes or zones of stress. Jointing is especially common in certain sedimentary rock types such as limestone, but it can be found in any rock that has been subjected to the stresses of volcanism and diastrophism (Fig. 13.3). Thus the processes of mountain building and of rock formation work with mechanical and chemical weathering to produce gradation.

Plants, animals, and humans all contribute to the mechanical disintegration of rocks. Plant roots wedge into cracks, and as their roots grow, they exert pressure on the rocks, eventually fragmenting them (Fig. 13.4). Burrowing animals, such as gophers and ground squirrels, also weaken rocks and cause fragmentation and disintegration. The actions of humans also encourage physical weathering; bulldozers, trail

FIGURE 13.4 Organic weathering: As the tree roots grow, the rock is wedged apart (Norway). (S. Brazier)

bikes, bombs, mining, quarrying, and even hikers all do their bit to shatter rocks into smaller fragments.

The freezing of water in cracks is one of the most important means of mechanical weathering. When water freezes it becomes less dense and expands in volume by about 11 percent, exerting tremendous outward force. Skiers who fail to fill their radiators with antifreeze before driving to the mountains find this an expensive lesson, and plumbing pipes in homes in cold-winter areas may burst if they are not well insulated.

Similarly, if water fills the cracks in rock surfaces, especially where these crevices are closed or confined, the expanding ice will wedge and split the rock apart (Fig. 13.5). Such **frost wedging** is most common where freezes are frequent and intense—in middle- and high-latitude regions and at high elevations. In low-latitude areas frost wedging does not occur because temperatures are not low enough. Frost wedging is an especially important weathering force in high latitudes because of the lack of other physical and chemical weathering processes, which depend directly or indirectly on warmer temperatures. Wedging is likewise frequent in high mountains above the timberline. The exposed bedrock of high mountains is splintered by frost wedging, and the shattered

FIGURE 13.6 Frost-shattered peaks with talus slopes at their base in the Dolomite Alps, Italy. (S. Brazier)

rocks, called **scree** or **talus,** tumble down the mountain slopes, where they form **talus cones** of angular shattered rock at their base (Fig. 13.6).

Mechanical disintegration or wedging also results from the growth of salt crystals. This commonly occurs in sandstone in arid and semi-arid areas. After rain falls, capillary moisture moves toward the surface of porous sandstone. Evaporation of this water leaves behind deposits of salt crystals. As these crystals grow, they wedge rock granules out of their sockets. Repeated over time, this action can create multitudes of niches and shallow caves in large sandstone cliffs; similar effects may be seen in granite rocks.

These caves or overhangs were important living space for Indian communities in the American Southwest, for example at Mesa Verde, Colorado, or Canyon de Chelly, Arizona (Fig. 13.7). When the Indian's cliff dwelling villages were facing toward the south, they were shaded when the sun was high in summer and warmed by the low-angle sun's rays in winter.

The intensive daytime heating and nighttime cooling of rock surfaces, such as occur in deserts and at high elevations, cause the expansion and contraction of most minerals and therefore of the rocks that they make up. The

FIGURE 13.5 Frost-shattered boulders, County Kerry, Ireland. (S. Brazier)

FIGURE 13.7 Canyon de Chelly, Arizona. Indian ruins in overhangs caused by salt crystal wedging. (S. Brazier)

alternating cycle of expansion and contraction may result in some weakening and fracturing of the rocks, which would, of course, encourage both physical and chemical weathering.

Granular disintegration breaks rock down into its constituent grains or particles, especially when rocks are composed of a variety of minerals, for example, granite, with its quartz, feldspar, and mica. These minerals expand and contract at different rates, setting up stresses and strains within the rock, which eventually may cause it to disintegrate.

It was once believed that extreme diurnal changes in temperature in desert regions were a *primary* cause of rock disintegration there. However, laboratory studies seem to refute this. It was also once believed that the exfoliation of rocks after forest fires was strictly a result of the rapid expansion that the rocks undergo in the intense heating. It seems more likely that the expansion fracturing is com-

pounded by the pressure of steam formed from the water in the rocks.

Exfoliation refers to the sheeting, slabbing, or breaking off of concentric or curved rock shells. It is especially common in granite and was once thought to be caused by temperature changes. However, exfoliation may actually be caused by both chemical and physical weathering processes. Shallow exfoliation producing sheets a millimeter or so thick is closely associated with hydration, a chemical weathering process. Apparently there is also some validity to the theory that the unloading of pressure on rocks can result in exfoliation. When upper layers of rock are removed through erosion, pressure on the rocks below is lessened, so that the rocks can minutely expand. As they do so, fractures develop, creating concentric shells of rock that are often compared to the layers of an onion. This characteristic fracturing is especially apparent in massive intrusive rocks like granite, where bedding planes do not exist to absorb the expansion as the overlying rocks have been removed. As rock layers are peeled away from large, solid rock formations, immense dome-shaped masses are left, called **exfoliation domes.** Yosemite Valley in California contains some of the world's best examples of exfoliation domes and sheeting, among which are Half-Dome and Royal Arches. Sugar Loaf overlooking Rio de Janeiro's bays is a similar exfoliated granite dome (Fig. 13.8).

CHEMICAL WEATHERING

Chemical weathering, or **decomposition,** prepares rocks for removal and transportation by the gradational agents in three ways. First, chemical alteration forms new minerals that are softer and/or finer and therefore are less resistant to erosion. Second, chemical weathering may create substances—through the addition of water or through chemical change—that are greater in volume than the original rock material. This expansion can fracture the rock, increasing the rate of both physical and chemical weathering by increasing the amount of surface area and by weakening the rock mass. Third, chemical weathering may dissolve minerals in water or in a weak acid, making them easy to remove and transport. As more and more minerals are removed, the number and size of pore spaces or other openings in the rock are increased, allowing the rate of weathering to increase.

Almost without exception, chemical weathering can take place only in the presence of water. Therefore the rate of chemical weathering is always increased by the addition of water, and chemical weathering is most rapid in humid climates. Even arid climates, however, have enough moisture to allow some chemical weathering to take place, evidence of which is found in the rounded boulders showing the exfoliation and granular disintegration common in those arid regions where granite rock is present (Fig. 13.9).

Chemical reactions are also more rapid at high temperatures. Consequently, low-latitude regions with high temperatures are subject to more chemical weathering than are places with lower temperatures. Subarctic and polar climates, for example, are subject to little chemical weathering. Therefore the effects of mechanical weathering are relatively more impor-

FIGURE 13.8 Sugar Loaf, Rio de Janeiro, Brazil, is a granite exfoliation dome. (S. Brazier)

FIGURE 13.9 Block disintegration of granitic rocks in Joshua Tree National Monument, California. Note how the rectangular jointing facilitates the weathering process and how the blocks become rounded or "spheroidal" where angles and corners are attacked by both physical and chemical weathering. (S. Brazier)

FIGURE 13.10 Artist's Palette in Death Valley, California. The varied shades here result from chemical weathering, especially oxidation of iron and other minerals. (S. Brazier)

tant and consequently more visible in those areas.

Thus in hot, humid regions such as the tropical rain forest, savanna, and monsoon climates, chemical weathering is more significant in affecting the shape of landforms than is physical weathering. The landforms and rocks of such climates show the importance of chemical weathering in their rounded shapes. In contrast, the landforms and rocks of regions in which mechanical weathering is relatively more important are sharp, angular, and jagged.

The principal processes of chemical weathering are oxidation, hydration and hydrolysis, carbonation, and solution. The most important agents performing these processes are water, oxygen, and carbon dioxide, all three of which are common elements found in soil, rock, precipitation, groundwater, and air.

OXIDATION   The chemical union of oxygen with another substance to form a new product is called **oxidation** (Fig. 13.10). The chemical compounds formed by oxidation are usually softer and finer than the original substances and often have a greater volume. One of the most common of the *oxides,* as these compounds are called, is iron rust, derived from the combination of iron and oxygen. Rust is composed of two iron oxides: the minerals hematite and limonite. The rusty stains visible on many rocks are a combination of hematite and limonite, both of which are softer and more easily removed from rocks than the original iron-bearing minerals from which they were formed.

HYDRATION AND HYDROLYSIS   Though both involve the addition of water to a chemical substance, hydration and hydrolysis are distinctly different processes. In **hydration,** water molecules are attached to the molecules of another substance without any chemical change occurring. Hydration produces expansion, which can in turn result in either granular disintegration or exfoliation. Hydration also weakens minerals, making them more susceptible to other chemical weathering processes, particularly oxidation and hydrolysis, as well as to physical weathering.

Unlike hydration, **hydrolysis** does involve a chemical change through the union of water with another substance. Many common rock minerals are susceptible to hydrolysis, particularly the silicate minerals that form igneous rocks. In many cases hydrolysis, like hydration, results in expansion that can lead to exfoliation. Hydrolysis is not limited to locations at or near the surface. In warm, humid climates it can take place 30 meters or more below the surface. This deep hydrolysis weakens and decays rocks at great depth, especially in tropical regions.

SOLUTION Though it does not actually involve chemical change, solution is an important process of chemical weathering. Once minerals are dissolved in water, they are easily removed and transported by ground, soil, or surface water flow. This is known as *chemical erosion*.

Some rock minerals, such as sodium chloride (salt) and calcium sulfate (gypsum), are soluble in neutral or acidic water. Mineral salts that are immediately soluble in water are called **evaporites**, since they are precipitated when the water becomes saturated with them, as during the evaporation process. Other minerals, the products of previous chemical weathering, are also soluble. Even silica becomes soluble in hot, wet climates. Removal of silica is a major feature of the laterization process (see Chapter 10).

Many minerals that are insoluble or only slightly soluble in pure water are easily dissolved by a slightly acidic solution. Rain and soil water, by absorbing carbon dioxide from the atmosphere and from decaying vegetation in the soil, often form a very weak solution of carbonic acid, which is capable of dissolving a variety of minerals, notably calcite or calcium carbonate, the ingredient of common limestone (Fig. 13.11). When acted on by carbonic acid, calcium carbonate forms calcium bicarbonate, a salt that is soluble in water. Thus the solution of limestone involves both carbonation (creation of calcium bicarbonate) and solution. In humid regions the leaching (or removal) of salts such as calcium bicarbonate can severely weaken rock by greatly increasing the size of pore openings.

In recent years the phenomena of acid rain and acid fogs, caused by industrial pollutants being dissolved into atmospheric moisture, have speeded up chemical weathering. Since many of the great monuments and sculptures of the world are composed of limestone (calcium carbonate) or marble (calcite), there is a growing concern for the damage to these treasures. The Parthenon in Greece, the Taj Mahal in India, and the Sphinx in Egypt are

FIGURE 13.11 Weathering of limestone in County Clare, Ireland. The furrows in the limestone surface are caused by solution of the calcium carbonate along the joints. (S. Brazier)

examples where chemical solution and salt buildup are damaging and rotting away monument surfaces (Fig. 13.12).

Carbonic acid is not the only acid active in chemical weathering by solution. Other acids, derived primarily from the decay of organic matter, are also present in groundwater and are capable of dissolving other rock minerals. While most organic weathering takes place below the soil surface, it can also affect exposed rock. Such decay is associated with the chemical activity of lichens and mosses, which secrete acid substances that assist rock disintegration, producing nutrients for these simple life forms. The colonization of bare rock by plants rooted in a soil starts with the activity of lichens and mosses.

FIGURE 13.12 The Parthenon, in Athens, Greece, is severely threatened by solution decomposition, which causes the marble to flake off. (S. Brazier)

DIFFERENTIAL WEATHERING

Different rock types are affected by different weathering processes and break down and are removed at different rates. Granite is altered by oxidation, hydration, and hydrolysis, and, though mechanically resistant, it is very weak chemically. Accordingly, it is usually covered by a deep weathered overburden. In most humid regions where solid granitic rock is visible at the surface, this overburden has been stripped off by glacial erosion. Limestone is removed quite rapidly by carbonation and solution. Shale is chemically inert but is mechanically weak and is susceptible to hydration, which converts it back to clay, its original material. Sandstone is only as strong as its cement, which varies from soluble calcite to inert silica.

Wherever a number of different rock types are associated in a landscape, some will be relatively *strong* or resistant to weathering and others will be *weak* or easily altered and removed. A rock that is strong in contrast to other rocks in one area may be weak relative to different rocks in another area. Likewise, rocks that are resistant in a climate dominated by chemical weathering may be weak where physical weathering processes dominate. In general, the more massive the rock (the fewer the joints and bedding planes), the more resistant it is to all types of weathering.

Differences in rock resistance to weathering are highly visible in the landscape. Resistant rocks stand out as cliffs, ridges, or mountains, while weak rocks are eroded away to form valleys, subdued hills, and gentle slopes. One of the most outstanding examples of *differential weathering and erosion* is the scenery at Arizona's Grand Canyon. In Arizona's dry climate, limestone is a resistant rock, as are certain sandstones and conglomerates. As elsewhere, shale is a relatively weak rock. Thus the canyon is a landscape of cliffs and ledges composed of limestone, sandstone, and conglomerate, separated by gentler slopes cut across the shale (Figs. 13.13, 13.14). The cliffs are undermined as the shale beneath is removed. At the base of the canyon, resistant and ancient metamorphic

FIGURE 13.13 Differential erosion in the Grand Canyon. The Tonto Platform is clearly visible, composed of Tapeats sandstone, which overlies early Precambrian schist, into which the river is now cutting. (S. Brazier)

rocks have produced a steep-walled inner gorge.

An equally good example of differential weathering and erosion may be seen in a cross section of the New Appalachians (Appalachian Ridge and Valley) region of the eastern United States (Fig. 13.15). The geologic structure here consists of sandstone, conglomerate, shale, and limestone folded into anticlines and synclines. These have been eroded so that the edges of the steeply dipping rock layers are exposed at the surface. In this humid area, forested ridges composed of resistant sandstone and conglomerate stand 1 to 2000 feet above agricultural lowlands excavated out of shale and soluble limestone.

We have used the expression *differential weathering and erosion* as a single term because, while it is weathering that breaks up rocks at differential rates, it is erosion, or removal, that produces the landforms we see. Thus, weathering and erosion cannot be separated when discussing the origin of landforms.

THE IMPORTANCE OF WEATHERING

The fragmented material (regolith) prepared by both chemical and physical weathering is not only available for erosion—either by direct gravitational transfer or as part of the load carried by rivers, glaciers, or wind—but it also forms the raw material for soil development. The regolith formed by weathering is the major source for the inorganic portion of soils, without which vegetation could not grow. And all animal forms of life depend on vegetation.

Weathering also plays an important role in breaking rocks down into their mineral components and in creating new compounds through chemical change. Valuable ores of iron, aluminum, tin, manganese, and uranium are concentrated through weathering processes operating over immense periods of time, during which soluble bases and even silica are removed, leaving behind increasingly rich residual concentrations of metallic oxides. This effect usually occurs under humid, tropical

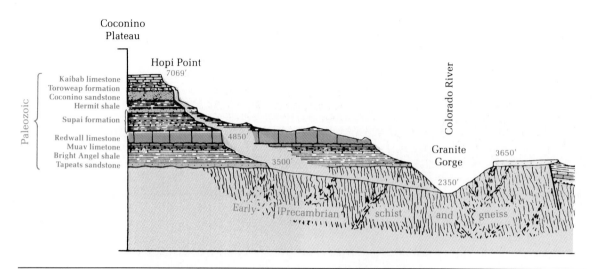

FIGURE 13.14 Arizona's Grand Canyon is a classic example of differential weathering and erosion in an arid climate. This cross section shows the relationship between rock type and surface form in the canyon. (U.S. Geological Survey)

conditions as part of the laterization process (see Chapter 10). Such secondary ores are to be distinguished from primary metallic ores, which are a part of the earth's crust as a consequence of volcanic action.

MASS MOVEMENT

Mass movement is a collective term for all the downslope movements of weathered rock materials that take place in direct response to the pull of gravity. All over the earth's surface and at all times gravity pulls objects toward the earth's center. On sloping surfaces the result of this force is a general downslope movement of loose material. Any material that does not have the stability to resist the force of gravity responds by rolling, falling, sliding, or flowing downslope, stopping only at the bottom of the slope or at a point where there is enough support to resist gravitational pull—the **angle of rest** for the material in question.

Mass movement occurs in a wide variety of ways. A single rock rolling and tumbling downhill is a form of gravitational transfer, as is an entire hillside flowing hundreds or thousands of feet downslope, burying homes, cars, and trees. Some mass movement is catastrophic in scale and violence, as in Figure 13.16. However, the more common type of mass movement is so slow as to be imperceptible, though tilted telephone poles, gravestones, and fenceposts indirectly reveal its presence.

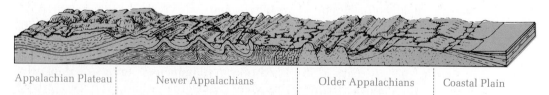

Appalachian Plateau | Newer Appalachians | Older Appalachians | Coastal Plain

FIGURE 13.15 The newer Appalachians (Ridge and Valley) region of the eastern United States is a magnificent display of the effects of differential weathering and erosion in a humid climatic region, as this block diagram shows. (A.K. Lobeck; map courtesy of Hammond Incorporated)

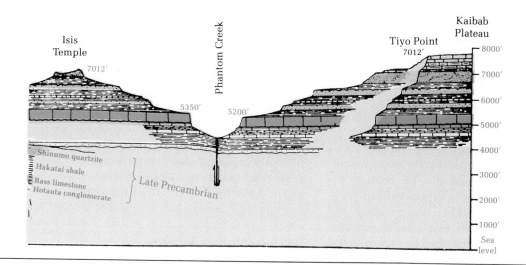

The combination of all forms of mass movement rivals the gradational work of water as a modifier of physical landscapes, because the pull of gravity is always present. Wherever there is regolith or soil on a slope of almost any inclination, gravity will be able to effect some movement downslope. Because gravitational force is stronger on steeper slopes, it pulls down the regolith faster there. Consequently, the regolith on steep slopes is apt to be thinner than it is on more gentle slopes and bedrock may be exposed.

Gravity is obviously the main factor causing mass movement. However, other elements are also important, especially in providing the initial instability leading toward movement. Water is the most important additional factor encouraging mass movement. Water fills pore

FIGURE 13.16 The giant San Jose Poaquil landslide set off by the Guatemala earthquake of February 4, 1976. (E.L. Harp, U.S. Geological Survey)

spaces, greatly increasing the weight of the material into which it has soaked. Water also lubricates material, helping particles to slip over each other more easily. Thus, especially in saturated material, water acts to minimize friction between particles of weathered material and thus to reduce the resistance to gravitational pull.

The shaking produced by earthquakes, explosions, wind, and even by the movements of heavy trucks or trains can be enough to shake material loose from a supported position, triggering movement. The undercutting of slopes by streams and by man's bulldozer are especially conspicuous triggers to mass movement. Groundwater, meltwater, and alternate periods of freezing and thawing and wetting and drying can contribute to mass movement, especially to slow mass movement. Even plants (particularly their root systems) and burrowing animals promote mass movement.

CREEP

Creep is the name given the slowest downhill movement of soil and regolith. Creep is so slow as to be imperceptible to the observer. Yet creep is the most widespread, persistent, and effective of all forms of mass movement, for it is going on at all times on nearly all slopes wherever there are weathered materials and soil available for movement.

For the most part, creep does not produce distinctive landforms. Instead, it gradually wears away slopes and deposits material at the base to be carried away by one of the gradational agents, usually running water. Sometimes the material that creeps downslope is not removed but accumulates in the basin, filling it in and leveling the bottom of the slope.

Several factors facilitate creep; in most locations these factors act in combination to move the regolith and soil slowly downslope. When soil water freezes, it expands in volume, pushing soil particles outward to make room for itself. Some of these particles are lifted vertically upward. When the ice thaws, the soil contracts. Soil particles sink back down, but the force of gravity causes them to be displaced downslope of their original position, as shown in Figure 13.17. The repeated cycles of freezing and thawing, which cause repeated lifting and downslope sinking of soil particles, result in downslope creep of the mass of the soil. The rate of movement is very slow, usually less than an inch per year.

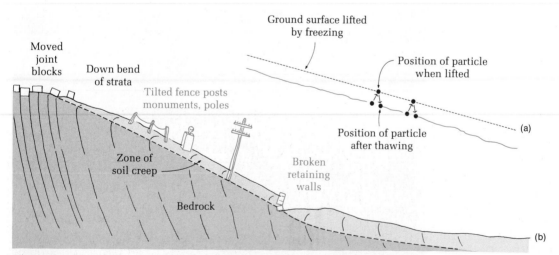

FIGURE 13.17 The relationship of creep to soil volume change. Diagram (a) shows the freeze-thaw mechanism that moves particles down slope. Diagram (b) presents some of the common visible effects of creep on natural and man-made objects. (After C.F.S. Sharpe)

Alternate periods of wetting and drying, which result in the alternate expansion and contraction of soil or regolith, also produce a net effect of downslope movement. This effect occurs because expansion tends to be greater on the downslope side of the material, in response to the ever-present pull of gravity. Likewise, there is less contraction on the downslope side.

Small burrowing animals such as gophers and ground squirrels are very effective earth movers. The tunnels they build help push soil downhill, and the excavated material they bring to the surface also tends to fall downslope. Plant roots, too, push soils outward. Even the trampling of soil and regolith by humans or herds of grazing animals tends to push surface material downhill a little. Every step up or down a steep slope or even across it pushes a little of the surface material downslope to a slightly lower position. Because most of the elements that encourage soil creep act near the surface, the rate of creep is greater there than below the surface. Therefore telephone poles, fence posts, and even trees, all of which are anchored at a level below the surface, exhibit a downslope tilt.

SOLIFLUCTION

Solifluction is the relatively slow downslope movement of soil and/or regolith that is saturated with water. Solifluction is most common at high latitudes or elevations where there is a layer of permafrost beneath the surface. During summer, when temperatures are higher, the top few meters of the soil thaw. However, the layer of permafrost beneath the surface prevents the downward percolation of thawed water. As a consequence, the unfrozen part of the soil becomes watersoaked, creating a soggy mass that sags slowly downslope in response to gravity until the next freeze arrives. The annual movement may amount to only several centimeters. Solifluction is encouraged by the sparseness of vegetation found in tundra regions. Plant roots on gentle slopes act to retard mass movements, but where permafrost exists, roots are very shallow. The portion of the soil that freezes and thaws annually and that moves when saturated is called the *active layer*.

Evidence of solifluction can be seen in tundra landscapes where irregular lobes of soil produce hummocky slopes (Fig. 13.18). Sometimes these slopes exhibit fractures and wrin-

FIGURE 13.18 Solifluction slippage marked by surface stripes, Jotunheimen Plateau, Norway. (S. Brazier)

kles from compression and tension during downslope movement. The general effect of both creep and solifluction is to produce rounded hillcrests and a landscape free of sharp angular features—the subdued landscapes usually associated with humid climates.

FORMS OF RAPID MASS MOVEMENT

There are several varieties of rapid mass movement. Some of these are landslides, rockslides, rockfalls, slumps, earthflows, and mudflows. They are distinguished from the slower types in that their rate of movement is visible and their effect on surface configuration is more dramatic. Of course, the speed of movement varies in each particular situation, depending on such variables as the quantity and composition of the material being moved, the steepness of slope, the nature of the vegetative cover, and the triggering factor.

Rapid mass movements usually leave a visible scar where material has been removed upslope and leave a deposit below consisting of debris that has slid, fallen, flowed, or rolled downslope. Despite the often spectacular nature of this type of mass movement both in force and in effect on the land, slow mass movement, especially creep, has a far greater cumulative effect on surface configuration.

LANDSLIDE   Though the term is often used to refer to any form of rapid mass movement, earth scientists give the name **landslide** only to the rapid downslope movement of a mass of material that moves as a unit and carries with it all the loose material above bedrock, and possibly great masses of bedrock as well. The movement is sudden, and the moving mass often attains very high velocities. Large landslides are rare but are often newsworthy because of their destructive qualities.

Landslides tend to occur more frequently on steep slopes in mountain regions during or after periods of heavy rain because of the additional weight of water and its lubricating qualities. Many landslides have been triggered by earthquakes. They are also more frequent on slopes that are undercut by streams.

While some landslides involve only the surface regolith, larger movements may detach huge masses of rock. Some rockslides are truly enormous, with volumes measured in cubic kilometers. Anything in their path is obliterated. More importantly, they may dam up valleys, which soon become filled with lakes. When the lakes become deep enough, they may wash out the dams, producing catastrophic sudden floods in the valleys downstream. Thus, immediately after such a major rockslide men must clear out the resulting dam and control the outlet of water trapped above it. This was done successfully after the Hebgen slide in southern Montana in 1959 (Fig. 13.19). This slide, one of the largest in North American history, was triggered by an earthquake and killed 26 people camped along the Madison River.

Major **rockslides** result from rock mass instability related to geologic structure and stream or glacial undercutting of slopes. Like landslides, major rockslides usually occur during exceptionally wet periods when the rock mass, or a sliding plane at its base, is well lubricated. Earthquakes have been associated with many, but not all, large historic rockslides. Today there are many locations in mountain regions where enormous slabs of rock supported by weak materials are poised on the brink of detachment, waiting only for an unusually wet year or a jarring earthquake to set them in motion.

ROCKFALL   When a small individual rock mass or several rocks fall nearly vertically downslope, clattering over other loose rock and debris, the movement is called a **rockfall.**

The rocks or rock fragments involved in rockfalls can vary greatly. A rockfall may consist of tiny granular particles skittering downslope or of a huge boulder bounding downhill and fragmenting along the way. In steep, rocky, mountainous areas, where rockfalls are common, cone-shaped accumulations of rock fragments build up at the base of the cliffs,

FIGURE 13.19 This large rockslide, which buried a section of the valley of the Madison River in Montana, just outside the boundaries of Yellowstone National Park, was set off by an earthquake. Over 20 vacationers were buried beneath the slide debris and a large lake was created by the incident. The lake is named Earthquake Lake. (U.S. Department of Interior, Bureau of Reclamation)

forming the talus slopes or cones mentioned earlier (see Fig. 13.6).

Rockfalls are particularly common in spring, when a traveler can encounter rocks scattered on mountain roads. Alternate periods of freezing and thawing at that time disturb precariously balanced rock masses, loosening them from their previously secure positions.

SLUMP AND EARTHFLOW Sometimes a mass of soil on a slope slips or collapses in a backward rotation, down at the top and up at

VIEWPOINT: LIVING IN HAZARD ZONES

Freeway collapse due to an earthquake in 1980 near Eureka, California. (California Department of Transportation)

The summer of 1982 was like most summers in southern California—hot, bright, sunny days with almost no rain. In the late summer and early fall, acres of the dry chapparal burned in the nearby mountains. This, too, was not unusual. In fact, it is estimated that about 100,000 acres burn every year in the Los Angeles area. Then the winter rains came, as they do every year, only in the winter of 1982–83 the torrential downpours were some of the worst in memory. The warm El Niño waters in the Pacific gave added strength to the winter storms. The result was that massive mudflows poured down the mountains, destroying everything in their path. People were killed, many others injured, and millions of dollars' worth of property was destroyed or damaged. What had happened? The sun-baked soils on the hills, unprotected by vegetation and unanchored by plant roots, had been unable to absorb the large amounts of rain that fell. Being largely unstable anyway, the lubricated and slippery soils simply poured off the hills—over cars, trees, through the windows and doors of expensive hillside homes. Those homes whose foundations were not built on firm ground were knocked over by the force of the flowing mud. Others were damaged, often beyond repair.

In the early morning hours of February 9, 1971, an earthquake literally shook people out of their beds in the San Fernando Valley in southern California. When it was over, 64 people were dead, and 800 homes, 65 apartment buildings, 4 hospitals, and 570 commercial buildings had been destroyed. Had the quake occurred a few hours later when children were in school and adults at work or shopping or driving on the freeways, the loss of life would probably have been in the thousands.

Then there is San Francisco—a jewel of a city with its hills and cable cars and bridges—and the potential site of a disaster that exceeds most people's imaginations. Seeing San Francisco as it stands today it is hard to believe that in 1906 a major earthquake and subsequent fire destroyed nearly the entire city. Since that time San Francisco has been rebuilt, and today the fabled "city by the bay" attracts tourists and new residents every day. However, much of this rebuilding has ignored the lessons of 1906. Skyscrapers have been built downtown within striking distance of the San Andreas Fault; an entire neighborhood has been built on fill that experts say would shake like a bowl of jello in the next quake; and outside the city rows of houses have been built on hills in what one earth scientist has called "essentially one big landslide area." It is estimated that if the 1906 earthquake were to strike San Francisco today, it would kill 3500 people (compared to 600 in 1906) and would injure 350,000 people. There is a saying in San Francisco that goes something like "We don't wonder *if* there will be another earthquake. We just wonder *when*."

Certainly California does not have a monopoly on the hazard zones of the world. The Atlantic coast of the United States, for example, continues to attract resort builders and summer home owners in spite of its proven record as a hurricane zone. Nor does California have a monopoly on people who choose to build and live in areas of high risk, but the state—with its ever-increasing population and its geologically unstable and earthquake-prone terrain—does offer classic examples of the growing tendency to build in hazard zones.

As more people have moved into California and as urban and suburban centers have spread, the pressures for land have increased accordingly. More and more homes are being built on hillsides, on unstable soils, on filled lands, and in canyons and ravines subject to sudden mudflows or floods. It is likely that many of the people who live in these homes are not fully aware of the potential dangers. There are certainly land developers and real estate people involved in selling houses in hazard zones who play down the potential dangers of a particular house site and play up the "spectacular views from this house of glass balanced on stilts above the city lights and smog." In many cases, however, it is not just the home buyer who does not know what dangers might befall him. Often the builder himself is unaware. In many situations, land development has proceeded at a pace far ahead of the geologic studies necessary to determine potential dangers and to make recommendations that could minimize damage in the case of a natural disaster. Then, too, there is the all-too-human tendency to believe that "disaster will always happen to the guy next door, not to me."

Probably one of the most important reasons people continue to move into and live in hazard zones is that many of these areas have positive features that apparently outweigh the possibilities of total destruction. For example, California's Mediterranean climate allows a freedom of lifestyle that obviously appeals to a great many people, and many of its more dangerous hillsides provide home owners with breath-taking views. These benefits—the climate, spirit, view—can be reaped everyday, whereas a landslide, a mudflow, or an earthquake is a once-in-a-lifetime possibility. As a seismologist said, speaking of his hillside home, "I guess this whole lot could wind up in our neighbor's pool down there someplace. We'd never move. We love the house and we love the area." Likewise those who choose to locate their vacation cottages in coastal resort towns like Biloxi, Mississippi, that are susceptible to the damaging waves and winds of a hurricane must find that the everyday advantages offset the risks.

It seems apparent that there is little we can do to convince people not to build in hazard zones and even less that we can do to convince them to move out of the hazard zones once they have settled there. And there seems little that can be done—at this point—about averting such natural disasters as floods, landslides, and earthquakes. Therefore, to minimize the personal and economic losses that are likely to result, we must focus our attention on other aspects of the problem.

Certainly a better understanding of the underlying structures of hazard zones, with mapping of soils, underlying rock features, and characteristic terrain and water drainage, is necessary. Increased knowledge of earthquakes and an ability to predict their occurrence would also sharply reduce losses in susceptible areas. In many cases, however, if people would simply apply some of the basic principles of geology, hydrology, and pedology that have been

known for years, many of the losses associated with hazard zone disasters could be averted.

In addition, there must be greater awareness on the part of the public of the possibilities of danger in these areas and of how and where to build to minimize damage. Building codes are another important part of reducing damage and loss of life in hazard zones. California has enacted some of the strictest such legislation in the country; the problem is enforcement. If California's building codes were enforced today, thousands of its buildings would have to be torn down or brought up to code. Structures that are critical to the functioning of an area hit by a disaster (such as hospitals, power stations, communications centers, fire and police headquarters) should be designed so that they can withstand the most extreme punishment. Finally, there should be well-thought-out plans and contingency programs for possible emergency situations.

Rockslide damage and burial of Highway 1 south of Big Sur, California. The slide resulted from severe winter storms of 1982–1983. (G. G. Wieczorek, U.S. Geological Survey)

393

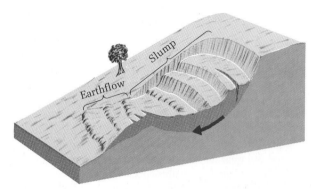

FIGURE 13.20 Cross section of a typical hillside slump and earthflow.

the base, so that what was once a portion of the hill slope ends up tilting backward (Fig. 13.20). The curved backward rotation of such a **slump** distinguishes it from a landslide. Slumps are likely to occur where moisture is concentrated at the base of a water-soaked mass of clay-rich soil. Slumps are conspicuous phenomena during exceptionally wet winters in California's San Francisco Bay region, where they frequently damage expensive hillside homes.

Earthflows are linear movements of moist (almost liquified), clay-rich regolith. Slumps often change into earthflows in a downslope direction. Some California earthflows are a mile or more long, and they generally have a tonguelike shape. Active earthflows move at rates of a few meters per hour, per day, or per month. Thus they are somewhat more rapid than solifluction but slower than landslides.

DEBRIS FLOW   The channeled movement of water and rock debris mixed in various proportions is called a **debris flow** or **mudflow.** It is more fluid than an earthflow and moves much more rapidly, its high water content helping to speed it along. In general, debris flows follow valleys rather than flow as sheets downslope.

A debris flow is the most fluid of all mass movements. Sometimes it is difficult to distinguish between a runny debris flow and a muddy stream. Debris flows result from torrential rainfall on steep, poorly vegetated slopes in dry regions such as the mountains of southern California (Fig. 13.21). The rains flush weathered debris into canyons, where it is mixed with flood waters. The result is a torrent of mud, gravel, boulders, plant debris, and water, which can knock out bridges and smash through or half bury buildings and homes. Such torrents frequently occur in wet winters following summers in which fires have destroyed hillside vegetation and left the soil vulnerable to removal by water runoff.

FIGURE 13.21 When torrential rains fall on poorly vegetated slopes, particularly those denuded by fire, the soil and loose regolith can produce devastating debris flows in the canyons and valleys below. (California Department of Transportation)

WEATHERING, MASS MOVEMENT, AND THE LANDSCAPE

In this chapter we have discussed only two phenomena associated with gradation: weathering and mass movement. Although both weathering and the slower forms of mass movement rarely attract human attention, they vitally affect the landscape. They can create distinctive landforms without help from other more familiar processes, and every slope reflects the local nature of weathering and mass wasting processes. The effect of rock structure on the landscape is expressed through the weathering characteristics of the material involved, and both the character of the weathering processes and the way that weathered regolith is affected by gravity are major determinants of how the landscape looks. Steep slopes reflect slow weathering; gentle slopes result from rapid weathering; rounded forms suggest chemical weathering and slow but general mass movement (creep or solifluction); and angular slopes are associated with physical weathering and mass movement in the form of rockfalls, rockslides, and debris flows. In the following chapters we will consider the forms produced by the more dynamic and generally visible agents of gradation: running water, groundwater, wind, glacial ice, and coastal waves.

QUESTIONS FOR DISCUSSION AND REVIEW

1. How does weathering differ from the transporting agents of gradation?

2. How are the joints and fractures in a rock related to the rate at which weathering takes place? How does physical weathering encourage chemical weathering in rock?

3. What are several ways in which expansion and contraction can affect the weathering of rock?

4. Under what environmental conditions do chemical decomposition and mechanical disintegration prevail? Where are these conditions found?

5. Why is chemical weathering more rapid in humid climates than in more arid climates?

6. In what three ways does chemical decomposition prepare rock for gradation? Give examples from instances of oxidation, hydration, and solution.

7. Compare the ways in which hydrolysis and carbonation work below the earth's surface.

8. What types of rocks best resist all types of weathering? How visible are they in the landscape, compared to less resistant rocks?

9. What function does weathering perform in shaping the land? What other functions does weathering perform?

10. What two factors encourage mass movement the most? How do they work together?

11. What factors facilitate creep? How does solifluction occur?

12. Describe the general effects of creep and solifluction on the landscape. With what climates are these effects usually associated?

13. What conditions encourage landslides and rockslides?

14. State the primary difference between an earthflow and a debris flow. What causes the development of each?

15. Suggest ways in which weathering and mass movement affect human lives.

LAND SCULPTURE BY UNDERGROUND WATER AND STREAMS

OCCURRENCE AND SUPPLY OF GROUNDWATER

While fresh water on or below the earth's surface, apart from that locked up as snow and ice, is most visible in streams and lakes, the greatest part of this resource is found beneath the ground surface as groundwater (also referred to as underground water). In the first part of this chapter we will look at the significance of groundwater for human beings, especially as it affects their domestic, agricultural, and industrial water supplies. And, in keeping with our continuing examination of those processes that help to shape the land, we will look at the gradational effects of groundwater. Although of minor gradational importance when compared with streams, groundwater gradation has a significant impact on landforms in certain areas.

Almost all groundwater is derived from precipitation. That which is so deep below the earth's surface that it has never been involved in the hydrologic cycle is called **juvenile water.** Other groundwater was once involved in the hydrologic cycle, but because of changes in the

earth's surface it has been locked out of that cycle for a long period of time. Such water, called **connate water,** is trapped in layers of sediment laid down by ancient rivers or seas. Future changes in the lithosphere could release these trapped waters and return them to the hydrologic cycle. Through volcanic activity these waters could be released in the form of geothermal energy (steam and hot water). The most obvious evidence of this activity is in the form of hot springs and geysers.

By far the largest proportion of groundwater is *meteoric,* meaning that it is derived from atmospheric sources. Meteoric groundwater forms the earth's primary supply of fresh water. It is brought to the surface by springs and wells, and it contributes significantly to standing bodies of water such as lakes and to running water in streams.

GROUNDWATER ZONES AND THE WATER TABLE

Groundwater includes water found within the soil as well as in the loose rock and bedrock below. Under conditions of modest precipitation

Bridge of Dee, Scotland. (S. Brazier)

and good drainage, as meteoric water sinks into the ground, it first passes through a level where air as well as water fills the spaces and pores within the soil and rock. This level is called the **zone of aeration.** Further movement downward brings the water to a second level, called the *zone of saturation,* where all the openings are filled with water. In between the two levels, and marking the upper limit of the zone of saturation, is an undulating surface called the **water table.** The water table does not remain fixed in position but fluctuates with the quantity of recent precipitation. After heavy precipitation, the water table will be higher. Since the water table reflects precipitation amount, it lies closer to the surface in humid regions than in arid regions.

In humid regions there are actually three groundwater zones (Fig. 14.1). The lowest zone is always saturated. The upper zone is almost never saturated. Between these two is a zone that is saturated under humid conditions and not saturated under drier conditions. It is through this middle zone that the water table fluctuates. Obviously, a well or spring originating within the permanently saturated zone will always have water, but one originating in the intermediate zone of fluctuation will run dry when the water table falls below it.

In many desert regions there is no saturated zone at all. Meteoric water evaporates in the soil rather than percolating downward to a water table. If groundwater is present, it may be very old, having accumulated during a past period of greater humidity. When such water is extracted in wells, it is not replaced from the atmosphere and the water table falls rapidly.

The water table in humid regions is not horizontal but instead tends to follow the general contours of the land. It is higher under hills and other high surfaces and lower beneath valleys and depressions. It is usually closer to the surface under low places than under high places. Because water tends to seek its own level and because it is affected by gravitational force, the groundwater under high land surfaces tends to flow downslope to a lower level as would a stream on the surface.

In humid areas of low relief the water table is often so high that it intersects the ground surface, producing lakes, ponds, or marshes such as those common in New England and along the Gulf Coast from Louisiana to Florida. Where the landscape is one of hills and narrow valleys the lowest points on the water table are controlled by the position of valley floors. As a stream deepens its valley, it makes a valley in the water table as well. Groundwater then

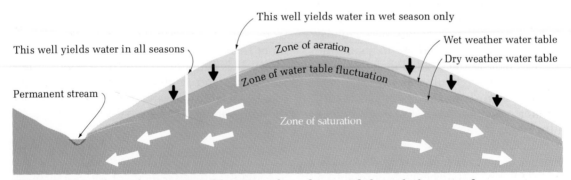

FIGURE 14.1 Groundwater zones. Water percolates downward through the zone of aeration, where air fills most of the openings between soil and rock particles. Eventually a level is reached at which all openings are filled with water. This level is the water table. Below it is the zone of saturation. The water table fluctuates in level as a consequence of varying input of water by precipitation. Permanent streams are maintained between rains by inflow from the zone of saturation.

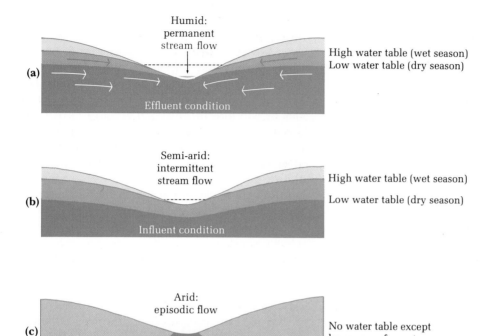

FIGURE 14.2   The influence of the water table on stream flow. (a) In humid areas groundwater flows into major stream channels throughout the year (effluent condition). Thus the streams flow continuously. (b) When a dry season occurs, the water table may drop below the stream bed, so that the stream dries up until the next wet season. (c) In desert regions the absence of a water table means that stream runoff occurs only during rains, and flow diminishes downstream (influent condition) because of seepage loss.

moves downslope in the subsurface and enters the stream. This **effluent** condition keeps the stream flowing between rains.

Streams in semiarid and arid regions flow only seasonally or immediately after rains. In the first case the water table lies below the stream bed during dry periods and rises to intersect the stream bed during wet periods. The stream is fed by groundwater only during wet phases but loses water by seepage during dry **influent** periods. In true desert situations, this influent condition is continuous, since there is no groundwater available to feed streams, and there is surface water flow only during and immediately after rains. Much of this downstream flow is eventually lost by seepage into the dry ground under the stream channel (Fig. 14.2).

FACTORS AFFECTING DISTRIBUTION OF GROUNDWATER

The amount of groundwater in an area depends on a variety of factors. Most fundamental is the amount of precipitation that falls in a given area and in areas that drain into it. Second is the rate of evaporation. A third factor is the amount and type of vegetation cover on the land. Although dense vegetation transpires great amounts of moisture back to the atmosphere, it prevents rapid runoff of rainfall, encourages percolation of water into the ground, and prevents rapid evaporation by providing shade. Thus the effect of forests in a humid area is to increase the supply of groundwater.

A fourth factor affecting the distribution of groundwater is the porosity of the soil and rocks. **Porosity** refers to the amount or proportion of space between the particles that make up the soil or rock. Thus some gravels, consisting of coarse particles that do not completely interlock, will be exceedingly porous and can hold large amounts of water. Dense interlocking crystalline rocks like granite, with virtually no pore space, can hold little water.

On the other hand, granite may indeed hold water within joints, which allow the passage of this water rather freely. They thus will be described as permeable. The **permeability** of a material, its ability to allow the passage of water through it, is related to the number and size of the spaces within the rock, which may be large pore spaces or joints, bedding planes, or other fractures. Porosity and permeability are not synonymous terms. A coarse sandstone, large-pored and jointed, will typically be porous and permeable: a clay, on the other hand, finely porous and unjointed, may contain significant amounts of water but be impermeable, since the water is locked around the tiny particles by soil moisture tension (Fig. 14.3). And a nonporous granite, by reason of its jointing, may be highly permeable. Porosity therefore affects the storage of groundwater and permeability affects the groundwater movement. Both of these factors affect the availability of water for wells and springs.

A rock layer like sandstone or limestone, which is porous and permeable, or a gravel or sand bed, can act as a container and transmitter of water and is called an **aquifer** (Latin: *aqua*, water; *ferre*, to carry). A rock like a shale, or clay, relatively impermeable and very finely porous, will restrict the passage of water and limit its storage, and therefore is called an **aquiclude** (Latin: *claudere*, to close off).

Sometimes a porous and permeable layer—an aquifer—will be found between two aquicludes. In this case water flows in the aquifer much as it would in a water pipe or hose. Water can pass through the aquifer and cannot escape outward through the aquicludes. Furthermore, soil water percolating downward may be prevented from reaching the zone of saturation by an aquiclude. An accumulation of such water above an aquiclude is called a **perched water table** (Fig. 14.4). Careless drilling can puncture the aquiclude supporting a perched water table, so that the water drains out. The well must then be deepened to reach the true water table.

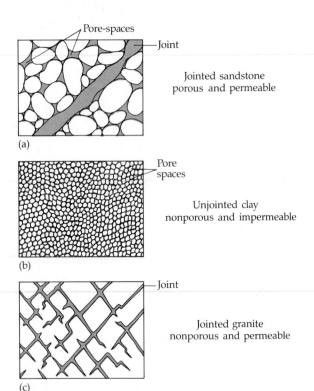

Porosity and permeability

FIGURE 14.3   The relationship between porosity and permeability in different kinds of rocks. (a) Jointed and porous sandstone: permeable. (b) Unjointed and virtually nonporous clay: impermeable. (c) Jointed and nonporous granite: permeable.

AVAILABILITY OF GROUNDWATER

SPRINGS

Springs are surface outflows of groundwater. They are caused by the landform configuration, the level of the water table, and the rela-

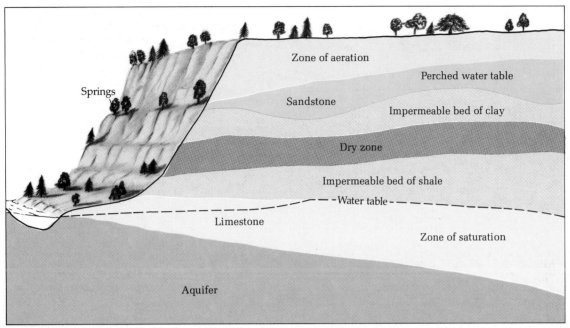

FIGURE 14.4 A perched water table occurs when an impermeable rock stratum is present. It is underlain by a dry zone, between impermeable beds of shale and clay. The water table in the permeable limestone bed carries water into the river valley.

tive position of various types of aquicludes. For example, a spring may occur along a valley wall where a stream or river has cut through the land to a level lower than a perched water table. There an impermeable layer of rock prevents further downward penetration of groundwater and thus forces the water to move horizontally. Such a spring is permanent if the water table always remains at a level above the valley floor; otherwise the spring is intermittent, flowing only when the water table is at a high level.

WELLS

Wells are artificial openings in the land surface dug or drilled below the water table. Water is extracted from wells by lifting devices ranging from water buckets to gasoline or electrically powered pumps. In shallow wells, the supply of water depends on fluctuations in the water table. In contrast, deeper wells that penetrate into lower aquifers provide more reliable

sources of water and are unaffected by seasonal periods of drought.

In an area where there are many wells, the drain on the supply of groundwater may exceed the intake of water that replenishes the supply. In most areas irrigated from wells the water table has fallen far below the reach of the original wells (Fig. 14.5). Progressively deeper wells must be dug (or the old ones extended) in order to reach the lower supply of water. In northern India in the Ganges Valley, the development of modern wells, deeper than the traditional hand-dug wells, has resulted in the lowering of the water table. In the southern High Plains of Texas, Oklahoma, and Nebraska, the drawdown of local aquifers is of serious concern to farmers. Costs of pumping are rising, causing higher crop farming costs, and in many instances dryland farming is replacing irrigated acreage.

In extreme cases of high groundwater demand, *subsidence,* or sinking of the land, occurs as the water pressure is reduced due to pump-

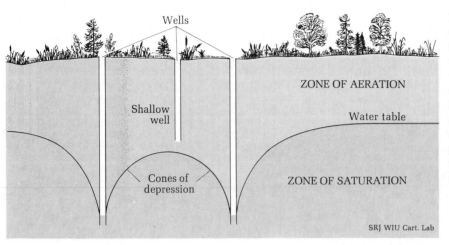

FIGURE 14.5 Cones of depression in the water table caused by pumped wells. In areas of many wells, adjacent cones of depression intersect, lowering the water table over the entire area, thus causing shallow wells to go dry. In extreme cases the ground surface may subside.

ing. The Central Valley of California, Mexico City, and Venice, Italy, all suffer from subsidence. In southern California, groundwater is artificially replaced by diverting rivers over permeable deposits. This process is known as groundwater recharge.

Because groundwater filters down through many layers of soil and rock before it reaches an aquifer, it is free of sediment but often carries a large chemical load dissolved from the materials through which it passes. Groundwater is thus said to be *hard* in comparison with *soft* rainwater. Moreover, just as increases in population, urbanization, and industrialization have resulted in the pollution of some of our surface waters, they have also resulted in the pollution of some of our groundwater supplies. The most recent danger to

groundwater supply has been from toxic waste seepage. In coastal regions, where pumping the groundwater has lowered the water table, salt water from the sea is able to seep in and replace the fresh water, since the pressure is no longer there to hold back the salt water. This salt water replacement has occurred in many localities, notably in southern Florida, Long Island (New York), and Israel.

ARTESIAN WELLS

In flowing **artesian wells** water rises to the surface and flows out under its own pressure, without pumping. In an artesian system water under pressure rises to a level above the water table. Certain conditions are prerequisite to an artesian structure (Fig. 14.6). First, a porous

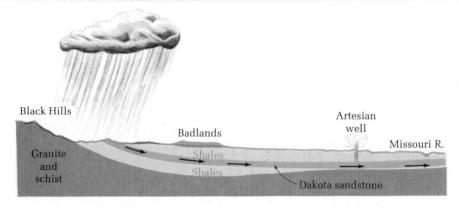

FIGURE 14.6 Conditions producing an artesian system. The Dakota sandstone, which is approximately 30 meters thick, transmits groundwater from western South Dakota to wells 320 kilometers away. (After Lobeck)

aquifer such as sandstone must be exposed at the surface in an area of high precipitation or infiltration. This aquifer must absorb water at the surface, must incline downward hundreds or thousands of feet below the earth's surface, and must be confined under an impermeable layer that prevents upward escape of the water. These conditions cause the aquifer to act as a pipe that conducts water through the subsurface. The water in the "pipe," provided with no exit, is under pressure from the water above it (closer to the area of intake at the surface). As a consequence of this pressure, water will move toward any available outlet. If that outlet happens to be a well drilled through the impervious layer and into the aquifer, the water will rise in the well, sometimes gushing out at the surface. The height to which the water rises depends on the amount of pressure exerted on the water. Pressure in turn depends on the quantity of water in the aquifer (more water, more pressure), on the angle of incline (steeper slope, more pressure), and on the number of other outlets, usually wells, available to the water (more outlets, less pressure). Sandstone exposed at the surface in Colorado and South Dakota transmits artesian water eastward to wells as far as 320 kilometers (200 mi) away. Other well-known artesian systems are found in Olympia, Washington, eastern Australia, and the western Sahara Desert. The word "artesian" was derived from the Artois region of France where such systems were once common.

LAND SCULPTURE BY UNDERGROUND WATER

The runoff of surface water is the major process that shapes landforms, but groundwater is also an agent of gradation. It is, of course, vital in the subsurface chemical weathering process, and, like surface water, it removes, transports, and deposits material prepared by weathering.

The principal mechanical effect of groundwater is to encourage mass movement through the lubrication of weathered material

and soil, which produces slumps, earthflows, and landslides. But it is by chemical action that groundwater contributes most to gradational processes. Through the removal of rock materials by solution and the deposition of those materials elsewhere, groundwater is an effective land-shaping agent, especially in areas where strata of limestone are present. Because limestone is easily dissolved by groundwater (which, as we have noted previously, is a weak solution of carbonic acid), distinctive landscapes evolve wherever groundwater can act on limestone.

KARST OR LIMESTONE LANDSCAPES

The distinctive features of a limestone landscape occur where limestone strata lie at or near the surface. The eastern Mediterranean region in particular exhibits features of limestone solution on a large scale. These are most clearly developed on the Karst Plateau along Yugoslavia's scenic Dalmatian Coast. Any landforms developed by solution in limestone are called **karst** landforms after this classic locality. Other Karst regions include areas of Kentucky and Tennessee, the Massif Central in France, parts of northern England, and vast areas of southwestern China.

Since limestone is a relatively common rock type, features created by the action of surface water or groundwater are found in many parts of the world. However, a true karst landscape, in which solution of limestone has been the dominant agent of land formation, is rather uncommon because of the special circumstances required.

A humid climate with plenty of precipitation is most conducive to karst development. Karst features are not developed in arid climates. However, some arid regions have karst features, which originated during a period when the climate was more humid.

A second important factor in the development of karst landforms is active movement of groundwater, so that the water saturated with dissolved calcium carbonate can be quickly replaced by unsaturated water. Vigorous movement of groundwater occurs when an outlet at a lower level is available, such as a deeply cut stream valley or a tectonic depression.

Limestone landscapes are typically marked by an absence of surface streams, since the permeability of the rock encourages the disap-pearance of surface water underground. Fre-quently the surface junction between imperme-able clays and shales and permeable limestones is marked by a series of **swallow holes** down which streams and rivers are "lost," or disap-pear: These seepage points may be widened

(a)

(b)

FIGURE 14.7 The south-eastern United States has a number of areas where sinkholes are common. (a) The nation's deepest sinkhole, with a depth over 50 meters, is in Al-abama. (U.S. Geological Survey) (b) A large sink-hole resulted in the city of Winter Park, Florida, when underground lime-stone caverns collapsed during a severe drought in May, 1981. (*Sentinel Star*, Orlando, Florida)

and deepened by solution into great vertical **potholes,** very often tens or even a few hundreds of meters deep.

Surface outcrops of limestone are pitted and pock-marked by chemical solution, especially along the joints, forming large, flat, furrowed exposures of limestone pavements. As this surface solution tends to be concentrated at joint intersections, circular depressions gradually develop called **sinkholes.** Often hundreds

of these may be conspicuously developed as in the southeastern United States from Florida to Kentucky (Fig. 14.7). Sinkholes that become plugged with clay or that have eaten through to a layer of insoluble shale produce lakes and ponds such as those found in central Florida.

In time such sinkholes, enlarged by continuing solution, may merge to form larger depressions called **dolines** (Fig. 14.8). Meanwhile the slowly circulating water below the

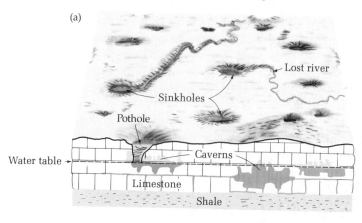

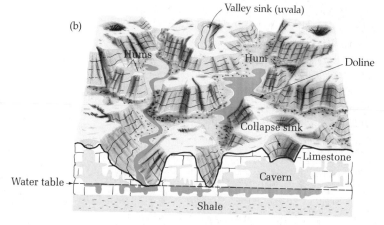

FIGURE 14.8 Karst development. (a) The cycle of erosion in soluble limestone begins with the incision of erosional valleys that lower the water table. Surface solution at joint intersections and along joints in the limestone causes the development of circular valley sinks, or dolines, and longer uvalas. Caverns form by underground solution and are evident by the disappearance of surface water and the appearance of collapse sinks and valley sinks. (b) Eventually the layer of limestone is divided into separate masses, and surface drainage reappears on insoluble rocks beneath. (c) In the last stages, limestone remnants, called haystack hills or hums, remain above the new land surface developed on other rock types.

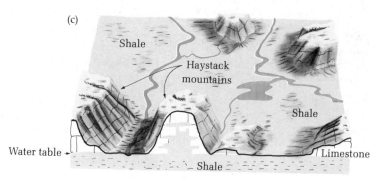

surface, traveling along joint and bedding planes, is also dissolving the limestone, so that **underground caves** and **caverns** develop, until the massive limestone complex becomes eaten into like a Swiss cheese. Collapsing subsurface cave roofs become **collapse sinks,** and the coalescence of sinks and potholes and dolines creates larger and larger depressions, often linearly arranged along the former underground water courses, to form valley sinks or **uvalas.** (These somewhat unusual terms are derived from Slavic languages used in Yugoslavia.)

In a mature karst landscape a complex underground drainage system all but replaces the normal surface patterns. The landscape may show large valleys with no streams at their base; these valleys were excavated by surface streams that were later diverted to underground paths by the development of swallow holes on the valley floor. This process is characteristic of the Mammoth Cave area in Kentucky. Some of the "lost rivers" may re-emerge as springs where they have cut down to impervious beds below the limestone.

In the late stage of karst development, especially in the wet tropics, only limestone remnants are left standing above the insoluble rock below. These remnants are usually in the form of small steep-sided and cave-riddled karst towers called **haystack hills** (or **hums**) (Fig. 14.8). Examples of this spectacular landscape are found in China and adjacent southeast Asia and on the islands of Puerto Rico, Cuba, and Jamaica (Fig. 14.9). These have been described as "egg-box" landscapes, since an aerial view of the pits and hums resembles the design of cardboard egg boxes.

Underground caverns are the most spectacular forms created by the solution of limestone. Groundwater, sometimes flowing as streams underground, carves out networks of caverns that later become filled with air and decorated with depositional forms. Examples of these in the United States are Carlsbad Caverns in New Mexico, Mammoth and Colossal Caves in Kentucky, and Luray and Shenandoah Caverns in Virginia. In fact, 34 states have caverns open to the public. Some are quite extensive, with rooms over 30 meters (100 ft) high and with kilometers of connecting passageways.

One factor necessary for cavern development in karst landscapes is the jointing of limestone strata. Groundwater solution works along jointing planes, for the joint pattern is evident in the rectangular pattern of the caverns that

FIGURE 14.9 Karst towers, Guilin, Southwest China. (S. Brazier)

are created. Streams do not run through all caverns, although the dry caverns often show evidence of previous stream activity, such as sedimentation of clay and silt on the cavern floor.

The depositional features of limestone caverns produce some of the most beautiful forms found in nature. Where water drips from the ceiling of a cavern and is evaporated (or partially evaporated), it leaves behind a deposit of calcium minerals called **travertine.** As these travertine deposits grow downward, they form icicle-like spikes called **stalactites** that hang from the ceiling. Calcium-saturated water dripping onto the floor of a cavern builds up similar but more massive structures called **stalagmites.** Stalactites and stalagmites often meet and continue growing to form columns or pillars (Fig. 14.10). A great variety of other depositional forms **(speleothems)** are also found, many of them very delicate in structure.

Not all caverns are alike. Some are well decorated with speleothems; some are not. Many have several levels and are almost spongelike in structure, while others are linear in pattern. The great variation in cavern forms indicates variations in mode of origin. Most large caverns appear to have developed either at the water table, where the rate of solution is most rapid, or below the water table in the zone of saturation. Subsequently, a decline in the level of the water table caused by the incision of surface streams, climatic change, or tectonic uplift has replaced the cavern water with air, allowing speleothem formation to begin. Underground rivers deepen some air-filled caverns, and collapse of their ceilings enlarges them upward. Many smaller caverns appear to have developed in the zone of aeration entirely above the water table.

Cavern development is a very complex process, involving such variables as rock structure, groundwater chemistry, hydrology, and regional tectonic and erosional history. The science of **speleology** (cavern studies) is a particularly challenging one, especially if we appreciate the fact that all that is known about caverns has been discovered by men and

FIGURE 14.10 Carlsbad Caverns is one of the largest and most impressive of the world's known caverns. This picture of one room in the enormous cavern system shows icicle-like stalactites, stalagmites, and pillars where the two have merged. (U.S. Department of the Interior, National Park Service)

women "spelunkers" crawling through mud and water in dark passages beset by unknown obstacles hundreds of feet underground.

Reference here should be made to another unique set of groundwater-related phenomena, namely **hot springs** and **geysers.** Here the water source is much deeper than in karst regions and indeed may well be connate water. So deep is its source that this water is heated and hence is referred to as geothermal water. It probably has been heated by contact with magmas deep below the surface and is accompanied by steam. Where the water bubbles out fairly continuously, this is a hot spring; where it is an intermittent and somewhat eruptive process, like Old Faithful in Yellowstone National Park, it is a geyser. Geysers appear to erupt when the steam pressure below reaches a critical level and forces the column of heated water out of the crust in an explosive manner.

Both hot springs and geysers contain significant amounts of calcium carbonate in solution, which becomes deposited in various forms, often as terraces or cones around the vent. These calcareous **tufa** and siliceous **geyserite** deposits are colorful and impressive forms (Fig. 14.11).

Geothermal groundwater phenomena are associated with areas of tectonic instability, often plate margins, such as in California, Mexico, New Zealand, and Italy. Iceland's use of such thermal waters for domestic heating and greenhouse fruit and vegetable gardens is well-known.

THE STREAM SYSTEM

Running water is the most important of the gradational agents that are constantly at work shaping the land. As an agent of gradation, running water performs both erosional and depositional functions. Streams are running waters that are normally contained in well-defined channels. Processes associated with the work of streams are known as **fluvial processes** (from Latin: *fluvius,* river).

The surface flow of all precipitation is called **runoff.** The amount of runoff in an area depends on several variables. Among these are the amount of precipitation, the evaporation rate, the permeability of surface materials, the density and type of vegetation cover, and the degree of surface slope. Of course, man can affect some of these variables and can greatly modify the natural runoff characteristics by urbanization, dam building, mining, logging, and agriculture.

Runoff occurs when there is more precipitation than is intercepted by the vegetation cover, evaporation, or infiltration into the ground. For short distances, runoff may occur in sheets (as **sheet wash** or unconcentrated flow), or it may subdivide into tiny rivulets or rills. Eventually this overland flow covers the land with a pattern of well-defined channels,

FIGURE 14.11 Geyserite deposits (siliceous) from hot springs, Yellowstone National Park. (S. Brazier)

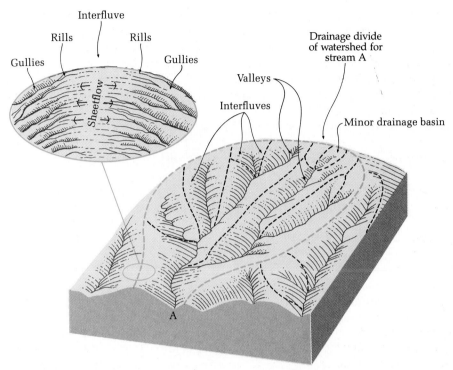

FIGURE 14.12   A small upland drainage basin showing component landforms and nature of water movement (inset). Dashed lines show drainage divides.

which join even larger ones and eventually form streams. The water in lakes can be considered runoff that is in temporary storage.

Each major stream and all its **tributaries** (the smaller streams that feed it) make up a **stream system.** The land surface drained by a stream system is called a **drainage basin** or **watershed** (Fig. 14.12).

Streams flow through valleys that are usually of the stream's own making. The higher lands separating one valley from the next are called *interfluves* or interfluvial ridges (from Latin: *inter*, between; *fluvius*, river). The term *ridge*, however, is misleading, for the higher land is not necessarily ridge-shaped. For example, it may actually be a high plain or tableland. On an interfluve that separates two stream systems, there is an imaginary line called a **divide.** On one side of the divide all surface runoff flows toward one stream system, while on the other side runoff flows toward another stream system. The North American *Con-*

tinental Divide separates runoff to the Pacific and Atlantic Oceans. Generally, it follows high ridges in the Rocky Mountains but also lies on gently sloping high plains and plateaus (Fig. 14.13).

Not all streams flow year round. In arid and semiarid lands streams often flow only after a heavy rain or during the rainy season, if there is one. Such streams are called **intermittent** streams. In contrast, **perennial** streams flow year round, though not always with the same volume or at the same velocity. Many streams overflow their banks on occasion, flooding the level land to either side. Such lands are called *floodplains.*

Total stream flow, called **discharge,** is the measure of the volume of water flowing past a cross section of the stream in a given unit of time (cubic meters per second or cubic feet per second). The United States Geological Survey maintains over 6000 gauging stations to measure stream discharge in the United States (Fig.

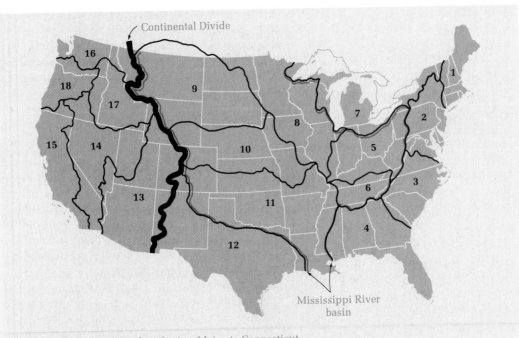

1 North Atlantic slope basins, Maine to Connecticut.
2 North Atlantic slope basins, New York to York River.
3 South Atlantic slope basins, James River to Savannah River.
4 South Atlantic slope and eastern Gulf of Mexico basins, Ogeechee River to Pearl River.
5 Ohio River basin except Cumberland and Tennessee River basins.
6 Cumberland and Tennessee River basins.
7 St. Lawrence River basin.
8 Hudson Bay and upper Mississippi River basins.
9 Missouri River basin above Sioux City, Iowa.
10 Missouri River basin below Sioux City, Iowa.
11 Lower Mississippi River basin.
12 Western Gulf of Mexico basins.
13 Colorado River basin.
14 The Great Basin.
15 Pacific slope basins in California.
16 Pacific slope basins in Washington and upper Columbia River basin.
17 Snake River basin.
18 Pacific slope basins in Oregon and lower Columbia River basin.

FIGURE 14.13   Major drainage basins in the contiguous United States.

14.14). On a world scale, the Amazon has by far the greatest discharge, while the Mississippi–Missouri system is ranked fourth (Table 14.1).

STREAM PATTERNS

Stream systems form various patterns of drainage. The two primary variables that influence the pattern developed by a particular system are geologic structure and the configuration of the land created by the tectonic and gradational processes. A **dendritic** (treelike) stream pattern (Fig. 14.15) is an irregular pattern with tributaries joining larger streams at acute angles (less than 90°). A dendritic stream pattern is by far the most common. Its existence is evidence of a lack of any specific geologic struc-

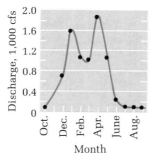

FIGURE 14.14 Hydrograph showing stream discharge measured at the Eldred, Pennsylvania, gauging station for the water year 1948–1949. (Data from Water Supply Paper 1143)

ture that might otherwise govern the pattern of stream drainage. A dendritic pattern also develops where the majority of rocks in the area have a similar resistance to weathering and erosion. In contrast, a **radial** pattern develops on dome- or cone-shaped geologic structures, such as volcanoes and domal uplifts. The opposite pattern is **centripetal** with the streams converging on a central point as in a basin of interior drainage. The **trellis** stream pattern consists of parallel streams linked by short, right-angled segments. The trellis pattern is usually evidence of parallel outcrops of erodible rocks, as in regions of folded sediments like the Appa-

lachians. Between the parallel stream segments one can assume the presence of more resistant ridge-forming rocks. The related **rectangular** pattern occurs where streams are guided by intersecting joints, usually in plutonic or metamorphic rocks.

All of these stream patterns so far described are related to, or *accordant* with, the structure and relief over which they flow. In some instances, however, stream and river systems show little relationship either to the underlying relief or geologic structure. Some rivers flow across fold structures or appear *discordant* with the rocks over which they flow. Such systems are probably either **antecedent** or **superimposed. Antecedent rivers** existed before a period of uplifting and folding but were able to successfully maintain their courses even while slow mountain building was taking place across their paths. The Columbia River in Washington and Oregon, cutting across the line of the Cascade Mountains, and the Brahmaputra River in India and Tibet (China), which flows through a mighty gorge across the Himalayan chain, probably originated in this way. Other rivers, such as those of the central Appalachians, for example, may have originated on earlier cover rocks, since stripped away by river erosion, so that the streams have

TABLE 14.1   Ten Largest Rivers of the World

	LENGTH		AREA		DISCHARGE	
	km	mi	sq km	sq mi	1000 m³/s	1000 cfs
Amazon	6276	3900	6133	2368	112–140	4000–5000
Congo (Zaire)	4666	2900	4014	1550	39.2	1400
Chang Jiang (Yangtze)	5793	3600	1942	750	21.5	770
Mississippi–Missouri	6260	3890	3222	1244	17.4	620
Yenisei	4506	2800	2590	1000	17.2	615
Lena	4280	2660	2424	936	15.3	547
Paraná	2414	1500	2305	890	14.7	526
Ob	5150	3200	2484	959	12.3	441
Amur	4666	2900	1844	712	9.5	338
Nile	6695	4160	2978	1150	2.8	100

Adapted from Morisawa: *Streams: Their Dynamics and Morphology.* New York, McGraw-Hill Book Company, 1968.

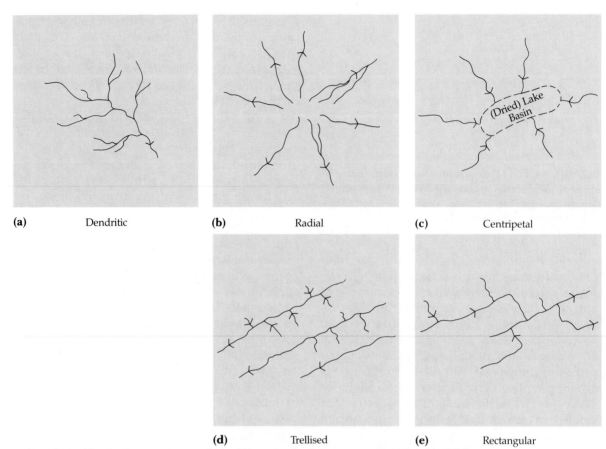

FIGURE 14.15 Drainage pattern often reflects geologic structure. (a) The dendritic pattern is found usually on homogeneous rocks. (b) The radial pattern indicates a central structural upland. (c) The centripetal pattern indicates a central structural lowland or basin. (d) and (e) The trellised and rectangular patterns indicate linear zones of weakness, either strips of erodible rock (trellis type) or well-developed jointing (rectangular type).

been **superimposed** on the rocks beneath. This sequence would explain why in many instances the rivers flow *across* the folds, creating water gaps like the Cumberland Gap or the gap formed by the Susquehanna River in Pennsylvania, both of which were important in early American history (Fig. 14.16).

EROSION BY STREAMS

Streams flow because of the earth's gravitational pull on water toward the center of the earth. The force of gravity has two components: One is the force that makes water flow downhill, and the other is the force that makes water exert pressure on the stream bed. The strength of each of these components, translated into the pressure exerted on the stream bed and the velocity of the stream, depends on (1) the steepness of the gradient of the stream bed and (2) the depth of water in the stream. Depth is related to the amount or volume of water in the stream, provided the stream's width remains constant. Every drop of water is a unit of potential energy for gravity to use. Thus, the greater the stream flow, the greater the amount of energy available to shape the land.

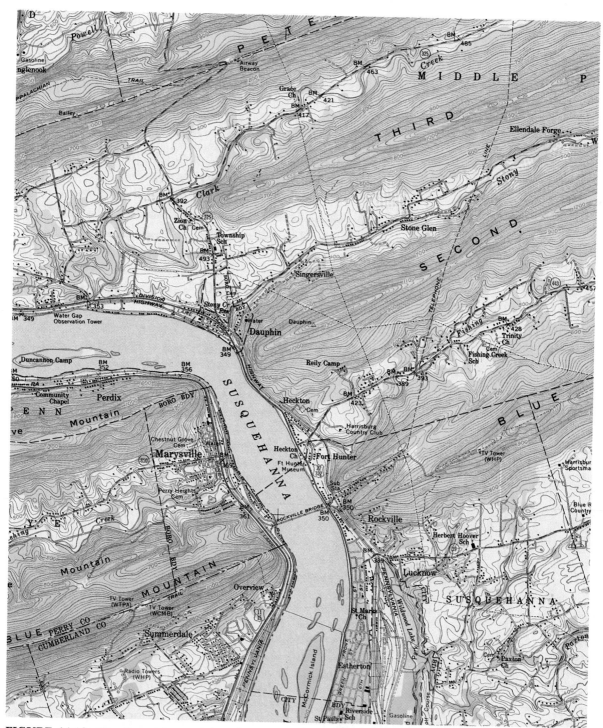

FIGURE 14.16 The Susquehanna River cuts across the folded Appalachian ridges (Harrisburg, Pennsylvania; scale 1:62,500; contour interval 20 feet). (U.S. Geological Survey)

A stream's energy is consumed continuously to overcome external friction (friction against channel walls) and to overcome internal friction (friction between water molecules, eddies, and currents). It is estimated that over 95 percent of a stream's energy is lost (consumed) in overcoming friction. Any excess energy, probably less than 5 percent, is used for vertical or horizontal erosion of the stream bed and banks, cutting the channel downward and laterally, and for transporting the sediment load.

Each stream achieves a tentative equilibrium or balance among these uses of its energy. This equilibrium is maintained by adjustments between channel slope, channel shape, channel roughness, availability of load, and particularly, velocity and volume of stream flow. These last two factors can change very rapidly, and partly as a consequence, they have a greatly magnified effect on the stream's equilibrium. This magnification is one reason why velocity and volume—or discharge—are such significant aspects of a stream's behavior, especially with regard to its ability to erode and to transport its sediment load (Fig. 14.17). This feedback mechanism of the stream's changing velocity was discussed in Chapter 1 (Fig. 1.5).

Volume of stream flow determines the size of the particles and the amount of sediment the water can carry. Friction of water in a stream significantly limits velocity and causes the water along the bottom and sides of a stream bed to slow down, so that maximum stream velocity is usually found in the center of the stream slightly below the surface.

The ability of a stream to pick up and carry materials is largely determined by the amount of *stream turbulence*. Turbulence is irregular or chaotic flow and is controlled by channel roughness and the velocity of stream flow. A very rough channel bottom will increase turbulence. A small increase in velocity will result in a significant increase in turbulence, which in turn greatly increases the rate of erosion as well as the load-carrying capacity of the stream (see again Fig. 14.17).

Volume of stream flow also affects the rate of erosion, as does the resistance of the rocks of the stream bed. Where rocks are highly resistant, little erosion of rock materials can take place.

Because the erosion rate is affected by so many variables (velocity, turbulence, volume of stream flow, resistance of rocks), erosion of the stream bed varies from place to place. All streams, however, have a **base level** below which they cannot erode. At the mouth of a stream—the location of its outlet into the ocean or lake—the level of the sea or lake is the stream's base level. The ultimate base level for virtually all stream gradation is sea level. Channel and valley deepening (downward erosion) occurs as long as there is enough slope to allow flow to the outlet.

A stream that begins to flow on recently uplifted land will gradually enlarge its channel in all directions by deepening, widening, and lengthening it through the erosion of material. Fluvial erosion is accomplished in several ways. First, simple **hydraulic action** will occur as the swirl of the river's currents wedges under loose slabs of rock on the bed or pounds away at the river banks and below waterfalls. **Plunge pools** at the base of waterfalls reveal this localized at-

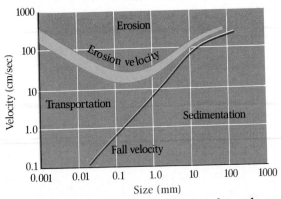

FIGURE 14.17 This classic diagram shows the relationship between stream velocity and ability to transport material of varying sizes. It will be noted that heavy gravels need a higher velocity to be moved, simply because of their size and weight. But very fine particles of silts and clays also take higher velocities for movement, since they stick together cohesively. Sand is relatively easily eroded and moved. (Morisawa, *Streams: Their Dynamics and Morphology*. New York, McGraw-Hill Book Co., 1968.)

tack point (Fig. 14.18). Hydraulic action also displaces loose debris and particles from the stream bed and channel walls and carries the material along in the current. These materials range in size from clay to silt, sand, gravel, and boulders.

As these particles and rocks bounce, scrape, and drag along the bottom and sides of the stream channel, they break off additional rock fragments. This form of erosion is called **abrasion** or **corrasion.** Under certain conditions abrasion makes round holes called **potholes** in the rock of the stream bed (Fig. 14.19). Potholes are formed only in special circumstances such as below waterfalls and in rapids, or at points of structural weakness such as joint intersections on the stream's bed. They range in size from many meters to a few centimeters across or deep. If you peer into a pothole, you can often see one or more round stones at the bottom. These are the *grinders.* The swirling movements of the stream water cause such stones to grind at the bedrock, making the pothole larger while the finely ground fragments are carried away. As potholes widen and join, the bed of the stream is deepened.

Some erosion is also carried out by the river's water chemically dissolving the bedrock. This so-called **corrosion** has a limited effect on

FIGURE 14.18   Plunge pool at the base of Vernal Falls, Yosemite National Park. (S. Brazier)

FIGURE 14.19   Potholes in bedrock of river, England. (S. Brazier)

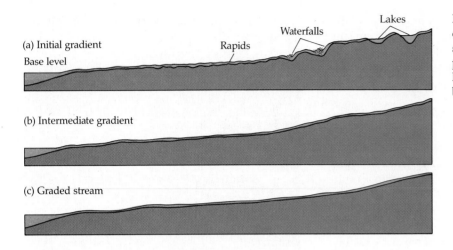

(a) Initial gradient
Base level
Rapids
Waterfalls
Lakes

(b) Intermediate gradient

(c) Graded stream

FIGURE 14.20 Idealized concept of the graded stream, in which the stream profile becomes more regular as the stream is graded to base level.

many rocks but may be significant in limestones and calcareous sandstones.

The size of the particles in the river's load is gradually reduced over distance. This wear and tear on the sediments as they bounce against each other and off the bed and banks of the stream is called **attrition.** This is the reason that rounded pebbles are often found in stream beds and that in the lower reaches of most large rivers the sediment load is composed primarily of dissolved materials and silt- or clay-sized particles.

The slope of a stream bed is its **gradient,** which is expressed as the vertical drop per unit of horizontal distance (meters per kilometer or ft per mi). A stream's gradient also affects the rate of erosion, for the steeper the gradient, the greater the velocity and turbulence of the stream. The gradient of a stream is usually greatest at its head and in new tributaries. However, the greater volume of flow farther downstream accelerates velocity even more than the high slope upstream, so that ability to pick up and carry materials is equivalent or even increased downstream.

As erosion continues, an idealized gradient is reached in which the loss of energy due to friction is counterbalanced by energy produced by the stream's flow. A stream in this condition is said to be in a graded condition. A **graded stream** has just the velocity necessary to remove the load eroded from the drainage basin. The graded stream is a theoretical balanced state averaged over a period of years. Near the mouth, stream erosion eats through to a graded condition much sooner than at the head, where the volume is lower. As the stream gradient is progressively reduced, grade is reached further and further upstream, eventually producing a smooth profile (Fig. 14.20).

Stream erosion also widens and lengthens the stream channel and valley. Lengthening occurs primarily at the source through headward erosion accomplished partly by the surface runoff that flows into the stream and partly by the undermining of the slope by springs feeding into the river. This lengthening of the river's course in an upstream direction is particularly important where soil erosion gulleys are eating back into agricultural land. Such gulleying may be counteracted by soil conservation practices that reduce rapid runoff. Channel lengthening also occurs as the stream becomes more sinuous and thus decreases its slope. Streams do not generally flow in straight lines but **meander** from side to side in the valley. This pattern is true of all streams that have floodplains, which are an indication of a graded or near-graded condition.

Valley widening is accomplished by erosion of the sides of the stream bed and commonly accompanies the outward expansion of stream loops or **meanders** (Fig. 14.21). The resultant undercutting by the stream may encourage the slump of material into the flowing stream. Channel widening is greatly encouraged by mass movement of material down the steep slopes created by a stream. Smaller streams flowing into a larger one also help to widen the valley through which the larger stream flows.

STREAM TRANSPORTATION

Streams transport material that they have eroded as well as a far greater mass of material brought to the stream by surface runoff and mass movement. This transported material is the **stream load.** Streams transport their load in several ways (Fig. 14.22). Some minerals are dissolved in the water itself—thus they are carried in *solution*. The finest particles are carried in *suspension* buoyed up by vertical turbulence. Such particles can remain in suspension as long as the force of the upward turbulence is stronger than the downward-settling tendency of the particles. Eroded fragments that are too large to be carried in suspension are moved along the bed of the stream by **traction.** When the particles are lifted from the stream bed, carried a short distance downstream, returned to the stream bed, lifted again, and so on, the process is called **saltation** (from French: *sauter*, to jump).

The load of a particular stream is measured by the weight of the material it is transporting. In most streams the largest portion of a load is the **suspended load**—the weight of the materials being carried in suspension. The **bed load** is the portion of the total load that is

FIGURE 14.21 The meandering Pastazi River, Ecuador, as it widens its valley through the Andes Mountains. (S. Brazier)

FIGURE 14.22 Transport of solid load in a stream. Clay and silt particles are carried in suspension; sand travels by suspension and saltation; larger material, from gravel to boulders, moves by traction.

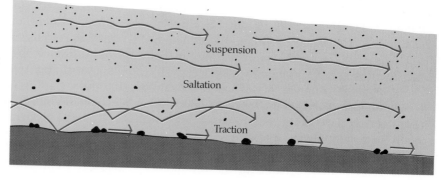

Suspension

Saltation

Traction

rolled, dragged, or bounced along the stream bed. The bed load and suspended load together constitute the solid load. The **dissolved load** is that portion that is in solution.

The relative proportion of each part of the load varies with the nature of the drainage basin and the rate of river flow. Some of the world's major rivers have huge suspension loads, the highest being that of the Huang He in North China. Known as the "Yellow River" because of the color of its fine sandy suspension load, it carries on average an annual load which makes up 10 percent of the total sediment carried by all the world's rivers.

The **capacity** of a stream refers to the maximum possible weight of its total solid load. A stream's **competence,** its ability to transport material, is measured by the diameter of the largest fragment it can carry as bed load. Both capacity and competence are increased significantly by small increases in stream velocity. Thus, a stream that doubles in velocity during a flood may increase its sediment load six to eight times. The huge boulders seen in many mountain streams arrived there during some past high-velocity flow that greatly increased stream competence and will be moved again when a flow of similar magnitude occurs. Thus rivers do most of their heavy earth-moving work during short periods of flood.

STREAM DEPOSITION

Since a stream's capacity to carry material depends primarily on stream velocity and discharge, a reduction in either of these elements will cause a stream to dispose of some of its load by depositing it. There are, of course, situations in which a stream is not carrying its maximum load. In such a situation a reduction in velocity and/or in discharge might not be sufficient to cause deposition.

The most common places where stream deposition occurs are on the inside of bends or meanders in the stream where velocity is reduced, on floodplains (or valley floors), at stream mouths (deltas), and at major breaks in slopes (e.g., where streams leave mountain areas). **Alluvium** is the name given to fluvial deposits, no matter what the type or size of material. Alluvium is usually recognized by the fact that it exhibits the characteristic sorting performed by streams. Changes in velocity cause a stream to sort fragments by size. As velocity fluctuates, the size of the particles that are picked up, transported, and deposited fluctuates accordingly. The velocity of a stream normally fluctuates with changes in discharge that occur in wet and dry seasons. The alluvium deposited by such a stream will exhibit these changes in velocity by alternating layers of coarse and fine material.

LAND SCULPTURE BY STREAMS

One way to examine some of the variety of landform features resulting from the gradational activities of streams is to look at a river course from its headwaters to its mouth. In the following discussion of upper, middle, and lower river courses you should note that erosion has greater significance in the upper course, while deposition is more important in the lower course.

FEATURES OF THE UPPER COURSE

In the upper course of a river the gradient is at its steepest. Here vertical erosion, particularly corrasion, works most strongly to reduce the stream to a graded condition. Erosion in the upper course cuts away at the land to create a V-shaped gorge or ravine. Flowing along the bottom of this ravine is the stream or river itself, cutting still deeper into the land. No floodplain is present, and the steep sides of such a valley encourage mass movement of material from the sides directly into the flowing stream. Valleys of this type, which are dominated by the down-cutting activity of the stream, are often called **youthful valleys** (Fig. 14.23).

Because erosion in the upper course is most important where the river comes in contact with bedrock, the effects of *differential erosion* are significant. Differential erosion occurs

FIGURE 14.23   Marble Canyon, a steep gorge cut at the bottom of the Grand Canyon of the Colorado River. (S. Brazier)

where a river must cut through rock layers of different resistance. Typically, rivers cutting through resistant rock have a steeper gradient than where they encounter nonresistant rock. This steep gradient gives the stream more energy. Rapids, cascades, and waterfalls indicate the resistant materials and steep gradients common in youthful valleys (Fig. 14.24). Where rocks are particularly resistant to weathering and erosion, valleys will be narrow and steep-sided; where rocks are less resistant, valleys will be more spacious.

Extremely young streams may spill from lake to lake, either over open land (like the Niagara River between Lakes Erie and Ontario) or through gorges. In either case the lakes are eventually eliminated by having their outlets lowered by erosion and by filling as deltas are built into them at the inlet point.

As a youthful stream reduces and smooths out its gradient further, and as its suspended load increases, vertical erosion becomes less significant and the lateral erosion of the channel sides assumes a more important role. Combined with mass movement, the effect is to displace the channel laterally, producing a wider valley floor with a smoother stream gradient. This lateral erosion expands the valley floor and eventually produces a narrow floodplain.

FEATURES OF THE MIDDLE COURSE

When a stream has reduced its gradient to a point where its sediment load is about equal to its load-carrying capacity as determined by its velocity, the stream reaches a *graded* condition. At this point further vertical erosion is impos-

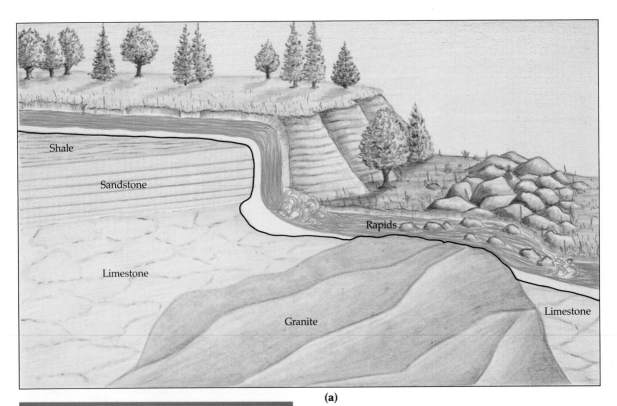

(a)

(b)

FIGURE 14.24   Effects of rock resistance on (a) stream profile and (b) valley form. Where rocks are resistant, steep-walled canyons, rapids, and waterfalls result and stream gradient increases significantly as shown (at right) in the Yellowstone Canyon. (R. Sager)

sible, since the stream has no excess energy available to pick up and transport more material. Consequently, a stream that has reached such a balance or equilibrium between load and velocity can erode only laterally, picking up material on one side and dropping it on the other.

It is in the middle course of most rivers that a graded condition exists. Here the river valley contains a floodplain but still maintains definite valley walls. The river is said to be *mature*. The mature river is sinuous, forming loops or meanders on the valley floor. Such a stream cuts away at the bank on the outside of each meander loop and deposits material on the inside of each loop (Fig. 14.25). Centrifugal force accelerates stream velocity on the outside of these bends, allowing sediment to be picked up, and decelerates velocity on the inside, allowing sediment to be deposited. Over time,

the size of the stream meanders is increased by this activity, widening the floodplain.

The land within the curves of a stream's meanders is relatively flat and covered with alluvium deposited by the stream, particularly during periods of falling velocity as floods subside. As lateral erosion continues and the size of the meanders is increased, the area of flat land covered with rich alluvium is also increased. Though the flooding of such land is always a potential danger, the richness of the floodplain's alluvial soils is an irresistible lure for farmers.

FEATURES OF THE LOWER COURSE

In the lower course of a river, deposition becomes more significant in shaping the landscape. The lower valley of a major river is wider than the meander belt (the area between

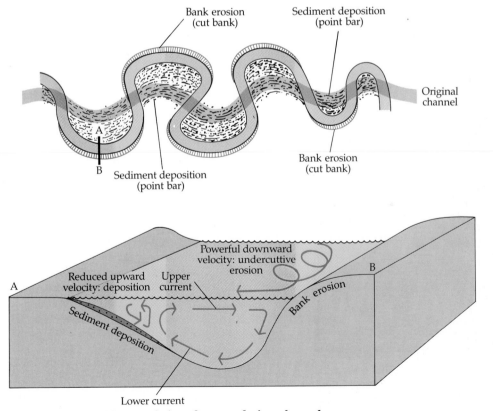

FIGURE 14.25 Characteristics of a meandering channel.

imaginary parallel lines connecting the outsides of meanders on each side of the valley) and shows evidence of shifts in the river's course (Fig. 14.26). In most cases this great valley width results from the fact that the lower valley is a vast floodplain of alluvium rather than an erosional plain. In other words, the alluvium of the lower valley is much deeper than the stream channel and has not been deposited by the meander process. These giant lower valleys are actually filled marine **estuaries** (river mouths drowned by the sea where salt and fresh water mix), as in the case of the Mississippi River, or subsiding tectonic depressions, like the Amazon and Congo Basins.

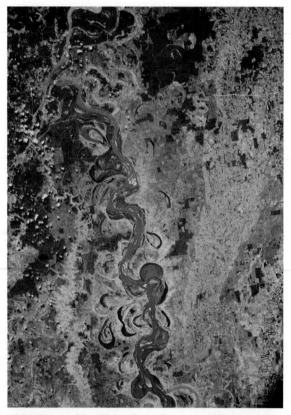

FIGURE 14.26   This photo from Apollo 9 shows a 160-km section of the lower Mississippi Valley. Note: meanders, oxbow lakes, yazoo streams, sand bars, and old channel remnants. (NASA)

During floods the floodplain of a river's lower course is inundated with sediment-laden water that deposits alluvium over it. These plains accordingly are called **alluvial plains** (Fig. 14.27).

A characteristic landform of the lower course, the **oxbow lake,** is itself evidence of the river's shifting course. The oxbow lake is formed, especially during flood periods, when a portion of the channel is cut off as the stream tends to seek a shorter, steeper, or straighter path. The cutoff section of the channel forms the oxbow lake (Fig. 14.28).

During flooding, a sharp reduction in a stream's velocity, due to friction at the channel's edge, results in heavy sedimentation along the river banks. Over time, the banks are built up into natural **levees** through successive deposition. Levees along the Mississippi River rise up to 5 meters (16 ft) above the rest of the floodplain.

As a river slows down or at times of reduced volume, deposition will occur in the channel itself. Thus in its lower course, a river sometimes raises the level of its channel bed. In rare cases, as in China's Huang He, the stream channel as well as its levees may be raised above the flat floodplains below. Flooding is an even greater danger in this situation because of the obvious drainage problems created in the so-called bottom lands. Sometimes humans aggravate such a situation still further by building up the levees artificially in an attempt to keep the river in its channel.

The presence of levees—both natural and artificial—prevents tributaries from joining the mainstream. The smaller streams are thus forced to run parallel to the larger one until a convenient junction can be found. These parallel tributaries are called **yazoo streams,** after the classic example of the Yazoo River, which parallels the Mississippi River for over 160 kilometers (100 mi) until it finally joins the larger river near Vicksburg, Mississippi.

When a stream has a heavy bed load in relation to its velocity and discharge, it deposits much of its load as sand and gravel bars in the stream bed. These obstructions break the

FIGURE 14.27   During floods the floodplain is inundated with sediment-laden water that deposits alluvium over it. These plains are called alluvial plains and have great agricultural potential due to the floodwaters bringing new parent material for soil. (S. Brazier)

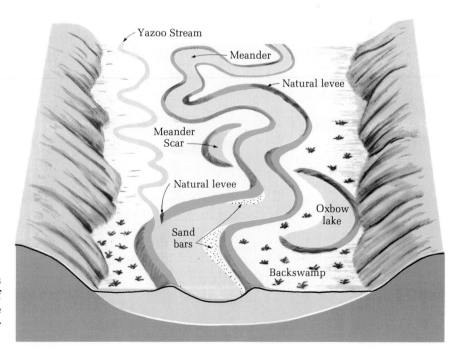

FIGURE 14.28   Features of a large floodplain of the type common in the lower portions of major rivers.

VIEWPOINT: FLOOD DANGER

Precipitation varies from place to place over the earth. Some locations receive too much water and others not enough. This irregularity can result in drought on the one hand and flooding on the other. Flooding is a natural process with streams. Under normal conditions a stream may overflow its banks once or twice a year in response to increased precipitation. Such flooding is a serious problem in areas where people have built their homes, factories, highways, farms, and cities without taking the potential for flooding into consideration.

Flooding results when the amount of precipitation received is more than the streams, lakes, and soil can absorb and transmit. Time is involved in flooding because the longer the time over which a given amount of rainfall occurs, the greater the amount of water that can be absorbed by the soil and carried away by the streams. The ability of the land to absorb water also plays a role in whether a given amount of rain will cause flooding. For example, flash floods occur in the semiarid American Southwest when a great deal of precipitation falls during a short period of time onto relatively impermeable soil and rocks. If this same heavy rainfall were to occur in a more humid region with a vegetation cover and permeable soil, a flood would be less likely.

Floods do occur in humid regions, however, when more precipitation falls than can be absorbed or carried away by the channels of streams. When precipitation is this heavy, streams rise over their banks, and water spreads over the adjacent low-lying land. A flood can be particularly extensive where open plains regions border a river, as in the Mississippi and Ohio Valleys.

Floods can result in great losses in human life. For example, the Yellow River (Huang He) in China, which is also known as "China's sorrow," has killed as many as a million people in a single flood. Floods can also cause extensive damage to crops, buildings, livestock, and soils. And when the waters recede, the mud-covered, waterlogged debris presents a formidable task of reclamation. A particularly devastating flood occurred in 1966 in Florence, Italy, when 19 inches of rain fell in 48 hours, causing the Arno River to inundate much of the city under 10 to 20 feet of water. Restoration of art treasures damaged by that flood is still going on. In 1972 the Susquehanna River overflowed its banks and flooded many towns, including Corning, New York, and Wilkes Barre, Pennsylvania. And in the spring of 1973 the giant Mississippi flooded in nine states, drowning almost four million acres of land and producing crop and livestock losses estimated at 36 million dollars. Economically, one of the worst results of this flood was the effect on subsequent crops. Because of the flooded lands, cotton, corn, soybeans, and sugarcane could not be planted at the normal time. The reduced supply of these commodities later caused their prices to rise nationwide.

Throughout history people have responded in a variety of ways to flooding and to the potential hazards of flooding. These include evacuation, both temporary and permanent, of hazard zones; build-

Floodwaters from the Susquehanna River inundate Harrisburg, Pennsylvania, after Hurricane Agnes in 1972. (J. L. Patterson, U.S. Geological Survey)

The James River flooding Richmond, Virginia, after Hurricane Agnes in 1972. (J. L. Patterson, U.S. Geological Survey)

ing levees to hold back abnormally high waters from floodplains; building houses on stilts so that flood water can flow beneath them; reinforcing existing structures and constructing new buildings to withstand the forces of a flooding stream; and moving personal property or inventories to upper stories or higher land. Probably the most obvious human responses, however, are those achieved through engineering.

Some of man's earliest engineering projects were attempts to gain some control over the surface waters of the earth: dams to hold back rivers and provide a supply of water for agricultural and domestic use during times of reduced precipitation, canals and new river channels to shorten the river and thereby minimize the area subject to flood damage, and bridges to span river-carved gorges. Today in the United States the Army Corps of Engineers is an effective agent of flood control, both riverine and coastal. First appointed by Congress before the Civil War to deal with the problems of flooding, the Corps of Engineers has built levees, cutoffs, dikes, reservoirs, dams, canals, and spillways and has deepened channels and diverted rivers as efforts at flood control. The engineers have also spent a great deal of time studying the problems of people living on floodplains. Their efforts have saved millions of lives and billions of dollars' worth of property and have opened up many new lands to agriculture, notably in the West and Southwest, by providing a dependable supply of water for irrigation in arid and semiarid regions.

To protect an area against potential floods the engineers first need to know something about the size of most floods, their rate of occurrence, and the size and rate of occurrence of maximum floods. Sev-

eral factors can be used to describe the extent of a flood. Among these are its height, how long it lasts (duration), the area it covers, the average speed of the flood waters (velocity), and the volume of water discharged per unit of time. These are affected, in turn, by such factors as the amount of precipitation received per unit of time, the time of year, the character of the soil (frozen, waterlogged, etc.), as well as the shape of the stream channel itself and the geomorphic features of the floodplain. For example, a swollen stream flowing through a narrow, steep-walled valley will not produce the same kind of flood as a river flowing through a wide, open floodplain.

There is another side to the picture, however. Many ecologists and others concerned with the environment feel that the engineers have a tendency to alter river systems without enough careful analysis of the possible effects their alterations could have on the other systems involved. For example, their attempts to straighten natural streams in Florida have had disastrous effects on the wildlife there. Critics of the Army Corps of Engineers point out, with some justification, that a dam is not always good for the environment as a whole, nor is it always necessary.

In spite of the flood-control projects of the Army Corps of Engineers and the billions of dollars spent by federal, state, and local governments to minimize damage in floods, the amount of damage continues to increase. There probably is no perfect solution for flood control. Certainly greater public awareness of the potential hazards of floodplain living would help. Perhaps this increased public awareness combined with the continued efforts of such groups as the Army Corps of Engineers—if tempered by more careful preplanning and analysis—can help to minimize losses during floods.

stream into strands that repeatedly separate and rejoin to give a braided appearance (Fig. 14.29). Such a stream is called a **braided stream.** This pattern may develop wherever sediment input into a stream is extremely high owing to weak sand and gravel stream banks or to unusual sediment sources such as melting glaciers and deserts. Braided streams are common on the Great Plains (the Platte River, for example), in the desert of the American Southwest, and in Alaska.

MAJOR DEPOSITIONAL FEATURES

ALLUVIAL FANS Where streams pass abruptly from narrow canyons to open plains, their channels flare out in width and become shallow. The greater external friction of the shallow channel quickly reduces stream velocity and carrying power. As a result, a large proportion of the eroded material carried by the stream from the higher lands is deposited at the base of the highland.

The point where the stream meets the plain is fixed by the canyon or ravine out of which it flows. As deposition occurs on the plain area, the stream, acting much like a

braided stream, shifts, subdivides, and continues to deposit material. Because the stream's entry point is fixed at the canyon mouth, this serves as a pivotal point. Deposition takes place around this point in the shape of a fan or cone, called an **alluvial fan** (Fig. 14.30).

An important characteristic of an alluvial fan is the sorting that usually occurs. Coarse boulders and pebbles are deposited near the fan's apex where the fan slope is steepest. Sediments become finer further away from the apex. Alluvial fans are especially common in arid regions because of the heavy sediment loads carried during brief periods of runoff. In such regions many alluvial fans are composed largely of debris flow deposits, which are not sorted like deposits produced by running water.

DELTAS When a stream flows into a relatively still body of water such as a lake or an ocean, stream velocity and load capacity are progressively reduced. Consequently, the stream deposits part of its load, larger fragments first and then finer ones. The resulting deposit is known as a **delta,** after the Greek letter, which some, like the Nile delta, resemble when seen on a map (Fig. 14.31).

FIGURE 14.29 The braided Nelchina River in Alaska. This condition results from excessive contributions of sediment from erodible valley slopes or bank materials. The sediment accumulates in gravel or sand bars, forcing the channel to divide repeatedly. (U.S. Geological Survey)

FIGURE 14.30 A series of alluvial fans form where streams initially confined in steep-channeled ravines abruptly issue onto open plains. As their velocity is checked, the streams deposit their loads and spread across the previously deposited alluvium. (U.S. Geological Survey)

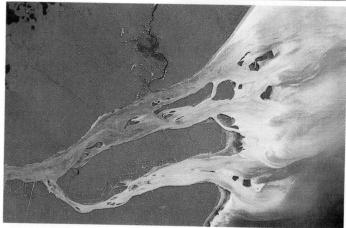

(a)

(b)

FIGURE 14.31 Satellite views of river deltas. (a) The Chang Jiang (Yangtze River) delta in central China. Note silt flowing into the ocean from the major distributaries. (b) The Nile Delta, the classic arcuate river delta, as it enters the Mediterranean. (EROS Data Center)

The normal pattern of delta formation resembles the formation of an alluvial fan. As stream deposition blocks the channel, the stream seeks another path. Once that new channel is clogged, the stream moves to another position. The multiple channels flowing away from the main stream, as it seeks the open water, are called **distributaries.** Natural levees build up along the sides of these distributary channels, even within the larger body of water. By continued deposition the stream extends the land far out into the water. Favored by the rich alluvial deposits and the abundance of moisture, vegetation takes a quick hold on this fertile new land and further secures its position. Delta lands such as those of the Mekong, Indus, and Ganges Rivers form the principal agricultural areas that feed the dense populations of Asia.

In general, deltas can attain great size only where the sediment supply is high and where wave, current, and tidal action are not strong. Most deltas have an arcuate (bow) shape, such as the Niger and Nile deltas.

The so-called bird's-foot delta of the Mississippi is a special type that results from delta subsidence after formation, leaving only the crests of the natural levees above sea level. The Mississippi delta changes rapidly over the years as new distributaries deposit alluvium that subsequently sinks, leaving only the levee crests as parallel fingers pointing seaward.

A final type of delta fills in at the heads of some marine *estuaries.* As previously noted, estuaries are river valleys that have been drowned by the sea either because the sea level has risen or because the land has been depressed (Fig. 14.32). Deltas form slowly in es-

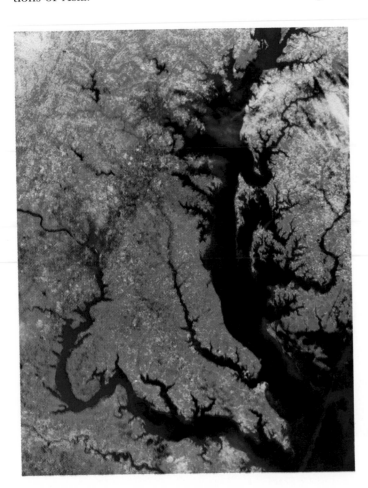

FIGURE 14.32 Chesapeake Bay is a major estuary produced by the submergence of the lower valley of the Susquehanna River as sea level rose at the end of the last glacial period. All along its margins are minor tidal estuaries. The Potomac River enters the main estuary at the bottom of the picture. Such environments have highly specialized life forms that are presently endangered by pollution. (NASA)

tuaries because of the counteraction of tides and sometimes because of low sediment input.

THE FLUVIAL CYCLE OF EROSION

Gradation by running water, combined with the effects of mass movement, creates a host of landforms. The famous American geographer William Morris Davis first suggested the existence of a fluvial cycle of erosion. By this he meant that the landforms that result from gradation by running water (particularly streams) pass through characteristic and recognizable evolutionary stages. In other words, recently uplifted land is eroded by streams and worn down so that, in successive stages and in the absence of further uplift, a relatively flat plain of low relief and elevation ultimately is formed. If this plain is uplifted again by tectonic forces, the erosional cycle will begin again.

In the following brief discussion of the three major stages of Davis's fluvial erosion cycle, you should keep three facts in mind. First, the cycle we will describe is an idealized model. In reality, the cycle may be interrupted at any stage by additional uplift. Second, uplifted regions do not go through all three stages—termed by Davis **youth, maturity,** and **old age**—at the same rate. The stages in the erosional cycle cannot be equated with specific amounts of time. Climate, vegetation, original elevation, and rock resistance are all important variables in the maturation of a landmass being reduced by stream erosion.

Third, the theoretical final stage of the erosion cycle has not been observed anywhere on earth. It would be a relatively flat, erosional plain worn down close to base level. Vast erosional plains at base level simply do not exist, although plains produced by deposition, such as the Amazon Basin, or areas of tectonic subsidence are common. When the cycle of erosion was described in the late nineteenth century by Davis, there was no information available to distinguish plains of erosion from those of aggradation (deposition).

YOUTH

After uplifting by tectonic forces, running water on the uplifted land begins strong vertical cutting down toward base level (Fig. 14.33). Rivulets carve rills that become gullies and ravines. Eventually, stream flow is concentrated into a few main streams that carve well-separated V-shaped valleys (Fig. 14.34). The depth of such valleys varies with the degree of original uplift above base level. The major portion of the land surface at this stage is not eroded to any great extent and remains at the original elevation created by uplift. Thus, because most of the land is still highly elevated, interfluves are broad and relatively flat. The steep gradients of the streams result in waterfalls and rapids. Lakes and marshes may develop on the uplifted surface, since a complete drainage system has not yet been developed.

MATURITY

The youthful stage of valley development gives way to maturity when the streams cease their strong vertical incision and when lateral stream erosion becomes the dominant process. In maturity most of the falls and rapids of youth have disappeared to be replaced by a smoother stream profile. When strong vertical erosion ceases, stream meanders begin to undercut the banks of the youthful valleys. As lateral erosion and deposition increase, the valley floor is widened. The narrow, V-shaped youthful valley is enlarged to a continuous plain in which stream meanders are free to migrate without obstruction. The plain is about the same width as the stream's meander belt.

There are many more streams and tributaries in the mature stage than there were in youth. By this time erosion has proceeded so that there is a well-developed stream system that has almost completely removed the original uplifted surface. A maturely dissected landscape can be recognized by the fact that its interfluves are narrow ridges, and most of the land surface, with the exception of floodplains, is in slope. As streams enlarge the valleys further through lateral erosion, the height and extent of the uplands are reduced still more.

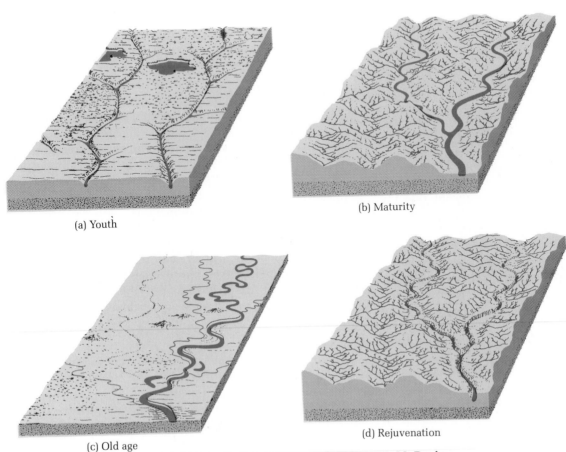

(a) Youth

(b) Maturity

(c) Old age

(d) Rejuvenation

FIGURE 14.33   Stages in the fluvial cycle of erosion, according to W. M. Davis.

OLD AGE

Old age is a theoretical stage reached when a landmass has been reduced by erosion almost to base level. In the old age of a landscape, streams flow in wide meanders over an open plain. Oxbow lakes, the evidence of changing stream courses because of erosion and deposition, are common. The major portion of the land is at low elevation and is relatively flat, with only a gentle slope in the direction of drainage. Such an erosional plain is called a **peneplain.** Sometimes the evenness of a peneplain may be interrupted by an occasional isolated hill. Such hills, or mountain remnants, remain standing long after the surrounding land is reduced by erosion because they are made of more resistant rock. Erosional remnants of this type are called **monadnocks** after the example of Mount Monadnock in New Hampshire.

Fewer streams and tributaries are present to provide drainage in the flat, old-age landscape than in a mature one. This lack of drainage, combined with the low gradients and ponding of drainage behind natural levees, can result in marshes and swamps, like those common in the lower Mississippi Valley.

REJUVENATION

In reality there are no peneplains as envisioned by Davis, since all erosional plains have more relief than the concept of the peneplain permits. Truly flat landscapes are instead all ag-

(a)

(b)

FIGURE 14.34   Examples of youthful terrain: (a) Heavy rains on the island of Kauai, accompanied by stream cutting and sliding of the soft basalt upland, cause the steep corrugated cliffs called "pali." (b) A steeply incised valley cut into the uplifting Colorado Plateau. (R. Sager)

FIGURE 14.35 The "goose necks" of Utah's San Juan River are a classic example of the rejuvenation of a meandering stream. Rejuvenation here has been a consequence of vertical uplift of the land surface. (U.S. Department of the Interior Bureau of Reclamation)

gradational. Nevertheless, there are many instances of erosion surfaces, probably flat in former times, that are either slightly or severely dissected as a consequence of uplift. When uplift interrupts the erosional cycle at any point, vertical incision toward base level is intensified, and waterfalls and rapids are created as valleys are deepened by erosion. The landscape and its streams are then said to be **rejuvenated.**

If new uplift occurs during maturity or old age, after the formation of large stream meanders, those meanders become **entrenched** by the rapid incision of the rejuvenated stream (Fig. 14.35). Now, instead of eroding the land laterally, with meanders migrating across a plain, the rejuvenated stream's primary activity is vertical incision.

It is important to note that all rivers reaching the sea were rejuvenated during the Pleistocene as a consequence of the lowering of sea level that resulted from continental glaciation. The accumulation of frozen water on the land lowered the sea 120 meters (400 ft), thus dropping base level for all coastal streams, which consequently incised. The subsequent melting of the ice elevated sea level—or base level—causing valleys to be filled with sediment. These effects have produced broad, flat, aggradational floodplains above buried channels cut far below sea level.

Sea level changes, as well as minor tectonic movements and climatic changes, often cause a relatively slight amount of rejuvenation in mature valleys, so that the valley is slightly deepened, the old valley floor being preserved in strips along its edges. These older valley floors are **stream terraces.** Some terraces are a consequence of successive periods of rejuvenation and aggradation (Fig. 14.36).

CRITICISM OF THE CYCLE OF EROSION

There have been many criticisms and challenges to the Davis fluvial cycle of erosion. The criticisms have been especially strong from geomorphologists dissatisfied with Davis's purely descriptive point-of-view. His idealized cycle was not based on precise measurements or observations and hence was a qualitative approach.

In defense of the Davis concept, one can state that it is still a useful tool for visualization of the fluvial processes and the landforms produced by streams. It is particularly valuable to beginning students of geomorphology who need a broad overview of the formation of fluvial landscapes. As long as we accept its major limitation—that it is not a useful research tool based on sound quantitative data—it serves its purpose. The Davis fluvial cycle of erosion was

(a)

(b)

FIGURE 14.36 (a) These stream terraces were formed by three phases of rejuvenation, which produced four successively lower valley floors. The oldest is at the first level; the youngest is at the fourth. (b) River terraces, Tien Shan Mountains, Sinjiang, China. Three clearly defined terraces indicate former floodplain levels. (S. Brazier)

the first well-accepted conceptualized model of landscape formation. It still serves as a basis for present and future studies, whether supportive or critical.

FLUVIAL GEOMORPHOLOGY TODAY

Modern fluvial studies in geomorphology are based on statistical analysis. Geomorphologists feel that the quantitative approach will produce detailed information that will help us better understand the complexity of the fluvial processes and the landforms produced. They will be able to define and describe streams and their resulting landscape in a rigorous manner, using the scientific method and the most modern research tools, from computers with visual displays to satellite-based remote sensors.

Since streams are dynamic systems in a constant state of change, innumerable types of quantitative data must be developed, tested, and used for descriptive and analytical purposes. A better understanding of fluid mechanics and energy flow must also be integrated with the geomorphic processes and landform descriptions.

A few of the more common mathematical parameters used in fluvial geomorphic studies are stream discharge, stream order, and drainage density. *Stream discharge*, which was briefly discussed earlier in this chapter, is probably the most important fluvial data that is collected and analyzed. It can be calculated by the simple equation $Q = AV$. Discharge (Q) is equal to the stream cross section area (A) times the stream velocity (V). This figure (measured in cubic meters per second or cubic feet per second) is most useful in comparative studies of streams and in the studies of energy available to streams in relation to erosive power and load capacity and competence.

For comparison of streams within a river basin or for comparison of river basins, **stream order** is used. Though there are several approaches to this subject, the first classification was done in the 1930s by an American hydraulic engineer, Robert E. Horton. First-order streams are the first headwater channels formed without tributaries. They can be seen in the field and are mapped on large-scale topographic maps. When two or more first-order streams join, the channel becomes larger, and it is classified as a second-order stream. When two or more second-order streams intersect, the channel is classified as a third-order stream, and so on. Generally, the higher the order, the larger and longer the channel, but the lesser the total number of streams in that order. The river basin order is based on the highest-order stream in the basin, the main stream (Fig. 14.37). The Mississippi River would be classed as a 10th-order stream, while the mighty Amazon would probably be a 13th-order stream.

Drainage density studies provide us with information on stream dissection of the basin area. The formula used is $D = L/A$. Drainage density (D) is equal to the total length of all the stream channels in the basin (L), divided by the total area of the drainage basin (A). The easily eroded Dakota Badlands may have an extremely high drainage density of over 200, while very resistant granite hills may have a drainage density of only 5 (i.e., 5 kilometers of stream channel for every square kilometer of basin area).

Future quantitative studies will not only help us to understand the origins and formational processes of fluvial landforms but will also help us predict water supplies and flood potential, estimate soil erosion, and trace toxic pollutant sources.

THE IMPORTANCE OF SURFACE WATERS

STREAMS

People have used streams and rivers for a variety of purposes throughout history. For example, the settlement and growth of the United States would certainly have been different had there been no Mississippi River with its far-reaching system of tributaries, a system that drains most of the area between the Appalachians and the Rockies. We have used the Missis-

FIGURE 14.37 **The hierarchy of stream ordering is illustrated by this fourth-order watershed.**

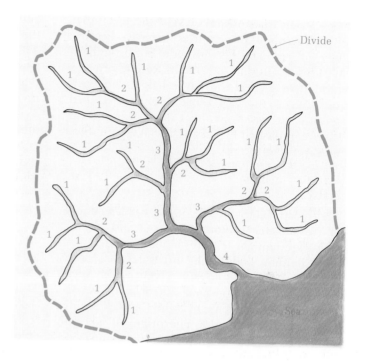

sippi, like many other rivers, for exploration, migration, and settlement. The number of major cities along the Mississippi (Minneapolis, St. Paul, St. Louis, Memphis, and New Orleans, to name a few) are evidence of people's tendency to settle along rivers. We have also used the Mississippi for the cheap transportation of bulk cargo. Today our major rivers still compete successfully with railroads and trucks as carriers of grain, lumber, and mineral fuels.

Large rivers like the Mississippi, as well as many smaller streams, have provided a source of power for saw mills and paper mills in earlier times and for hydroelectric plants more recently. Rivers have also provided water for irrigation, and the rich alluvial soils along their banks have often become productive agricultural lands. We have used streams as a source of food and water, as a depository for industrial wastes, and for recreation (boating, fishing, water skiing, swimming).

Obviously, there are many benefits to living near streams and rivers. However, such riverside settlement has its penalties, particularly in the form of floods. The variability of flow in

stream channels constitutes the greatest hazard and impediment to their use. Stream channels can apparently carry the maximum flows that can be anticipated once in every year or two. Larger flows, which may occur once in every 5, 10, 100, or 1000 years, spill out of the channel and cover the surrounding land, often with disastrous results. Similarly, exceptionally low flows may produce crises in water supply.

Because stream flow is so variable, few developed areas today use completely natural water supplies. Dams trap the potentially devastating flood flows of wet periods and make the water available during dry years. Many river systems now consist of a series of lakes, or reservoirs, impounded behind great man-made dams. Whole river basins have been remodeled in this way. The outstanding example is the Tennessee River Valley, which was "tamed" during the 1930s. A look at a hydrographic map will demonstrate that most of our great rivers, such as the Missouri, Columbia, and Colorado, have been transformed by dam construction and consist largely of a string of reservoirs. Unfortunately, the life of these reservoirs, like that of any lake, will be rather short—a couple of centuries at most—because they will gradually fill with sediment brought in by the streams entering them. Recently, it was decided to protect the natural flow and environment of the few remaining undeveloped streams and rivers in the United States. This was accomplished by passing the Wild and Scenic Rivers Act.

LAKES

The most visible portion of the earth's supply of fresh water is stored in lakes and ponds, which are familiar features in many landscapes. Any standing body of inland water is a **lake.** Most of the world's lakes are fresh water, such as Lake Superior or Lake Victoria. However, some are salt water, such as the Caspian Sea or Great Salt Lake (Utah). The origin of lakes is controlled by topography, since they must occupy a depression on the land surface. The majority of the world's lakes are products of glaciation such as North America's five Great Lakes and Minnesota's ten thousand lakes. Rivers, groundwater, tectonic activity, volcanism,

and humans also produce lakes. Lake Baikal in the Soviet Union is the world's deepest lake. This long Siberian lake reaches over 1525 meters (5000 ft) deep and was caused by a fault depression. Crater Lake, Oregon, North America's deepest lake, was caused by volcanism.

Lakes and ponds (small shallow lakes) are among the most temporary of the natural features of the earth's surface. Few have been in existence more than 10,000 years, a mere twinkling in geologic time. As soon as they are formed, their destruction begins through two processes, sedimentation and biological activity.

Entering streams wash sediment—sand and silt—into lakes, forming deltas, which eventually fill the entire lake basin (Fig. 14.38). At the same time, plant and animal organisms, using the mineral nutrients brought into the lake by streams, occupy the shore and build up communities on the floor, thereby helping the shoreline advance farther into the lake area. As the lake becomes smaller and shallower, plants take root on its floor, and it becomes a marsh. Eventually the marsh is filled and becomes a meadow. This aging process is inevitable and is accompanied by progressive changes in the

FIGURE 14.38 The temporary nature of lakes can be seen in this glacial lake in the Sierra Nevada of California. This lake is being filled by sediments, and eventually vegetation will completely cover the lake bed. (S. Brazier)

FIGURE 14.39 Norris Dam was the first dam built by the Tennessee Valley Authority (T.V.A.) and was completed in 1936. It is located on the Clinch River in Tennessee. This dam is multipurpose, supplying electrical power, water storage, and recreational aspects as well as flood control. (Tennessee Valley Authority)

lake's plant and animal life. The first occupants are forms adapted to water that is cold and deep. As the lake ages, there is a shift toward life forms adapted to progressively shallower and warmer water. Eventually, toads hop where pike and trout formerly swam.

Lakes are important to man for more than their obvious scenic appeal and their value for fishing and recreational activity. Lakes, like oceans, affect the climates around them, particularly by reducing daily and seasonal temperature ranges. Major fruit-producing areas have formed near lakes in Florida, New York, Michigan, and Wisconsin because of the moderating effects of lakes.

The benefits of lakes are such that humans have produced tens of thousands of them artificially by the construction of dams (Fig. 14.39). These artificial lakes, or reservoirs, are some of man's most ambitious construction projects. However, while the benefits of lakes have always been obvious, the necessity of protecting them from destruction by chemical, thermal, and biological pollution due to human activities has only recently been realized. The Great Lakes of the eastern United States and Canada, and Lake Erie in particular, are an instructive example of the damage that can be done to a large complex natural system by human misuse over a short period of time.

QUESTIONS FOR DISCUSSION AND REVIEW

1. What is the water table? How is it related to climate?

2. Describe a sinkhole. What is the formation process involved? What type of climate would be most conducive to its formation?

3. Name the two primary variables that influence a stream system's pattern. How do the various patterns reflect these variables?

4. Which variables affect the erosion rate of a stream bed? What determines the base level to which a stream bed will erode?

5. What constitutes the largest portion of a

stream's load? State the ways a stream can transport its load.

6. How does a stream sort alluvium? Explain the relationship of this sorting to velocity changes in a stream.

7. Where in the course of a stream is the gradient steepest? How do rapids and waterfalls in stream beds relate to rock resistance and steepness of gradient?

8. Describe the development of natural levees and the relationship of levees to yazoo streams.

9. Why do alluvial fans develop? In what ways does the formation of a delta resemble that of an alluvial fan?

10. Describe the stages of Davis's fluvial erosion cycle. What are some of the unique aspects of the "old age" landscape?

11. What are stream terraces? How are they formed? How did changes in sea level during the Pleistocene cause stream terraces to form on land?

12. Discuss the positive and negative aspects of building dams.

13. What kinds of processes help form lakes? Why are lakes such short-lived features of the earth's surface?

15

LANDFORMS OF DESERT REGIONS

WATER IN THE DESERT

The scenery of desert regions is unlike that of any other environment. Desert landforms are usually angular, intricate, and colorful, with bare rock widely exposed. These landforms attract the eye much more than does the sparse vegetative cover with which they are associated. In actuality, however, it is the absence of a continuous vegetative cover—rather than the direct influence of the desert climate—that gives desert landforms their unique character. Without a vegetative canopy to break the force of raindrops and without the binding effects of root networks, a blanket of moisture-retentive soil cannot accumulate on slopes. A grain of rock loosened by desert weathering is swept away by the first rainfall. The presence of a mantle of weathered regolith and soil on slopes in more humid areas gives such slopes their subdued, rounded form. Remove the vegetative cover and erosion will accelerate; the permeable soil cover will be lost; and desert-like landforms will be created, regardless of the climate.

Every feature of the desert landscape reflects a deficiency of water, yet the effects of running water are visible everywhere—on slopes as well as in valley bottoms. Much of the rain that falls, sparse as it is, encounters impermeable, soil-free surfaces and runs off immediately, eroding the land. A long-established fallacy is that desert landforms are produced by wind erosion. Wind erosion does occur in deserts, but it is entirely subsidiary to water erosion. The effects of wind erosion are confined to poorly consolidated or loose material, and in most deserts, such as those of the southwestern United States, wind rarely leaves a mark on solid rock. The main effect of wind is to transport fine material brought into desert basins by running water. Nevertheless, since the absence of vegetation permits wind gradation to operate on a far grander scale in deserts than in any other environment, we will examine wind-produced, or **eolian,** landforms in detail in this chapter.

CLIMATE AND VEGETATION

The importance of water and wind as gradational agents in arid regions is strongly related to the climate of those areas, in the past as well as the present. First, temperature and precipi-

Monument Valley, Arizona. (Courtesy of R. Mona)

tation are important factors in determining the amount of runoff available for gradational work and for stream flow. Second, winds are an element of weather or climate, and their strength, direction, and velocity are all determined by climatic factors. Third, climate, particularly precipitation, affects the amount of vegetative cover in arid regions. Where there is little or no vegetation, the effects of both water and wind on the shape of the land are greatly increased.

Arid lands have higher daily temperature ranges than humid regions at comparable latitudes. The excessive daytime heating and nighttime cooling that take place in deserts is caused by the lack of moisture in the air, which is reflected in the clear desert skies.

Even more characteristic of desert climates is the small amount of precipitation received. Sometimes a desert location may go years without receiving any rain, though such situations are exceptional. Most desert locations receive some precipitation each year, although the amount and timing are highly unpredictable. The rains that do fall are often of a convectional nature, caused by the intense heating of the land during the day. Such rains, though brief and limited in their coverage, are intense, often producing rapid runoff and flash floods over the desert landscape. The most important aspect of desert rainfall, in terms of landform development, is that when it does occur, much of it falls on impermeable surfaces, producing runoff capable of performing as a powerful agent of erosion.

Although running water is a highly effective agent of land formation in deserts today, it is only an occasional agent. In most parts of the desert, running water is active only during and shortly after rainstorms. Certain evidence, however, indicates that many desert regions have not always had the arid climates they have today.

It is believed that many areas were much wetter in the past, most recently during the Pleistocene Epoch. At the same time that glaciers were advancing in the high latitudes and in mountain regions, precipitation seems to

have increased in the middle and subtropical latitudes where deserts are found today. Evidence of this **pluvial** (rainy) period in today's deserts are lake deposits and wave-cut shorelines found in presently dry lands (see Fig. 16.26), twenty-thousand-year-old pollen grains, fossilized animal excrement indicative of a vegetation cover that could only have existed under more humid conditions, and neolithic cave paintings and rock drawings (petroglyphs) in these areas depicting animals associated with more humid climates.

Geomorphologists who study the landforms in arid regions believe that in all likelihood water was once not only the dominant gradational agent but may also have been a continuous one. They thus attribute certain desert landforms that seem incompatible with the present climate to the work of water under earlier pluvial climatic conditions.

STREAMS IN THE DESERT

Most desert streams are *intermittent;* that is, they flow only during and shortly after heavy rains or a particularly rainy period. During the rest of the time, the beds of these streams lie exposed and dry. There is no groundwater input to sustain a **base flow** between periods of surface runoff. The Mojave River in California is an excellent example of an intermittent stream that loses water volume by seepage into the ground rather than gaining volume from groundwater inflow as do streams of humid regions.

Since desert streams seldom have enough volume to make the trip to the sea before evaporating or seeping into their dry beds, they often terminate in interior depressions, where they form temporary lakes. These, too, eventually evaporate and disappear, only to reappear when future rains supply more inflow. In arid regions there are many such lake basins that have not been filled with water since the pluvial period of the Pleistocene.

Where surface runoff drains into interior basins rather than to the sea, the sea does not govern erosional base level as much as it does in humid lands. When deposition raises the surface of such a basin, the base level for the streams that flow into it is also raised. If tectonic activity lowers the basin, erosional base level drops and stream erosion is rejuvenated.

Some of the streams found in deserts originate in more humid regions nearby. Most of these streams, however, have insufficient volume to sustain flow across a large arid region. Without tributaries or groundwater inflow to replenish their losses caused by evaporation and underground seepage, the streams dwindle and finally disappear. The 465-kilometer-long Humboldt River in Nevada is an outstanding example. Only a few large streams that originate in humid uplands have sufficient volume to survive the long journey across hundreds of kilometers of desert to the sea (Fig. 15.1). The classic examples of such **exotic streams,** as they are called, are the Nile (Egypt), Tigris-Euphrates (Iraq), Indus (Pakistan), Huang (China), Murray (Australia), and Colorado (United States) Rivers. Only the exotic streams, the streams that come into the desert from outside and survive the desert to reach the sea, erode toward a base level governed by the level of the sea. All other desert streams—during their infrequent flows—erode toward the base level set by their particular terminal basin.

WATER AS A GRADATIONAL AGENT IN ARID LANDS

When rain falls in the desert, running water deeply erodes the unprotected surface. Sheets of water run down slopes, picking up loads of material. Channels become filled with flooding streams of muddy water. All the material removed by surface streams and runoff is carried along, just as in humid lands, until velocity and/or volume of stream flow decreases sufficiently for deposition to occur. Eventually these streams disappear because their seepage and evaporation losses exceed their flow. The huge amounts of debris are deposited along the way as the stream loses volume and velocity. The processes of erosion, transportation, and deposition by running water are the same in both arid and humid lands. However, the landforms that result differ, partially because of the intermittent nature of desert runoff, the lack of drainage flowing through to the sea, and the lack of a vegetation cover to protect surface materials against rapid erosion.

FIGURE 15.1 A satellite view of the Nile River crossing the Sahara Desert. Dark irrigated croplands contrast sharply with the sandy desert. The Nile is an outstanding example of an exotic stream, rising in the wet Ethiopian Highlands and East African rift lakes and crossing barren desert to reach the Mediterranean Sea. (NASA)

FIGURE 15.2 This stream cut wash is caused by rushing waters after rains in arid regions. A high flash-flood risk, such features are called arroyos in the American Southwest and wadis in the Middle Eastern deserts. (S. Brazier)

LANDFORMS OF ARID FLUVIAL EROSION One of the most common desert landforms created by the erosional activities of surface runoff are the channels of the intermittent streams themselves. Created by the rushing surface waters of heavy rainstorms, these vertical-walled streambeds, usually cut in unconsolidated alluvium, are called **washes** or **arroyos** in the Southwest United States, **barrancas** in Mexico, and **wadis** in North Africa and Southwest Asia (Fig. 15.2). The flash floods to which washes or wadis are prone make them bad risk areas for desert camping. Though it may sound bizarre, people have drowned in the desert during such floods.

Where steep slopes are underlain by weak clays and shales, rapid runoff produces a close network of V-shaped gullies, creating a severe, rugged configuration of ridges and dry ravines. The early French fur trappers in the American West called such areas in the Dakotas "badlands to cross" (Fig. 15.3). The phrase stuck,

FIGURE 15.3 The Badlands of South Dakota. Impermeable clays that lack a soil cover produce rapid runoff, leading to intensive gully erosion. (U.S. Dept. of Interior, National Park Service)

and those regions are still called the Badlands. Any similar area of rugged, complex, barren topography is now called a **badland.** Other examples can been seen at Death Valley and Pinnacles, both national monuments in California (Fig. 15.4). Such landscapes are not formed in humid climates because there vegetation usually stimulates soil development, increasing slope permeability and decreasing runoff and drainage density (number of streams within an area). Removal of vegetation from clay or shale areas will, however, cause badland topography to form in humid areas even more quickly than in arid areas.

Where the desert surface has been uplifted by geologic forces, exotic streams and their tributaries respond by cutting steep-sided canyons. Where the canyon walls consist of alternating layers of resistant and erodible rocks, erosion of the weaker formations (usually shale) will cause the canyon walls to retreat quite rapidly. The walls will be terraced with near-vertical ledges marking the resistant layers (ordinarily sandstone or limestone). Eventually only flat-topped, steep-sided mesas, capped by resistant layers, will remain. **Mesas** are a relatively common part of the landscape in the Col-

orado Plateau of Utah, Arizona, New Mexico, and Colorado. After additional erosion from all sides, a mesa will be reduced to a smaller remnant, lacking a flat summit, called a **butte.** The existence of mesas and buttes in a landscape are evidence that uplift occurred in the past and that fluvial erosion of the uplifted land has been extensive since that time. Monument Valley, in the Four Corners country of Utah, Arizona, New Mexico, and Colorado, is an exquisite example of the late stages of such a development (Fig. 15.5).

The erosion of mountain slopes fringing a desert basin or plain is carried on by sheetwash and gully erosion. As such slopes gradually retreat under erosion, a new and far more gently sloping surface of bedrock is created, called a **pediment** (Fig. 15.6). Characteristically, there is a sharp break in gradient between that of the hills or mountains, which rise at angles of 20° to 30° and that of the pediment, whose slope is usually only 2° to 7°.

Where alluvium has been deposited on the surface of a pediment, the surface appearance of the water-*eroded* form may closely resemble a water-*deposited* alluvial fan. In some situations, in fact, it may not be possible to de-

FIGURE 15.4 Gully erosion causing badland topography at Zabriski Point, Death Valley National Monument, California. (S. Brazier)

FIGURE 15.5 Mitten Butte, Monument Valley, Utah. (S. Brazier)

termine the existence of such an underlying pediment without excavating through the alluvium at the surface. In the case of an alluvial fan, the alluvium will be tens or even hundreds of meters thick. In contrast, the layer of alluvium overlying a pediment is only a thin sheet, often no more than a meter deep.

Geomorphologists do not all agree as to precisely how a pediment is formed. Clearly it is left behind as a surface of transportation for the debris being eroded from the receding

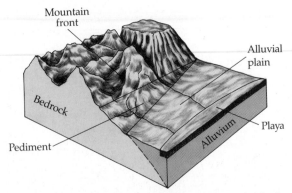

FIGURE 15.6 The pediment is cut into the bedrock at the foot of eroding mountain slopes in arid regions. In some places it will be covered with alluvial deposits.

mountain front. However, there is general agreement that the pediment, the sloping surface of bedrock that separates mountain slopes from basins in arid regions, is an erosional form created by the action of running water.

LANDFORMS OF ARID FLUVIAL DEPOSITION Deposition is as important as erosion in creating landform features of the desert, and in many areas the aggradation carried on by water certainly does as much to level the land as does erosion. For example, alluvial fans achieve their greatest development where intermittent streams laden with debris flow out of a mountainous region onto arid plains. Therefore, alluvial fans are especially associated with landscapes resulting from fault blocks, as in the Great Basin of Nevada and Utah. Here streams periodically erupt from canyons in the uplifted blocks and deposit their loads of debris in the adjacent grabens (Fig. 15.7).

There are several reasons for the development of alluvial fans in such settings. First, highland areas that fringe a desert are subject to a great deal of erosion, primarily because of their thin vegetation cover and the torrential convectional and orographic downpours that occur over mountains. Thus, even to begin

FIGURE 15.7 An alluvial fan, seen from the slopes above Death Valley. The dark alluvial deposits contrast with the light saline deposits of the valley floor. (S. Brazier)

with, these streams probably carry a larger load than comparable streams in more humid regions. As the streams pass from confined canyons in the highlands into the open desert, they deposit a large amount of their coarser materials near the mouth of the canyon. As streams flow out into the desert itself, their depth decreases and their volume is significantly reduced through seepage into the dry ground. Not far from the canyon, the stream itself may disappear. (In humid areas streams tend to continue in restricted valleys after leaving high-

lands, the water loss is not nearly so significant, and groundwater inflow and entering tributaries sustain the stream flow. As a result most highland streams that flow into the lower lands of humid regions do not create large alluvial fans.)

Along the bases of many highland areas in arid regions adjacent alluvial fans are so large that they join together to form an undulating, ramplike surface called a **bajada** (Fig. 15.8). A bajada is composed of relatively steep, sloping fans composed of coarse debris. Where

FIGURE 15.8 A *bajada* is formed when a series of alluvial fans join, forming an alluvial slope along the front of an eroding mountain range. (GeoPhoto Publishing Company, Alan L. Mayo, University of Colorado, Colorado Springs)

the associated streams are large, fans are less steep, less permeable, and larger in area. Where such fans coalesce they form a **piedmont alluvial plain,** like that along the western front of Utah's Wasatch Mountains, extending north and south of Salt Lake City.

Piedmont alluvial plains may be richly supplied with plant nutrients. If they can also be supplied with sufficient water, these plains make fertile agricultural lands, particularly where the finer materials have been deposited. Supplying water to a piedmont alluvial plain is not as difficult as it would be in most other parts of the desert. First, the associated mountain streams may have perennial flow above the fan and can be used for irrigation. Gravity helps distribute mountain water over the gradually sloping alluvial plain. Second, the porous character of the alluvium and the gentle downward slope of the plain provide an especially good setting for wells, and, in some instances, for artesian systems.

Because of their rich soils and relatively accessible water supplies, piedmont alluvial plains have been transformed into highly productive farmlands. In Utah the sloping alluvial plain of the Salt Lake City oasis is used to grow such crops as wheat, sugar beets, alfalfa, vegetables, and soft fruits. The farmlands near Phoenix, Arizona, produce citrus fruits, dates, cotton, alfalfa, and vegetables.

Desert basins surrounded by mountains are sometimes called **bolsons.** They were formed by faulting, the same activity that uplifted the fringing highlands. The lowest part of a bolson commonly is occupied by a dry lake bed, known as a **playa.** Floods occasionally transform the playa into a lake in the space of a few hours. The lake may be gone the next day, or it may persist for weeks (Fig. 15.9). Playa lakes almost never rise high enough to spill over into an adjacent basin. Consequently, these lakes lose most of their volume through seepage and evaporation in the dry desert air. The repeated cycle of inflow and evaporation leaves behind thick deposits of minerals that crystallize from evaporating brines: chlorides, sulfates, and carbonates (for example, rock salt,

FIGURE 15.9 Badwater, Death Valley, a playa lake at the lowest elevation in North America, 86 meters *below* sea level. (S. Brazier)

FIGURE 15.10 Salt deposits on the surface of a salt flat or salina in Death Valley. (S. Brazier)

gypsum, and lime, respectively). When the playa lake dries out, its bed may be encrusted with sparkling white salt deposits. Then it is known as a **salt flat** or **salina** (Fig. 15.10). Many playas are merely clay pans, indicating that their lakes leak away in the subsurface rather than evaporating at the surface.

Playas are useful to man in several ways. For one, we mine the rich deposits of evaporite minerals such as potash, salt, borates, and sodium nitrate, all important industrial chemicals that have been deposited in playa beds (Fig. 15.11). The flat, hard surface of some playas also makes them suitable as racetracks and air strips. Utah's famous Bonneville Salt Flat is not a typical playa but the bed of a long-extinct Pleistocene pluvial lake. The western portion, where world land speed records are set, does not usually flood under existing climatic conditions.

From the foregoing discussion one might assume that desert scenery always includes cliffs or canyons or mountain walls overlooking either alluvial fans or pediments. This is not strictly true. In addition to the dune land-

FIGURE 15.11 The extraction of evaporite minerals from the beds of extinct desert lakes and playa surfaces is a major industry in California's Mojave Desert. This aerial view is of the open pit mine at Boron, California, operated by U.S. Borax Corporation. It is the world's principal source of borate minerals. (U.S. Borax)

scapes, called **ergs,** which we will discuss shortly, and which in fact make up a relatively small proportion of the world's desert landscapes, there are vast desert areas that consist only of featureless plains. Some of these are expanses of bedrock known as **hamada;** others are gravel- or pebble-covered surfaces called **reg** or **serir.** The Syrian, Gobi, and Australian Deserts are largely reg and hamada types.

WIND AS A GRADATIONAL AGENT

As a gradational agent, wind is less effective than running water, waves, groundwater, moving ice, and mass movement. However, under certain conditions, wind is a significant gradational force and contributes visibly to the shaping of the land. Landforms—whether in the desert or elsewhere—that are created by wind are called *eolian landforms* (after Aeolus, the god of the winds in classical mythology).

The two primary conditions for wind to become an effective gradational agent are absence of complete vegetation cover and the presence of dry, fine material at the surface. These two conditions are most widespread in arid regions and beaches, though they are also found on or adjacent to dry river beds, areas of recent alluvial or glacial deposition, newly plowed fields, and overgrazed lands.

A continuous vegetation cover reduces wind velocity near the surface, absorbs the force of the wind and prevents it from being directed against the land surface, and holds materials in place with its root network. Without such a protective cover, fine-grained and sufficiently dry surface materials are subject to removal by any strong gust of wind that occurs. However, if surface particles are damp, they will adhere together in wind-resistant aggregates.

The gradational activities of the wind are similar in most ways to those of running water. Running water and wind detach and remove materials by similar means. Both gradational agents transport material by traction, saltation, and suspension. However, unlike running water, wind cannot transport material in solution. Furthermore, wind has hardly any lateral or vertical limitations on movement. The result is that the dissemination of material by the wind can be far more widespread than that by streams, and deposited material is not necessarily concentrated in surface depressions.

The most important similarity between the actions of wind and running water is that the size of the particles they can pick up and carry (their **competence**) is controlled by their velocity. The result is that wind erosion selects the finer particles for transportation and leaves behind the larger, coarser particles that the wind is incapable of lifting. Likewise, wind deposits are stratified according to changes in velocity.

EROSION BY WIND

Strong winds frequently blow through arid regions, whipping up loose surface materials and transporting them within turbulent air currents. Wind removes or erodes surface materials by two processes (Fig. 15.12). The first of these is **deflation,** which is similar to the hydraulic force of running water. As wind velocity increases, the first particles to be affected are the finest ones present at the surface—microscopic bits of clay and silt—essentially, dust. The finest particles transported by the wind are carried in suspension, buoyed by vertical currents. Such particles will remain in suspension as long as the strength of the upward currents of air exceeds the tendency of the particles to fall to the ground. If the wind velocity surpasses 16 kilometers per hour (10 mph), surface sand grains will be put into motion. The lighter particles that are nevertheless too large to be carried in suspension are bounced along the ground by the transportation process of **saltation.** As these particles are bounced along, they dislodge other particles that are then added to the wind's load or are driven forward on the ground as surface sand creep. The particles carried in suspension by the wind make up its suspended load. The particles that bump

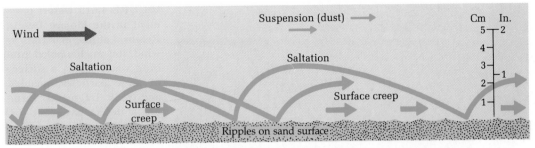

FIGURE 15.12 Movement of sand by saltation and surface creep due to impact of saltating grains. When sand is blowing, forward ripple movement is rapid enough to be seen. (After R. A. Bagnold, 1941)

and jump along the ground make up the wind's bed load.

The second way in which the wind erodes is by **abrasion.** This process is analogous to the abrasion produced by the transported load of streams, breaking waves, and glacial ice, but it operates on a much more limited scale. In order for the wind to remove surface materials by abrasion, it must carry cutting tools that are harder than the particles of the abraded surface. As these materials hit against a surface of weaker material, they break off bits of the weaker rock. Quartz sand is the most effective of the wind's abrasive agents. Yet sand grains are relatively large and heavy and rarely are lifted higher than 1 meter above the surface. Consequently, the effect of natural sandblast is limited to a zone close to ground level.

Where fine dust particles predominate on the land surface, they will be picked up and carried in suspension by the strong winds. The result is a thick, dark, swiftly moving cloud of dust, swirling over the land. Dust storms can be so severe that visibility drops to nearly zero and almost all sunlight is blocked. They can also be highly destructive, removing layers of surface materials, and depositing them elsewhere, sometimes in a thick, choking layer, all within a matter of a few hours. Sand storms are similar to dust storms but occur in areas where sand is the dominant surface material. Because sand is heavier than dust, most sandstorms are confined to a low level near the surface, often not more than a meter above the ground. Evidence

of the restricted height of desert sandstorms can be seen on automobiles that have traveled through the desert. After a sandstorm, the pitting, gouging, and abrading effects of natural sandblast may be visible only on the lower portions of the vehicle, leaving the upper part unaffected.

Erosion by deflation can produce hollows or depressions in a barren surface of unconsolidated materials. These depressions, which vary in size from a few centimeters to a few kilometers across, are called **deflation hollows** or **blowouts.** In the Kalahari Desert deflation hollows collect rainwater and thus attract game animals and their hunters, the Bushmen. Often blowouts form where there was already a slight depression in the surface, such as in silty playa deposits.

Deflation has been thought to produce a close-fitting mosaic of rock fragments called **desert pavement** (or *reg* in North Africa and *gibber* in Australia) common in many desert regions. Because erosion by deflation is selective, the smaller clay and silt particles from an area of materials of mixed sizes are selectively eroded and transported to another location; sometimes even the larger and heavier grains of sand are removed as well. Such selective removal leaves behind the larger particles, pebbles, and rock fragments, which together form the cobbled surface of desert pavement (Fig. 15.13). This erosional feature is widespread in parts of the Sahara Desert, interior Australia, the Gobi in central Asia, and the American

FIGURE 15.13 Desert pavement in the Mojave Desert. Removal of fine materials by wind and unchanneled running water leaves a surface of large particles, pebbles, and rock fragments forming a desert pavement. (R. Sager)

Southwest. Some recent research indicates that desert pavement may also be a product of unchanneled running water. Regardless of its origin, desert pavement is important for the protection it affords the material below the top layer of coarse pebbles and rocks. Pavement formation stabilizes desert surfaces by preventing continuous widescale erosion. Unfortunately, off-road recreational vehicles may disturb this stability, thus damaging desert ecological systems.

Where abrasion or sandblast action is at work on rocks of varying resistance, differential erosion results in the etching away of softer sections of rock while the more resistant rock remains. The rock face becomes honeycombed or latticed in intricate designs (Fig. 15.14).

Eolian abrasion also produces **ventifacts** (wind-fashioned rocks). A ventifact is commonly a rock fragment that has been trimmed flat on one side by sandblast, after which erosion of its support has caused it to turn over, so

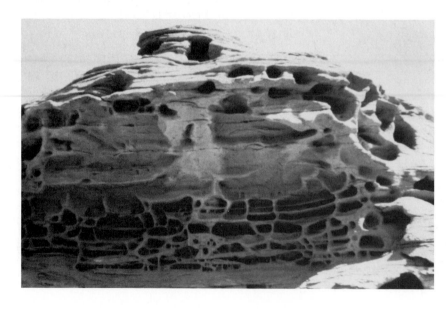

FIGURE 15.14 Differential erosion due to sandblast abrasion can create a honeycombed or latticed design in bedrock. (S. Brazier)

FIGURE 15.15 This desert rock is a ventifact, showing the effect of natural sandblast. Note its smooth and polished look. (U.S. Geological Survey)

that a second side is faceted by eolian abrasion. Often three flat faces are produced, which meet along sharp edges (Fig. 15.15). This highly distinctive rock form is an indication of sandblast. While not extremely common, ventifacts are plentiful where conditions are appropriate for their formation.

Eolian abrasion works much too slowly to be as effective as deflation in changing the shape of the land. However, in favorable locations major abrasional forms are produced. The largest are the long grooves, called **yardangs,** cut in ancient lake-bed silt deposits.

The best examples of yardangs are seen in Iran and Afghanistan.

Another feature often attributed to wind abrasion is the pedestal or balancing rock, commonly and incorrectly thought to form where sandblast attacks the base of an individual rock so that the larger top part appears balanced on a thinner pedestal below. Actually, such forms result from salt crystallization and weathering processes in the damper environment at the base of an outcrop, and are unrelated to sandblast (Fig. 15.16).

Where the land surface has been eroded to bedrock, wind abrasion will polish the rock surface but will not significantly erode it. The speed of eolian erosion in arid regions depends primarily on the character of the materials exposed. Only where they are mechanically weak can abrasion and deflation be effective.

DEPOSITION BY WIND

All material transported by wind is deposited somewhere in some characteristic manner. The coarser material is often deposited in drifts in

FIGURE 15.16 Mushroom Rock, Death Valley. A pedestal rock caused by desert weathering processes. (S. Brazier)

the shape of hills or ridges, called **dunes.** The finer material settles far from its source area in the form of a blanket covering the existing topography. Fine, silty material deposited in this manner is called **loess.**

SAND DUNES   To many people the word desert evokes the image of endless sand dunes, blinding sandstorms, a blazing sun, mirages, and an occasional palm oasis. Although there *are* such deserts, particularly in Arabia and North Africa, many others have rocky or gravelly surfaces, some scrubby vegetation, and no sand dunes of any consequence. Nevertheless, sand dunes are certainly the most spectacular features of wind deposition, whether they occur as "sand seas," or *ergs,* in the Sahara, or as hills behind a Cape Cod beach (Fig. 15.17).

Dune topography is almost infinite in its variation. For instance, dunes in the great ergs of the Sahara and Arabia look like continuously rolling sea waves. Others are shaped like individual crescent rolls sitting on a plate of sand. Sometimes eolian sand forms "sand sheets" with no dune formation at all. Laboratory experiments indicate that the different formations are the result of the amount of sand available, the strength and direction of the dominant winds, and the amount of vegetation cover. As wind saturated with sand encounters obstacles or changes in the topography that decrease its velocity, sand drops out and piles up in drifts. These formations further decrease wind velocity so that the dunes grow larger, until an equilibrium is reached between dune size and the ability of the wind to feed sand to the dune.

Sand dunes may be classified as *live* or *fixed* (Fig. 15.18). Live dunes change their shape *and* advance downwind. Dunes may change their shape with changing wind direction and/or wind strength. Dunes move forward as the wind erodes their windward slope. This causes sand ripples to migrate up to the crest and deposit their load on the steep leeward slope, or **slip face,** which is at the angle of repose for dry sand (about 35°, which is the steepest slope dry sand can retain without slipping or falling). When wind direction and velocity are relatively constant, a dune can move forward while maintaining its form. The speed at which live dunes move downwind varies. Large dunes travel extremely slowly, while smaller ones may move up to 40 meters a year.

A dune whose shape and position are maintained over time is said to be fixed. Dunes

FIGURE 15.17   Desert sand dune scenery, Death Valley. (S. Brazier)

(a)   (b)

FIGURE 15.18 These two photographs contrast live dunes (a) with fixed dunes covered by vegetation (b). Live dunes normally have a sharp crest, with a gentle back slope that undergoes erosion, and a steep advancing slip face at the angle of repose of the dry sand. Fixed dunes usually have rounded crests and give no indication of recent erosion or deposition. (U.S. Dept. of Interior, National Park Service)

are normally fixed by vegetation, by the position of a wind-breaking obstacle, or by back-and-forth movement of the crest under the influence of opposing winds. Where vegetation lies in the path of a live dune, the dune may move over plants and drown them in sand. However, depending on the size and extent of the vegetation, dune movement may be impeded. Vegetation is able to fix a sand dune if plants can gain a foothold and send roots down to moisture beneath the dune. This task is difficult for most plants, because the sand itself offers little in the way of nutrients or moisture. In places where a sufficient cover of vegetation has been able to develop, a sand dune invasion may be halted. One such place is in the Sand Hills of Nebraska, where giant dunes, probably formed during an interglacial period, have been fixed by a cover of grasses that now serve as grazing lands. Similar stabilized dunes

are found along the southern edge of the Sahara Desert, which clearly extended farther toward the equator in the recent geologic past. Both instances involve changes in climate that affected sand supply and wind patterns.

TYPES OF SAND DUNES Many dunes are similar enough so that some basic types can be described (Fig. 15.19). **Barchans** are crescent-shaped isolated dunes. Their windward slope is the convex curve of the crescent, a gentle slope up which sand is moved. The concave leeward slope is at the angle of repose. The two horns of the crescent point downwind. Barchans are formed in areas of low sand supply where moderate winds blow from a constant direction. Although they form as isolated dunes, barchans often appear in swarms.

Parabolic dunes are somewhat similar to barchan dunes but have a reverse orientation.

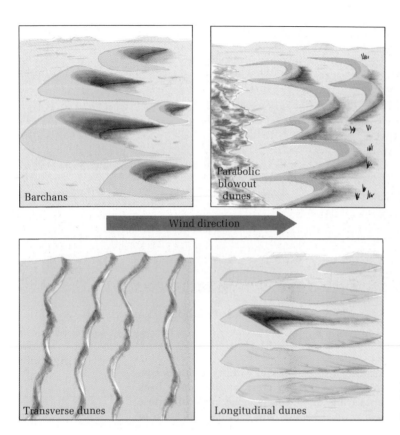

FIGURE 15.19 Principal dune types. Wind direction is the same in all figures as indicated by the arrow.

Here the points of the curving dune trail behind, while the crest advances most rapidly. Such dunes often resemble hairpins that are open in the upwind direction. Often their "tails" are stabilized by vegetation. Parabolic dunes commonly appear along beaches and in nondesert environments.

Transverse dunes form where light to moderate winds blow from a constant direction and there is an abundant supply of sand covering the entire landscape. Transverse dunes take the shape of a series of crests and troughs whose peaks run perpendicular to the direction of prevailing winds (hence the name *transverse*). These dunes look like sea waves. The windward slope of transverse dunes, like that of the barchans, is gentle, while the leeward slope is at the angle of repose.

Longitudinal dunes (Fig. 15.13) are long, narrow, parallel dunes that are aligned with the prevailing wind direction. A small sand supply

and strong winds are important factors contributing to the formation of longitudinal dunes. There is no consistent distinction between the back slopes and slip faces of these dunes, and their summits may be either rounded or sharp. Longitudinal dunes cross vast areas of interior Australia where they are known as *sand ridges*. A type of longitudinal dune called a **seif** (from the Arabic, *sword*) is found in the deserts of Arabia and North Africa. Seifs are huge, sharp-crested dunes, sometimes hundreds of kilometers long, whose troughs are almost clear of sand (Fig. 15.20). They may reach 180 meters (600 ft) in height.

The **blowout dune** is most commonly formed on beaches where sand supply is abundant, where winds are moderate and blow from a constant direction, and where vegetation has partially fixed the sand. The shape of the blowout dune is that of an elongated sand hill with a deflation hollow on the windward side.

FIGURE 15.20 A satellite view of Saharan longitudinal dunes, which are over 100 kilometers in length. (NASA)

Where vegetation has insufficiently fixed a blowout dune, its leeward side gradually encroaches upon the land as a parabolic dune.

LOESS The wind can carry dust-sized particles of clay and silt, resulting from deflation, for hundreds or thousands of miles before depositing them. Eventually these particles settle down to form a tan or gray blanket of **loess** that covers the existing topography over widespread areas. These deposits vary in thickness from a few centimeters to over 100 meters. In Northern China on the margins of the Gobi Desert, the loess is 30 to 90 meters (100 to 300 ft) thick (Fig. 15.21).

Loess may originate from deserts, dry river floodplains, or other unvegetated surfaces. The extensive loess deposits of the

FIGURE 15.21 A steep-sided gully eroded into the extensive loess deposits of northern China. (S. Brazier)

VIEWPOINT: YELLOW EARTH AND YELLOW RIVER

China's huang-t'u or "yellow earth" is the largest extent of loess in the world. It covers much of northern China, especially the region across which the middle Huang He or "Yellow River" makes its huge right-angled loops through the Ordos Plateau. The silt deposit is so fine that it has the feel of talcum powder, and its brownish-yellow color dominates the landscape. Blown for millennia by the cold winter winds moving out of Central Asia, this dust was lifted from the Mongolian steppes and Gobi desert and carried southeast. There it buried earlier surface topography to depths of an average of 15–30 meters (50–100 ft), and in some areas as deep as 76–91 meters (250–300 ft). It created a landscape of khaki-colored slopes and cliffs, pale brown river waters, and a sky that seems perpetually dusty. Peasants working in the fields sometimes wear gauze masks over their mouths and noses to avoid breathing in dust, and dust storms will carry the yellow earth as far as Beijing and Shanghai during the winter.

Being wind-borne, the loess is generally unstratified and lacks the horizontal bedding planes of water-deposited segments. It is soft and extremely porous and permeable. It also has vertical cleavage that is reflected in steep cliffs and river banks. These last two characteristics arise from the fact that the loess region was typically a steppe-grass vegetation zone: As new deposits of the loess smothered the earlier grasses, they continued to grow upwards. Their former decayed stems left a series of irregular vertical tubes through the continuing accumulation of dust. Thus water falling on its surface is rapidly absorbed like a sponge. This percolating moisture dissolves minerals and is frequently brought to the surface by capillary action in the hot summers. When evaporated it leaves a hard crust, which strengthens the surface and walls of vertical cleavage.

Beneath this crusting, however, the soil is unconsolidated and soft. Occasional heavy summer rains tear through the crust, and streams pour through the hills, scouring out gullies in the unprotected surface. The loess region becomes a dissected landscape, with alluvial valleys and plains separated by gullied ridges and eroded barren hills.

In spite of these seemingly negative characteristics, nowhere else in the world has loess played such a key role in the history of people. The Wei Valley, tributary to the Huang He, was the cultural hearth of early Chinese civilization. Not only could its steppe and light woodland cover be readily cleared, but this soft soil was easily worked with primitive tools and made productive. It is probable that the need for water control and irrigation en-

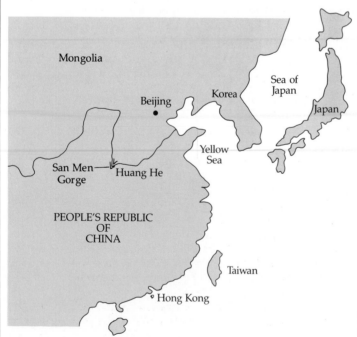

The location of the Huang He Basin, China.

Homes carved into loess deposits (left). Cultivated loess valleys (center). The fertile loess plains (right). (S. Brazier)

couraged the development of a highly organized and cooperative society, which has been the hallmark of the Chinese to the present day. Shared labor has always been needed to keep the silting channels cleared in order to water crops. The ancient capital of Chang'an (near the modern city of Xian) existed as early as 1122 B.C. Here the great tombs of the 'Chin emperors—constructed 22 centuries ago and now so famous for their vast terracotta armies—were discovered by a peasant digging a well in his loess field in the early 1970s.

Settlement was located in the fertile alluvial valleys and also on the loess plateau itself. Since the loess was unconsolidated, it was easy to dig into. Through the centuries houses and whole villages have been constructed as caves hollowed out in the faces of cliffs. Inside the cave dwellings it is well insulated, cool in the hot summers and warm in the cold winters. Hollow brick beds have fires lit under them for winter heating. At Yenan in the 1930s, Communist soldiers under Mao Tse-tung found refuge here in the war against Japan. Mao himself lived in a cave and consolidated his leadership there.

The fact that this area is prone to earthquakes has meant disastrous losses of life as the cave homes collapsed. In the 1920 earthquake, over 250,000 people were killed either by burial in their homes or by the famine and starvation that followed.

Through the centuries roads and farm tracks have cut down into the soft loess: Once the traditional wooden-wheeled carts have broken through the crusted surface, wind and rain carries the dust away. It is possible to travel on these incised tracks with only an occasional glimpse of the surrounding countryside above.

While wheels and animal hoofs throughout the centuries have eroded into the surface, millennia of

tillage also have destroyed the natural vegetation and contributed to catastrophic erosion here. Farmers cultivating dry wheat and millet on the loess surface plowed up and down slopes and removed all vestiges of woodland for firewood. Gullies and ravines grew larger and the area became one of chronic poverty.

With the communal mobilization of production brigades in the 1950s and 1960s, great efforts were made to deal with the earlier mismanagement of agriculture. A local village drew up long-range schemes for their area.

> High and remote hills, into forest.
> Low hills, gentle slopes, into terraced fields.
> Gullies, into orchards.
> River beds, into farm gardens.

Throughout the plateau, gullies were filled in, dams were built to hold the silt runoff, slopes were terraced, contour plowing and strip cropping were introduced, and tree planting programs were initiated.

In its passage through the loess plateau, the Huang He becomes a transporter of massive amounts of suspended load. Before it enters the loess region, the Huang He carries about 1 percent sediment by weight: After its passage around the two great bends through the loess, it carries 40 percent. It has the consistency of a yellow porridge, caused by constant runoff from the huang t'u.

The Huang He leaves the hilly loess region and enters the North China Plain through San Men Gorge (see the map). As its velocity slows, a vast sediment load is deposited. In the geologic past the North China Plain was a large marine bay, with its coastline along the foothills of the mountains. Millions of tons of sediment have built a huge delta whose margin was pushed eastward by the slow-flowing river.

But what fertility! This onetime desert dust, carried first by the wind and then by stream and river, now forms one of the greatest agricultural regions in the world—but not without its problems. As more and more silt was deposited, the river's bed rose higher and higher and floods became more frequent. The river changed its course many times. And for 4000 years the Chinese farmers built the dikes higher and higher. In the 1940s, it was estimated that the river level at flood stage was 8 meters (25 ft) higher than the surrounding plain. No wonder the river in this region is known as "China's Sorrow," as millions have drowned throughout Chinese history owing to its flood waters.

In spite of the fact that this is probably one of the most elaborately measured and analyzed rivers in the world, it still presents challenges. The great San Men Dam and reservoir, completed in 1960, lost 40 percent of its capacity by siltation within only four years. New techniques and reconstruction of the dam now ensure more effective sluicing out of sediment at peak flow, but at the cost of losing electrical-generating capacity. Even the propellers of vessels plying the river wear down faster than on any other of the world's rivers because of the silt. Somewhat tamed in its lower course, the Huang He finally reaches its destination, the Hwang Hai (Yellow Sea), a sea colored by the transported loess.

American Midwest and of Europe were derived from the glacial deposits of retreating continental ice sheets. As winds blew across barren till and outwash plains, they picked up a large load of fine sediment that formed the loess deposits of downwind regions.

Certain interesting characteristics of loess affect the shape of the land where it forms the surface material. For example, though fine and dusty to the touch, loess maintains vertical walls when cut through naturally by a stream or artificially by a road. Sometimes slumping will occur down these steep faces. This slumping gives a steplike profile to many loess bluffs. Furthermore, loess is easily eroded because of its fine texture and unconsolidated character. As a result, loess-covered plains that are unprotected by vegetation often become gullied. Where loess covers hills, both gully erosion and slumping are conspicuous. A particularly severe erosion problem is currently causing the collapse of the high loess bluffs along the Mississippi River at Vicksburg, Mississippi.

Due to its high calcium carbonate content and young unleached characteristics, loess is the parent material for many of the earth's most fertile agricultural soils. Extensive loess deposits are found in northern China, the Pampas of Argentina, the North European Plain, the Soviet Ukraine, Central Asia, and the Midwestern Plains and the Mississippi Valley of the United States (Fig. 15.22). Most of these areas now form extremely productive farming regions.

THE EROSION CYCLE IN DESERTS

Theoretically, the cycle of erosion in arid lands is similar in many ways to the cycle of erosion in humid regions, except that the formation of pediments rather than peneplains is emphasized and wind plays a more important role. The cycle in humid regions consists of a continuous lowering of the soil-covered land surface by the progressive flattening of slopes. In arid regions the removal of uplands is accomplished by the horizontal retreat of rocky cliffs or escarpments. The presumed final product in humid areas is the nearly flat peneplain near base level; in arid areas it is called a **pediplain** (composed of coalescing pediments). Any residual relief present in the late stages of the arid erosion cycle occurs in the form of **inselbergs** (island mountains) that rise abruptly above surrounding pediments (Fig. 15.23).

In this arid cycle of erosion, W. M. Davis applied the same principles originally developed for humid landscapes to those of deserts. His familiarity with American deserts, with

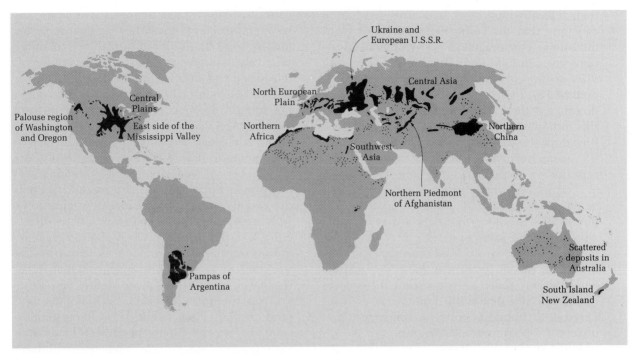

FIGURE 15.22   Major loess regions of the world. Most loess deposits are peripheral to deserts and recently glaciated regions.

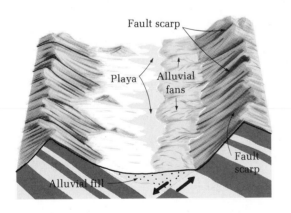

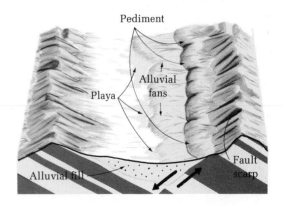

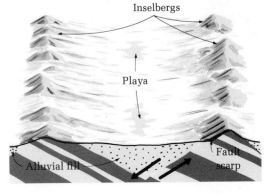

FIGURE 15.23   The cycle of erosion in mountainous deserts (after Davis). Alluvial fans develop and basins fill with debris as fault block movement uplifts mountain ranges. After mountain uplift ceases, range fronts wear back, producing pediments. Eventually erosion may reduce the mountains to isolated remnants called inselbergs, which rise above flat, windswept plains.

their typically fault block structures, led him to conclude that the young horsts or fault block mountains encouraged the development of orographic rainfall. Thus, fluvial processes with their erosion of the fault scarps, the cutting of pediments, and the building of alluvial fans dominated the earlier phases of the cycle. As the mountains were reduced by erosion, the fluvial processes became less important, and the eolian forms became more dominant, except during occasional convectional storms.

However, whether these same steps would be equally applicable to all desert landscapes is doubtful. The older, more geologically stable Australian and Saharan deserts were not block-faulted like the North American deserts; thus the initial fluvial phase may have been different. Further, there are significant differences in erosion and deposition under arid as compared with humid conditions, which result in different topographic expressions.

First, interior basins—instead of sea level—may provide base levels for erosion. The result is that erosional plains in arid lands may be found at elevations either well above or below sea level. In addition, instead of remaining relatively constant, the local base levels are often raised through deposition in the basins or lowered through faulting. Second, infrequent but intense precipitation plus little or no protective vegetative cover results in extremely rapid and powerful surface runoff, which is uncommon in humid lands. Third, the absence of a cover of vegetation and soil minimizes creep as a slope-forming process and permits the varying degrees of bedrock resistance to be fully expressed in the landscape. Thus differential weathering and erosion attain their maximum development in deserts. Fourth, the weakness of long-range transportational forces results in conspicuous loose debris accumulation, in the form of talus, alluvial fans, and sand dunes. Thus, while it is interesting to compare the evolution of desert landscapes with those of humid areas, arid landforms have their own unique characteristics.

QUESTIONS FOR DISCUSSION AND REVIEW

1. What are eolian landforms? What gradational agent most significantly affects the desert landscape?

2. How do climate and vegetation affect landforms in the desert?

3. Have deserts always been dry? Explain your answer.

4. What is an exotic stream? How does such a stream differ from an intermittent stream?

5. Why do landforms produced by erosion in the desert differ from erosional landforms in more humid regions?

6. Define the following terms: *badland, mesa, butte,* and *pediment.*

7. How is a bajada formed?

8. Why do lakes occasionally form in bolsons? How are such lakes related to salinas?

9. Why are most sandstorms confined to a low level? How do they differ from dust storms?

10. What is deflation? How does it affect the desert landscape?

11. Describe the formation of a ventifact.

12. What is the difference between a barchan and a transverse dune?

13. Why is loess an important natural resource? What problems might you encounter if you tried to cultivate loess?

14. Compare the cycle of erosion in deserts with that of humid lands.

16

GLACIAL SYSTEMS

One can hardly imagine greater natural beauty than in the Swiss Alps, Canadian Rockies, or the coastal fiord country of Norway and Alaska. Rugged mountain peaks rise high above deep lake-filled valleys, or narrow deep sea lanes, creating the ultimate in scenic appeal for many people. Masses of moving ice, known as **glaciers,** have transformed the appearance of high mountains, as well as large portions of continents, into unique landscapes. These great icy currents are one of the most effective and spectacular geomorphic agents on the earth's surface.

GLACIER FORMATION AND THE HYDROLOGIC CYCLE

Glaciers are moving masses of ice that have accumulated on land in areas where more snow falls during a year than melts. As the snow falls, it is a hexagonal crystal of intricate beauty and variety. Once on the land surface, however, it is soon transformed into a more compacted mass of smaller rounded grains. As the air space is lessened by compaction and melting, the grains become more dense. With further melting, refreezing, and increased weight from newer snowfall above, the snow reaches a granular recrystallized stage between snow flakes and ice, known as **firn.** With additional time, pressure, and refrozen meltwater from above, the firn will be recrystallized into glacial ice. The small firn granules will become larger, interlocked blue ice crystals. When the ice is thick enough, usually over 30 meters (100 ft), the weight of the snow and firn above will cause the crystals to become plastic and flow outward or downward from the area of snow accumulation.

Glaciers are open systems, with snow entering the system and mainly meltwater leaving the system in a constant cycle. The glacial system is controlled by two basic climatic conditions, precipitation in the form of snow and freezing temperatures. First, there must be sufficient snowfall to exceed the annual loss through melting, evaporation, sublimation, and calving. **Calving** is when the glacier loses solid chunks as icebergs to the sea or large lakes. Mountains along middle-latitude coastlines, and even at the equator, can support glaciers

The Columbia Glacier, Alaska. (Larry Mayo, U.S. Geological Survey, 76 M3-51)

due to heavy orographic snowfall, despite intense sunshine and warm surrounding climates. Yet, some very cold polar regions in subarctic Alaska and Siberia, and a few valleys in Antarctica, have no glaciers due to a dry climate.

A second climatic condition is temperature. Summer temperatures must not be high for too long or all the snowfall from the previous winter would melt. Surplus snowfall is essential, since it allows for the pressure of accumulated snow over the years to transform older buried snow into firn and glacial ice and to create depths great enough for the ice to flow.

Glaciers are part of the earth's hydrologic cycle and are second only to the oceans in the total amount of water contained. About 2 percent of the earth's water is presently frozen as ice. Two percent may be a deceiving figure, however, since over 80 percent of the world's *fresh* water is locked up as ice in glaciers, with the majority of it in Antarctica. The total amount of ice is even more awesome if we estimate the water released upon the melting of the world's glaciers. Sea level would rise over 70 meters (230 ft). This would change the geography of the planet considerably. In contrast, should another ice age occur, sea level would drop drastically. During the last ice age, sea level dropped over 120 meters (400 ft).

When snow falls on high mountains or in polar regions, it may become part of the glacial system. Unlike rain, which returns rapidly to the sea or atmosphere, the snow that becomes part of a glacier is involved in a much more slowly moving system. Here water may be stored in ice form for hundreds or even hundreds of thousands of years before being released again into the liquid water system as meltwater. In the meantime, however, this ice is not stagnant, but is moving slowly across the land with tremendous energy, carving into even the hardest rock formations. The glacier reshapes the landscape as it engulfs, pushes, drags, and finally deposits rock debris in places far from its original location. And so even after the land is long released from its icy entombment, a tremendous variety of glacial land-

forms remains as a reminder of the energy of the glacial system.

Throughout most of the earth's history, glaciers did not exist. When a period of time occurs when significant areas of the earth are covered by glaciers, we call it an **ice age.** At the present time about 10 percent of the earth's land surface is covered by glaciers. Present-day glaciers are found in Antarctica, Greenland, and at high elevations on all the continents except Australia. In the recent past, from about 2 million to about 10,000 years before the present, nearly a third of the earth's land area was covered by ice thousands of meters thick. In the much more distant past other ice ages have occurred.

TYPES OF GLACIERS

There are two major types of glaciers: **alpine glaciers** and **continental ice sheets.** Alpine glaciers have their source in mountain areas, usually accumulating in depressions initiated by stream erosion. Those that are confined by the rock walls of the valley they occupy may also be called **valley glaciers** (Fig. 16.1).

Another variety of alpine glacier is the **piedmont glacier** (Fig. 16.2). This variety forms where two or more valley glaciers coalesce and move together over flatter land at the base of a mountainous region. Whereas piedmont glaciers result from glacial flow beyond the limits of confining valleys, some alpine glaciers do not reach the valleys below the zone of high peaks. Instead, they occupy distinctive high amphitheaters, called **cirques,** and are known as **cirque glaciers** (Fig. 16.3).

Alpine glaciers create the characteristic rugged scenery of the high mountains of the world. They can presently be found in the Rockies, Sierra Nevada, Cascades, Olympic, Coast Ranges, and numerous Alaskan ranges of North America. They are also found in the Andes and Alps and in the Himalayas, Pamirs, and other large Asian mountain ranges. Small glaciers are even found in tropical Africa on Mount Kenya, Kilimanjaro, and the Ruwenzori Range. The largest alpine glaciers in existence

FIGURE 16.1 The South Cascade Glacier, Washington. This is a small valley glacier with a clearly defined firn line separating the lighter upper accumulation zone from the darker lower ablation zone. (Austin Post, U.S. Geological Survey, FR6025-60)

FIGURE 16.3 Several small cirque glaciers surrounding Mt. Assiniboine in the Canadian Rockies. A bergschrund, or great crevasse, can be seen around the edge of the cirque glacier in the center of the photo. (Austin Post, U.S. Geological Survey, 61-F2-79)

FIGURE 16.2 The upper portion of the Malaspina Glacier, Alaska. This large piedmont glacier formed when several valley glaciers coalesced over the flatter land at the base of the mountain range. (Austin Post, U.S. Geological Survey, 6611-157)

today are found in Alaska and the Himalayas, where some reach lengths of over 100 kilometers (62 mi).

The second and larger type of glacier is the continental ice sheet, a far more significant formation than the valley glacier. Continental ice sheets, at one time, covered as much as 30 percent of the land. They still blanket Greenland and Antarctica, and smaller ice caps are present on some Arctic mountains. In contrast to alpine glaciers, continental ice sheets are unconfined, flowing over even the higher portions of the land. Thus they are not elongated but flow outward in all directions from their source area.

FEATURES OF AN ALPINE GLACIER

Alpine glaciers can be divided into two parts, a zone of accumulation and a zone of ablation (Fig. 16.4). The upper portion of the glacier, where snowfall exceeds

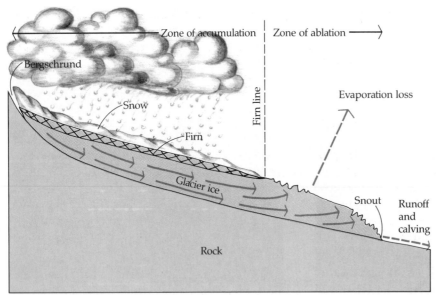

FIGURE 16.4 Cross-sectional diagram showing the principal features of a valley glacier.

ablation (losses through melting, evaporation, and sublimation), is termed the **zone of accumulation.** The lower portion of the glacier, where ablation exceeds snowfall, is termed the **zone of ablation.** This lower zone is able to develop only because, under the stress of gravity, glacial ice flows constantly from the zone of accumulation down to the zone of ablation. Downslope movement is seen in the development of a great crack or crevasse around the head of the glacier. This crack is known as the **bergschrund** (see center of photo, Fig. 16.3). It shows that the ice mass is pulling away from the confining rock walls of the cirque and is moving outward and downward. The end of a glacier, its **terminus** or **snout,** marks the lowest advance of the zone of ablation. If the snout reaches the sea, it will calve to form icebergs.

The **firn line,** which separates the two zones of a glacier, represents an equilibrium point between snowfall and ablation. The firn line tends to coincide with the snow line. Several factors influence the level of the snow line (see center of photo, Fig. 16.1). Latitude and elevation, which are temperature controls, are important factors. Equally important is the amount of snowfall received during the winter. In general, the colder the temperature and the greater the snowfall, the lower is the snow line. Other factors cause variation in the level of the snow line, such as the amount of insolation. A shady mountain slope, or one where there is a high percentage of cloud cover, will have a lower snow line than one that receives more insolation. Wind is another factor, since it produces snow drifts on the lee side of the mountain ranges. Thus, in the midlatitudes of the Northern Hemisphere the snow line is lower on the north (shaded) and east-facing (leeward) slopes of mountains. Consequently, the most significant glacier development is on these slopes.

Some alpine glaciers are not caused primarily by snowfall on the glacier itself, but by the accumulation of snow blowing and drifting into protected areas. The Colorado Rockies and the Ural Mountains in the Soviet Union provide good examples of alpine glaciers formed by snow drift.

EQUILIBRIUM AND THE GLACIAL BUDGET

When the snout of a glacier does not move, the glacier is said to be in a state of dynamic equilibrium—that is, a balance in the system has been achieved between accumulation and ablation of ice and snow. As long as this rare condition is maintained, the glacier's snout will remain in the same location, although the glacial ice continues to flow forward.

Let us assume that for several years exceedingly heavy amounts of snow are received, and the glacier's equilibrium is upset. Under the pressure of this surplus snow accumulation, more ice will be produced, and the snout will advance until it reaches a new point of equilibrium where receipt of ice and snow equals wastage. A deficit of snow in the glacial budget will cause the snout to retreat or recede, by melting, until the dynamic equilibrium between receipt and wastage is again achieved. An increase or decrease in temperature will also cause an advance or retreat of the ice mass.

However, even in a retreating glacier the ice is constantly flowing forward. Forward movement stops only when the ice becomes too thin to flow at all. Glaciers fluctuate constantly

(Fig. 16.5). From about 1890 to 1960, most Northern Hemisphere glaciers were in steady retreat. However, in the late 1960s stabilization and some local readvances had become apparent. The upper parts of many glaciers actually have been thickening since about 1950.

HOW DOES A GLACIER FLOW?

The mechanics by which an alpine glacier flows are complex, and explanations are still theoretical. Most of the movement, however, is caused by gravity. It is believed that a glacier flows because of a combination of factors, one of which is basal slip over its rock floor. Steep slopes, which allow gravity to work, and meltwater, which reduces friction by lubricating, must both be involved. Basal slip is particularly important in middle-latitude glaciers and during summer when much of the glacier is near its melting point and meltwater is available. During winter and in cold-climate glaciers, with lit-

FIGURE 16.5 Three congruent glaciers in Alaska show the variability of glacial flow. The blunt-ended glacier on the left has stopped advancing. The middle glacier with the great amounts of morainal material on its terminus is retreating. The debris-free snout of the glacier on the right is advancing. (Austin Post, U.S. Geological Survey, F4-138)

tle meltwater available, the glacier may freeze onto the bedrock and basal slip may be prevented.

Internal plastic deformation is another means by which ice flows. At depth the tremendous weight of the ice causes the individual ice crystals to arrange themselves in parallel layers and then slide over each other, much like a deck of cards. This type of flow only seems to occur below about 30 meters (100 ft) of ice, in an area known as the **zone of plastic flow.**

The upper surface of the glacier is brittle and does not have plastic deformation. Motion of the ice in this upper **zone of brittle flow** occurs by fracturing and faulting. Here the ice breaks and cracks as a solid material. These cracks, called **crevasses,** are common where

the ice mass is stretched, particularly along the margins and snouts of alpine glaciers (Fig. 16.6). Where glaciers drop over steep slopes they form an **ice fall.** Here, intersecting crevasses break the ice into a morass of unstable ice blocks. This is an extremely dangerous area for mountain climbers and scientists who work on the ice. The snout of the glacier, where it meets seawater, can also be dangerous, as large blocks of ice calve off and topple into the water, creating large waves.

The speed of glacial flow varies from an imperceptible fraction of a centimeter per day to as much as 30 meters (100 ft) per day. In addition, the speed of an individual glacier may change from time to time because of changes in the dynamic equilibrium and from place to

(a)

(b)

FIGURE 16.6 The upper surface of a glacier is brittle and as it breaks it forms cracks. These cracks, as in (a), are called crevasses. Crevasses are particularly numerous along the margins and snout of the glacier (b) where the ice is stretched. (Both photos, R. Sager)

place because of changes in gradient or even in the amount of friction encountered with adjacent rock. The speed of glacial flow is greatest where there is a steep slope, where the ice is thickest, and where temperatures are warmest. For example, midlatitude coastal alpine glaciers flow much faster than the Antarctic Ice Sheet.

Even within a small section of a glacier, glacial flow varies. In the middle of the upper surface, where friction is least, speed of flow is greatest. On the sides and bottom surfaces of the glacier friction slows the rate of flow.

Sometimes a glacier's velocity will increase many times its normal rate, causing the glacier to travel hundreds of meters per year. The reasons for such enormous **surges,** as these velocity increases are called, are not completely clear, although lubrication of the glacial bed by pockets of meltwater explains some of them.

GLACIERS AS AGENTS OF GRADATION

As a glacier moves over the land, whether in the shape of an ice sheet or as an alpine glacier, it scrapes along the land surface, picking up and carrying along with it boulders and rock fragments. This erosive process of lifting and incorporating rock and soil into the glacial ice is called glacial **plucking** or **quarrying.** Plucking is encouraged by weathering processes, particularly the freezing of water in joints and fractures in the bedrock, which breaks rock fragments loose.

Glaciers also drag rock fragments along their undersides. Many of the rock fragments traveling with a glacier become tools of erosion themselves as they scrape and gouge stationary surface rocks. The **striations** (gouges, grooves, and scratches) produced by such glacial **abrasion** (as this erosion process is called) are often cited as evidence of previous glaciation in areas devoid of glaciers today (Fig. 16.7). Striations indicate the direction of flow long after a glacier has disappeared. Rock surfaces subjected to intense glacial abrasion are typically smoother and more rounded than those produced mainly by plucking or quarrying. The typical landform of such rock surfaces is the **roche moutonnée**—a bedrock hill that is smoothly rounded on the upstream side most subject to abrasion, with some plucking evident on the downstream side (Fig. 16.8).

Plucking and abrasion provide glaciers with a load of rock fragments of all sizes, from the finest ground **rock flour** to giant boulders.

FIGURE 16.7 Glacial abrasion produces smooth rock surfaces with scratches and grooves parallel to the direction of ice movement. (U.S. Geological Survey)

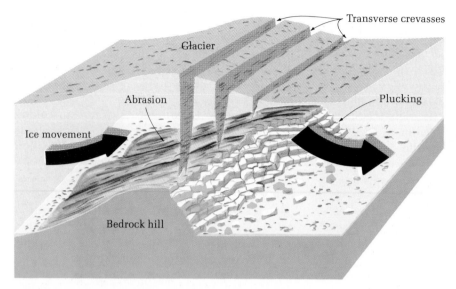

Transverse crevasses

Glacier

Abrasion

Plucking

Ice movement

Bedrock hill

FIGURE 16.8 The characteristic asymmetric form produced by glacial erosion of a bedrock hill. Arrows show the direction of ice flow. Such forms are known as roches moutonnées or sheep-backed rocks.

Much of the load is concentrated near the surfaces of the glacier that are in contact with land surfaces—the source of most of a glacier's load. More material, particularly in the case of alpine glaciers, is transported on the upper glacial surface where it has been eroded from the valley sides. Even more material joins it from mass movement of mechanically weathered materials due to ice wedging on the high unglaciated slopes above. Where valley glaciers join, these materials cause the characteristic stripes on the glacier surface (Fig. 16.9). The volume and ve-

FIGURE 16.9 Where valley glaciers join, the eroded materials from glacial erosion, weathering, and mass movement form the characteristic stripes on the glacier surface. (Austin Post, U.S. Geological Survey, 80R3-022)

locity of a glacier do not determine the size of the particles it can erode and transport, as is the case with running water. As a consequence, the sorting of sediment by size that occurs as running water changes speed does not occur with flowing ice.

Ice is especially aggressive in removing decayed or nonresistant rock, often producing erosional depressions. The bottom ice currents (but not the ice surface) may move obstructions even if the ice has to move uphill for awhile to drag material out of the depression. The fact that ice currents (unlike flows of water) can move upward means that after the ice has melted a glaciated surface may have many depressions. After the retreat of a glacier these depressions are often filled with water to form lakes.

EROSIONAL LANDFORMS DUE TO ALPINE GLACIATION

Ice thickness and erosion are greatest at the head of alpine glaciers. Glacial erosion works headward even as the glacier flows downslope. The headward erosion of a glacier produces a valley head shaped like a steep-sided bowl or amphitheater. Glacial undercutting of the rock walls above the ice level, combined with mechanical weathering, causes mass movement and increases the steepness of the bowl's walls, forming a cirque. If a lake forms in the cirque depression it is called a **tarn** (Fig. 16.10a).

Often two or more cirque glaciers will form near the top of the same mountain. As the cirques of two such glaciers are enlarged, the wall of rock between them will be shaped into a jagged, sawtooth spine of rock, called an **arête** (Fig.16.10b). Where three or more cirques meet at a mountain summit, they form a characteristically pyramid-like peak called a **horn** (Fig. 16.10c). The Matterhorn, in the Swiss Alps, is the world's classic example. A **col** is a notch or pass formed where two cirques have intersected to produce a low saddle between high peaks.

Unlike streams, which initially erode V-shaped valleys, glaciers erode characteristically steep-sided U-shaped valleys called glacial troughs. In addition, a glacier's tendency to move straight ahead rather than to meander causes it to straighten out the original valley by eroding away interlocking spurs.

By eroding decayed or weak rock on the valley floor, the glacier often creates a sequence of rock steps and excavated basins. During glaciation, ice falls will be present at the steps. The result is a "glacial stairway." When the ice retreats, rock-bound lakes may fill the basins, often looking like beads connected by a glacial stream flowing down the glacial trough. Such lake chains are called **paternoster lakes.**

A large glacier often has tributaries that merge with it. These tributary glaciers, like the main ice stream, carve U-shaped channels. However, because they have less volume than the main glacier, the rate of erosion in these tributaries is less rapid. As a result, their troughs are smaller and not as deep as those of the main glacier. Nevertheless, during the peak glacial phases the top surface of the ice that flows from the smaller glacier is at the same level as the ice in the larger glacier. Not until the two glaciers begin to wane does the difference in height between their trough floors become apparent. The higher trough of the old tributary glacier is called a **hanging valley.** A stream that flows down such a channel will drop down to the lower channel by a high waterfall or chain of cataracts. Yosemite Falls and Bridalveil Falls in Yosemite National Park are excellent examples of hanging valley waterfalls. Yosemite Valley, itself, is a classic example of a glacial trough (Fig. 16.10d).

Once a landscape has been degraded by an alpine glacier, it shows a sharp contrast between the glacial trough scoured smooth by ice flow and the jagged peaks above the former level of the ice. The rugged quality of these upper surfaces is caused primarily by mechanical weathering above the ice surface and by undercutting at the head of the ice mass. Glacier National Park, Montana, is an excellent site of such rugged alpine glacial terrain (Figs. 16.11

FIGURE 16.10 (a) A cirque depression and tarn lake formed by headward erosion of a glacier in the Canadian Rockies. (R. Gabler) (b) Jagged sawtooth spines of rock, such as these in the French Alps, are called arêtes. (R. Sager) (c) Mt. Assiniboine in the Canadian Rockies is a classic example of a horn. Horns are caused by several glaciers cutting headward into a mountain peak. (R. Sager) (d) Glaciers carve steep-sided U-shaped valleys called glacial troughs. Yosemite Valley, California, is a classic example of a glacial trough. (R. Sager)

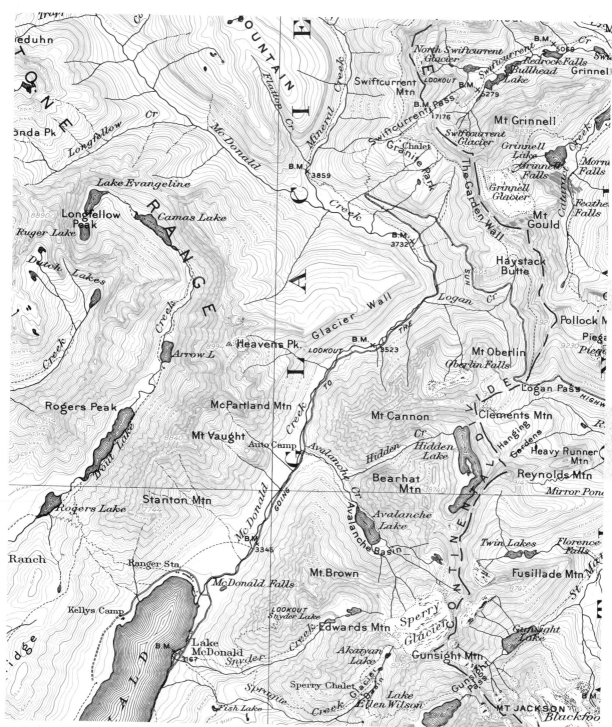

FIGURE 16.11 The glacial topography of a portion of Glacier National Park as shown on a topographic map. (Chief Mountain, Montana, scale 1:125,000, contour interval 100 feet, U.S. Geological Survey)

FIGURE 16.12 Glacier National Park, Montana, is an excellent example of rugged alpine terrain. Cirques, tarns, arêtes, and horns form spectacular scenery. (Austin Post, U.S. Geological Survey, 61F2-33)

and 16.12). The sequence in development of alpine glacial landforms is illustrated in Figure 16.13.

In areas where mountainous regions lie near the coast, alpine glaciers may reach the sea and produce icebergs. This situation is presently found along the coasts of British Columbia, southern Alaska, Chile, Greenland, and Antarctica, and was formerly characteristic of Scotland, Norway, Iceland, and New Zealand. When such a glacier disappears, the sea invades the former glacial trough, creating a deep, inland finger of the sea called a **fiord.** This invasion by the sea is possible primarily because—unlike streams, which can erode only to base level—glaciers can erode far below sea level. Because of its density, ice must be nine-tenths submerged before it will float. Thus the sea may enter deep glacial channels once the ice has melted. In addition, fiords that were

formed during periods of large-scale glaciation were later submerged as the sea level rose with the melting of the glaciers (see again Fig. 8.13).

DEPOSITION BY ALPINE GLACIERS

Alpine glaciers carry debris on their surfaces, frozen in their interiors, and dragged along at the bottom. As mentioned previously, the material they carry includes boulders, rocks, and fragments plucked by the glaciers themselves from their channel sides and floor as well as the smaller fragments and particles produced by abrasion. Gravitational rockfalls from steep trough walls may also supply a glacier with debris. Glaciers deposit huge chunks of bedrock, fine rock flower, layers of pollen, dead plants and insects, soils, and volcanic dust:

All glacial deposits are included within the general term **drift,** whether they are unsorted

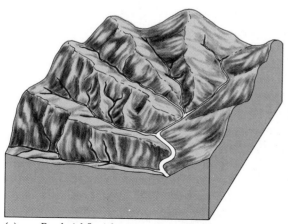

(a)   Preglacial fluvial topography

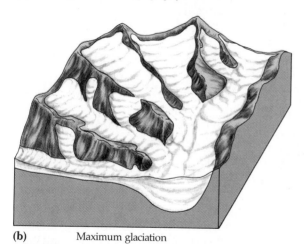

(b)   Maximum glaciation

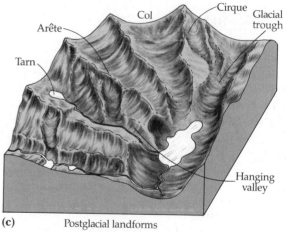

(c)   Postglacial landforms

FIGURE 16.13 This sequence of diagrams illustrates the development of glacial landforms in an alpine region. Mountain topography (a) prior to glaciation, (b) maximum valley glaciation, and (c) postglacial landforms are shown.

and unstratified ice deposits or the orderly deposits of meltwater streams issuing from the glacier. To differentiate these two types of deposits, the term **till** is applied to unsorted drift laid down by ice. Meltwater deposits are called **glaciofluvial deposits.**

Glaciers deposit a portion of their load when their capacity is reduced. Linear glacial deposits called **moraines** occur along the margins of glaciers. The deposits laid down along side margins are called **lateral moraines** (Fig. 16.14a). When two glaciers join together, their interior lateral moraines form a **medial moraine** in the center of the new glacier. At the

snout of the glacier all the debris carried forward by the "conveyor belt" of ice and pushed ahead of the glacier is deposited in a jumbled heap of rocks and fine material, called an **end moraine.** End moraines marking the farthest advance of the snout are **terminal moraines.** End moraines deposited as a consequence of a halt in snout retreat, followed by a stabilization of the ice front prior to further retreat, are called **recessional moraines.** Glaciers also deposit a great deal of till along the floor of their channel as they retreat, particularly near the snout where melting is greatest, where the ice becomes thinner, and where load capacity is

(a)   (b)

FIGURE 16.14   The long even-crested ridges in (a) are lateral moraines built up by valley glaciers, which advanced down from California's Sierra Nevada range. (John S. Shelton) In (b) meltwaters from the snout of an Alaskan glacier have formed a braided stream and deposits of glacial outwash. (U.S. Geological Survey)

consequently reduced. This glacial till deposit is called **ground moraine.**

Braided streams of meltwater, laden with sediment, commonly issue from the glacier terminus. The sediment, called **glacial outwash,** is deposited beyond the terminal moraine, larger rocks and debris first and then progressively finer particles. Often resembling an alluvial fan confined by valley walls, this depositional form is called a **valley train.** Valleys in glaciated regions may be filled to depths of several hundred feet by outwash, producing extremely flat valley floors (Fig. 16.14b).

CONTINENTAL ICE SHEETS

Continental ice sheets differ from alpine glaciers, especially in size and shape. However, as agents of landform gradation, ice sheets are similar in most ways to alpine glaciers, and much of what we have dis-

cussed about alpine glaciers applies as well to continental ice sheets. The differences in the gradational activities of the two types of glaciers are primarily differences in scale, attributable to the enormous disparity in size between the two.

EXISTING ICE SHEETS

As we have noted, glacial ice covers about 10 percent of the earth's land area, and alpine glaciers can be found in mountain regions on most continents. However, in area and mass, alpine glaciers are insignificant in comparison to the ice caps of Greenland and Antarctica, which account for 96 percent of the area covered by glaciers today. Ice caps resembling those of Greenland and Antarctica, but on a much smaller scale, are also present in Iceland, the arctic islands of Canada and the USSR, and in Alaska and the Canadian Rockies.

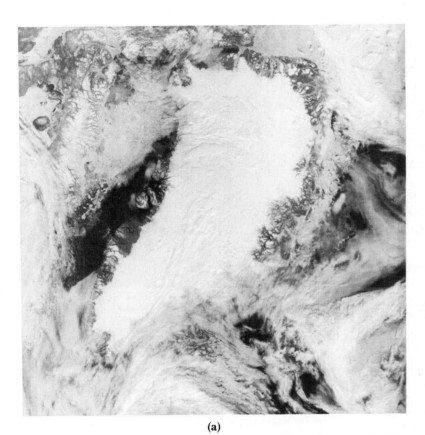

(b)

(a)

FIGURE 16.15 Except for the mountainous edges, the Greenland Ice Sheet almost completely covers the world's largest island. (a) Ice thickness in the center of the island is over 3 kilometers thick and depresses the bedrock below sea level. The small map (b) shows the extent of the ice sheet. (Satellite image produced from USAF DMSP [Defense Meteorological Satellite Program] film transparencies archived for NOAA/ NESDIS at the University of Colorado, CIRES/National Snow and Ice Data Center)

The Greenland ice cap covers the world's largest island with a layer of ice that is over 3 kilometers thick in the center. The only land exposed in Greenland is a narrow, mountainous strip along the coast (Fig. 16.15). Where the ice does reach the sea, it usually does so through fiords. These ice flows to the sea resemble alpine glaciers and are called **outlet glaciers.** When the ice reaches the sea, huge chunks are broken off by melting and the action of waves and tides. The resulting **icebergs** are a hazard to vessels in the North Atlantic shipping lanes south of Greenland. Tragic maritime disasters, such as the sinking of the

Titanic, have been caused by collisions with these huge irregular chunks of ice, nine tenths of which are invisible below the sea surface. Today, by using radar, satellites, and the ships and aircraft of the International Ice Patrol, these sea disasters are minimized.

The Antarctic ice cap covers some 13 million square kilometers (5 million sq mi), an area almost seven and one-half times the size of the Greenland ice cap, which covers 1.7 million square kilometers (650 thousand sq mi). Little is known about the land beneath the thick layer of ice in Antarctica (Fig. 16.16). Like Greenland, little land is exposed in Antarctica, and

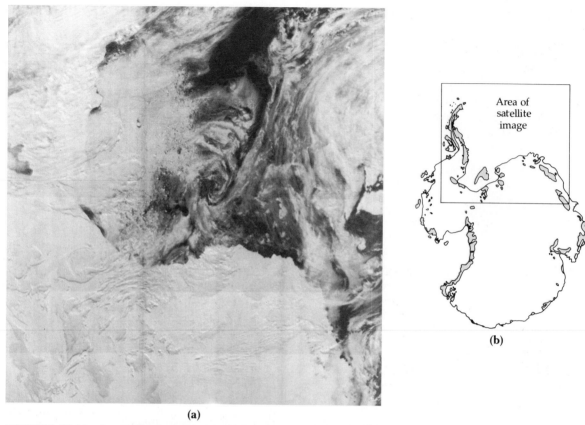

(a)

(b)

FIGURE 16.16   A satellite image of a small portion of the Antarctic Ice Sheet and Antarctic Peninsula. (a) The world's largest ice sheet covers over 13 million square kilometers of land and reaches a thickness of over 4 kilometers. The small map (b) shows the extent of the ice sheet and the location of the satellite image. (Satellite image produced from USAF DMSP [Defense Meteorological Satellite Program] film transparencies archived for NOAA/NESDIS at the University of Colorado, CIRES/ National Snow and Ice Data Center)

the weight of the 5-kilometer-thick ice in some interior areas has depressed the land well below sea level. Where the ice reaches the sea, it floats in enormous, flat-topped plates called **ice shelves.** These are the source of icebergs in Antarctic waters. They do not have the irregular shape of Greenland's icebergs, and because they do not float into heavily used shipping lanes, these huge tabular Antarctic icebergs are not as much of a hazard to navigation (Fig. 16.17). They do, however, add to the problem of access to Antarctica for scientists. The huge wall of the ice shelf itself, the broken-up and

melting sea ice, and the extreme climate combine to make Antarctica inaccessible to all but the hardiest individuals and equipment. This icy continent serves, however, as a natural laboratory for scientists to study the ice ages.

THE PLEISTOCENE ICE AGE

The Pleistocene Epoch or ice age is the name given that period of geologic history during which there were at least four major advances and retreats and a vastly greater number of minor movements of continental ice sheets over

FIGURE 16.17 The Antarctic Ice Sheet produces enormous flat-topped tabular icebergs. Due to the density difference between ice and seawater, most of the iceberg is hidden below the sea surface. (Official U.S. Coast Guard Photograph)

large portions of the world's land (Fig. 16.18). Scientists believe the Pleistocene Epoch began about 2 million years ago. Since that time, ice sheets have spread outward over the land from centers in Canada, Scandinavia, and the eastern Soviet Union, and also from high mountain ranges to cover nearly a third of the earth's land surface. At the same time, sea ice expanded equatorward. In the Northern Hemisphere, sea ice was present along the coasts as far south as Delaware in North America and Spain in Europe. Between each glacial advance a warmer **interglacial** occurred, during which time the enormous continental ice sheets and sea ice retreated and almost completely disap-

peared. An examination of glacial deposits has determined that within each major glacial advance there were many minor retreats and advances, which may reflect small changes in global temperature and precipitation.

The gradational effects of the last major glacial advance, the Wisconsinan stage, which ended about 10,000 years ago, are the most visible in landscapes today. The glacial landforms created during the Wisconsinan stage are relatively new and have not been destroyed to any great extent by the other gradational agents. Consequently, we are able to derive a fairly clear picture of the extent and actions of the ice sheets at that time.

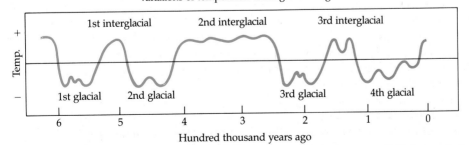

FIGURE 16.18 Only slight drops in the earth's temperature may bring on an ice age. During the Pleistocene, temperature variations caused ice advances (1st interglacial, Nebraskan; 2nd, Kansan; 3rd, Illinoian; 4th, Wisconsinan) and ice retreats (1st interglacial, Sangamonian; 2nd, Yarmouthian; 3rd, Aftonian).

The land covered during the major glacial advances in North America and Eurasia are depicted in Figure 16.19. Continental ice sheets in North America extended as far south as the Missouri and Ohio Rivers and covered nearly all of Canada and much of the northern Great Plains, the Midwest, and the northeastern part of the United States. In New England, the ice was thick enough to overrun the highest mountains, including Mount Washington with an elevation of 2063 meters (6288 ft). The ice was more than 2 kilometers deep in the Great Lakes area. In Europe, glaciers spread over what is now most of Great Britain, Scandinavia, Germany, Poland, and the western USSR. Although there apparently was some glaciation in Siberia, it was too dry there (as it is today) to allow for such massive snow and ice accumulation as occurred in North America and northwestern Europe.

Also, during each advance of the ice sheets, alpine glaciers were much more extensive and massive in highland areas than they are today. In fact, it was the examination of landforms in the Swiss Alps that led Louis Agassiz in 1840 to publish the theory of past glaciations. Though at first considered radical, the theory today is acknowledged as fact and is supported by countless detailed studies of glacial deposits and erosional forms in various parts of the world.

Where did the water locked up in all the ice and snow come from? Its original source was the oceans. During the periods of great glacial growth there was a general lowering of sea level, exposing large portions of the continental shelf. The recent melting and glacial retreat would have raised the oceans a similar amount—about 120 meters (400 ft). Evidence of the most recent rises in the sea level can be seen along many submerging coastlines around the world.

MOVEMENT OF THE CONTINENTAL ICE SHEETS

It is a popular misconception that all continental ice sheets originate at the poles and spread toward the equator. Actually, the great centers of Pleistocene glacial accumulation (aside from highland areas, Antarctica, and Greenland) were in the upper midlatitudes, in the vicinity of Hudson's Bay in Canada, on the Scandinavian Shield, and in eastern Siberia. The accumulated ice in these centers flowed in all directions. As with valley glaciers, the initial flow direction of advancing ice sheets is determined largely by the path of least resistance, found in

FIGURE 16.19   Maximum spread of the continental ice sheets and highland ice caps during the Pleistocene. Much of North America and Eurasia was under thousands of meters of ice during that time of ice advance and retreat.

preexisting valleys and belts of softer rock. These huge ice sheets, however, also invaded and submerged some fairly rugged terrain, such as New York's Adirondacks, New England's White Mountains, and much of Scandinavia.

Continental ice sheets began to flow outward in all directions from a central zone of accumulation when the ice was thick enough to allow for plastic flow. The radial expansion is analogous to the spreading of pancake batter in a frying pan. If all of the batter is poured into the center of the pan, it will eventually spread outward to cover the entire bottom of the pan. Like an alpine glacier, an ice sheet flows outward from a zone of accumulation to a zone of ablation. Like alpine glaciers, ice sheets retreat and advance with small changes in temperature and snowfall and will thin and disappear when ablation exceeds accumulation.

ICE SHEETS AND EROSIONAL FORMS

Ice sheets erode the land in the same way as alpine glaciers but on a much larger scale. As a result, landforms created by ice sheet erosion are far more extensive than those formed by alpine glaciation stretching over millions of square kilometers of North America, Scandinavia, and Siberia. As ice sheets flowed out over the land, they gouged the earth's surface, enlarging valleys that already existed, scouring out rock basins, and smoothing off existing hills. The eroding ice sheets removed most of the soil and then attacked the bedrock itself.

Today these ice-scoured plains are areas of low, rounded hills, lake-filled depressions, and wide exposures of bedrock. Because the ice sheets plowed through and totally disrupted the former stream patterns, and because glaciation has been so recent that new drainage systems have not had time to form, ice-scoured plains are characterized by thousands of lakes, marshes, and areas of muskeg (poorly drained areas grown over with vegetation). The major characteristic of glacially eroded lands, such as

those found in Canada and Finland, is the great expanses of exposed gouged bedrock and standing water.

ICE SHEETS AND DEPOSITIONAL FORMS

Again, scale makes ice sheet deposition different from that of alpine glaciers. Though terminal and recessional moraines, ground moraines, and glaciofluvial deposition are found, these deposits form significantly larger features in the landscapes caused by ice sheets (Fig. 16.20).

TERMINAL AND RECESSIONAL MORAINES Terminal and recessional moraines form belts of low hills and ridges crossing the land in areas of glacial deposition. These features are rarely more than 60 meters (200 ft) high (Fig. 16.21). The last major Pleistocene glacial advance through New England left a terminal moraine running the length of New York's Long Island and formed the offshore islands of Martha's Vineyard and Nantucket. Cape Cod and Lake Michigan's rounded southern end were formed by recessional moraines. End moraines are usually arcs convex toward the direction of ice flow. Their pattern and placement indicate that the ice sheets did not maintain an even front but spread out in tongues or lobes channeled by the previous terrain (Fig. 16.22). The positions of the terminal and recessional moraines give evidence of more than simply the direction of ice flow. In addition, the character of the deposited material can be examined to detect the pattern and sequence of advances and retreats of each successive ice sheet.

TILL PLAINS In the zone of ice sheet deposition, massive accumulations of unsorted glacial till accumulated, often to depths of 30 meters or more. Because of the uneven nature of the deposition, the configuration of the land today covered by till varies from place to place.

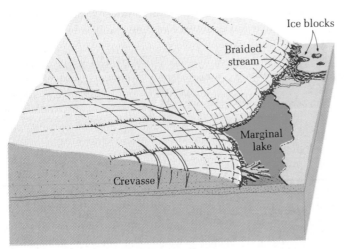

Ice blocks

Braided
stream

Marginal
lake

Crevasse

(a)

FIGURE 16.20 The alteration of landscape by ice sheet deposition. Many of the
features associated with ice stagnation
in (a) are further modified by glacial meltwater as the ice sheet retreats (b).

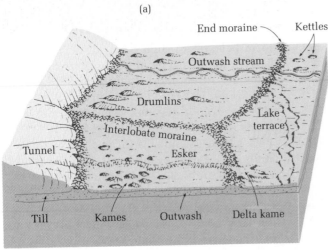

End moraine   Kettles

Outwash stream

Drumlins

Lake
terrace

Interlobate moraine

Tunnel   Esker

Till   Kames   Outwash   Delta kame

(b)

In some areas the till is too thin to hide the original contours of the land. In other regions the thick deposits of the till form broad, rolling plains of low relief. Small hills and slight depressions, some filled with water, characterize most **till plains** and reflect the uneven glacial deposition. Some of the best agricultural land of the United States is found on the gently rolling till plains of Illinois and Iowa. These young chernozem and prairie soils (mostly mollisols), developed on the till, are extremely fertile.

OUTWASH PLAINS Beyond the belts of hills that are the terminal and recessional moraines lie the **outwash plains.** These are extensive, relatively smooth plains covered with the sorted deposits carried forward by the meltwater from the ice sheets. The outwash plains, which may cover hundreds of square kilometers, are analogous to the valley trains of alpine glaciers.

Some outwash plains are marked by occasional water-filled pits, called **kettle holes,** which were formed when a block of ice became detached from the glacial terminus and was buried in the stratified outwash deposits. Eventually, where the block of ice had melted, a hole remained. Today countless kettle holes form lakes in the outwash plains.

FIGURE 16.21 The hilly topography of an end moraine in the American Midwest. (John S. Shelton)

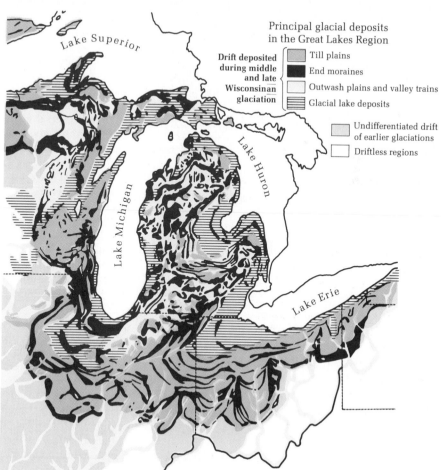

Principal glacial deposits in the Great Lakes Region

Drift deposited during middle and late Wisconsinan glaciation

Till plains

End moraines

Outwash plains and valley trains

Glacial lake deposits

Undifferentiated drift of earlier glaciations

Driftless regions

Lake Superior

Lake Michigan

Lake Huron

Lake Erie

FIGURE 16.22 Glacial deposits of the Great Lakes region. Note the curved pattern of the end moraines. (Trewartha, Robinson, and Hammond, *Fundamentals of Physical Geography*, 2nd ed., New York, McGraw-Hill, 1968)

VIEWPOINT: A FUTURE ICE AGE?

Drought in the Soviet steppes reduces wheat harvests; starvation and death come to Ethiopia and other African nations bordering the Sahara desert as water holes dry up; and Florida oranges are glazed with ice in an abnormally cold winter—all these news items have appeared on television and in newspapers within the last few years. Is something happening to the world's climates? Are weather patterns really changing? Or are these merely minor fluctuations in our "post-Pleistocene," that is, "post-glacial," era?

Certainly, such fluctuations have been on record since the maximum extent of the ice sheets about 18,000 years ago. Archeological records tell us of American Indians moving out of the Mesa Verde and Chaco Canyon areas of the American West as their crops failed because of increasing aridity. History tells us that the River Thames in London was frozen solid for six weeks in 1684, while in 1666, the same river barely provided any water. The migration patterns of the Vikings suggest warmer years when ice was less frequent in subarctic waters.

Previously, these variations have been seen as just that—minor fluctuations in an otherwise generally mild "post-glacial" era, with the Ice Age a thing of the past. Since the 1960s, however, this picture has been seriously challenged by paleoclimatologists (who study former climates), geographers, and geophysicists. The old "classic" notion of a Pleistocene Ice Age with four major *glacials* separated by warmer *interglacials* is slated for modification. Analysis of deep ice cores in Greenland and Antarctica and of deep sea sediment cores reveals a series of

Glaciers of the Mount Blanc Massif, Chamonix, France. (R. Sager)

glacials, perhaps seven or eight, that occurred during the Pleistocene and lasted about 100,000 years each, with interglacials persisting for only 10,000 years or so. Further, the global climate we experience today appears to be at the end of a brief warm period within a series of multiple glaciations and therefore is relatively *untypical* of the Pleistocene Epoch we probably still inhabit. The Pleistocene norm consists of cooler and more arid conditions, to which we may return in a few thousand years or less. Even more surprisingly, such marked climatic changes from interglacial to glacial conditions appear from the ice-sheet records to take place with amazing rapidity—perhaps within less than 100 years.

What is the evidence for such astonishing new interpretations? Much of it has emerged from ice-core studies carried out in Greenland and Antarctica by a Danish physicist, Willi Dansgaard, of the University of Copenhagen. Using a machine called a *mass spectrometer,* which can accurately measure oxygen atoms, he is able to analyze how much heavy oxygen (O_{18}) is locked up in the Greenland ice. Findings indicate climatic changes based on the fact that the less heavy oxygen there is in an ice layer, the colder the climate was at the time this ice was forming. Techniques of dating such as carbon-14 and magnetic reversal analysis can date these ice layers. It appears that about 90,000 years ago the global climate, like our own today, was comparatively mild, and suddenly in less than 100 years, it switched to conditions of extreme glaciation; within about another 1000 years the climate reverted to warmer conditions.

Other research in a quite different part of the world appears to support the ice core findings. Core samples of the muds at the bottom of the Caribbean Sea were collected to analyze the proportion of heavy oxygen in minute marine fossils within them. Work with these and other ocean bed cores shows the same patterns of cooling and warming. Tree-ring dating (dendrochronology) and pollen analysis tell us of changing vegetation growth patterns and distributions, as climates have fluctuated through warmer and then drier, colder periods. The investigation of windblown dust, called loess, deposited during periods of increased aridity and powerful winds circulating around the glacial high-pressure regions, also confirms essentially the same picture— a Pleistocene Ice Age of seven (or eight) glacials. Our own present interglacial period has already lasted for 10,000 years, about the length of interglacial periods of warmth. The implication seems to be that we may be facing a renewed glacial period in the not-too-distant future.

But what are the reasons for such significant and relatively sudden climatic changes? Many hypotheses have been suggested for the cooling of global climates in the past. Some scientists have suggested enormous volcanic eruptions, which spew out so much volcanic dust that sunlight is filtered out and the earth cools. However, volcanic debris should surely be visible within the ice and marine sediment cores, and there appears to be no correlation between periods of cooling and volcanic dust layers.

Recent work on plate tectonics might suggest a link between earlier positions of the plates and periods of cooling. Certainly the distribution of land and sea surfaces is highly significant in the patterns of world pressure and wind systems. It is particularly important that large landmasses be near the poles for an ice age to occur. Ice ages do not occur when continents are all far from the poles. This is true for the Pleistocene and occurred 200 to 300 million years ago when what are now the Southern Hemisphere continents were glaciated. At that time Africa was located where Antarctica is today. But neither volcanic theories nor plate tectonics can explain the regularity of the 100,000-year glacial/10,000-year interglacial cycle that seems to be emerging from the current research.

Can we look for such regular cyclic variations in the output of energy from the sun? Is there a kind of sunspot cycle that reduces the solar constant for regular periods of time? So far we hardly have sufficient or long enough records to hazard a guess. However, a new research satellite launched by the United States, known as a *Solar Max,* may help answer this question in the near future.

We *do* know that there have been periods of variability in our earth's *receipt* of solar energy owing to variations in its movement with respect to the sun. A Yugoslavian geophysicist, Milutin Milankovitch, drew attention to them as long ago as the 1920s. He referred to three changes in the earth's motions: (1) variations in the orbit of the earth around the sun, from almost circular to elliptical and back again. This cycle or "stretch" of the orbit takes 90,000 to 100,000 years; (2) variation in the "wobbles" of the earth's axis as it orbits around the sun, rather like a top that wobbles as it spins. The rhythm of this movement is about 24,000 years; (3) variations in the tilt of the axis. This tilt varies between 21.8° (more upright) and 24.2° (more tilted), and one complete "roll" (as a ship rolls) takes 40,000 years. The combination of "stretch," "wobble," and "tilt" can cause significant changes in seasonal energy receipt. The analysis of their total rhythmic effect is complicated, but they do seem to match surprisingly well the new dating of glacials and interglacials emerging from the recent research.

What does seem certain is that a new glacial period is not an impossibility. The fact is that a decrease in the mean global temperature of merely 5°C (9°F) would indeed plunge the world into another glaciation. And evidence suggests, as we've seen, that this could happen quite abruptly. The sobering thought is that we live, technologically, in a "balmy weather" world. Our irrigation systems, our farming patterns, and our transport systems are geared to generally *good* weather conditions. Can we face the onset of colder climates—or more arid ones—since the two appear to go together? What would our energy, water, and food supplies be like in an approaching new ice age?

Icebergs from Portage Glacier, Alaska. (R. Sager)

DRUMLINS A **drumlin** is a streamlined hill, usually a half kilometer in length and less than 50 meters high, molded in glacial drift on the till plains (Fig. 16.23a). Drumlins are oddly shaped, resembling half an egg or the convex side of a teaspoon, and are usually found in swarms, with as many as 100 or more clustered together. Glaciologists do not yet understand exactly how drumlins were formed. The most conspicuous feature is their elongated shape, which follows the direction of ice flow. Their broad, steep noses face the direction from which the ice advanced, while their gently slop-

ing, narrow tails point in the direction of ice flow. Thus their geometry is the reverse of that of roches moutonées. They are well developed in Ireland and in the states of New York and Wisconsin. Bunker Hill, near Boston, is probably the most historical of American drumlins.

ESKERS An **esker** is a narrow, winding ridge composed of glaciofluvial gravels (Fig. 16.23b). Sometimes eskers are as long as 200 kilometers but usually do not exceed several kilometers. It is believed that most eskers were formed by streams of meltwater flowing in ice tunnels at

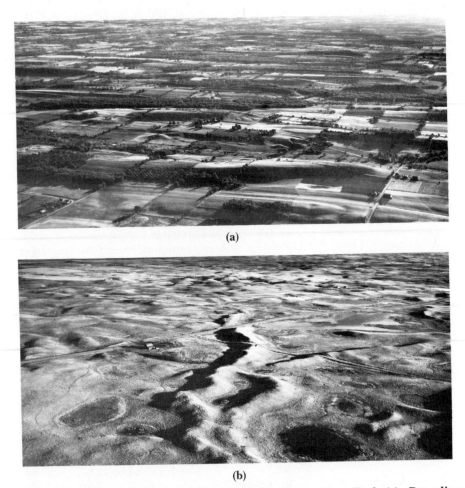

(a)

(b)

FIGURE 16.23 A drumlin field east of Rochester, New York (a). Drumlins are streamlined hills, elongated in the direction of ice flow. An esker east of Aberdeen, South Dakota (b). Eskers are ridges of sand and gravel formed by alluvium in a tunnel under the ice during melting. (John S. Shelton)

the base of ice sheets. Eskers are a prime source of clean gravel and sand for the construction industry. Being natural embankments, they are frequently used in marshy, glaciated landscapes as highway and railroad beds. Eskers are especially well developed in Finland.

KAMES   Conical hills composed of sorted glaciofluvial deposits are called **kames.** They are presumed to have formed in contact with glacial ice when sediments accumulated in ice pits, crevasses, and among jumbles of detached ice blocks. Kames, like eskers, are excellent sources of useful sand and gravel, and are especially common in New England. **Kame terraces** are landforms resulting from accumulations of glaciofluvial sand and gravel along the margins of ice tongues melting in valleys in areas of hilly relief. They occupy the position of the lateral moraines of alpine glaciers. Examples of kame terraces can be seen in New England and New York.

ERRATICS   Scattered on the surface of the glacial drift may be boulders that differ from the local bedrock. Often the source regions of such rocks, called **erratics,** can be identified, indicating the direction of ice flow during the Pleistocene. In England, erratics may be from a source region as far away as Norway. Before the theory of an ice age was proposed, many hypotheses were developed to explain the existence of erratics. Among these explanations was one based on the belief that the Biblical flood transported rocks from one place to another. The term *drift* was originated in connection with the flood hypothesis. However, a flood would not account for the striations present on the erratics and could not move large boulders hundreds of kilometers.

GLACIAL LAKES AND LAKE PLAINS

The ability of ice sheets to form lake basins has already been mentioned. Many of the resulting lakes have disappeared since the retreat of the last continental ice sheets from Eurasia and North America. However, there are still millions of glacier-created lakes in existence.

Ice sheets created some lakes by scooping out huge elongated basins along former stream valleys. If the glacier built a morainic dam at one end of such a basin, a meltwater lake would remain after the retreat of the ice sheet. New York's beautiful Finger Lakes are the classic example of moraine-dammed lakes in ice-deepened elongate basins. Alpine glaciers can produce similar results, as in the case of Washington's Lake Chelan and Lakes Maggiore, Como, and Garda of the Italian Alps. Although many of the lakes formed by the ice sheets are no longer in existence, the **glaciolacustrine** (from *glacial,* ice; *lacustrine,* lake) deposits that they laid down prove their former existence and size.

Some lakes formed where the disruption of drainage by glacial deposition prevented depressions from being drained of meltwater. This situation usually developed where water was trapped between a large end moraine and the ice front or where the land sloped toward, instead of away from, the ice front. In both situations, temporary **ice-marginal lakes** were formed from meltwater. They were drained and ceased to exist when the retreat of the ice front uncovered an outlet route.

During their existence the floors of these ice-marginal lakes accumulated layers of fine sediment. The result of this sedimentation can be seen in the extremely flat surfaces that characterize most glaciolacustrine plains. The outstanding example of such a plain is the valley of the Red River in North Dakota and Minnesota. This plain is the flattest landscape in the United States and is of great economic importance, since it is well adapted to the growing of wheat. The wheat belt and the flat plains of the Red River continue north into Canada, where the river eventually flows into Lake Winnipeg. These plains are the result of deposition in a vast Pleistocene lake held between the front of the retreating continental ice sheet on the north and the morainic dams and higher topography to the south. This lake has been called Lake Agassiz, after the Swiss scientist who espoused the theory of an ice age. Lake Winnipeg is the last remnant of Lake Agassiz, lying in the deepest part of the ice-scoured and sediment-filled lowland.

Another ice-marginal lake produced much more spectacular landscape features but not in the area of the lake itself. In northern Idaho a glacial lobe, moving southward from Canada, blocked the valley of a major tributary of the Columbia River and caused the formation of an enormous ice-dammed lake known as Glacial Lake Missoula. This lake covered almost 7800 square kilometers (3000 sq mi) and was 610 meters (2000 ft) deep at the ice dam. Eventually the ice dam failed and Lake Missoula emptied in a stupendous flood that engulfed much of eastern Washington. The racing floodwaters scoured the basaltic terrain, producing Washington's channeled scabland, consisting of intertwining steep-sided troughs **(coulees),** dry waterfalls, scoured-out basins, and other features quite unlike those associated with normal stream erosion.

The Great Lakes of the eastern United States and Canada are the world's largest lake system. Lakes Superior, Michigan, Huron, Erie, and Ontario lie in former river valleys that were vastly enlarged and deepened by glacial erosion. All of the lake basins except that of Lake Erie have been gouged out to depths below sea level and have irregular bedrock floors lying beneath thick blankets of glacial till.

The history of the Great Lakes is very complex, resulting from the back-and-forth movement of the ice front, which produced many changes of lake levels and overflow in varying directions at different times (Fig. 16.24). The earliest lakes appear to have emerged near the southern tip of the Lake Michigan Basin and the western end of the Erie Basin. These lakes drained westward to the Mississippi through the Illinois and Wabash Rivers.

Ice retreat exposed the southern fringe of the Ontario Basin, which was occupied by a lake with an eastern outlet through New York's

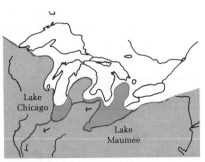

(a) Glacial retreat

(b) Port Huron glacial advance

(c) Glacial retreat (Post Valderan)

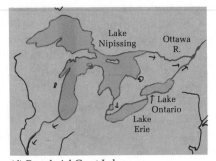

(d) Postglacial Great Lakes

FIGURE 16.24   Stages in the formation of the Great Lakes of North America during glacial retreat at the close of the Pleistocene.

Mohawk and Hudson Valleys. By this time the basin of Lake Huron was emerging, with an outlet westward across Michigan through the Grand River to Lake Michigan (Pleistocene Lake Chicago). This outlet also channeled the overflow from the Erie Basin (Pleistocene Lake Maumee) as the Wabash route ceased to function.

Further ice retreat exposed the western portion of the Superior Basin (Pleistocene Lake Duluth), which overflowed westward to the Mississippi through the St. Croix River in Minnesota. Lakes Michigan, Huron, Erie, and Ontario (Pleistocene Lake Iroquois) were now linked to overflow eastward to the Mohawk and Hudson valleys, with Lakes Michigan and Huron spilling along the ice front into Lake Iroquois rather than following their present route through Lakes St. Clair and Erie.

About 9000 years ago the St. Lawrence outlet was exposed by ice retreat and the Great Lakes formed a single system, emptying to the east—the upper lakes entering the St. Lawrence by way of the Ottawa River. This low outlet permitted the lakes to diminish well below their present levels and areas. However, complete deglaciation caused the earth's crust to slowly rise (rebound) because of unloading and that raised the outlet. This process so enlarged the lakes that the old Illinois River outlet of Lake Michigan began to function again, and a new connection formed between Lakes Huron and Erie through Lake St. Clair. Eventually the St. Lawrence outlet was lowered, the Illinois river link was abandoned, and crustal rise terminated the Ottawa River link between the upper and lower lakes. With these developments the modern lake system was finally established.

A totally different type of Pleistocene lake related to Pleistocene climatic change rather than to glacial action must also be mentioned. Such a lake is known as a **pluvial lake.** During the Pleistocene, decreases in evaporation caused by climatic cooling, and increases in precipitation in formerly dry regions, caused pluvial lakes to grow in desert basins. Some of the largest of these were in the Great Basin region of the United States. The principal pluvial

lakes were Lake Bonneville in Utah and Lake Lahontan in Nevada (Fig. 16.25). About 120 smaller lakes also formed in the arid western United States, from southern Oregon to the Mexican border. Lake Bonneville covered some 52,000 square kilometers (20,000 sq mi) and reached a depth of more than 300 meters (1000 ft). The present Great Salt Lake is a small remnant of the vast lake that once existed west of what is now Salt Lake City. Lake Bonneville remained at a high level long enough to inscribe its shoreline on mountain flanks all around the lake's perimeter. Several lake levels are very well marked, in the form of wave-cut cliffs, beaches, spits, and sand bars up to 300 meters (1000 ft) above the dry desert floor (Fig. 16.26). Similar features may be seen around the margins of extinct Lake Lahontan and in many other desert basins.

OTHER CONSEQUENCES OF PLEISTOCENE GLACIATION

Beyond the ice sheets themselves other changes occurred. The cool Pleistocene midlatitude arctic and subarctic conditions created periglacial landscapes (*periglacial* means beyond the ice margins). The permafrost conditions caused landforms peculiar to tundra climates, such as ice wedges, patterned ground, and smoothed hill slopes due to solifluction.

The weight of the ice depressed the land surface several hundred meters. As the ice retreated, the land started to rise slowly in a process known as **isostatic rebound.** Areas such as Hudson Bay and the Baltic Sea may someday emerge above sea level. Measurable isostatic rebound raises elevations of areas such as Sweden, Canada, and some of the Soviet Union up to 2 centimeters per year (1 in). Should Greenland and Antarctica lose their ice sheets someday, their depressed central land areas would also rise to reach isostatic balance.

The rapidly changing climate conditions of the ice age also forced migration of animals and changes in plant life, in some cases causing extinctions. A major question which remains today is the cause of the mass extinction of the

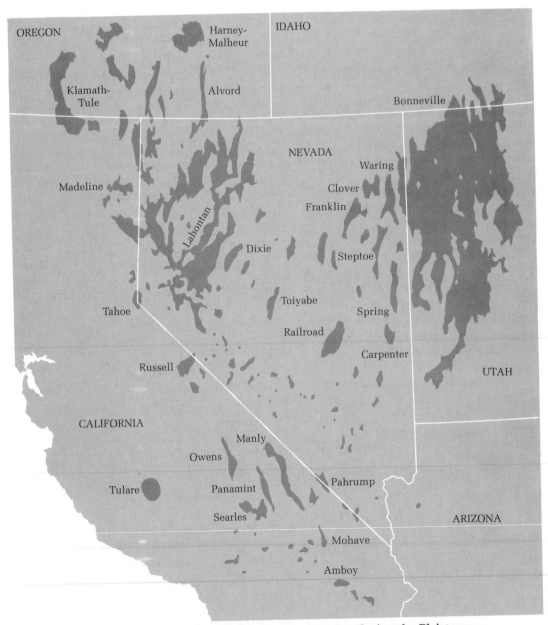

FIGURE 16.25 Lower temperatures and higher precipitation during the Pleistocene produced increased runoff in the arid Great Basin of the southwestern United States. The excess water collected in more than 100 lakes in desert depressions. Today, only a few shrunken descendants of these lakes survive.

large Pleistocene mammals, such as the sabre-toothed cat, woolly mammoth, mastodon, and giant ground sloth. Why did these animals survive throughout most of the Pleistocene and then suddenly disappear? The rapidly changing climate also had a major effect upon humans, by creating land bridges and encouraging migrations. This drastically changed the geographic distribution of early humans.

What would be the effects of another glacial advance upon modern humans? What caused the ice ages? These are two questions

FIGURE 16.26 The foothills of Utah's Wasatch Mountains above Salt Lake City are cut by horizontal terraces. These are wave-cut benches marking the shoreline levels of giant Pleistocene Lake Bonneville. (U.S. Geological Survey)

which remain to be answered (see Viewpoint "A Future Ice Age?"). We must also question whether humans can help cause another ice age through pollution, nuclear war (the "nuclear winter" theory), or other environmental change. Though we cannot answer all of these questions, we do know that glacial systems are dynamic and that we now live in a rare time in the earth's history when conditions are favorable for the return of the ice sheets—bringing to an end our present existence in an interglacial.

QUESTIONS FOR DISCUSSION AND REVIEW

1. Describe the various types of glaciers and explain how each is formed.

2. Diagram and label the characteristic parts of a glacier.

3. How does a glacier maintain its budget in a state of equilibrium?

4. Explain how a glacier moves.

5. How does a glacier accumulate its load? Cite examples of evidence of glacial erosion and movement.

6. Define the following terms: *moraine, till, cirque, arête, horn,* and *outwash.*

7. Distinguish glacial valleys from stream valleys in mountain areas.

8. How are hanging valleys formed?

9. Describe the major types of deposition by alpine glaciers.

10. Why do you think the major stages of glaciation in North America were given the names they were given?

11. How has ice sheet erosion altered the landscape? What kinds of landscapes are produced after continental glaciers deposit their loads and recede?

12. What are the differences between eskers, drumlins, and kames?

13. Explain the relationship of glaciation to the origin and history of the Great Lakes.

14. How does the present extent of glaciation compare to the maximum extent of Pleistocene glaciation?

15. What are some theories as to the cause of the Ice Age? What seems to be the most accepted theory?

17

COASTAL LANDFORMS AND THE GLOBAL OCEAN

INTRODUCTION TO THE OCEANS

The oceans started forming over 3 billion years ago. The first beginnings of life, the development of the first molecules that could reproduce themselves, probably began in these prehistoric oceans. In the late 1920s the British biologist Haldane hypothesized that life began in a "warm dilute soup" of organic compounds such as sugars and amino acids, which are the building blocks of proteins. Scientists have since met with some success in recreating the conditions of a primordial "soup" that would allow for the development of life under certain conditions. It is theorized that as life evolved in this medium over hundreds of millions of years, the character of the medium also changed. By the time of the Cambrian age (some 600 million years ago), seawater was probably very much as it is today.

The oceans contain over 97 percent of the earth's water and cover about 71 percent of the earth's surface. In fact, the name for our planet, "earth," is a misnomer. It should probably be called "oceanus" or "hydro" since it is the only water planet in our solar system. Yet our knowledge of the oceans is not extensive because they present a hostile environment to man as a land-dwelling creature. Consequently, we have been slow to learn about them. Only since the voyage of the *H. M. S. Challenger* in 1872 have scientists seriously begun to explore the ocean's complexity. Only during this century has oceanography become an important science. Most of our knowledge of the ocean has developed since World War II with the refinement of sonar devices (which use sound waves to determine the topography of the ocean floor), deep diving vehicles, coring devices, and satellites.

The oceans actually form a single, large, continuous body of water that surrounds all the land masses of the earth. It is commonly accepted, however, that the "global ocean" is composed of three or four subsidiary oceans. These are separated largely on the basis of their geographic locations, although they do differ somewhat in certain other characteristics as well.

The three principal oceans are the Pacific, the Atlantic, and the Indian Oceans. The approximate area of the Pacific Ocean is 166 million square kilometers (64 million sq mi). The

The Kilauea Peninsula, Kauai, Hawaii. (R. Sager)

Atlantic is half that size, or 83 million square kilometers (32 million sq mi), and the Indian Ocean is about 73 million square kilometers (28 million sq mi). In comparison the entire continent of North America is only about 23 million square kilometers (9 million sq mi) in area. In fact the total land area of the earth, 149.5 million square kilometers (57.5 million sq mi), is smaller than the area of the Pacific Ocean, which is the largest geographic feature on the earth's surface.

The oceans vary greatly in depth as well as in size. The Pacific Ocean has an average depth of 4200 meters (14,000 ft), while the Atlantic and Indian Oceans have average depths of 3900 meters (13,000 ft). However, these figures are somewhat misleading, as the ocean floors are not flat plains but have mountains, trenches, and basins that vary the depths considerably. For instance, the Pacific reaches a maximum depth of over 11,000 meters (36,000 ft).

The Arctic Ocean is far smaller at 13 million square kilometers (5 million sq mi) and shallower, averaging 1220 meters (4000 ft), than the other three oceans. For this reason, it is sometimes referred to as a *sea*. Seas are salt-water bodies that are smaller than oceans and are somewhat enclosed by land. Unlike lakes, which can be fresh or salt water, seas always interchange water with the oceans. Some of the larger seas of the earth are the Mediterranean, Baltic, Bering, Caribbean, East China, North, Black, Yellow, and Red Seas.

CHARACTERISTICS OF OCEAN WATERS

The ocean waters are a very dilute solution of many salts. By weight, 3.5 percent (35 parts per thousand [ppt]) is solid matter. Of this matter, the most common substance by far is sodium chloride, or common table salt (see Table 17.1). Other elements found as salts in the ocean are magnesium, sulfur, calcium, potassium, and bromine. The oceans also contain the dissolved gases of the atmosphere, especially oxygen and carbon dioxide. Some elements, such as chlorine, bromine, boron, and sulfur, are found in greater quantities in the oceans than on land.

The composition of the oceans varies over time and from place to place, but there is little doubt that ocean waters contain every element found on land. For example, there are about 40 pounds of gold per cubic mile of ocean water and 200 pounds of lead per cubic mile. It is now possible to extract salt (sodium chloride) and magnesium from the sea, and in the future the extraction of other substances may also prove to be worthwhile.

The concentration of dissolved salts in water, referred to as its salinity, varies throughout the oceans and seas of the earth. The average salinity of *all* oceans is 3.5 percent (35 ppt), but this can vary in the open ocean from 3.2 to 3.8 percent (32 to 38 ppt). The variation is much greater in enclosed or partially enclosed seas.

Several factors affect salinity, primarily the amount of precipitation and the rate of evaporation. In humid regions, where precipitation is high, fresh water tends to dilute the seawater and reduce salinity. Moreover, the flow of rivers into the sea tends to lower the concentration of dissolved salts. On the other hand, in arid and semiarid areas where precipitation is low and the evaporation rate is high, salts are more concentrated and salinity is therefore higher.

TABLE 17.1 Composition of Seawater

ELEMENT	PARTS PER THOUSAND	ELEMENT	PARTS PER THOUSAND
Chlorine (Cl)	18.98	Potassium (K)	0.38
Sodium (Na)	10.56	Bromine (Br)	0.065
Magnesium (Mg)	1.27	Carbon (C)	0.028
Sulfur (S)	0.88	Strontium (Sr)	0.013
Calcium (Ca)	0.40	Boron (B)	0.005

Temperature in the oceans varies and, together with salinity, affects density. Density is the mass per unit volume of a substance and is measured in kilograms per cubic meter or pounds per cubic foot. The density of fresh water at about 4°C (39°F) is 1000 kilograms/cubic meter (62.4 lb/cu ft). The greater the salinity, the denser the water. The salinity of ocean water raises water density to about 1025 kilograms/cubic meter (64 lb/cu ft). Above 4°C/39°F in fresh water (lower in salty water), an increase in temperature results in a decrease in density. The density of ocean waters affects circulation because differences in density cause gravitational displacement of the water. Since warm water is less dense than cold, it tends to float above colder, denser water. Cold or highly saline water near the surface sinks and is replaced by warmer or less saline water. This effect produces patterns of surface water movement that vary with the seasons.

SURFACE VARIATIONS

The ocean surface shows significant variations in salinity and temperature. Salinity is highest in subtropical regions near 30° N and S because of the low precipitation, scant stream flow, and high evaporation rate in that area. Salinity decreases toward the equator because of the abundant rainfall, heavy stream flow, and lower evaporation rate due to increased cloud cover. For example, salinity is only 2 percent (20 ppt) near the equatorial coast of Brazil because of the diluting influence of the Amazon River as well as the abundant rainfall and reduced evaporation rate. Lowest salinity is found in polar regions because of the extremely low evaporation rate and because of fresh water stored on land in the form of snow and glacial ice, which contributes continuous fresh meltwater inflow during the warm season. Salinity is higher in middle latitudes than in polar regions due to warmer ocean temperatures and greater evaporation. Highest salinities are found in enclosed seas in dry, hot regions, where the evaporation rate is high and stream flow low. The Red Sea, for example, has a salinity of more than 4 percent (40 ppt). Des-

ert basin lakes have the highest salinity; the Dead Sea is so saline, 23.8 percent (238 ppt), that only primitive life forms can survive in it (thus its name). The Great Salt Lake in Utah has a maximum salinity of 22 percent (220 ppt).

In general, surface temperatures decrease with increased latitude; thus, the coldest temperatures are found in polar regions. Surface temperatures can be as low as −1.8°C (28.7°F) and reach a maximum of 32°C (90°F) in the Persian Gulf in summer.

The pattern of ocean currents affects the distribution of surface temperatures. Where ocean currents flow away from tropical latitudes, water temperatures are higher, and where currents move from high to low latitudes, temperatures are lower. Because of the pattern of ocean currents, the eastern sides of the oceans (west coasts of continents) in the middle latitudes tend to be relatively cool, while the western sides (east coasts of the continents) are relatively warm. This pattern has a significant effect on marine life, weather, and climate.

VARIATIONS WITH DEPTH

The vertical temperature distribution in the oceans is due to insolation heating the surface water. The heat is mixed downward by conduction and wave action. The result is a surface layer of essentially uniform temperature, below which there is a transition zone to the uniformly cold waters of the deep ocean. The change from the heated surface layer to the cold, deeper water occurs in a well-defined zone that has a depth of 15 to 60 meters (50 to 200 ft). This zone is called the **thermocline.** The thermocline is most apparent in the warm season when the surface water is most strongly heated.

Pressure increases with depth. The effect is comparable to the increase on atmospheric pressure as we descend down a mountain slope to a lower elevation. Water pressure is usually measured in pounds per square inch. It increases by the weight of one atmosphere (14.7 pounds per square inch) for every 10 meters

(33 ft) of water depth. At a depth of 300 meters (1000 ft), water pressure is 445 pounds per square inch. At 600 meters (2000 ft), it is 892 pounds per square inch, and at 1800 meters (6000 ft), it is 2685 pounds per square inch. At the base of the deepest ocean trench, it exceeds 16,000 pounds per square inch. Such pressure strains human organs such as eardrums, lungs, and eyes even during shallow descents into the ocean. For this reason the deepest scuba dive has been only about 100 meters (330 ft). Deep-diving vehicles, such as bathyscaphes and submarines, can descend thousands of feet while maintaining a more normal pressure for humans inside (Fig. 17.1).

MOVEMENTS OF OCEAN WATERS

The oceans exhibit three major types of movement. First is the movement of surface water pushed by the winds through the main ocean bodies. The moving streams are called **currents.** Second is the rise and fall of **tides,** evident along seacoasts. Finally, there are the oscillatory motions of the ocean surface called **waves,** which are generated by the force of winds.

OCEAN CURRENTS

Ocean currents are the primary means by which both water and heat are transported horizontally and vertically in the ocean. The movements that result involve both the surface currents and the much slower-moving subsurface circulation—thus, motion is three-dimensional. The physical geographer's primary interest is in the surface currents because they influence weather, climate, navigation, and the character and quantity of marine life.

The major ocean currents and drifts that affect the surface layers of water are caused primarily by the winds. Other controls are the Coriolis effect and the size, shape, and depth of the sea or ocean basin. Local currents may be caused by differences in density, which are

FIGURE 17.1 The deep-diving submersible AL-VIN has taken scientists to the ocean bottom to study the midocean ridges and made numerous scientific discoveries. Inside *Alvin* crew members are under normal atmospheric pressure, while outside, the submersible is under tremendous pressure from the ocean depths. (John Porteous, Woods Hole Oceanographic Institution)

due to variations in temperature and salinity, tides, and wave action.

Generally speaking, warm currents move poleward as they carry tropical waters into the cooler waters of higher latitudes, as in the case of the Gulf Stream or the Brazil Current. Cool currents deflect water equatorward, as in the California Current and the Humboldt Current. Warm currents tend to have a humidifying and warming effect on the east coasts of continents along which they flow, while cool currents tend to have a drying and cooling effect on the west coasts of the land masses.

The surface currents of the oceans are related to the wind and pressure systems of the atmosphere. The major ocean currents move in broad circulatory patterns, called **gyres,** around the subtropical highs (see Chapter 4).

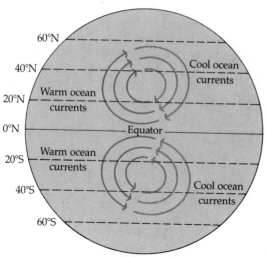

FIGURE 17.2 The major ocean currents flow in broad gyres in opposite directions in each hemisphere.

Due to the Coriolis effect, the gyres flow clockwise in the Northern Hemisphere and counterclockwise in the Southern Hemisphere. As a general rule, the surface currents do not cross the equator (Fig. 17.2).

Waters near the equator in both hemispheres are driven toward the west by the tropical easterlies or the trade winds. The current thus produced is called the Equatorial Current. At the western margin of the ocean, its warm tropical waters are deflected poleward along the coastline. As these warm waters move into higher latitudes, they move through waters cooler than themselves and are identified as warm currents (Fig. 17.3).

Because of the Coriolis effect, these warm currents, like the Gulf Stream and the Kuroshio Current, are deflected more and more to the right (or east). At about 40°N, the westerlies begin to drive these warm waters eastward across the ocean, as in the North Atlantic Drift and the North Pacific Drift. Eventually, these currents run into the land at the eastern margin of the ocean, and most of the waters are deflected toward the equator. By this time, these waters have lost much of their warmth, and while moving equatorward into the subtropical latitudes, they are cooler than the adjacent waters. They have become cool currents. These waters complete the circulation pattern when they rejoin the westward-moving Equatorial Current.

On the eastern side of the North Atlantic, the North Atlantic Drift moves into the seas north of the British Isles and around Scandinavia, keeping those areas warmer than their latitudes would suggest. Some Norwegian ports north of the Arctic Circle remain ice-free because of this warm water. Cold polar water flows southward into the Atlantic and Pacific along their western margins as the Labrador and Oyashio Currents.

The circulation in the Southern Hemisphere is comparable to that in the Northern except that it is counterclockwise. Also, because there is little land poleward of 40°S, the West Wind Drift (or Antarctic Circumpolar Drift) circles the earth across all three major oceans as a cool current, almost without interruption. It is cooled by the influence of the Antarctic ice cap (Table 17.2).

The general circulation described below is consistent throughout the year, although the position of the currents follows seasonal shifts in atmospheric circulation. In addition, in the North Indian Ocean, the direction of the circulation reverses seasonally according to the monsoon winds.

The cold currents along west coasts in subtropical latitudes are frequently reinforced by **upwelling.** As the trade winds in these latitudes drive the surface waters offshore, colder waters rise from lower levels to replace them. This upwelling of cold waters adds to the strength and effect of the California, Humboldt (Peru), Canary, and Benguela Currents.

On occasion the current systems break from their normal directional flow pattern. A periodic alteration in the Pacific Ocean which has the greatest worldwide effect is known as **El Niño.** Though the ultimate cause is still unknown, its devastating effects are well remembered from the particularly severe El Niño of 1982–1983. The trade winds become weak when atmospheric pressures at the opposite sides of the Pacific Ocean become reversed.

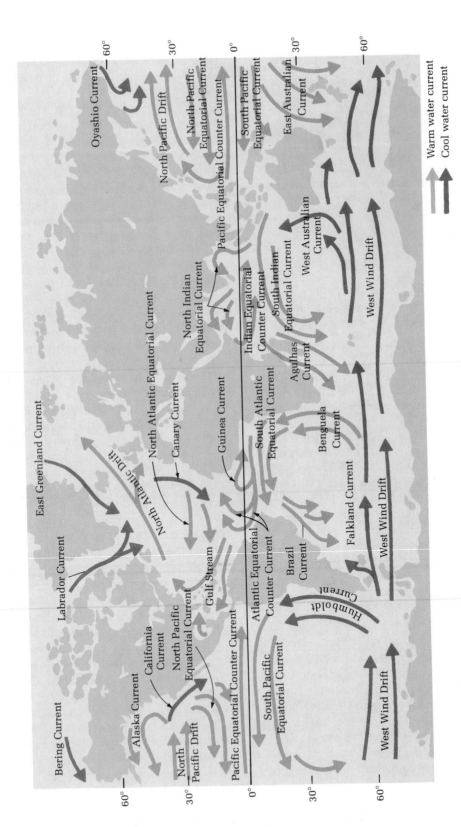

FIGURE 17.3　Map of the major world ocean currents, showing warm and cool currents.

TABLE 17.2   Primary Ocean Currents and Temperature Characteristics

PACIFIC OCEAN		ATLANTIC OCEAN		INDIAN OCEAN	
Oyashio	Cool	East Greenland	Cool	North Indian monsoon currents	
Bering	Cool	Labrador	Cool	(reverse seasonally with the	
North Pacific Drift	Warm	North Atlantic Drift	Warm	monsoon winds)	
Kuroshio (Japan)	Warm	Gulf Stream	Warm	North Indian Equatorial	Warm
Alaska	Warm	Canary	Cool	Indian Equatorial Counter	
California	Cool	Guinea	Warm	Current	Warm
North Pacific Equatorial	Warm	North Atlantic Equatorial	Warm	South Indian Equatorial	Warm
Pacific Equatorial Counter		North Atlantic Equatorial		Agulhas (Mozambique)	Warm
Current	Warm	Counter Current	Warm	West Australian	Cool
South Pacific Equatorial	Warm	South Atlantic Equatorial	Warm		
Humboldt (Peru)	Cool	Brazil	Warm		
East Australian	Warm	Benguela	Cool		
West Wind Drift	Cool	Falkland	Cool		
(Antarctic Circumpolar					
Drift; also present in South					
Atlantic and South Indian					
Oceans)					

The normal east-flowing Pacific equatorial currents also reverse as the trade winds weaken. This brings warm tropical waters, which have piled up on the western side of the Pacific, flowing back toward the shores of South America and sometimes as far north as California.

As the water temperatures rise in the eastern Pacific along South and North American coasts, upwelling is stopped. The cool Peru and California currents are deflected away from shore or are forced under the warmer water. Marine life dies or migrates to cooler waters, fisheries are destroyed, and unusual weather patterns occur. In 1982–1983, normally dry coastal Peru and southern California had powerful storms and floods, Tahiti and Hawaii experienced rare hurricanes, while Australia, on the other side of the Pacific, recorded the worst drought and brushfires in its history. As earth scientists sort through all the data to discover the causes of El Niño, they have again been reminded of the close association between the atmosphere and the hydrosphere and the complex relationship between these earth systems.

TIDES

The periodic rise and fall of ocean waters in response to external forces acting on them are called the tides. For the waters to rise in one place as a high tide, they must be pulled away from another part of the earth, resulting in a low tide there. The gravitational pull of the moon and sun, and the force produced by rotation of the earth-moon system, are the major causes of the tides (Fig. 17.4). Tides are complex phenomena; in order to explain them we must first consider the relationship between the earth and the moon.

Since the moon revolves around the earth every 29.5 days, one might assume that the center of its revolution would be at the center of the earth. However, both the moon and the earth actually revolve around a common center of gravity that lies within the earth and that is always on the side of the earth that faces the moon. Because of the rotation of the earth-moon system around this axis, centrifugal force causes the earth and the moon to tend to fly away from each other. However, this tendency is blocked because the centrifugal force is exactly balanced by the gravitational attraction between the earth and the moon. Nevertheless, the centrifugal force does cause the earth's fluid hydrosphere to bulge out on the side opposite the moon. This effect is one of the external forces that raises the water surface enough to produce a high tide. The ocean waters on

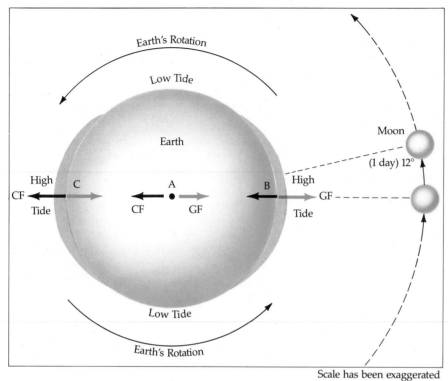

A. Gravitational force (GF) and centrifugal force (CF) are equal. Thus separation between earth and moon remains constant.

B. Gravitational force exceeds centrifugal force causing ocean water to be pulled toward moon.

C. Centrifugal force exceeds gravitational force causing ocean water to be forced outward away from moon.

Scale has been exaggerated

FIGURE 17.4 The tides are a response to gravitational and centrifugal forces. The gravitational attraction of the moon, periodically reinforced or opposed by the sun, pulls a bulge of water toward it, while the centrifugal force of rotation of the earth-moon system forces an opposing mass of water to be flung outward on the opposite side of the earth. The earth rotates through these two bulges every 24 hours; thus, there are two high and two low tides each day. Actually a "tidal day" is 24 hours and 50 minutes, since the moon moves 12° eastward each day along its monthly orbit around the earth.

the side of the earth facing the moon respond to the moon's gravitational force by bulging toward the moon. This bulge of water is directly opposite the bulge produced by centrifugal force, and the earth rotates through two high tides daily. Because they are 180° of longitude apart, they are separated by some 12 hours (half a day, because 180° is half the earth's circumference). Between these two tidal bulges, the waters recede as they are pulled toward the areas of high tide. Accordingly, there are two low tides, midway (90° of longitude) between the high tides.

The earth's 24-hour rotation, together with the moon's daily movement of 12° east-ward, along its monthly 360° path around the earth, means that theoretically coastlines will experience two high tides and two low tides every 24 hours, 50 minutes (the length of a tidal day). The time between two high tides is called the **tidal interval,** and it averages 12 hours, 25 minutes. However, the ideal tidal pattern does not occur everywhere, though the most common tidal pattern does approach the ideal model of two high tides and two lows in a day. This *semidiurnal* tidal regime is characteristic along the Atlantic coast of the United States. In bodies of water that have restricted access to the open ocean, such as the Gulf of Mexico or the Caribbean Sea, the tidal pattern

FIGURE 17.5 The extreme tidal range of the Bay of Fundy on Canada's east coast. (Canadian Government Travel Bureau)

may show only one high tide and one low during a day. This type of tide is called *diurnal* and is not as common as the semidiurnal. A third type of tidal pattern consists of two high tides of unequal height and two low tides, one lower than the other. The waters of the Pacific Coast of the United States exhibit this *mixed pattern.*

Tidal regimes are extremely complex and variable, even over short distances. We have described the tides as though they were bulges of water through which the earth rotates. In actuality, this "wave of equilibrium" theory does not account for the many variations in tidal regimes. A somewhat different theory explains the tides as an oscillatory movement of the hydrosphere set up as a consequence of the forces we have discussed. This "stationary wave theory" describes the tides as a back-and-forth sloshing of water, like the waves produced if a tray of water is tipped first one way and then the other. A circular motion, due to the earth's rotation, is superimposed on this up-and-down or rocking movement.

The difference in sea level at high tide and low tide is called the **tidal range.** Tidal range varies from place to place in response to factors such as the shape of the coastline, the depth of the water, access to the open ocean, and the topography of the ocean floor. The average tidal range along open ocean coastlines like the Pacific Coast of the United States is 1.5 to 3 meters (5 to 10 ft). In restricted or partially enclosed seas, such as the Baltic or Mediterra-

nean Sea, the tidal range is usually 0.7 meters (2 ft) or less. Funnel-shaped bays, such as the Bay of Fundy on Canada's east coast, are apt to produce extremely high tidal ranges. The Bay of Fundy is famous for its enormous tidal range, which averages 15 meters (50 ft) and may reach as much as 21 meters (70 ft) (Fig. 17.5).

SPRING AND NEAP TIDES The sun acts as a tidal influence on the ocean waters, but because it is so much farther away, its tidal effect is less than half (46 percent) that of the moon. When the sun, moon, and earth are lined up, as they are when there is a new or full moon, the additional influence of the sun on the ocean waters causes abnormally high and low tides, increasing the tidal range. This situation occurs every two weeks and is called **spring tide** (*spring*, here, does not refer to the season). A week after a spring tide, when the moon has revolved a quarter of the way around the earth, its gravitational pull on the earth is exerted at an angle of 90° to that of the sun. At this time the forces of the sun and moon tend to counteract one another. The moon's attraction, at the time of the first quarter and last quarter moons, is diminished by the counteracting force of the sun's gravitational pull. Consequently, the high tides are not as high, and the low tides are not as low. This moderated situation, which also occurs every two weeks, is called **neap tide** (Fig. 17.6).

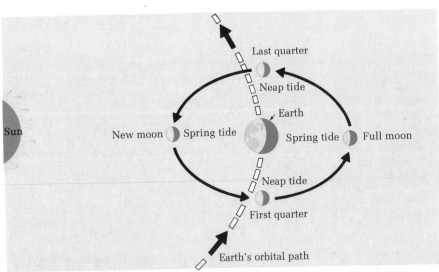

FIGURE 17.6 The highest tidal ranges (spring tides) occur when the moon and sun are aligned on the same side of the earth, or on opposite sides at full moon or new moon. The lowest tidal ranges (neap tides) occur when the gravitational forces of the moon and sun are acting at right angles to each other, at first- or last-quarter moon.

Based on astronomical information, long periods of observation of tidal patterns, and up to 40 local factors, tide tables can be prepared years in advance. Tide-predicting machines, ingenious devices of gears and pinions, have been used since the late 1800s. Today, computers provide tide tables of local coastal areas, which are useful to sailors, fishermen, surfers, and even beach joggers.

HORIZONTAL TIDAL CURRENTS The tidal movement that we have been describing is a vertical rise and fall of ocean waters. There are also horizontal currents, especially in bays or sounds, called **tidal currents.** The **flood tide** is the incoming current that accompanies rising tide; the outgoing **ebb tide** accompanies falling tide. The time period between flood and ebb, with no current, is called **slack water.** The velocity of tidal currents may be high enough to affect shipping and to erode the shoreline. The tidal current through the Golden Gate, the entrance to San Francisco Bay, is 4 knots (4.6 mph) during both flood and ebb tides. In some of the channels between islands of the Inside

Passage of the Alaska-British Columbia coast, there are tidal currents of up to 10 knots (11.5 mph). These currents can be hazardous to ships trying to navigate between the islands.

When incoming flood tides are opposed by the strong currents of large rivers, the tide may develop a steep wavelike front, known as a **tidal bore.** A few rivers famous for their tidal bores are the Amazon, Yangtze, and Seine. In exceptional instances, as in the Amazon, the tidal bore may be 5 meters (16 ft) in height.

WAVES

Waves are undulations of the surface layers of both large and small bodies of water. Contrary to appearance, waves do not transport water horizontally from one place to another, except where they break and run up on a beach. Rather, the form or shape of waves and their energy are transmitted *through* water. The movement of waves is similar to the movement of stalks of wheat, as wind blows across a wheat field and causes wavelike ripples to roll across its surface. The wheat returns to its original po-

FIGURE 17.7 The submerged bow of a Navy destroyer during a typhoon. The steep choppy waves within a storm are called "sea." They become whitecaps when the wind is strong enough to blow the tops off the waves. (Official U.S. Navy Photo)

sition after the passage of each wave. Water, too, returns to its original position (or close to it) after transmitting a wave. Another image that may serve as an analogy for the idea of a wave moving through water is the movement of a wave transmitted along the length of a snapped rope.

Most natural water waves are initiated by wind; the few that are not are produced by seismic or volcanic activity (tsunamis) or are an effect of tidal activity (tidal bores). When wind blows across water, friction arises between them. Some of the wind's energy is thus transmitted to the water. Waves are the result of this energy transfer (Fig. 17.7).

In a sufficiently deep, open body of water, waves appear as undulations on the surface. In Figure 17.8 you can see what happens to individual parts of the surface water during the passage of a wave. First, the upward movement of water produces a **wave crest.** The subsequent sinking of the water surface produces a **wave trough.** Water particles rise and fall, producing an endless series of waves passing along. The actual movement of water particles is circular, so that there is a small amount of

forward movement during each rise. This circular or oscillatory pattern of movement is the reason why ocean waves are called **waves of oscillation** (Fig. 17.8). The height of a wave is measured vertically from trough to crest. This height is the diameter of the orbit of surface water particles during wave passage. The horizontal distance between two wave crests (or two troughs) is the **wave length. Wave period** (usually measured in seconds) is the time it takes for successive crests to pass a fixed point. A "sea state" report would read something like this: 3- to 4-foot waves from the northwest, at 10-second intervals. This type of report is useful to sailors, fishermen, and surfers.

The factors that determine wave size are (1) velocity of the wind, (2) duration of wind, and (3) fetch. **Fetch** refers to the distance of open water across which the wind can blow without interruption. An increase in any of these three factors produces an increase in the size (height and wavelength) of waves.

Gentle breezes form *ripples* on the sea surface, providing roughness necessary for the wind to "grip" the water. If the wind increases, the ripples are soon transformed into larger

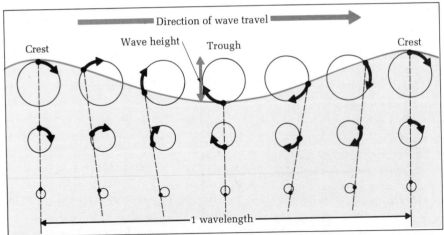

FIGURE 17.8 Orbital paths of water particles cause oscillatory wave motion in deep water. Wave height is the diameter of the surface orbit, which is the vertical distance from trough to crest. Wavelength is the distance between successive crests. (After Hydrographic Office Publication No. 11)

waves. In a storm, steep, choppy, chaotic waves called *"sea"* occur. They become *whitecaps* if the wind is strong enough to break the tops off the waves.

When the waves travel out of the generation area, or when the wind dies out, the chaotic sea is transformed into more gentle, longer-period waves called *swells*. Swells can travel thousands of miles and completely cross an ocean. For instance, large southwest swells arriving on the California coast in summer have been generated by winter Antarctic storms occurring south of New Zealand. The regular rhythm of the swells is the type of wave motion usually associated with sea-sickness.

CATASTROPHIC WAVES

Tsunamis are a train of waves caused by undersea seismic activity, such as earthquakes and volcanic eruptions, or undersea landslides. In deep ocean water they travel at speeds of up to 725 kilometers (450 mi) per hour and may pass beneath a ship unnoticed. However, they are extremely dangerous, rising over 30 meters (100 ft) as they meet the shore, engulfing and devastating entire coastal settlements, and often causing great loss of life and property.

In 1946, an Alaskan earthquake caused a tsunami in Hilo, Hawaii, which attained a height of over 10 meters (33 ft) and killed 150 people. When Krakatoa erupted in 1883, it generated a tsunami that reached over 40 meters (130 ft) and killed over 37,000 people in nearby Indonesian Islands. Tsunamis are sometimes called "tidal waves," even though they have no relationship to tidal activity.

Similarly, **storm surges,** which are sometimes called "storm tides," are not related to tides, although they may coincide with high tide. Storm surge is the name given to the rise in sea level beneath a severe storm (low air pressure), and to the waves driven onshore by the strong winds accompanying hurricanes or typhoons. During Hurricane Camille, in 1969, the sea level was over 8 meters (25 ft) above normal along the Gulf Coast.

Like tsunamis, these high seas may be enormously destructive. In 1900, Galveston, Texas, was destroyed and several thousand lives were lost owing to a hurricane-produced storm surge. In 1970 a storm surge produced by a typhoon in the Indian Ocean drowned over 200,000 people in the Ganges Delta area of Bangladesh, and in 1985 thousands of more lives were lost in a similar storm.

WAVES AS GRADATIONAL AGENTS

Waves, and their resulting local currents, are the gradational agents associated with large bodies of water: the oceans, seas, and major lakes. Their effect is felt in the narrow zone where land, air, and water meet. The line of this meeting is called the **shoreline.** Landform changes, however, are made both on the landward and seaward sides of the shoreline. Furthermore, the position of the shoreline fluctuates with the tides, and with long-term rises and falls in sea (or lake) level, as well as with diastrophic movements of the land itself. The **coastal zone** includes the area on land as well as areas presently submerged under water, through which the shoreline boundary fluctuates with time.

Waves, like streams, both aggrade and degrade the land. By removing, transporting, and depositing material, waves constantly work on the narrow strip of land with which they are in contact. Although the gradational effects of waves are multiplied many times over during storms when the largest waves are developed, all waves that reach the coastal zone do some work in shaping the land (Fig. 17.9).

As waves pound away at the shore, their general effect is to straighten and smooth the shoreline. Peninsulas or headlands that extend further into the water than other parts of the land are gradually cut back by wave action. Bays and inlets, on the other hand, are gradually filled in.

THE BREAKING OF WAVES

As long as the water is deep enough, and waves can roll along without disturbing the bottom, the ocean (or lake or sea) floor will remain unaffected. However, as the water depth decreases to half the distance between wave crests, the orbital motion of the water is retarded by friction with the bottom (Fig. 17.10). Speed of wave transmission slows, wavelength decreases, and wave height increases. Eventually wave height reaches a point of instability and the wave curls and collapses, producing surf, or breakers. If the waters far from shore are shallow, the waves will break at some distance offshore. If, however, water remains deep right beside the coast, the waves are forced to break against the land.

FIGURE 17.9 Large breakers pounding the shore at Point Lobos, California. (R. Sager)

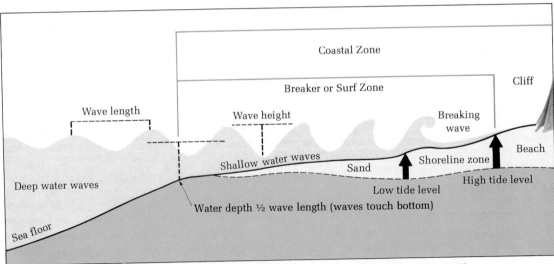

FIGURE 17.10 Waves begin to touch bottom when the water depth becomes one half the distance between wave crests. Then the wave velocity and wavelength decrease, and the wave height increases until breaking occurs.

If the wave crashes directly against solid material, the impact is often sufficient to break the material apart, whether it is jointed rock or a man-made structure. The direct impact of storm waves erodes back coastal cliffs and headlands. More commonly, the waves break before striking the land directly; then the water surges forward as foaming **swash** or *uprush,* which picks up sand and carries it onto a beach. The force of gravity eventually overcomes the waning momentum of the water, which finally reverses direction and drains back down the beach slope as **backwash**—taking some sand with it. The material returned to the water in the backwash is flung back during the breaking of the next wave. When the waves are large, chunks of rock and sand serve as abrasive tools similar to those carried as stream bed load. Thus abrasion is an important factor in wave erosion, just as in fluvial erosion. The power of waves, combined with the buoyancy of water, enables waves to carry large rocks, even boulders weighing tons. Such pieces of rock, when thrown against cliffs, act like cannonballs. Wave erosion of the coastal zone is also carried on by the process of solution, as is the case with surface and groundwater, as well as by hydraulic

action from the sheer physical force of the pounding water and the explosive effect of the air compressed between breaking waves and cliff faces. The crystallization of salts from evaporating ocean spray also helps detach mineral grains from cliffs, helping to push back the land.

WAVE REFRACTION

Wave refraction refers to the bending of waves as they approach a shore. An important consequence of wave refraction is that wave energy becomes concentrated along some parts of the shoreline and is greatly reduced in others. To see how this happens, imagine an irregular shoreline of bays and headlands (Fig. 17.11). Offshore waves are essentially parallel to each other, approaching the shore either directly or obliquely. However, as they approach land, the waves reach shallow water in some places sooner than in others. As a rule, the shallow waters off headlands will be reached by a wave before the shallow waters of a bay. As a result, the part of a wave that approaches a headland will be slowed down and forced to break before the part that is approaching the bay. Because

FIGURE 17.11 Wave refraction is important in straightening shorelines. Wave energy is concentrated on headlands, eroding them back, while in bays, wave deposition causes beaches to grow seaward.

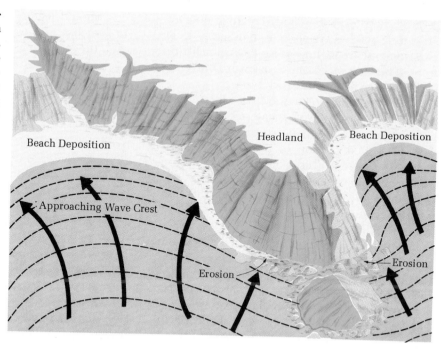

one part of the wave is slowed down before another part, the wave is bent or refracted.

Consequently, wave energy is directed more toward the headlands and less toward bays and coastal indentations. Because of the energy concentrated at headlands, wave erosion is more intensive there. The debris produced by wave erosion at headlands joins with fluvial material brought to the coast by streams and is shunted toward areas where there is less wave energy, accumulating in the bays between headlands. This effect is the reason waves tend to reduce shoreline irregularity: The concentration of erosion on headlands wears them back, while deposition in bays fills them seaward.

COASTAL LANDFORMS

As waves erode the coastal zone, they create distinctive landforms (Plate 21). **Sea cliffs** are created where waves pound against steeply sloping land or highly resistant rocks. The erosive action of the waves gradually eats away at the land, creating steep cliffs to the water (Fig.

17.12). Where the rocks are well jointed but cohesive, wave erosion may create **sea caves** along lines of weakness. **Arches** result where two caves meet from either side of a headland. When the top of an arch collapses, or a sea cliff retreats and a resistant pillar is left standing, it is called a **stack.**

Waves frequently cut notches in a cliff by erosive action. Notches are particularly common where limestone cliffs border the sea. The seawater dissolves the limestone, just as

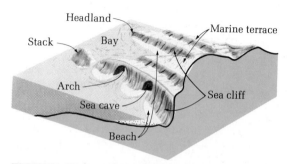

FIGURE 17.12 Diagram of the major coastal erosional landforms due to wave activity. Photographs of specific erosional features appear on Plate 21.

groundwater does on land, creating notches. However, notches are also cut in other rocks by abrasion and are an important factor in cliff retreat. The rate of coastal erosion is controlled by both wave action and rock type.

The presence of a wave-cut cliff implies removal of a large mass of material. The cliff is the vertical portion of a notch cut into the land. The horizontal portion of the notch is below water level and consists of a broad abrasion platform or **wave-cut bench** (Fig. 17.13). The material removed to produce the sea cliff and wave-cut bench comes to rest in the beaches of the coastal indentations.

Much of the material eroded from the cliff or beach is carried to sea by backwash and coastal currents. Because the turbulence important in transport of debris is less seaward, coarse debris accumulates and is battered into smaller particles close to the land. Finer particles move out into the sea. Thus marine sedimentation, like fluvial deposition, is organized by size and weight, with the finest particles being laid down farthest from shore. As these deposits accumulate, a **wave-built terrace** is created just seaward of the wave-cut bench.

Should tectonic activity uplift these wave-cut benches and wave-built terraces above sea level, they are then called **marine terraces.**

The most visible evidence of deposition are the **beaches.** They reflect the balance between input of material by swash and removal by backwash, which *combs* the beach. Not all beaches are made of sand. Some are formed from gravel, cobble, and even silt. The nature of beach material depends on its source: Granite, basalt, shale, conglomerate, and coral all result in beaches that differ in type, color, and texture. When particle sizes are large and wave energy is high, beaches will be much steeper than where only fine material is present and wave energy is low. Since wave height due to storms is greater in winter than in summer in the midlatitudes, winter beaches are generally narrower, steeper, and composed of coarser material than are summer beaches. Winter waves are relatively destructive, while the smaller summer waves are constructive. In winter, beach deposits may be removed entirely by destructive storm waves. Summer beaches are generally temporary accumulations of sand deposited over winter beach materials.

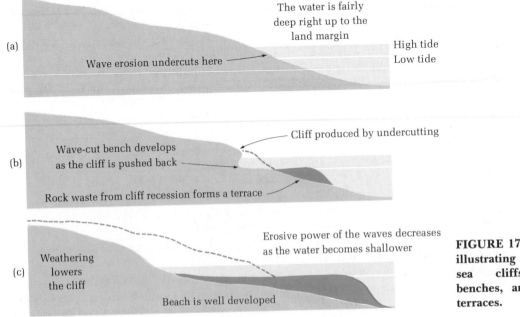

FIGURE 17.13 Diagram illustrating the origin of sea cliffs, wave-cut benches, and wave-built terraces.

FIGURE 17.14 Longshore currents carry beach material along the coast. The building of a jetty traps beach material on one side, while it starves the beach downcurrent. The longshore current in this aerial photo of East Hampton, New York, moves from left to right. (R. Sager)

LONGSHORE CURRENTS The movement of water and material by obliquely approaching waves along a shoreline is called a **longshore current.** Where waves come in at an angle and break, the swash pushes material ahead of it toward shore. Though the waves and consequent swash strike the shore at an angle, the backwash, responding to gravity, moves directly downslope perpendicular to the shoreline (Fig. 17.14). The result is that material pushed diagonally up the beach by swash is not returned to its original position by backwash. Repeated over time, the result is the mass transport of materials along the shore. As a general rule, longshore currents carry tons of material southward on both the Atlantic and Pacific coasts of the United States.

Rip currents are strong, seaward-moving currents found near the shore and are caused by the channeled return of water and sediments from large waves that have broken against the land. Rip currents (also known as *undertow* and *rip tides*) can be dangerous to swimmers, for the backwash can pull them out to sea. Swimmers caught in such undertow should not try to swim against it, but should

strike out parallel to the coast, for the rip current is seldom wide. Rip currents are usually visible as streaks of foamy turbid water flowing perpendicular to the shore.

Beaches are in equilibrium when income and outgo of sand are in balance. We can increase the size of a beach by building an obstruction to the longshore current, which prevents removal, while sand input remains the same. The obstruction may be accomplished by constructing a **groin** or **jetty,** a concrete or rock wall perpendicular to the beach. Of course, this obstruction starves the next beach area, which now has no input but still has the usual rate of removal (Fig. 17.14). Beach stability is frequently engineered to keep harbors free of sediment and to encourage beach growth. Every change in the natural process has effects that extend well beyond the area of concern, with results that are seldom beneficial in the long run.

Beaches are transitory by nature. When man upsets the natural sand supply by damming rivers or building groins to slow beach migration, the beach's temporary nature may be accelerated. In Florida and New Jersey,

VIEWPOINT: PROTECTING OUR COASTAL RESOURCES

Waikiki on Oahu, Hawaii. (R. Sager)

Our Atlantic, Gulf, Pacific, and Great Lakes coastlines are under tremendous pressure from conflicting uses and rapidly rising land values. Some of our coasts have been extensively modified and polluted.

Over 50 percent of the American population now lives in counties along our coasts. This has created conflicting uses for residential development, industry, energy production, agriculture, commercial fishing, recreation, and environmental protection.

Especially near our large cities, coastal modification and pollution have taken their greatest toll. Dredging for harbors and port facilities and filling of wetlands for airports, marinas, and industrial sites have completely changed the urban shoreline. Pollution from chemicals, construction runoff, and sewage have altered coastal waters.

Oil pollution from tankers and offshore rigs is the biggest offender, fouling beaches and endangering birds, seals, and lobsters and their habitats. The U.S. Department of the Interior is opening more offshore areas on the outer continental shelf for oil leases as pressure to find new oil and gas deposits grows. Already, Texas, Louisiana, California, and Alaska have extensive offshore oil drilling activity, with more to follow. Interest is now also directed at our Atlantic Coast, to Baltimore Canyon off the New Jersey shore and Georges Banks off the New England coast.

Special consideration must be given to the most sensitive coastal environments, specifically, barrier islands, beaches, and wetlands. America's barrier islands form a long slender chain running parallel to our Atlantic and Gulf coasts from New York to Texas. Fronted by the ocean on one side and marshes or bays on the other, they are made up of unconsolidated sediments, mainly sand, and are thus in a constant state of change. Their beaches and sand dunes migrate, their marshes fill in, but most of all, sudden changes from storms can wipe out whole sections of a barrier island or completely overwash it. Some barrier islands such as Atlantic City, Galveston, and Miami Beach are urban areas. Others are undeveloped, inaccessible, or set aside as National Seashores or National Wildlife Refuges. Most, however, are considered prime real estate property with coastal access. Thus, as real estate profits soar, development continues, in spite of the risk to life and property. On some barrier islands, population growth has been so rapid that evacuation in the path of a hurricane would now be almost impossible.

hundreds of millions of dollars are being spent to replenish the sand beaches. The beaches not only serve the obvious recreational needs but are also necessary to protect settlements from storm waves.

SPITS, BARS, AND BARRIER ISLANDS

Wave erosion of nonresistant materials produces a great deal of sediment. Where rivers flow into the sea, they often bring additional sediment. Material not carried out into deeper waters or deposited on the wave-built terrace is transported along the shore from the area of intense wave action to a more protected area where deposition can take place. When a shoreline is relatively straight, except for a bay, the material carried by longshore currents will be deposited as a **spit** that continues the shoreline into the bay mouth (Fig. 17.15). If the spit is joined by one extending from the opposite side of the bay or if the spit grows completely across the bay, the result is called a **baymouth bar.**

The formation of a baymouth bar changes the bay into a protected lagoon. The salinity of such a lagoon will vary from that of the open sea, depending on such factors as river flow and climate. The salinity in turn will affect the organic life of the lagoon. A river that flows into the waters of a bay protected by a spit or bar will be able to build up a delta. Eventually, the entire lagoon may be transformed into salt marsh or coastal wetland by

Beach access, crowding, and "hotel row" development have become prime coastal issues. Public access to beaches is one of the most critical issues in coastal management and planning. For instance, there are more than 22 million Californians and 1100 miles of coastline. Over 50 percent of the coastline is private property (in some states this figure is over 90 percent), and a significant portion of the remaining coast is off-limits owing to military installations or its rugged inaccessible terrain. Coastal planning must address the public access issue and protect coastal zones from unplanned development in the future. Responsible citizens must strike a delicate balance between private property rights and public access rights and at the same time maintain public safety and natural habitat protection.

A coastal wetland is a life zone of shallow water, generally where salt and fresh water mix. Most persons consider these wetlands useless and uninteresting, yet biologically they are the most productive ecosystem in the world. They play a vital role in the food-energy web and serve as a nursery for fish and shellfish as well as a wildlife refuge for millions of coastal and migratory birds. The wetlands also provide the coastline with storm protection and serve as a pollution filtration system. They are being modified and destroyed at an alarming rate. Over 40 percent of America's coastal wetlands have disappeared under industrial sites, airports, subdivisions, and city dumps. Other coastal wetlands, such as those in Chesapeake Bay, are suffering from the effects of agricultural herbicides.

Barrier island and lagoon near Georgetown, South Carolina. (NOAA, National Ocean Service, photo 84-ZP-1074, February 16, 1984)

To help save our coasts from becoming an endangered resource and to regulate conflicts between environmental protection and development, laws must be enacted and enforced. Two examples of coastal protection legislation are the Federal Coastal Zone Management Act of 1972 and the California Coastal Act of 1976.

The coast is one of the earth's most productive and beautiful natural environments. To preserve our coastline, we must change our view of it from a commodity to be sold to the highest bidder to a vital resource necessary for the future of this planet.

river deposits on one side trapped behind wave deposits on the other. If wave deposition connects spits between the mainland and islands close to the shore, such spits are called **tombolos** (Plate 23).

On gently shelving coasts or along coastlines where waves are forced to break at considerable distance from the shore, the churning motion of the water builds submerged **offshore bars** parallel to the coast. As these bars grow and emerge above sea level they form barrier beaches and the larger **barrier islands** (Plate 23). Usually there is a sand dune zone and a shallow lagoon between the barrier island and the mainland. Barrier islands are the most common type of beach deposit on low-relief

coastlines. The Atlantic and Gulf coasts from New York to Texas are dominated by barrier islands. Some excellent examples of large barrier islands are Fire Island (New York), Cape Hatteras (North Carolina), Cape Canaveral and Miami Beach (Florida), and Padre Island (Texas).

CLASSIFICATION OF COASTS

Though there are numerous ways to classify coastlines, there is no universally accepted classification system. Two of the most popular classifications of coasts were proposed by Johnson and Shepard. In 1919, Douglas Johnson, a student of William Morris Davis, distinguished

FIGURE 17.15 Diagram (a) shows the major coastal depositional features. A spit (b) is a beach deposit connected at one end to the coast, while a baymouth bar (c) is connected to the coast at each end of a bay. (Both photos, R. Sager) Barrier islands (d), such as this one seen in a satellite view of Cape Hatteras, North Carolina, are common from New York to Texas. (NASA) Other coastal depositional features are shown in Plate 23.

four major shoreline types: shorelines of emergence, shorelines of submergence, compound shorelines, and neutral shorelines.

Shorelines of emergence occur where the water level has been lowered or the land has risen in the coastal zone. In either case, land emerges that was once covered by seawater, and features created by wave action such as wave-cut terraces, seacliffs, stacks, and beaches

are found above the level of the present shoreline. The position above the present shoreline of wave-created landforms serves as evidence of emergence. Shorelines of emergence were probably common during Pleistocene glaciation, prior to 10,000 years ago, for the formation of glaciers would have lowered the level of the oceans over 100 meters. Features of emergence are presently best developed along tectonically active coasts such as that of California, where marine terraces are found as much as 370 meters (1200 ft) above sea level.

As the Pleistocene glaciers melted, the seas gradually rose, creating **shorelines of submergence** in which features of the coastal zone were submerged beneath the seas all around the world. Shorelines of submergence also occur where the level of the land has been lowered by tectonic forces, as is the case in San Francisco Bay. The features of a new shoreline of submergence are related to the character of the coastal lands prior to submergence. Plains, for instance, will produce a far more regular shoreline than will a mountainous region. Two special types of submerged shorelines are ria shorelines and fiord shorelines. **Rias** are created where river valleys are "drowned" by a relative rise in sea level or a sinking of the coastal area. The valleys themselves become estuaries, and the interfluves form peninsulas. *Fiords,* which are drowned glacial valleys, form scenically spectacular shorelines (see again Fig. 8.13). A fiord shoreline is highly irregular with deep, steep-sided arms of the sea penetrating far inland in the troughs originally deepened by glaciers. Tributary streams cascade down the canyonlike fiord walls, which may be several thousand feet high.

The configuration of submerged shorelines in areas of mountainous relief varies depending on the orientation of geologic structure with respect to the shoreline. Where shorelines cut across the grain of fold or fault structures, the coast is a ragged one of long peninsulas, offshore islands, and deep bays. The coasts of Greece and western Turkey are outstanding examples of such shorelines. However, where the geologic structures parallel the coastline, as in California, the shoreline is rather regular, with only occasional shallow bays and few good harbors.

Submerged shorelines in areas of considerable relief produce the more spectacular marine erosional and depositional features mentioned earlier: wave-cut cliffs, arches, stacks, tombolos, baymouth bars, etc. When areas of low relief and weak rocks are submerged, they are simply eroded back in a regular manner, with the eroded material forming barrier islands such as those found almost continuously along the Atlantic and Gulf Coasts of the United States. The evolution of submerged coastal plains is so much like that of emergent coasts that in such instances the entire concept of submergence and emergence is of little relevance.

Neutral shorelines, such as coral reefs and river deltas, cannot be classified as either submerging or emerging. More accurately, these shorelines are pushing outward-prograding.

Actually, most shorelines show evidence of more than one type of development, largely because the level of the land and the level of the seas have changed many times during the geologic history of the earth. For this reason, most shorelines are considered to be compound. **Compound shorelines** are characterized by features of both submerged and emerged shorelines and often of neutral shorelines as well.

Since coastlines are really shaped by both terrestrial and marine geomorphic processes, a better classification system may be one first proposed by Francis P. Shepard in 1937. Shepard recognized only two types of coasts. **Primary coastlines** are formed mainly by land erosion and deposition processes. Such formation is due to tectonic activity or sea level changes that cause the shoreline to change its level too rapidly for the marine processes to shape the coast. The major types and good examples of primary coastlines are drowned river valleys (Chesapeake Bay), glacial erosion coasts (fiords of Alaska), glacial deposition coasts (north shore of Long Island), river deltas (Mississippi delta), volcanic coasts (Hawaii), and faulted coasts (California).

Secondary coastlines are those formed mainly by the marine geomorphic agents, especially waves, and by marine organisms. Marine erosional coasts are dominated by such features as sea cliffs, arches, stacks, and sea caves. Marine depositional coasts have such features as barrier islands, spits, and bars. One example of coasts built by marine organisms is the coral reef. Mangrove trees and salt-marsh grasses also trap sediments to build new land areas in shallow coastal water.

THE OCEAN FLOOR

It is only in this century that scientists have been able to explore the ocean floor and discover the details of its contours. The invention and continuing refinement of sonic depth-finding devices, which can make continuous recordings of ocean depths by use of reflected sound waves, have helped enormously, as have specialized research ships designed for deep-sea drilling, such as the *Glomar Challenger,* and deep-diving vehicles like the *Trieste,* which has descended to a depth of 10,912 meters (35,800 ft) below sea level. General advances in technology have produced materials and engineering designs able to withstand the enormous pressures found far below the ocean surface (Fig. 17.1).

Before the most recent explorations of the sea, it was believed that most of the ocean floor consisted of flat plains. We now know that this is far from the truth. Beneath the ocean waters are irregularities on the seabed surface comparable in size and extent to those found on the continents (Fig. 17.16). Mountains, basins, plateaus, ridges, volcanic cones, escarpments, canyons, and trenches all are present beneath the ocean (Plate 22).

The oceans actually spill over the rim of their basins and onto the edges of the continental platforms. Thus, some of the sea floor is geologically a part of the continents. As a result, there are two major topographic subdivisions of the ocean basins: (1) the continental shelf and continental slope, which are part of the continental landmasses, and (2) the deep ocean floor (Fig. 17.17).

CONTINENTAL SHELF

The **continental shelf** consists of the rims of the continents flooded by the oceans. This part of the sea floor slopes gently from sea level to an average maximum depth of 200 meters (656 ft) beneath the ocean waves. The continental shelf varies a great deal in width. In some places it is almost nonexistent, while in others it is as wide as 1300 kilometers (800 mi). Generally, where broad continental plains slope to meet the ocean, the continental shelf is quite wide. Where mountains are found along the edge of the continent, the shelf is narrow. These two extremes are found on the two coasts of the United States. Beyond the plains of the Atlantic and Gulf Coasts, the continental shelf extends outward as much as 480 kilometers (300 mi). On the Pacific Coast, with its mountainous margins, the continental shelf is so narrow as to be practically nonexistent in some places. It is, of course, not surprising that the character of the continental shelf is related to the character of the nearby land, especially when we remember that the shelf is an extension of that land and geologically is not a part of the ocean basins.

The continental shelf area is "the treasure chest of the sea." The supply of light and the supply of nutrients washed into these waters

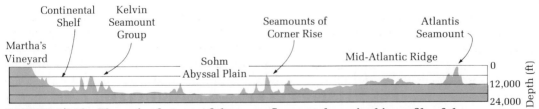

FIGURE 17.16   The major features of the ocean floor are shown in this profile of the North Atlantic Ocean from the United States east coast to the Mid-Atlantic Ridge. (After Yasso, *Oceanography*, New York, Holt, Rinehart, and Winston, 1965)

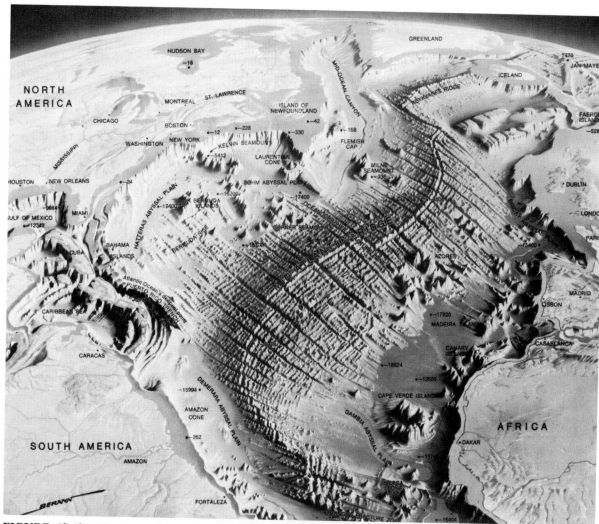

FIGURE 17.17 Physiographic diagram of the floor of the North Atlantic Ocean showing features such as the Mid-Atlantic Ridge, seamounts, trenches, and the continental shelf and slope of the surrounding continents. (From a painting by Heinrich Berann, courtesy of ALCOA)

from the continents allow marine plants and animals to grow and thrive. Thus it is in the shallow waters above the continental shelf that 90 percent of the fish we eat are caught. Virtually all the lobster, crab, shellfish, and shrimp live in these rich waters. Vast untapped quantities of gas and oil are stored in the shelf sediments, as well as other minerals including diamonds, tin, and gold.

The topography of the continental shelf is not as smooth as we might expect from the effects of constant wave abrasion and from the settling of sediments (sand, silt, and clay) brought from the adjacent land. Actually, the shelf, though it is a *relatively* smooth, sloping plain, has ridges, depressions, hills, valleys, and canyons. Some of the higher features break the surface of the water and appear as islands, such as Long Island and Martha's Vineyard. During periods of low sea level in the Pleistocene, when much water ordinarily stored in the seas was held on the land as glacial ice, the continental shelves were exposed to the atmosphere and were grazed by animals. The teeth of these Pleistocene land mammals are occasionally dredged up by fishermen.

Submarine canyons are probably the most striking features of the continental margin seabed. These canyons are usually steep-sided erosional valleys cut into the sediment and rock of the shelf and slope. They resemble the canyons cut by rivers on land (Fig. 17.18). Some are larger than the Grand Canyon. V-shaped in cross section, these submarine canyons are often found opposite the mouths of major rivers such as the Congo or Hudson. This distribution has led to the hypothesis that rivers have helped form the canyons. Yet there are canyons that are not located opposite either present-day or ancient river mouths. Many theories have been advanced as to how the submarine canyons were formed. The most widely accepted explanation is that they were produced by **turbidity currents**—periodic, mas-

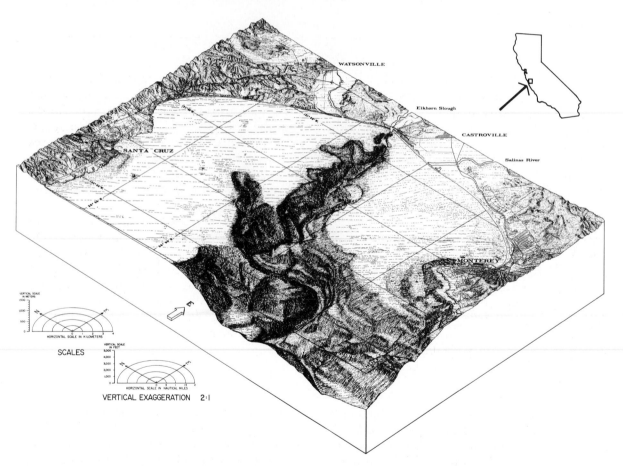

OBLIQUE MAP OF MONTEREY BAY, CALIFORNIA

PHYSIOGRAPHY BY TAU RHO ALPHA
1978

FIGURE 17.18 Physiographic diagram of a submarine canyon. Monterey Canyon off the coast of California is an example of a submarine canyon that is deeper than the Grand Canyon of the Colorado River. (Tau Rho Alpha, U.S. Geological Survey)

FIGURE 17.19 The major features of the continental margin.

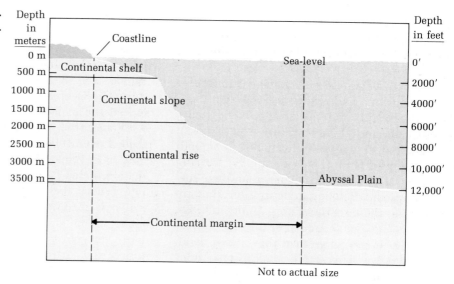

Not to actual size

sive, submarine flows of sediment that stream from the continental shelves to the deep ocean floor. Apparently such flows have a high capacity for eroding the ocean floor. The upper portions of the canyons may have developed by normal erosional processes during the Pleistocene, when the low sea levels exposed much of the shelf areas.

CONTINENTAL SLOPE

Marking the outer edge of the continental shelf is a relatively steep drop, usually 3000 to 3600 meters (10,000 to 12,000 ft), to the ocean floor (Fig. 17.19). This drop, called the **continental slope,** forms the margin of the adjacent continent and the side of the ocean basin. The landward boundary of the continental slope, where the land drops off abruptly, usually occurs where the waters are somewhere between 120 to 180 meters (400 to 600 ft) deep.

The continental slope is actually not a steep incline, since it descends at an angle of 15° at the most, though it definitely slopes more sharply than does the shelf. What is most characteristic of the continental slope is its great descent, usually some 3600 meters (12,000 ft), but sometimes as much as 9000 meters (30,000 ft), to the deep ocean floor or to the submarine trenches.

Far less sediment is deposited on the continental slope than on the shelf, partly because of the increased distance from the continents. Sediment is transported across the slope by turbidity currents in submarine canyons. Also, sea life is far less abundant on the continental slope because the supply of nutrients from the land is scanty and light for photosynthesis cannot penetrate the deeper waters.

The gently sloping surface at the base of the continental slope is known as the **continental rise.** It has an average slope of less than 1°. Although the continental rises are well developed in the Indian and Atlantic Oceans, they are almost nonexistent in the Pacific owing to the presence of trenches at the edges of the continental margins.

FEATURES OF THE DEEP OCEAN FLOOR

The deep ocean floor lies at an average depth of 3600 to 3900 meters (12,000 to 13,000 ft) below sea level. Until recently it was believed to be a relatively smooth plain. Now, with the use of sophisticated hydrographic profiling devices, we have discovered that the topography of the deep ocean floor is almost as irregular as that of the land. The ridges and depressions of the ocean floor rival and even surpass some of those found on land, both in size and complexity of pattern. On the other hand, large areas blanketed by sediment are virtually featureless plains.

Smoothing and leveling agents are important in shaping land above sea level (see Chapters 13–16). Running water, glaciers, wind, weathering, and gravity all work, sometimes in concert and sometimes individually, to smooth and level the land. They do this by wearing down landforms that are higher than their surroundings and by filling in depressions that are lower than their surroundings. Under the sea, these leveling agents are either missing entirely or are much less active than they are on land. Consequently, many of the landforms under the sea, created by volcanic action and by breakage and bending of the earth's crust, remain basically in their original form. The two most impressive features of the deep ocean floor are the oceanic ridges and the submarine trenches. Other major features of oceanic topography are abyssal plains, fault scarps, seamounts, and guyots.

RIDGES The oceanic ridges are interconnected chains of mountains found in all three major oceans (Fig. 17.17). The best known of these is the Mid-Atlantic Ridge, which extends from Iceland almost to Antarctica before it swings eastward around Africa toward the Indian Ocean. The Mid-Indian Ridge forms an inverted Y shape that extends into the Red Sea. Its southerly arms link the Mid-Atlantic Ridge to the Pacific Ridge and on to the East Pacific Rise. This 64-000-kilometer-long (40,000 mi-long) continuous chain of mountains averages about 1600 kilometers (1000 mi) wide and rises an average of 1500 to 3000 meters (5000 to 10,000 ft) above the ocean floor. In some places its highest peaks rise above the surface of the water as islands. The Azores, Ascension Island, and the island of Tristan da Cunha are some of the highest peaks of the Mid-Atlantic Ridge. The Azores rise 8100 meters (27,000 ft) above the ocean floor.

From the results of recent oceanographic research, especially ocean floor drilling and direct observation of the oceanic ridges from the manned U.S. submersible *Alvin*, earth scientists now know that new material is being added to the earth's crust by undersea volcanic activity (Fig. 17.1).

The Deep Sea Drilling Project (DSDP) operated the drillship *Glomar Challenger* between 1968 and 1983 (Fig. 17.20a). *Glomar Challenger* drilled over 1000 holes in all the major oceans. The most remarkable discovery was the confirmation of sea floor spreading. The addition of new crustal rock material along the midocean ridges is pushing the older parts of the crust apart. Drilling of the ocean floor now seeks more detailed data on the hidden three quarters of the earth's geology. The Ocean Drilling Program (ODP) will continue to explore the ocean bottom with the drillship *JOIDES Resolution* (Fig. 17.20b).

TRENCHES **Oceanic trenches** represent the deepest parts of the ocean. Usually long, narrow, arc-shaped, and steep-sided, these depressions are aptly called *trenches*. Most trenches are found not in the middle of the ocean basins, as we might expect, but near their margins. Trenches are usually found adjacent to areas that have a great deal of volcanic and seismic or earthquake activity and are most common on the seaward (convex) side of curving chains of volcanic islands (called **island arcs**) such as the Aleutians.

Most oceanic trenches—the deepest ones—are found around the rim of the Pacific Ocean. Challenger Deep, in the north Pacific's Marianas Trench, the deepest known part of the ocean, reaches 10,915 meters (35,810 ft) below sea level (Fig. 17.21). In 1960 Jacques Piccard and Donald Walsh descended in the bathyscaphe *Trieste* to the bottom of that trench. Placed in this trench, Mount Everest, the highest mountain on earth, would still have a mile of water above its summit. Other major trenches in the Pacific are the Kuril, Japan, Philippine, Tonga, and Peru-Chile Trenches (Fig. 17.22). The deepest part of the Atlantic is the Puerto Rico Trench, while the Java Trench is the Indian Ocean's deepest point.

(a)

(b)

FIGURE 17.20 (a) Named after the first vessel outfitted for an oceanographic expedition, H.M.S. *Challenger* (1872-1876), the Deep Sea Drilling Project's (DSDP) drillship *Glomar Challenger* has crisscrossed the world's oceans, boring into the seabed in search of evidence for the origin, age, and history of the ocean basins. (Scripps Institution of Oceanography) (b) The new drillship *JOIDES Resolution* will continue the exploration of the ocean bottom under the Ocean Drilling Program (ODP). A computer-controlled positioning system maintains the ship over a specific location while drilling in water depths of over 8000 meters. (John Beck, Ocean Drilling Program)

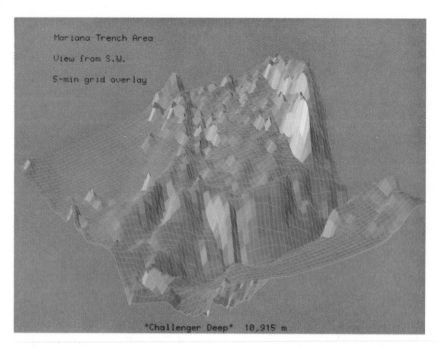

FIGURE 17.21 A computer-produced digital terrain model of the Mariana Trench area of the western North Pacific Ocean. Vertical exaggeration is about 22:1. Challenger Deep, the deepest part of the Mariana Trench, is the deepest known place on earth. It reaches depths of 10,915 meters, deep enough to swallow Mt. Everest with over a mile of water depth above its summit. (Courtesy of Peter W. Sloss, National Geophysical Data Center, NOAA, from ocean depth data supplied by the U.S. Navy and the Defense Mapping Agency)

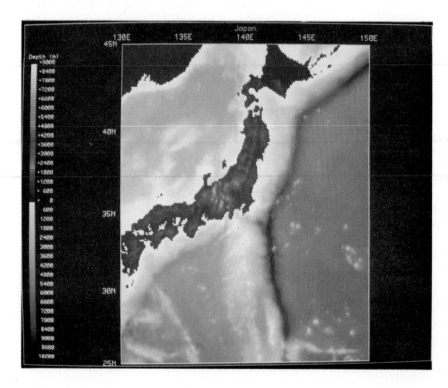

FIGURE 17.22 A computer-produced color-scaled contour map of the northwest Pacific Ocean and Japanese islands. The deep blue represents the depths of the Japan Trench, the dark blue represents the Northwest Pacific Abyssal Basin, while the light blue and white show the shallow waters of the undersea ridges and continental shelf of Japan. As the Pacific Plate moves northwest, it is forced down toward the mantle, forming the Japan Trench. (Courtesy of Peter W. Sloss, National Geophysical Data Center, NOAA, from ocean depth data supplied by the U.S. Navy and the Defense Mapping Agency)

As with the midocean ridges, the trenches play a major role in the earth's geologic evolution. As noted in Chapter 12, earth scientists believe that oceanic plates are descending below continental plates and are being recycled to the earth's interior by way of the trenches. This process, known as *subduction,* is a key concept in the theory of plate tectonics. Ocean floor drilling has shown that no ocean floor rock is older than 200 million years. The rock is being created at the midocean ridges and destroyed in the trenches by subduction. Thus, it appears that the oceanic crust is recycled over many millions of years.

ABYSSAL PLAINS **Abyssal plains** of very low relief lie at depths of 3000 to 6000 meters (10,000 to 20,000 ft) and cover about 40 percent of the ocean floor. They are caused by the deposition of thick masses of marine sediments that bury the ocean floor relief. Much of this sedimentary blanket consists of fine brown and red clays, contributed by turbidity flows from the continental slopes. A significant portion also consists of the remains of microscopic marine organisms. Such organic deposits are known as **ooze.** The famous white chalk cliffs of Dover are uplifted ancient ocean floor ooze deposits.

SEAMOUNTS AND GUYOTS One of the more common features of ocean topography is the **seamount,** an underwater volcanic mountain. Unlike the continuous chains of mountains that form the oceanic ridges, seamounts are relatively isolated mountains or groups of mountains, usually with an elevation above the ocean floor of 900 meters (3000 ft) or more (Fig. 17.18). Often steep-sided with small summits, seamounts are the remains of once-active volcanoes that developed above randomly distributed centers of high temperatures within the mantle. Though most seamounts are not nearly as impressive in size as the midocean ridges, some are high enough to break the ocean surface and appear as islands.

Guyots, discovered during World War II, are seamounts with flat instead of peaked tops,

found at depths of a few thousand meters below sea level. The origin of their flat tops at such depths is not certain. It seems most probable that they are volcanoes whose summits have been planed off by wave erosion. Later they subsided to their present depths, possibly during lateral movement of the sea floor away from the oceanic ridges and "hot spots" anchored in the mantle (see Chapter 12).

ISLANDS AND CORAL REEFS

There are three basic types of islands— continental, oceanic, and coral. **Continental islands** are found on the continental shelves. They are geologically part of the continent and are separated from them because of sea level changes or tectonic activity. The world's largest islands, such as Greenland, New Guinea, Borneo, and Great Britain, are continental. Smaller continental islands include Washington's San Juan Islands, New York's Long Island, and California's Channel Islands. The hundreds of barrier islands along the Gulf and Atlantic coasts of the United States are also continental.

Oceanic islands are volcanoes that rise from the deep ocean floor. They are not geologically related to the continents. Most of the oceanic islands occur in **island arcs** along the edges of the trenches, such as the Aleutians, Kurils, and Marianas. Others are peaks of the midocean ridges rising above sea level, such as the Azores. Many oceanic islands occur in chains or lines, such as the Hawaiian Islands. These **island chains** are caused by the oceanic crust sliding over a stationary "hot spot" in the mantle. The exact cause of these "hot spots," where new volcanic material rises to the oceanic crust surface, is not known. Yet we can predict, for instance, that the Hawaiian Islands will move with the Pacific plate toward the northwest, and the islands will slowly sink into the thin oceanic crust. Hence, the islands to the northwest will submerge to become seamounts or guyots. A new volcanic island will form to the southeast. Evidence of this motion is indi-

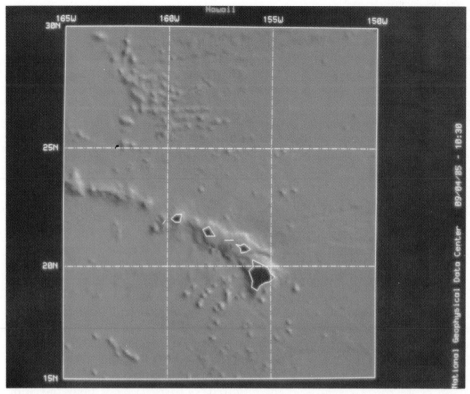

FIGURE 17.23   A computer-produced shaded-relief plot of the Pacific Plate in the Hawaiian Islands area. As the Pacific Plate moves northwest across a hot spot in the underlying mantle, magma rises through the crust, forming volcanic islands. The big island of Hawaii is still active and over the hot spot. Older islands have moved to the northwest and become inactive. To the far northwest most of the islands have become submerged seamounts. (Courtesy of Peter W. Sloss, National Geophysical Data Center, NOAA, from ocean depth data supplied by the U.S. Navy and the Defense Mapping Agency)

cated by the fact that the youngest islands of the Hawaiian chain, Hawaii and Maui, are to the southeast, while the older islands, such as Kauai and Midway, are located to the northwest (Fig. 17.23).

Coral reefs and atolls are formed by an accumulation of skeletal remains of tiny sea animals called polyps that secrete a limy skeleton of calcium carbonate. Reef corals need special conditions to grow—clear and well-aerated water, water temperatures above 20°C (68°F), plenty of sunlight, and normal salinity. These conditions can be found in the shallow waters of tropical regions such as Hawaii, the Florida Keys, and the U.S. Virgin Islands. Today, increasing pollution, dredging, souvenir coral collecting, and other man-made stresses threaten the survival of many coral reefs.

A **fringing reef** is a coral reef built along a coast. Fringing reefs tend to be wider where there is more wave action, bringing a continuous supply of well-aerated water and additional nutrients for increased coral growth. They are usually absent where there is a river mouth, because the corals cannot grow where the waters are laden with sediment or where river water lowers the salinity of the marine environment. Sometimes the coral forms a **barrier reef,**

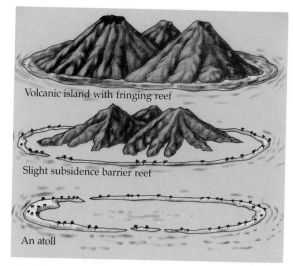

Volcanic island with fringing reef

Slight subsidence barrier reef

An atoll

FIGURE 17.24 The theory of coral reef development based on an explanation by Charles Darwin. Successive reef forms develop from island subsidence and coral reef upbuilding. First (a) the fringing reef is attached to the shore; then as the island subsides (b) the barrier reef forms; further subsidence causes the coral to build upward, while the volcanic center of the island is completely submerged, forming (c) an atoll.

which lies offshore, separated from the land by a shallow lagoon. The Great Barrier Reef of Australia is over 1930 kilometers (1200 mi) long.

An **atoll** is a ring of coral reefs encircling a lagoon that has no inner island. Figure 17.24 indicates the manner in which atolls develop. In response to the subsidence of volcanic islands, the fringing reef grows upward as fast as subsidence occurs, becoming a barrier reef and finally an atoll. This explanation of atolls was developed by Charles Darwin in the 1830s and has been proved correct by drilling into several Pacific atolls. Drilling indicates that there has been as much as 1200 meters (4000 ft) of subsidence and an equal amount of reef development in the past 60 million years (Fig. 17.25).

Sometimes small islands are formed on atolls from storm wave action, which erodes the reef and heaps the coral debris above sea level. Such atoll islands are most common in the Pacific Ocean. These islands pose several problems as environments for people. First, they have a low elevation above sea level and provide no defense against huge storm waves or

FIGURE 17.25 Satellite photograph of atolls of the Tuamotu Archipelago in French Polynesia. (NASA)

tsunamis, which can inundate an entire island, drowning all its inhabitants. Second, little vegetation can survive in the lime-rich rock and soil of which the atoll islands are composed. The coconut palm is one exception to this rule, and therefore the coconut is depended upon to fulfill many of the needs of atoll island inhabitants in Polynesia and Micronesia.

MARINE ECOSYSTEMS

The living organisms of the ocean can be divided into three groups according to where or how they live in the ocean. The lowest-order group is called **plankton.** Plankton is made up of the ocean's smallest—usually microscopic—plants and animals. These tiny plants and animals float freely with the movements of ocean water, are true "drifters," and form the basis of the oceanic food chain. The second group is composed of the animals that swim in the water. This group is called **nekton** and includes fish, sea mammals such as whales and seals, and invertebrates such as squid. The third group is the plants and animals that live near or on the ocean floor. This group is called the **benthos.** It includes coral and such burrowing or crawling animals as the barnacle, crab, lobster, and oyster, and plants such as turtle grass.

Life in the ocean depends on the sun's energy and on the nutrients available in the water. The plant plankton, or **phytoplankton,** are the most important link in the ocean food chain. They take the dissolved nutrients in the water and, through the process of photosynthesis, produce foods usable by animal plankton, or **zooplankton,** and by the smallest nekton. Phytoplankton is the only food source for these animals, who in turn form the food source for larger carnivorous fish and sea mammals. These in turn are prey for still larger animals. The *food chain* just described is part of a full food *cycle* in the ocean, for through the excretions of animals and the decomposition of both plants and animals in the ocean, chemical nutrients are returned to the water and are again made available for transformation by phytoplankton into usable foods. The phytoplankton also play a major role in the production of oxygen, a most essential gas in our atmosphere.

The uneven distribution of nutrients in the oceans and the fact that sunlight can penetrate only to a depth of about 120 meters (400 ft), depending on the clarity of the water, means that the distribution of marine organisms is also variable. Most organisms are concentrated in the upper layers of the ocean, where the most solar energy is available. In deep waters where the ocean floor lies below the level to which sunlight penetrates, the benthos depend on whatever nutrients, plants, and animals filter down to them. For this reason benthos animals are scarce in deep and dark ocean waters. They are most common in shallow waters near coasts where there is sunlight and a rich supply of nutrients, and where phytoplankton and zooplankton are abundant as well.

The waters of the continental shelf have the highest concentration of marine life. The supply of chemical nutrients is greater in waters near the continents, where nutrients are washed into the sea from the land. Marine organisms are also concentrated where waters from the bottom rise to the surface layers, where sunlight is available. Such vertical exchanges are sometimes the result of variations in salinity or density. A similar situation occurs where convection causes bottom layers of water to rise and mix with top layers, as is the case in cold polar waters and in midlatitude waters during the colder winter months. Marine life is also abundant in areas where there is a mixing of cold and warm ocean currents and in regions of upwelling.

During the 1977 dives of the manned submersible vessel *Alvin* off the East Pacific Rise, scientists for the first time observed abundant sea life on the floor of the ocean at depths of over 2500 meters (8100 ft). It was previously presumed that these cold (2°C/46°F), dark waters were a virtual biological desert.

The undersea volcanic mountain range produces vents of warm, mineral-rich waters, which nourish bacteria and large colonies of crabs, clams, mussels, and the newly discovered giant 3-meter (10-ft) red tube worms. This entirely new deep ocean ecosystem, which exists without the benefit of sunlight, has caused scientists to rethink old theories about the ocean and its chemistry. This unusual "chemosynthetic" vent community has more recently been observed at several other Pacific Ocean Ridge sites.

THE GLOBAL OCEAN AS A RESOURCE

The oceans have always been used by humans as sources of food and salt and as avenues of commerce. Ocean beaches have served as recreation areas for millions of people who live and visit the edges of the land. More recently, however, new oceanic resources have become critically important to the world's peoples. Foremost among these are the petroleum and natural gas deposits of the continental shelves.

The ocean provides us with renewable and nonrenewable resources. The ocean's *renewable resources* are those that are replenished at a rate equal to or greater than the rate of consumption, such as water, fish, or sea mammals. *Nonrenewable resources* are those that are never replenished or are replenished at a much slower rate than the consumption rate. These include oil, natural gas, and manganese. Renewable resources can become nonrenewable if used at such a rate that exhaustion or extinction of the resource occurs. This situation has occurred with some species of whales, for instance.

BIOLOGICAL RESOURCES FROM THE OCEANS

Today most countries do not depend heavily on marine food sources. Yet as world popula-

tion continues to grow, more attention will be directed toward the oceans as a source of protein. The oceans cannot, however, help to solve world food shortages, without an end to rapid population growth.

Five nations—Japan, China, Peru, the USSR, and Norway—catch over 75 percent of the world's fish tonnage. Japan and the USSR, in particular, have modern worldwide fishing fleets that fish all of the oceans. The most important fishing areas are off the coasts of northwestern Europe and the North Sea, eastern North America, western North and South America, and east Asia. Over half of the world catch, 53 percent, comes from the Pacific, 40 percent from the Atlantic, and only 5 percent from the Indian Ocean.

Because foreign fishing fleets were overfishing the areas off our coasts, the United States passed the Fishing Conservation and Management Act of 1976. This law proclaimed a 322-kilometer (200-mi) exclusive economic zone to protect our coastal fish resources. The U.S. Coast Guard enforces the law with fines, seizure, and imprisonment for violations. With careful management and cooperation, we should be able to continue to extract fish from our oceans far into the future.

The hunting of sea mammals, especially whales and seals, has been a commercial enterprise for centuries. Whales, the largest mammals on earth, have been hunted for food, ambergris (perfume), bone products, and oils for the manufacture of soap, margarine, and cosmetics. Other sea mammals were usually hunted for their fur. Modern whaling, with deadly accurate harpoons, has depleted many species. The International Whaling Commission (I.W.C.) was formed to protect whale populations by establishing whaling quotas for each species. Today whales are still hunted by a few nations, but most nations now support a total ban on all whaling. The Japanese continue to operate their whaling fleets under considerable protest from environmental organizations.

Mariculture or *aquiculture* is the farming of marine animals and plants as commercial crops. The Japanese have farmed the seas for centuries, raising oysters, shrimp, mussels, carp, and edible seaweed. Mariculture has great potential, both for raising sea animals as a direct food source and for growing sea plants used for the production of vitamins, drugs, food additives, and energy. Mariculture programs in the United States presently produce abalone, scallops, lobster, algae, and kelp (a giant seaweed) on a commercial basis.

PHYSICAL RESOURCES FROM THE OCEANS

Today petroleum is the single most eagerly sought-after offshore resource in this energy-hungry world. Petroleum probably develops from the constant rain of organic debris in the nutrient-rich waters of the continental shelves. To be preserved and ultimately changed into petroleum, this debris must accumulate in poorly ventilated (oxygen-poor) bottom waters. Thus not all areas of the continental shelves can be expected to yield oil. Nevertheless, vast deposits have been found, for example, in the North Sea; off the coasts of southern California, Texas, Louisiana, and Alaska; off the coasts of Peru, Venezuela, and Argentina; in the Mediterranean Sea and Persian Gulf; around Africa's Gulf of Guinea; in the seas of Indonesia; and off the west and south coasts of Australia. Exploration for oil and gas deposits on the continental shelves goes on at a feverish pace and has produced international controversies over the ownership of the shelf areas. It is projected that by 1990 over one third of the world's oil will come from the seabed, since the sea is the only remaining area where great expansion of production seems possible (Fig. 17.26).

Besides covering oil and natural gas, the ocean waters and their motions are a vast potential energy source. The ocean is the largest solar collector, and it also stores gravitational energy in the tides.

Tidal power is presently tapped for electrical energy in France and the USSR. The largest operating tidal power plant is at St. Malo, on the Brittany (Atlantic) coast of France. Sites for the construction of future tidal power plants are limited, since a large tidal range of 9 meters (30 ft) is necessary for efficient operation. The only areas in the United States with potential tidal power capability are the coasts of Maine and Alaska.

Other sea motions, such as waves and currents, could also possibly be tapped to produce electricity. These ideas are in an early stage of development. Presently, large-scale energy production from waves and currents is not feasible.

Temperature differences between the warm surface waters and the deep cold waters below may be used to drive electric generators. This method is known as *ocean thermal energy conversion* (OTEC). The first pilot OTEC plant is now being tested off the island of Hawaii.

As the cost of conventional energy sources rises and the supply falls, the ocean will come into more prominence as a future source of energy production.

In addition to salt, magnesium and bromine have long been extracted from seawaters. Bromine is an essential constituent of the gasoline made from petroleum. Magnesium is the most valuable element extracted from seawater in the United States. The mineral wealth of the seas even includes metals. Every metallic element is present to some degree in seawater, though the amount is very small. A cubic mile of the ocean contains about 90 million tons of chlorine and 50 million tons of sodium but only 47 tons each of zinc, iron, and aluminum, 14 tons of copper and tin, 1 ton of silver, 200 pounds of lead, and 40 pounds of gold. None of these widely dispersed elements are economically recoverable at present.

However, waters and sediments heavily saturated with such metals as zinc, copper, lead, silver, and gold have been discovered in volcanic regions of the oceanic ridge systems, especially in the Red Sea. These rich metallic brines may someday be exploited, but the technology for doing so has not yet been developed. Of more immediate significance are minerals precipitated as nodules on the sea floor.

Plate 21 Coastal Landforms of Wave Erosion

The rugged terrain of the wave-eroded and uplifted California coastline.

Wave erosion is concentrated on this volcanic headland on the island of Maui, as the waves reduce shoreline irregularities.

These rugged sea cliffs at Pt. Lobos, California, are created as waves pound directly against resisting rocks.

This sea stack off Santa Barbara Island, Channel Islands National Park, California, was formed when a resistant pillar of rock was left standing as the sea cliff retreated.

Sea arches, such as this one shown at low tide near Homer, Alaska, form as sea caves cut completely through a headland.

Marine terraces are uplifted wave-cut benches found in tectonically active regions, such as the Azores over the Mid-Atlantic Ridge.

The floor of the Oceans. Copyright © Marie Tharp.

...are the most visible evidence of wave ...n, as demonstrated at Laguna Beach, ...

All beaches are not made of sand, as evidenced by these glacial gravels and cobbles at Montauk Point on New York's Long Island.

...e of beach material depends upon its ...Coral reefs are the source for these fine-...ight-colored beaches, common to most ...islands, such as Tutuila, American

Black Sand Beach on the island of Hawaii was formed from deposits of volcanic materials.

...lo is formed when wave-deposited mate-...nect a nearby island with the mainland, ...n by this example in Baja California,

On gentle shelving coasts, wave deposition wil form barrier islands that extend for many miles parallel to the original shore. The Hamptons or

FIGURE 17.26 Drilling for offshore oil in the Gulf of Mexico. (American Petroleum Institute)

The most abundant of these are phosphorite and manganese nodules. Phosphorite nodules are not in strong demand, because there are extensive phosphate sources on land. Manganese is a strategic mineral that is in short supply. Manganese nodules, precipitated on pebbles, shark's teeth, and bones, virtually cover vast areas of the deep ocean floor. Their origin is a puzzle to oceanographers. Many schemes for their collection have been proposed, but three problems cause considerable difficulty. The first is the metallurgical separation problem presented by the unique mineral compounds present in the nodules. The second problem relates to the massive disturbance of sea-floor ecosystems that would result from the large-scale dredging or vacuuming systems proposed. The third problem concerns the question of who owns the sea floor and how rights to exploit it should be allocated.

Of course, the greatest resource the oceans have to offer is their own water—free of its salt content. As the world population grows, so does the demand for agricultural, domestic, and municipal water. In many climatically attractive coastal regions like Israel and southern California, fresh water is in short supply. It is almost nonexistent along desert coasts. But beyond the shore lies an inexhaustible supply—unfortunately contaminated by salt. Removal of the salt is the obvious answer, but *desalinization* is a very expensive procedure and the cost of desalinized water so greatly restricts its use that it is foolish to anticipate irrigating deserts with desalinized sea water. Nevertheless, hundreds of desalinization plants are in operation or under construction around the world, with steady growth projected. One of the pioneer countries in desalinizing seawater was the desert sheikdom of Kuwait. Kuwait had energy supplies to spare in the form of petroleum and was desperate for water to service its growing population based on the oil industry. Today Kuwait alone has more than 50 desalinization plants. Key West, Florida, is the first United States city to derive its municipal water supply from the sea through desalinization. The cost of desalinized water is expected to decline as technology improves, especially where desalinization plants can be linked to solar energy production.

MANAGING THE GLOBAL OCEAN RESOURCE

The ocean itself is a resource. One traditional use of the ocean has been as a depository for the waste products of civilizations. However, the current production of waste, particularly in the form of insecticides, municipal and industrial fallout, and sewage, has surpassed the level at which it can be diluted, degraded, and dispersed by natural systems. The ocean, like the other components of the environment, is showing the strain of man's use.

Oil spills have created the greatest public concern over ocean pollution. Oil spills from supertanker accidents and offshore oil well *blowouts* are the most sensational of all forms of pollution. Besides the spectacular oil spills, there is a continual oil discharge by hundreds of oil tankers cleaning their tanks after unloading their liquid cargo. Oil slicks and floating lumps of oil tar may be encountered anywhere on the seas today, not just in coastal waters (Fig. 17.27).

Toxic substances such as DDT, radioactive waste, mercury, and lead derived from gasoline are found in alarming concentrations far from the shore. Pollution of the seas has already had an impact on fishing industries. Localized studies have shown that production of marine plankton has been reduced as a consequence of small DDT and mercury concentrations. Sometimes, even a temperature increase of a few degrees (*thermal pollution*) can affect marine plant and animal communities adversely. Nevertheless, if we ruled out the use of the ocean waters as a depository for waste materials, the accumulation of municipal, industrial, and toxic waste products on the land would be both phenomenal and disastrous.

The effects of ocean pollution on the food chain, which is based on plankton, are not clear yet but can hardly be beneficial. Life, which probably began in the sea, is now being threatened at its fundamental level in the sea by our activities on the land and in the oceans themselves. The management and conservation of the resources of the oceans is one of the greatest challenges facing mankind. It becomes particularly important as the pollution dangers increase and our technological ability to exploit the ocean's resources expands.

There have been conflicts between nations for centuries over territorial and resource claims to the oceans. In 1609 the Dutch jurist

FIGURE 17.27 The sinking oil tanker *SS Argo Merchant* after she ran aground off New England in December, 1976. Over 7 million gallons of oil cargo caused a major oil spill. (Official U.S. Coast Guard Photograph)

Hugo Grotius introduced the concept of *Mare Liberum*—that the oceans are the common heritage of mankind. The modern legal background for the management of the world's oceans is still based on the fact that the high seas belong to all nations.

It was not until 1980 that the nations of the world completed a draft of *A Law Of The Sea* (LOS) treaty after years of frustrating negotiations. The 1980 Law Of The Sea involves 150 nations and contains over 400 separate articles. The major issues addressed are: (1) establishment of a 19-kilometer (12-mi) territorial sea for all coastal states with rights of innocent passage for ships of other nations; (2) recognition of a 322-kilometer (200-mi) exclusive economic zone, which gives the coastal state control over its offshore oil and fishing resources; (3) provision of strong international pollution controls and penalties; and (4) acceptance of a common heritage seabed beyond all national jurisdictions. This allows both private and United Nations sponsored mining interests to exploit deep ocean resources such as man-

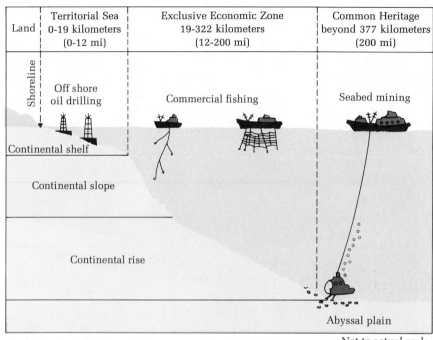

Land	Territorial Sea 0-19 kilometers (0-12 mi)	Exclusive Economic Zone 19-322 kilometers (12-200 mi)	Common Heritage beyond 377 kilometers (200 mi)

Shoreline

Off shore oil drilling

Commercial fishing

Seabed mining

Continental shelf

Continental slope

Continental rise

Abyssal plain

Not to actual scale.

FIGURE 17.28 Major oceanic areas involved in the Law of the Sea (LOS) treaty.

FIGURE 17.29 The oceans are the common heritage of mankind, and the Law of the Sea (LOS) recognizes the fact that the high seas belong to all nations. (R. Sager)

ganese nodules with revenues paid to the poor developing nations of the world (Fig. 17.28). Hopefully all the nations of the world, including the United States, will sign the treaty in the near future. This will establish, by treaty, a rule of law for an area that covers three-quarters of the globe.

A common Law Of The Sea is the only proper approach to protecting our oceans (Fig. 17.29). If all of humankind is to benefit from the resources of the oceans it is also the respon-sibility of all humans to protect the oceans. Surely, history has taught us the aesthetic value of perpetuating the waters that surround our continents. Our fascination with the oceans can be seen in the painting, poems, and music created by artists. The oceans have found their way into the human soul. The restlessness of the oceans, their ever-changing changelessness, their power and vastness all speak to the human spirit and imagination.

QUESTIONS FOR DISCUSSION AND REVIEW

1. List the major chemicals and minerals present in the oceans. Which ones are economically recoverable for human use?

2. Why and how do salinity and temperature affect the density of water? Where is ocean water least saline? Where is it least dense?

3. Explain the general pattern and effect of the major oceanic currents.

4. Describe the major factors that produce tides. What are some of the variations in tidal patterns?

5. What factors can cause ocean waves? Which is most important? What determines wave height?

6. Describe the general effect that waves have on a shoreline. What part does wave refraction play in this process?

7. Explain the relationship between the sorting of marine sediment and the development of beaches. Why do beaches change seasonally?

8. What are the ways by which man can alter a coastline, and with what effects?

9. Why do baymouth bars develop? What effect do they have on bays?

10. Compare the Atlantic Coast with the Pacific Coast of the United States. How do they differ, and why?

11. Describe the important topographic features of the ocean floor. Why have these features remained relatively unchanged since they were first formed?

12. How are features of the continental shelf related to adjacent landmasses? What are some of the major topographic features of the continental shelf?

13. Describe the development of coral reefs and atolls. Why are coral islands of limited value for human habitation?

14. Describe the food chain of the oceans. Which link in the chain is most important?

15. What are the major mineral resources of the oceans? Discuss the problems associated with the recovery of these resources.

16. What are some of the major types of pollution affecting the oceans? What are some of the effects produced by this pollution?

GLOSSARY

ablation loss of glacial ice and snow cover by means of melting, evaporation, and sublimation.

abrasion (corrasion) erosion of a stream by the grinding and rolling of rock particles and boulders carried by a stream, or by wave action, wind, or glacial ice.

absolute humidity mass of water vapor present per unit volume of air, expressed as grams per cubic meter, or grains per cubic foot.

abyssal plains deep low-relief ocean floor, generally covered with a blanket of marine sediments.

acid rain rain with a pH value of less than 5.6, the pH of natural rain; often linked to the pollution associated with the burning of fossil fuels.

adiabatic heating and cooling change of temperature *within* a gas because of compression (resulting in heating) or expansion (resulting in cooling); no heat is added or subtracted from outside.

advection horizontal heat transfer within the atmosphere; air masses moved horizontally, usually by wind.

advection fog fog produced by the movement of warm moist air across a cold sea or land surface.

aggradation process of leveling the land whereby low places or depressions are filled in.

air mass large portion of the atmosphere, sometimes subcontinental in size, that may move over the earth's surface as a distinct, relatively homogeneous entity.

air mass analysis explanation of weather phenomena by a study of the actions and interactions of major portions of the atmosphere.

albedo proportion of solar radiation reflected back from a surface, expressed as a percentage of radiation received on that surface.

Aleutian low center of low atmospheric pressure in the area of the Aleutian Islands, especially persistent in winter.

alluvial fan fan-shaped aggradational feature where a stream emerges from a mountain channel onto a flat plain and deposits material, generally found in arid regions.

alluvium fragmented earth materials deposited by a river or stream.

alpine glacier moving glacial ice accumulated in high sheltered mountain valleys; also called *valley glacier*.

altitude heights of points above the earth's surface.

angle of inclination tilt of the earth's polar axis at an angle of 23½° from the vertical to the plane of the ecliptic.

angle of rest angle of slope created by a bed of loose sand, gravel, or rock.

annual range of temperature difference between the mean daily temperatures for the warmest and coolest months of the year.

Antarctic circle parallel of latitude at 66½°S; the northern limit of the zone in the Southern Hemisphere that experiences a 24-hour period of sunlight and a 24-hour period of darkness at least once a year.

anticline arch, crest, or upfold in a wave of crustal folding.

anticyclone an area of high atmospheric pressure, also known as a *high*.

aphelion position of the earth's orbit at farthest distance from the sun during each earth revolution.

aquiclude rock layer that restricts flow and storage of groundwater; it is impermeable and nonporous.

aquifer rock layer that is a container and transmitter of groundwater; it is both porous and permeable.

Arctic circle parallel of latitude at 66½°N; the southern limit of the zone in the Northern Hemisphere that experiences a 24-hour period of sunlight and a 24-hour period of darkness at least once a year.

arête jagged, sawtooth spine or wall of rock separating two expanding cirque basins.

artesian well groundwater that flows to the surface under its own pressure.

asthenosphere thick, plastic-like layer within the earth mantle that theoretically flows in response to convection and instigates the surface movement of tectonic plates.

atmosphere blanket of air, composed of various gases, that envelops the earth.

atmospheric air pressure (barometric pressure) force per unit area that the atmosphere exerts on any surface at a particular elevation.

atmospheric controls geographic features that affect climate and weather patterns; e.g., distance from the ocean, wind direction, altitude.

atmospheric disturbance refers to variation in the secondary circulation of the atmosphere that cannot correctly be classified as a storm; e.g., front, air mass.

atmospheric elements components of weather; e.g., temperature, precipitation, pressure, wind.

atoll ring of coral reefs and islands encircling a lagoon, with no inner island.

autotroph organism which, because it is capable of photosynthesis, is at the foundation of a food web and is considered a basic producer.

axis an imaginary line between the geographic North Pole and South Pole, around which the planet rotates.

azonal soils major soil order of the Russian-American System; soils in this order, such as recent alluvium, show little horizon development because of their immaturity.

Azores high see Bermuda high.

backing wind shift change in wind direction counterclockwise around the compass; e.g., from east to northeast, to north, to northwest.

badlands region of rugged, barren topography with sharp ridges and ravines; caused by gully erosion of soft materials.

bajada continuous series of alluvial fans forming a gently sloping, low-relief area along the base of a mountain range.

barchan crescent-shaped sand dune with tips that point downwind.

barometer instrument for measuring atmospheric pressure.

barrier island long, narrow, wave-built island separated from the mainland by a lagoon, formed on low-relief coastlines.

base level elevation below which a river or stream cannot erode; although sea level is the ultimate base level, basins or lakes may be local base levels.

batholith largest of the deep-seated igneous masses generally known as plutons.

beach coastal region of unconsolidated sediments between the low tide line and the upper limit of wave action.

bedrock solid rock layers of the earth's crust that underlie soil and other unconsolidated earth materials.

benthos the ocean bottom and the plants and animals that live on the sea floor.

bergschrund large crevasse or crack around the head of a valley glacier, formed by the downslope movement of ice away from the valley rock wall.

Bermuda high persistent, high atmospheric pressure center located in the subtropics of the north Atlantic Ocean.

biomass amount of living material or standing crop in an ecosystem or at a particular trophic level within an ecosystem.

biome one of earth's major terrestrial ecosystems, classified by the vegetation types that dominate the plant communities within the ecosystem.

biosphere the life-forms, human, animal, or plant, of the earth that form one of the major earth subsystems.

bolson desert basin, surrounded by mountains, with no drainage outlet.

bora cold downslope wind in Yugoslavia (see katabatic wind).

boreal forest coniferous forest dominated by spruce, fir, and pine found growing in subarctic conditions around the world north of the 50th parallel of latitude.

braided stream stream channel with multiple subchannels that form a braided pattern flowing through alluvial deposits.

butte isolated erosional remnant of a tableland with a flat summit, often bordered by steep-sided escarpments. Buttes are usually found in arid regions of flat-lying sediments and are smaller than mesas.

calcification soil-forming process of subhumid and semiarid climates. Soil types in the pedocal category, the typical end products of the process, are characterized by little leaching or eluviation and by the accumulation of both humus and mineral bases (especially calcium carbonate, $CaCO_3$).

caldera collapsed summit area of a stratovolcano,

thought to be caused by the expulsion or withdrawal of supporting magma.

calíche   hardened layers of lime ($CaCO_3$) deposited at the surface of a soil by evaporating capillary water.

calving   the formation of icebergs by a mass of ice breaking away from the snout of a glacier at its junction with the sea or a lake.

Campos   tropical savanna region of the southern interior of Brazil, characterized by mixed grass and thornbush vegetation.

Canadian high   high atmospheric pressure area that tends to develop over the central North American continent in winter.

capillary water   soil water that clings to soil peds and individual soil particles as a result of surface tension. Capillary water moves in all directions through the soil from areas of surplus water to areas of deficit.

carnivore   animal that eats only other animals.

cartography   the science of map-making.

catastrophism   once-popular theory that all the earth's landforms developed in a relatively short time in a catastrophic fashion.

Celsius (or centigrade) scale   temperature scale in which 0° is the freezing point of water and 100° its boiling point at standard sea level pressure.

centrifugal force   force that pulls a rotating object away from the center of rotation.

chaparral   sclerophyllous woodland vegetation found growing in the Mediterranean climate of the western United States; these seasonal, drought-resistant plants are low-growing, with small, hard-surfaced leaves and deep, water-probing roots.

chinook   dry warm wind on the eastern slopes of the Rocky Mountains (see foehn wind).

cinder cone   volcano formed primarily from the expulsion of cinders, ash, and other solid rock fragments from a central vent or system of vents.

circle of illumination   line dividing the sunlit (day) hemisphere from the shaded (night) hemisphere; experienced by individuals on the earth's surface as sunrise and/or sunset.

cirque   deep, sometimes steep-sided amphitheater formed at the head of an alpine valley by glacial ice erosion.

cirque glacier   glacial ice limited to a cirque basin and not entering the alpine valley itself.

cirrus   high, detached clouds consisting of ice particles. Cirrus clouds are white and feathery or fibrous in appearance.

classification   process of systematically arranging phenomena into groups, classes, or categories based on some established criteria.

clastic rock   sedimentary rock formed by the compaction and cementation of clay, silt, sand, or other rock debris.

climate   accumulated and averaged weather patterns of a locality or region; the full description is based upon long-term statistics and includes extremes or deviations from the norm.

climatology   scientific study of climates of the earth and their distribution.

climax vegetation   final succession of natural vegetation within a plant community that has reached equilibrium or balance with the surrounding environment.

climograph   graphic means of giving information on mean monthly temperature and rainfall for a select location or station.

closed system   system in which no substantial amount of energy and/or materials can cross its boundaries.

cloud   mass of suspended water droplets (or at high altitudes, ice particles) in air above ground level.

cold front   leading edge of a relatively cooler, denser air mass that advances upon a warmer, less dense air mass.

Comprehensive Soil Classification System (CSCS)   also known as the Seventh Approximation System; a system of soil classification developed by the Soil Conservation Service of the USDA, it is based on the composition and specific features, such as distinctive horizons, that characterize various soils. (Descriptions of the major soil orders of the CSCS may be found in Chapter 10.)

condensation   process by which a vapor is converted to a liquid during which energy is released in the form of latent heat.

condensation nuclei   minute particles in the atmosphere (e.g., dust, smoke, pollen, sea salt) on which condensation can take place.

conduction   transfer of heat within a body or between adjacent matter by means of internal molecular movement.

connate water   groundwater trapped in the pore spaces of sedimentary rock at the time it was first deposited; water locked out of the hydrologic cycle in sedimentary rocks.

Continental Divide   line of separation dividing runoff between the Pacific and Atlantic Oceans. In North America it generally follows the crest of the Rocky Mountains.

continental ice sheet   thick ice mass that covers a major portion of a continent and buries all but the highest mountain peaks; it usually flows from one or more areas of accumulation outward in all directions.

continental islands islands that are geologically part of a continent and are usually located on the continental shelf.

continental rise gently sloping depositional surface at the base of the continental slope.

continental shelf gently sloping submarine surface extending from the coast to the steep continental slope.

continental slope steeply sloping submarine surface that is seaward of the continental shelf.

contour map (topographic map) map that uses contour lines (lines of equal elevation to show differences in elevation (topography).

convection process by which a circulation is produced within an air mass or fluid body (heated material rises, cooled material sinks); also, in tectonic plate theory, the method whereby heat is transferred to the earth's surface from deep within the mantle.

convectional precipitation precipitation resulting from condensation of water vapor in an air mass that is rising convectionally as it is heated.

convergent wind circulation pressure-and-wind system where the airflow is inward toward the center, where pressure is lowest.

coral reef ridge of limestone built up by the accumulation of skeletal remains of tiny sea animals.

core extremely hot and dense, innermost portion of the earth's interior; the molten outer core is 2400 km (1500 mi) km thick; the solid inner core is 1120 km (700 mi) thick.

Coriolis effect effect of the earth's rotation on horizontally moving bodies, such as wind and ocean currents; such bodies tend to be deflected to the right in the Northern Hemisphere and to the left in the Southern Hemisphere.

coulee snaking, steep-sided channel cut through lava formations by glacial meltwater.

creep slow downslope movement of soil and regolith caused by the pull of gravity, also refers to slow fault zone displacement.

crevasse stress crack commonly found along the margins and at the snout or terminus of a glacier.

crust relatively shallow, approximately 8–64 km (5–40 mi) thick, low-density surface layer of the earth.

cumulus globular clouds, usually with a horizontal base and strong vertical development.

cyclone center of low atmospheric pressure, also known as a *low*.

cyclonic precipitation see frontal precipitation.

debris flow channelled movement of earth material mixed with water usually following drainage; a *mudflow* is a rapidly moving debris flow with the consistency of mud.

decomposer organism that promotes decay by feeding on dead plant and animal material and returns mineral nutrients to the soil or water in a form that plants can utilize.

deflation surface erosion and removal of fine earth materials by the wind.

degradation process of leveling of land whereby raised portions are worn down and removed.

delta depositional landform where a river flows into a still body of water, such as a sea or lake.

dendritic term used to describe a drainage pattern that is treelike with tributaries joining the main stream at acute angles.

deposition accumulation of earth materials at a new site after they have been dropped by the transporting agents: water, wind, or glacial ice.

desert pavement desert surface accumulation of pebbles and stones, the finer materials having been removed by wind and/or water erosion.

detritivore animal that feeds on dead plant and animal material.

dew tiny droplets of water on ground surfaces, glass blades, or solid objects. Dew is formed by condensation when air at the surface reaches the dew point.

dew point the temperature at which an air mass becomes saturated; any further cooling will cause condensation of water vapor in the air.

diastrophism distortion of the solid earth crust by bending, folding, warping, or fracturing (faulting).

differential weathering (and erosion) process whereby different types of bedrock weather (and erode) at varying speeds due to differing resistance to the weathering (and erosion) processes; such differing resistance often produces distinctive landform features.

dike igneous intrusion that forms a vertical rock mass after molten material has been forced through a crustal fracture and cools at right angles to flat-lying rock layers.

dip the angle that a stratum of rocks or fault makes with the horizontal plane.

discharge (stream discharge) rate of stream flow; measured as the volume of water flowing past a cross section of a stream per unit of time (cubic meters or cubic feet per second).

distributary branching stream that flows away from the main stream, common on deltas; the opposite of a tributary.

diurnal (daily) range of temperature difference between the highest and lowest temperatures of the day (usually recorded hourly).

divergent wind circulation pressure-and-wind system where the airflow is outward away from the center, where pressure is highest.

divide line of separation between drainage basins; generally follows high ground or ridge lines.

doldrums zone of low pressure and calms along the equator.

drainage basin (watershed) total land surface area drained by a stream system.

drainage wind see katabatic wind.

drift all material deposited by glacier; includes both unsorted and unstratified material and sorted debris deposited by meltwater.

drizzle fine mist or haze of very small water droplets with a barely perceptible falling motion.

drumlin streamlined, elongated hill composed of glacial drift. Drumlins are usually found in swarms, with as many as 100 or more clustered together; their elongated shapes indicate the direction of ice flow.

dry adiabatic rate rate at which a rising mass of air is cooled by expansion when no condensation is occurring (10°C/1000 m or 5.6°F/1000 ft).

dune (sand dune) mound of sand and other coarse materials deposited and shaped by the wind.

dynamic equilibrium constantly changing relationship among the variables of a system, which produces a balance between the amount of energy and/or materials that enter a system and the amount that leave.

earthflow linear movement downslope of moist, clay-rich soil and regolith, usually exhibiting a tongue-like shape.

earthquake series of vibrations or shock waves set in motion by the sudden movement of crustal blocks along a fault.

earth system set of interrelated components or variables (e.g., atmosphere, lithosphere, biosphere, hydrosphere), which interact and function together to make up the earth as it is presently constituted.

easterly wave trough-shaped, weak, low-pressure cell that progresses slowly from east to west in the trade-wind belt of the tropics; this type of distur-

bance sometimes develops into a tropical hurricane.

ecological niche combination of role and habitat as represented by a particular species in an ecosystem.

ecology science that studies the interactions between organisms and their environment.

ecosystem community of organisms functioning together in an interdependent relationship with the environment which they occupy.

effective precipitation actual precipitation available to supply plants and soil with usable moisture; does not take into consideration storm runoff or evaporation.

elevation vertical distance from mean sea level to a point or object on the earth's surface.

ellipsoid of rotation a rotating, near-sphere with an elliptical (oval-shaped), rather than pure circular, plane or cross section.

eluviation removal by gravitational water of fine soil components from the surface layer (A horizon) of the soil.

empirical classification classification process based on statistical, physical, or observable characteristics of phenomena; it ignores the causes or theory behind their occurrence.

end moraine accumulation of rocks and fine glacial material at the terminus or snout of a glacier.

environment surroundings, whether of man or of any other living organism; includes physical, social, and cultural conditions that affect the development of that organism.

eolian referring to the work of wind; associated with wind erosion, transportation, and deposition.

epicenter point on the earth's surface directly above the focus of an earthquake.

epipedon surface soil layer that possesses specific characteristics essential to the identification of soils in the Comprehensive Soil Classification System. (Examples of epipedons may be found in Table 10.2.)

equator great circle of the earth midway between the poles; the zero degree parallel of latitude that divides the earth into the Northern and Southern Hemispheres.

equatorial low zone of low atmospheric pressure centered more or less over the equator where heated air is rising. (See also doldrums.)

equilibrium state of balance between the interconnected components of an organized whole.

equinox one of two times each year (approximately March 21 and September 23) when the position of the noon sun is overhead (and its vertical rays strike) at the equator; all over the earth day and night are of equal length.

erg desert region of active sand dunes, most common in the Sahara.

erosion pick-up and removal of earth materials by water, wind, or glacial ice.

erratic large rock or boulder transported and deposited by a glacier above bedrock of different composition.

esker narrow, winding ridge composed of glacio-fluvial gravels; believed to have been formed by streams of meltwater flowing in tunnels of a stagnant ice sheet, or on a melting glacial surface.

estuary coastal waters where salt and fresh water mix.

evaporation process by which a liquid is converted to the gaseous (or vapor) state by the addition of latent heat.

evaporite mineral such as common salt or gypsum that is soluble in groundwater and accumulates when water evaporates in arid climates.

evapotranspiration combined water loss to the atmosphere from ground and water surfaces by evaporation and, from plants, by transpiration.

exfoliation progressive breaking off of concentric slabs or sheets from the exposed portions of massive rocks due to weathering.

exotic stream (or river) stream that originates in a humid region and has sufficient water volume to flow across a desert region.

extratropical disturbance convergence of cold polar and warm subtropical air masses over the middle latitudes.

Fahrenheit scale temperature scale in which 32° is the freezing point of water, and 212° its boiling point, at standard sea level pressure.

faulting movement of adjacent crustal blocks along joints, or fracture planes, in bedrock.

fault scarp the steep cliff or exposed face of a fault where one crustal block has been displaced vertically relative to another.

feedback sequence of changes in the elements of a system, which ultimately affects the element that was initially altered to begin the sequence.

fetch distance over open water that winds blow without interruption.

fiord deep, glacial trough along the coast invaded by the sea after the removal of the glacier.

firn compact granular snow formed by partial melting and refreezing due to overlying layers of snow.

firn line boundary between the zones of ablation and accumulation on a glacier, representing the equilibrium point between net snowfall and ablation.

fluvial term used to describe landform processes associated with the work of streams and rivers.

focus point within the earth crust where an earthquake occurs.

foehn wind warm, dry, downslope wind on lee of mountain range, caused by adiabatic heating of descending air.

fog mass of suspended water droplets within the atmosphere that is in contact with the ground.

foliation process whereby metamorphic rocks tend to develop parallel banding or platy structures during formation.

front sloping boundary or contact surface between air masses with different properties of temperature, moisture content, density, and atmospheric pressure.

frontal lifting lifting or rising of warmer, lighter air above cooler, denser air along a frontal boundary.

frontal precipitation precipitation resulting from condensation of water vapor in an air mass that is rising over another mass along a front.

frost frozen condensation that occurs when air at ground level is cooled to a dew point of 0°C (32°F) or below; also any temperature near or below freezing that threatens sensitive plants.

frost wedging (frost shattering) breaking apart of bedrock by the expansive power of water freezing, melting, and refreezing in joints, cracks, and crevices.

galactic movement movement of the solar system within the Milky Way galaxy.

galeria forest jungle-like vegetation extending along and over streams in tropical forest regions.

genetic classification classification process based on the causes, theory, or origins of phenomena; generally ignoring their statistical, physical, or observable characteristics.

geography study of earth phenomena; includes an analysis of distributional patterns and interrelationships among these phenomena.

geomorphic processes various movements, changes, and interactions that have taken place within the earth's crust to produce landforms.

geostrophic winds upper-level winds in which the Coriolis effect and pressure gradient are balanced, resulting in a wind flowing parallel to the isobars.

glaciofluvial deposit sorted glacial drift deposited by meltwater.

glaciolacustrine deposit sorted glacial drift deposited by meltwater in lakes associated with the margins of glaciers.

glaze translucent coating of ice that develops when rain strikes a freezing surface.

gleization soil-forming process of poorly drained areas in cold, wet climates. The resulting soils have a heavy surface layer of humus with a water-saturated clay horizon directly beneath.

graben depressed landform or crustal trough that develops when the crust between two parallel faults is lowered relative to blocks on either side.

gradational processes processes that derive their energy indirectly from the sun and directly from earth gravitation and serve to wear down, fill in, and level off the earth's surface.

graded stream stream where slope and channel size provide velocity just sufficient to transport the load supplied by the drainage basin; a theoretical balanced state averaged over a period of many years.

gravitational water meteoric water that passes through the soil under the influence of gravitation.

great circle any circle formed by a full circumference of the globe; the plane of a great circle passes through the center of the globe.

greenhouse effect warming of the atmosphere that occurs because short-wave solar radiation heats the planet's surface, but the loss of long-wave heat radiation is hindered by atmospheric elements such as carbon dioxide.

Greenwich mean time (G.M.T.) time at zero degrees longitude used as the base time for the earth's 24 time zones; also called Universal Time or Zulu Time.

ground-inversion fog see radiation fog.

ground moraine glacial till deposited on the earth's surface beneath a melting glacier.

groundwater (underground water) all subsurface water, especially in the phreatic zone (zone of saturation).

guyot flat-topped seamount thought to be formed by the slow subsidence of a volcanic island.

habitat location within an ecosystem occupied by a particular organism.

hail form of precipitation consisting of pellets or balls of ice with a concentric layered structure usually associated with the strong convection of cumulonimbus clouds.

hanging valley tributary trough that enters a main glaciated valley at a level high above the valley floor.

hardpan dense, compacted, clay-rich layer occasionally found in the subsoil (B horizon) that is an end product of excessive illuviation.

heat energy budget relationship between solar energy input, storage, and output within the earth system.

heat island mass of warmer air overlying urban areas.

herbivore animal that eats only living plant material.

heterotroph organism that is incapable of producing its own food and that must survive by consuming other organisms.

high see anticyclone.

horizon the visual boundary between earth and sky; see also soil horizon.

horn pyramid-like peak created where three or more expanding cirques meet at a mountain summit.

horst raised landform that develops when the crust between two parallel faults is uplifted relative to blocks on either side.

humidity amount of water vapor in an air mass at a given time.

humus organic matter found in the surface soil layers that is in various stages of decomposition as a result of bacterial action.

hurricane severe tropical cyclone of great size with nearly concentric isobars. Its torrential rains and high-velocity winds create unusually high seas and extensive coastal flooding; also called willy-willies, tropical cyclones, baguios, and typhoons.

hydration attachment of water molecules to molecules of other elements or compounds without chemical change.

hydrologic cycle circulation of water within the earth system, from evaporation to condensation, precipitation, runoff, storage, and re-evaporation back into the atmosphere.

hydrolysis union of water with other substances involving chemical change and the formation of new compounds.

hydrosphere major earth subsystem consisting of the waters of the earth, including oceans, ice, fresh-water bodies, groundwater, and water within the atmosphere and biomass.

ice age period of earth history when much of the earth's surface was covered with massive continental glaciers. The most recent ice age is referred to as the Pleistocene Epoch.

iceberg free-floating mass of glacier broken off by tidal and wave action.

ice cap small ice sheet found in highland areas that usually covers all but the highest mountain peaks.

ice fall portion of a glacier moving over and down a steep slope, creating a rigid white cascade, crisscrossed with deep crevasses.

Icelandic Low center of low atmospheric pressure located in the north Atlantic, especially persistent in winter.

ice-marginal lake temporary lake formed by the disruption of meltwater drainage by deposition along a glacial margin, usually in the area of an end moraine.

ice sheet mass of glacial ice thousands of feet thick that is of continental proportions and covers all but the highest points of land. The sheet usually flows from one or more areas of accumulation outward in all directions.

ice shelf large flat-topped plate of ice from the Antarctic ice cap, which overlies Antarctic waters and is a source of icebergs.

igneous rock one of the three major rock types; formed from the cooling and solidification of molten earth material.

illuviation deposition of fine soil components in the subsoil (B horizon) by gravitational water.

inselberg remnant residual hill rising above an arid or semiarid plain; produced by stream erosion of a former mountainous area.

insolation incoming solar radiation, i.e., energy received from the sun.

instability condition of air when it is warmer than the surrounding atmosphere and is buoyant with a tendency to rise; the lapse rate of the surrounding atmosphere is greater than that of *unstable* air.

interglacial warmer period between glacial advances, during which continental ice sheets and many valley glaciers retreat and disappear or are greatly reduced in size.

intermittent stream stream that flows part of the time, usually only during, and shortly after, a rainy period.

International Date Line line roughly along the 180 degree meridian, where each day begins and ends; it is always a day later west of the line than east of the line.

Intertropical Convergence Zone (ITC) zone of low pressure and calms along the equator, where air carried by the trade winds from both sides of the equator converges and is forced to rise.

intrazonal soils major soil order of the Russian-American classification system; soils in this order have well-developed profiles that exhibit the strong influence of local environmental factors such as drainage, slope, or parent material.

inversion see temperature inversion.

isarithm line on a map that connects all points of the same numerical value, such as isotherms, isobars, and isobaths.

island arc curved row of volcanic islands along a deep oceanic trench; found near tectonic plate boundaries where subduction occurs.

isobar line drawn on a map to connect all points with the same atmospheric pressure.

isostasy theory which holds that the earth crust *floats* in hydrostatic equilibrium in the denser plastic layer of the mantle.

isotherm line drawn on a map to connect all points with the same temperature.

jet stream high velocity upper air current with speeds of 120–640 kph (75–250 mph).

joints cracks or systems of cracks revealing lines of weakness in bedrock.

juvenile water groundwater derived from molten rock and not previously a part of the hydrologic cycle in the form of atmospheric or surface water.

kame conical hill composed of sorted glaciofluvial deposits; presumed to have formed in contact with glacial ice when sediment accumulated in ice pits, crevasses, and among jumbles of detached ice blocks.

kame terraces landform resulting from accumulation of glaciofluvial sand and gravel along the margin of a glacier occupying a valley in an area of hilly relief.

karst topography landforms developed as a result of the dissolving of limestone bedrock by groundwater.

katabatic wind downslope flow of cold, dense air that has accumulated in a high mountain valley or over an elevated plateau or ice cap.

kettle hole water-filled pit formed by the melting of a remnant ice block left buried in drift after the retreat of a glacier.

laccolith massive igneous intrusion that bows overlying rock layers upwards in a domal fashion as it forces its way toward the surface.

land breeze air flow at night from the land toward the sea, caused by the movement of air from a zone of higher pressure associated with cooler nighttime temperatures over the land.

landslide mass of earth material, including all loose debris and often portions of bedrock, moving as a unit rapidly downslope.

lapse rate see normal lapse rate.

latent heat of condensation energy release in the form of heat, as water is converted from the gaseous (vapor) to the liquid state.

lateral moraine moraine deposited along the side margin of an alpine glacier or lobe of a continental ice sheet.

laterite iron, aluminum, and manganese rich layer in the subsoil (*B* horizon) that can be an end product of laterization in the wet-dry tropics (tropical savanna climate).

laterization soil-forming process of hot, wet climates. Latosols, the typical end product of the process, are characterized by the presence of little or no humus, the removal of soluble and most fine soil components, and the heavy accumulation of iron and aluminum compounds.

latitude angular distance (measured in degrees) north or south of the equator.

lava molten earth material expelled at the surface from volcanoes or earth fissures. From this material extrusive igneous rock is formed.

leaching removal by gravitational water of soluble inorganic soil components from the surface layers of the soil.

leeward located on the side facing away from the wind.

levee natural raised alluvial bank along margins of a river on a floodplain; artificial levees may be constructed along river banks for flood control.

liana woody vine found in tropical forests that roots in the forest floor but uses trees for support as it grows upward toward available sunshine.

life support system interacting and interdependent units (e.g., oxygen cycle, nitrogen cycle) that together provide an environment within which life can exist.

lightning visible electrical discharge produced within a thunderstorm.

lithification the combined processes of compaction and cementation that transform clastic sediments into sedimentary rocks.

lithosphere solid crust of planet earth that forms one of the major earth subsystems. In a more technical definition related to tectonic plate theory, the lithosphere consists of the earth crust and the uppermost rigid zone of the mantle, which is divided into individual plates that move independently on the plastic material of the asthenosphere.

Llanos region of characteristic tropical savanna vegetation in Venezuela, located primarily in the plains of the Orinoco River.

loam soil soil with a texture in which none of the three soil grades (sand, silt, or clay) predominate over the others.

loess wind-deposited silt; usually transported in dust storms and derived from arid or glaciated regions.

longitude angular distance (distance measured in degrees) east and west of the prime meridian.

longshore current current flowing parallel to the shore within the surf zone, produced by waves breaking at an angle to the shore.

long-wave radiation electromagnetic radiation emitted by the earth in the form of waves more than 4.0 microns in amplitude, which includes heat re-radiated by the earth's surface.

low see cyclone.

magma melt or molten earth material, situated beneath the earth's surface, from which plutonic and intrusive igneous rock is formed.

mantle moderately dense, relatively thick (2880 km/1800 mi) middle layer of the earth's interior that separates the earth crust from the earth core.

maquis sclerophyllous woodland and plant community, similar to North American chaparral; can be found growing throughout the Mediterranean region.

mariculture (aquiculture) sea farming, or the cultivation of marine plants and animals for commercial purposes.

mass wasting (mass movement) collective movement of surface materials downslope as a result of earth gravitation.

medial moraine central moraine in a large valley glacier; formed when two smaller valley glaciers come together to form the larger glacier and their interior lateral moraines merge.

Mercator projection mathematically produced, conformal map projection showing true compass bearings as straight lines.

mercury barometer instrument measuring atmospheric pressure by balancing it against a column of mercury.

meridian one half of a great circle on the globe

connecting all points of equal longitude; all meridians connect the North and South Poles.

mesa flat-topped erosional remnant of a tableland characteristic of arid regions with flat-lying sediments; typically bordered by steep-sided escarpments and may cover large areas.

mesopause upper limit of mesosphere, separating it from the thermosphere.

mesosphere layer of atmosphere above the stratosphere; characterized by temperatures that decrease regularly with altitude.

metamorphic rock one of the three major rock types; formed from other rock within the crust by change induced by heat and pressure.

meteoric water groundwater derived from atmospheric sources.

meteorology study of the patterns and causes associated with short-term changes in the elements of the atmosphere.

microclimate climate associated with a small area at or near the earth's surface; the area may range from a few inches to several miles in size.

millibar unit of measurement for atmospheric pressure; one millibar equals a force of 1000 dynes per square centimeter; 1013.2 millibars is standard sea level pressure.

mineral homogeneous naturally occurring inorganic substance that possesses fairly definite physical characteristics and nearly constant chemical composition.

mistral cold downslope wind in southern France (see katabatic wind).

Mohorovicic discontinuity (Moho) interface between the earth's crust and the more dense mantle.

monsoon seasonal wind that reverses direction during the year in response to a reversal of pressure over a large land mass. The classic monsoons of Southeast Asia blow onshore in response to low pressure over Eurasia in summer and offshore in response to high pressure in winter.

moraine unsorted glacial drift deposited beneath and along the margins of a glacier.

mountain breeze air flow downslope from mountains toward valleys during the night.

muskeg poorly drained vegetation-rich marshes or swamps usually overlying permafrost areas of polar climatic regions.

natural resource any element, material, or organism existing in nature that may be useful to humans.

natural vegetation vegetation that has been allowed to develop naturally without obvious interference from or modification by man.

nimbus term used in cloud description to indicate precipitation; thus cumulonimbus is a cumulus cloud from which rain is falling.

normal lapse rate decrease in temperature with altitude under normal atmospheric conditions; approximately 6.6°C/1000 m (3.6°F/1000 ft).

northeast trades see trade winds.

occluded front boundary between a rapidly advancing cold air mass and an uplifted warm air mass cut off from the earth's surface; denotes the last stage of a midlatitude cyclone.

ocean current horizontal movement of ocean water, usually in response to major patterns of atmospheric circulation.

oceanic islands volcanic islands that rise from the deep ocean floor.

oceanic ridge (midocean ridge) linear seismic mountain range that interconnects through all the major oceans; it is where new molten crustal material rises through the oceanic crust.

oceanic trench (trench) long narrow depression on the sea floor usually associated with an island arc. Trenches mark the deepest portions of the oceans and are associated with subduction of oceanic crust.

omnivore animal that can feed on both plants and other animals.

open system system in which energy and/or materials can freely cross its boundaries.

orogeny period in earth history associated with major tectonic or mountain-building activity.

orographic precipitation precipitation resulting from condensation of water vapor in an air mass that is forced to rise over a mountain range or other raised landform.

outcrop bedrock exposed at the earth's surface with no overlying mantle of regolith or soil.

outwash glacial drift deposited beyond an end moraine by glacial meltwater.

outwash plain extensive, relatively smooth plain covered with sorted deposits carried forward by the meltwater from an ice sheet.

oxbow lake crescent-shaped lake or pond formed on a river floodplain in an abandoned meander channel.

oxidation chemical union of oxygen with other elements to form new chemical compounds.

ozone gas with a molecule consisting of three atoms

of oxygen, (O_3); forms a layer in the upper atmosphere that serves to screen out ultraviolet radiation harmful at the earth's surface.

Pacific high   persistent cell of high atmospheric pressure located in the subtropics of the North Pacific Ocean.

parallel   circle on the globe connecting all points of equal latitude.

parallelism   tendency of the earth's polar axis to remain parallel to itself at all positions in its orbit around the sun.

parent material   residual (derived from bedrock directly beneath) or transported (by water, wind, or ice) mineral matter from which soil is formed.

paternoster lakes   chain of lakes connected by a postglacial stream occupying the trough of a glaciated mountain valley.

ped   soil aggregate or mass of individual mineral particles with a distinctive shape that characterizes a soil's structure.

pediment   gently sloping bedrock surface, usually covered with fluvial gravels, located at the base of a stream-eroded mountain range in an arid region.

pediplain   desert plain of pediments and alluvial fans; the presumed final erosion stage in an arid region.

peneplain   theoretical plain of extreme old age; the last stage in the cycle of erosion, reached when a landmass has been reduced to near base level by stream erosion in a humid region.

perihelion   position of earth at closest distance to sun during each earth revolution.

permafrost   permanently frozen layer of subsoil and underlying rock found in midlatitude subarctic and polar climates where the season is too short for summer thaw to penetrate more than a few feet below ground level.

permeability   characteristic of soil or bedrock that determines the ease with which water moves through the earth material.

pH scale   scale from 0 to 14 that describes the acidity or alkalinity of a substance and which is based on a measurement of hydrogen ions; pH values below 7 indicate acidic conditions; pH values above 7 indicate alkaline conditions.

photosynthesis   the process by which carbohydrates (sugars and starches) are manufactured in plant cells; requires carbon dioxide, water, light, and chlorophyll (the green color in plants).

piedmont glacier   glacier that forms where two or more valley glaciers coalesce to cover lower lands at the base of a mountainous region.

plane of the ecliptic   plane of the earth's orbit about the sun and the apparent annual path of the sun among the stars.

plankton   passively drifting or weakly swimming marine organisms, including both phytoplankton (plants) and zooplankton (animals).

plant community   variety of individual plants living in harmony with each other and the surrounding physical environment.

plate tectonics   theory that superseded continental drift and is based on the idea that the lithosphere is composed of a number of discrete segments or *plates* that move independently of one another, at varying speeds, over the earth's surface.

playa   dry lake bed in a desert basin.

plucking   see quarrying.

plug dome   a particularly steep-sided, explosive type of volcano with its central vent or vents plugged by the rapid congealing of its highly acidic lava.

pluton   extensive mass of igneous rock formed by the cooling of magma deep within the earth's crust.

pluvial   rainy time period, usually pertaining to glacial periods when deserts were wetter than at present.

podzolization   soil-forming process of humid climates with long cold winter seasons. Podzols, the typical end product of the process, are characterized by the surface accumulation of raw humus, strong acidity, and the leaching or eluviation of soluble bases and iron and aluminum compounds.

polar easterlies   easterly surface winds that move out from the polar highs toward the subpolar lows.

polar front   shifting boundary between cold polar air and warm subtropical air, located within the middle latitudes and strongly influenced by the polar jet stream.

polar highs   high-pressure systems located near the poles where air is settling and diverging.

polar jet stream   high-velocity air current within the upper air westerlies.

pollution   alteration of the physical, chemical, or biological balance of the environment that has adverse effects on the normal functioning of all lifeforms, including humans.

porosity   characteristic of soil or bedrock that relates to the amount of pore space between individual peds or soil and rock particles and which determines the water storage capacity of the earth material.

potential evapotranspiration hypothetical rate of evapotranspiration if at all times there is a more than adequate amount of soil water for growing plants.

prairie almost treeless tall grasslands in middle latitudes.

precipitation water in liquid or solid form that falls from the atmosphere and reaches the earth's surface.

pressure belts zones of high or low pressure that tend to circle the earth parallel to the equator in a theoretical model of world atmospheric pressure.

pressure gradient rate of change of atmospheric pressure horizontally with distance, measured along a line perpendicular to the isobars on a map of pressure distribution.

prevailing wind direction from which the wind for a particular location blows during the greatest proportion of the time.

prime meridian (Greenwich meridian) half of a great circle that connects the North and South Poles and marks zero degrees longitude. By international agreement the meridian passes through the Royal Observatory at Greenwich, England.

productivity rate at which new organic material is created at a particular trophic level.

pyroclastic material solid rock material (cinders, ash, and rock fragments) thrown into the air by a volcanic eruption.

quarrying (plucking) process whereby active glaciers break away and carry forward weathered and fractured bedrock.

radiation emission of waves that transmit energy through space (see *short-wave radiation* and *long-wave radiation*).

radiation fog fog produced by cooling of air in contact with a cold ground surface.

rain falling droplets of liquid water.

rain shadow dry, leeward side of a mountain range, resulting from the adiabatic warming of descending air.

recessional moraine end moraine deposited behind the terminal moraine, marking pauses in the retreat of a valley glacier or ice sheet.

reg desert surface of gravel and pebbles with finer materials removed; common to large areas in the Sahara.

regolith mantle of weathered earth materials that usually covers bedrock.

relative humidity ratio between the amount of water vapor in air of a given temperature and the maximum amount of water vapor that the air could hold at that temperature, if saturated; usually expressed as a percentage.

remote sensing mechanical collection of information about the environment from a distance, usually from aircraft or spacecraft, e.g., photography, radar, infrared.

remote sensing devices variety of techniques by which information about the earth can be gathered from great heights, typically from very high-flying aircraft or spacecraft.

revolution (earth) motion of the earth along a path, or orbit, around the sun. One complete revolution requires approximately 365¼ days and determines an earth year.

rhumb line line of true compass bearing (heading).

ribbon falls high, narrow waterfall dropping from a hanging glacial valley.

Richter scale scale that measures the energy released by an earthquake, used by earth scientists to rate and compare the magnitude and potential destructive power of earthquakes.

rift valley major lowland that forms in a graben or down-faulted crustal block.

rime ice crystals formed along the windward side of tree branches, airplane wings, etc., under conditions of supercooling.

rip current strong, narrow surface current flowing away from shore. It is produced by the return flow of water piled up near shore by incoming waves.

roche moutonnée bedrock hill subjected to intense glacial abrasion on its upstream side, with some plucking evident on the downstream side.

rockfall nearly vertical drop of individual rocks or a small rock mass caused by the pull of gravity on steep slopes.

rock flour rock fragments finely ground between the base of a glacier and the underlying bedrock surface.

Rossby waves horizontal undulations in the flow of the upper air winds of the middle and upper latitudes.

rotation (earth) turning of the earth on its polar axis; one complete rotation requires 24 hours and determines one earth day.

runoff flow of water from the land surface, generally in the form of streams and rivers.

salinization soil-forming process of low-lying areas in desert regions; the resulting soils are characterized by a high concentration of soluble salts as a result of the evaporation of surface water.

Santa Ana very dry foehn wind occurring in southern California. (See foehn wind.)

saturation (saturated air) point at which sufficient cooling has occurred so that an air mass contains the maximum amount of water vapor it can hold. Further cooling produces condensation of excess water vapor.

savanna tropical vegetation consisting primarily of coarse grasses, often associated with scattered low-growing trees or patches of bare ground.

scale ratio between distance as measured on the earth and the same distance as measured on a map, globe, or other representation of the earth.

sea breeze air flow by day from the sea toward the land; caused by the movement of air toward a zone of lower pressure associated with higher daytime temperatures over the land.

sea floor spreading movement of oceanic crust in opposite directions away from the midocean ridges, associated with the formation of new crust at the ridges and subduction of old crust at ocean margins.

seamount submarine volcanic peak rising from the deep ocean floor.

sedimentary rock one of the three major rock types; formed by the accumulation, compaction, and cementation of fragmented earth materials, organic remains, or chemical precipitates.

seismograph scientific instrument utilized to read the passage of vibratory earthquake waves produced by crustal faulting.

selva characteristic tropical rain forest comprised of multistoried, broad-leaf evergreen trees with significant development of lianas and relatively little undergrowth.

sextant navigation instrument used to determine latitude by star and sun positions.

shield volcano low-lying volcano formed by the cooling and accumulation of successive fluid lava flows extruded from a central vent or system of vents.

short-wave radiation radiation energy emitted by the sun in the form of waves of less than 4.0 microns (1 micron equals one ten-thousandth of a centimeter); includes X-rays, gamma rays, ultraviolet rays, and visible light waves.

Siberian high intensively developed center of high atmospheric pressure located in northern central Asia in winter.

sill igneous intrusion that forms a horizontal rock mass after molten material has been forced between rock layers and subsequently cools.

sinkhole surface depression, or pit, produced by the dissolving of limestone bedrock by groundwater.

slash-and-burn agriculture also called swidden or shifting cultivation; typical subsistence agriculture of primitive societies in the tropical rain forest. Trees are cut, the smaller residue is burned, and crops are planted between the larger trees or stumps before rapid deterioration of the soil forces a move to a new area.

sleet form of precipitation produced when raindrops freeze as they fall through a layer of cold air; may also, locally, refer to a mixture of rain and snow.

slope aspect direction a mountain slope faces in respect to the sun's rays.

slump mass of soil and regolith that slips or collapses downslope with a backward rotation.

small circle any circle that is not a full circumference of the globe. The plane of a small circle does not pass through the center of the globe.

smog combination of chemical pollutants and particulate matter in the lower atmosphere, typically over urban-industrial areas.

snout (terminus) end portion of a glacier that marks its farthest advance at a particular time, where ablation equals net snowfall plus the ice received from the zone of accumulation.

snow precipitation in the form of ice crystals.

snow line elevation in mountain regions above which summer melting is insufficient to prevent the accumulation of permanent snow or ice.

soil grade classification of soil texture by particle size: clay (less than 0.002 mm), silt (0.002–0.05 mm), and sand (0.05–2.0 mm) are soil grades.

soil horizon distinct soil layer characteristic of vertical zonation in soils; horizons are distinguished by their general appearance and their specific chemical and physical properties.

soil profile vertical cross section of a soil that displays the various horizons or soil layers that characterize it; used for classification.

solar constant rate at which insolation is received just outside the earth's atmosphere on a surface at right angles to the incoming radiation.

solar energy see insolation.

solifluction slow movement or flow of water-saturated soil and regolith downslope due to gravity; causes characteristic lobes on slopes in permafrost areas where only the top few feet of earth material thaws in summer and drainage is poor.

solstice one of two times each year when the position of the noon sun is overhead at its farthest distance from the equator; this occurs when the sun is overhead at the Tropic of Cancer (about June 22) and the Tropic of Capricorn (about December 22).

source region nearly homogeneous surface of land or ocean over which an air mass acquires its temperature and humidity characteristics.

southeast trades see trade winds.

specific humidity mass of water vapor present per unit mass of air, expressed as grams per kilogram of moist air.

spit beach feature attached to the mainland and built partially across a bay or inlet by the depositional action of longshore currents.

spring any surface outflow of groundwater, generally where the water table intersects the ground surface.

squall line narrow line of rapidly advancing storm clouds, strong winds, and heavy precipitation; usually develops in front of a fast-moving cold front.

stability condition of air when it is cooler than the surrounding atmosphere and resists the tendency to rise; the lapse rate of the surrounding atmosphere is less than that of *stable* air.

stationary front frontal system between air masses of nearly equal strength; produces stagnation over one location for an extended period of time.

steppe middle-latitude semiarid vegetation, treeless and dominated by short bunch grasses.

stock individual deep-seated igneous mass of limited size; it is often associated with other igneous masses, known generally as plutons.

storm local atmospheric disturbance often associated with rain, hail, snow, sleet, lightning, or strong winds.

storm surge rise in sea level due to wind and reduced air pressure during a hurricane or other severe storm.

storm track path frequently traveled by a cyclonic storm as it moves in a generally eastward direction from its point of origin.

strata distinct layers or beds of sedimentary rock.

stratus uniform layer of low sheetlike clouds, frequently grayish in appearance.

stratopause upper limit of stratosphere, separating it from the mesosphere.

stratosphere layer of atmosphere lying above the troposphere and below the mesosphere, characterized by fairly constant temperatures and ozone concentration.

stratovolcano (composite cone) volcano formed of alternating layers of lava and pyroclastic material expelled from a central vent or system of vents.

stream load amount of material transported by a stream at a given instant; includes bed load, suspended load, and dissolved load.

striations gouges, grooves, and scratches produced in bedrock by rock fragments and boulders imbedded in a glacier.

strike the compass direction taken by a rock stratum or fault plane, which is at right angles to their dip.

structure nature, arrangement, and interrelationships of the bedrock layers and other earth materials underlying a region.

subduction process associated with plate tectonic theory whereby an oceanic crustal plate is forced downward into the mantle beneath a lighter continental plate when the two converge.

submarine canyon steep-sided erosional valley cut into the continental shelf or continental slope.

subpolar lows east/west trending belts or cells of low atmospheric pressure located in the upper middle latitudes.

subsurface horizon buried soil layer that possesses specific characteristics essential to the identification of soils in the Comprehensive Soil Classification System. (Examples of subsurface horizons may be found in Table 10.2.)

subtropical highs cells of high atmospheric pressure centered over the eastern portions of the oceans in the vicinity of 30°N and 30°S latitude; source of the westerlies poleward and the trades equatorward.

subtropical jet stream high-velocity air current flowing above the sinking air of the subtropical high-pressure cells; most prominent in the winter season.

succession progression of natural vegetation from one plant community to the next until a final stage of equilibrium has been reached with the natural environment.

surface of discontinuity three-dimensional surface with length, width, and height separating two different air masses; also referred to as a *front*.

surge sudden shift downslope of glacial ice; possibly caused by a reduction of basal friction with underlying bedrock.

swell regular longer-period sea wave traveling a significant distance from the area of generation by the wind.

syncline trough, depression, or downfold in a wave of crustal folding.

system group of interacting and interdependent units that together form an organized whole.

taiga term used to describe the northern coniferous forest of subarctic regions on the Eurasian landmass.

taku cold downslope wind in Alaska (see katabatic wind).

talus rock debris in a cone-shaped deposit at the base of a steep slope or escarpment; usually a result of frost wedging and individual rockfalls with debris accumulating at the angle of rest.

tarn lake located within a glacial cirque.

tectonic processes processes that derive their energy from within the earth's crust and serve to create landforms by elevating, disrupting, and roughening the earth's surface.

temperature degree of heat or cold and its measurement.

temperature gradient rate of change of temperature with distance in any direction from a given point; refers to rate of change horizontally; a vertical temperature gradient is referred to as the *lapse rate*.

temperature inversion reverse of the normal pattern of vertical distribution of air temperature; in the case of inversion, temperature *increases* rather than decreases with increasing altitude.

terminal moraine end moraine that marks the farthest advance of an alpine glacier or ice sheet.

terra rosa characteristic calcium-rich (developed over limestone bedrock) red-brown soils of the climatic regions surrounding the Mediterranean Sea.

thermocline vertical zone of ocean water where there is a sharp change in temperature with depth.

thermosphere highest layer of atmosphere extending from the mesopause to outer space.

thunder sound produced by the rapidly expanding, heated air along the channel of a lightning discharge.

thunderstorm intense convectional storm characterized by thunder and lightning, short in duration and often accompanied by heavy rain, hail, and strong winds.

tide periodic rise and fall of sea level in response to the gravitational interaction of the moon, sun, and earth.

till unsorted glacial drift, characterized by variation in size of deposit from clay particles to boulders.

tombolo wave depositional beach feature (bar or spit) connecting an island to the mainland.

tornado small, intense, funnel-shaped cyclonic storm of very low pressure, violent updrafts, and converging winds of enormous velocity.

trade winds consistent surface winds blowing in low latitudes from the subtropical highs toward the intertropical convergence zone; labeled northeast trades in the Northern Hemisphere and south-east trades in the Southern Hemisphere.

transpiration transfer of moisture from living plants to the atmosphere by the emission of water vapor, primarily from leaf pores.

transportation movement of earth materials from one site to another as a result of the transporting power of water, wind, or glacial ice.

travertine calcium carbonate (limestone) deposits resulting from the evaporation in caves or caverns and near surface openings of groundwater saturated with lime.

tree line elevation in mountain regions above which cold temperatures and wind stress prohibit tree growth.

trophic level number of feeding steps that a given organism is removed from the autotrophs (e.g., green plant—first level, herbivore—second level, carnivore—third level, etc.).

trophic structure organization of an ecosystem based on the feeding patterns of the organisms that comprise the ecosystem.

Tropic of Cancer parallel of latitude at 23½°N; the northern limit of the migration of the sun's vertical rays throughout the year.

Tropic of Capricorn parallel of latitude at 23½°S; the southern limit to the migration of the sun's vertical rays throughout the year.

tropopause boundary between the troposphere and stratosphere.

troposphere lowest layer of the atmosphere, exhibiting a steady decrease in temperature with increasing altitude and containing virtually all atmospheric dust and water vapor.

trough elongated area or "belt" of low atmospheric pressure.

tsunami ocean wave produced by submarine earthquake, volcanic eruption or landslide; not noticeable in deep ocean waters, but building to dangerous heights in shallow waters.

tundra treeless vegetation of polar regions and very high mountains, consisting of mosses, lichens, and low-growing shrubs and flowering plants.

turbidity current submarine flow of sediment-laden water.

uniformitarianism widely accepted theory that the

earth's landforms have developed over exceedingly long periods of time as a result of processes that may be observed in the present landscape.

upper air westerlies system of westerly winds in the upper atmosphere, flowing in latitudes poleward of 20°.

upwelling upward movement of colder, nutrient-rich, subsurface ocean water, replacing surface water that is pushed away from shore by winds.

valley breeze air flow upslope from the valleys towards the mountains during the day.

valley glacier see alpine glacier.

valley train outwash deposit from glacial meltwater, resembling an alluvial fan confined by valley walls.

variable one of a set of objects and/or characteristics of objects, which are interrelated in such a way that they function together as a system.

veering wind shift the change in wind direction, clockwise around the compass; e.g., east to southeast, to south, to southwest, to west, and northwest.

ventifact wind-fashioned rock produced by wind abrasion (sandblasting).

volcanism movement of molten material either within or at the surface of the earth's crust.

warm front leading edge of a relatively warmer, less dense air mass advancing upon a cooler, denser air mass.

warping broad and general uplift or settling of the earth crust with little or no local distortion.

wash (arroyo, wadi, barranca) generally steep-walled channel of an intermittent stream in an arid region; the stream bed is characteristically choked with coarse alluvium.

water budget relationship between evaporation, condensation, and storage of water within the earth system.

water table upper limit of the zone of saturation, below which all pore spaces are filled with water.

water vapor water in its gaseous form.

wave-cut bench gently sloping surface produced by wave erosion at the base of a sea cliff.

wave refraction bending of waves as they approach a shore, aligning themselves with the bottom contours of the surf zone.

weather atmospheric conditions, at a given time, in a specific location.

weathering mechanical (physical) fragmentation and chemical decomposition of rocks and minerals in the earth's crust.

westerlies surface winds flowing from the polar portions of the subtropical highs; carrying fronts, storms, and variable weather conditions from west to east through the middle latitudes.

wet adiabatic rate rate at which a rising mass of air is cooled by expansion when condensation is taking place. The rate varies but averages 6.0°C/1000 m (3.2°F/1000 ft).

wind air in motion from areas of higher pressure to areas of lower pressure; movement is generally horizontal, relative to the ground surface.

windward location on the side that faces toward the wind and is therefore exposed or unprotected; usually refers to mountain and island locations.

xerophyte vegetation type that has genetically evolved to withstand the extended periods of drought common to arid regions.

yazoo stream a stream tributary that flows parallel to the main stream for a considerable distance before joining it.

zonal soils major soil order of the Russian-American classification system; soils in this order have well-developed profiles that exhibit the strong influence of vegetation and climate. (Descriptions of zonal soil types from the Russian-American system may be found in Table 10.1.)

zone of ablation lower portion of a glacier, below the firn line, where melting, evaporation, and sublimation exceed net snow accumulation.

zone of accumulation subsoil or *B* horizon of a soil, characterized by deposition or illuviation of soil components by gravitational water, also the upper portion of a glacier, above the firn line, where net snow accumulation exceeds the melting, evaporation, and sublimation of snowfall.

zone of aeration upper groundwater zone above the water table where pore spaces may be alternately filled with air or water.

zone of depletion top layer, or *A* horizon, of a soil, characterized by the removal of soluble and insoluble soil components through leaching and eluviation by gravitational water.

zone of saturation zone immediately below the water table, where all pore spaces in soil and rock are filled with groundwater.

List of Plates

550

INDEX